Optimization Methods Applied to Power Systems II

Optimization Methods Applied to Power Systems II

Editors

Francisco G. Montoya
Raúl Baños Navarro

MDPI • Basel • Beijing • Wuhan • Barcelona • Belgrade • Manchester • Tokyo • Cluj • Tianjin

Editors
Francisco G. Montoya
University of Almeria
Spain

Raúl Baños Navarro
University of Almeria
Spain

Editorial Office
MDPI
St. Alban-Anlage 66
4052 Basel, Switzerland

This is a reprint of articles from the Special Issue published online in the open access journal *Energies* (ISSN 1996-1073) (available at: https://www.mdpi.com/journal/energies/special_issues/optimization_ii).

For citation purposes, cite each article independently as indicated on the article page online and as indicated below:

LastName, A.A.; LastName, B.B.; LastName, C.C. Article Title. *Journal Name* **Year**, *Volume Number*, Page Range.

ISBN 978-3-0365-0358-5 (Hbk)
ISBN 978-3-0365-0359-2 (PDF)

Contents

About the Editors

Francisco G. Montoya, professor in the Engineering Department and the Electrical Engineering Section at the University of Almeria (Spain), received his M.S. from the University of Malaga and his Ph.D. from the University of Granada (Spain). He has published over 75 papers in JCR journals and is the author or coauthor of books published in different editorials. His main interests are power quality, smart metering, smart grids and evolutionary optimization applied to power systems, and renewable energy. Recently, he has become passionately interested in geometric algebra as applied to power theory.

Raúl Baños Navarro is an Associate Professor in the Department of Engineering at the University of Almeria (Spain). He received his M.S. in Computer Science at the University of Almeria and his second M.S. in Economics from the National University of Distance Education (UNED). He wrote his Ph.D. dissertation on computational methods applied to the optimization of energy distribution in power networks. His research activity includes computational optimization, power systems, renewable energy systems, and energy economics. He has been a research visitor at Napier University (Edinburgh, UK) and at the Universidade do Algarve (Portugal). As a result of his research, he has published more than 150 papers in peer-reviewed journals, books, and conference proceedings.

Preface to "Optimization Methods Applied to Power Systems II"

Electrical power systems are large and complex systems that cover the generation, transmission, distribution, and use of electrical energy. The inherent complexity of these systems is augmenting due to the increase in the generation capacity from renewable energy sources. Therefore, the analysis, design, and operation of current and future electrical power systems require an efficient approach to different optimization problems, including load flow, fault location, contingency analysis, system restoration after blackouts, economic dispatch, unit commitment, islanding detection, filter design, and smart grids control. The fast evolution of this field and the difficulty of managing large systems demand the design and validation of advanced techniques for supporting decision-making processes. Currently, power system analysis, design, operation and management tasks are being supported by advances in computational optimization methods, from classical optimization techniques such as linear and nonlinear programming and integer and mixed-integer programming to recent advances in meta-heuristic optimization, including bio-inspired meta-heuristics which have allowed researchers and practitioners to obtain quality solutions in reduced response times. Taking into account the high dimensions and complexity of real-life power systems, efficient network planning, operation, or maintenance requires a permanent research effort.

Therefore, this book includes recent advances in the application of optimization techniques that directly apply to electrical power systems so that readers may familiarize themselves with new methodologies directly explained by experts in this scientific field.

Francisco G. Montoya, Raúl Baños Navarro
Editors

Article

Online Economic Re-dispatch to Mitigate Line Overloads after Line and Generation Contingencies

Oswaldo Arenas-Crespo [1], John E. Candelo-Becerra [1,*] and Fredy E. Hoyos Velasco [2]

[1]   Departamento de Energía Eléctrica y Automática, Facultad de Minas, Universidad Nacional de Colombia, Sede Medellín, Carrera 80 No. 65-223, Campus Robledo, Medellín 050041, Antioquia, Colombia; oarenasc@unal.edu.co

[2]   Escuela de Física, Facultad de Ciencias, Universidad Nacional de Colombia, Sede Medellín, Carrera 65 No. 59A-110, Medellín 050034, Antioquia, Colombia; fehoyosve@unal.edu.co

*   Correspondence: jecandelob@unal.edu.co; Tel.: +57-4-425 50 00

Received: 15 February 2019; Accepted: 9 March 2019; Published: 13 March 2019

Abstract: This paper presents an online economic re-dispatch scheme based on the generation cost optimization with security constraints to mitigate line overloads before and after line and generation contingencies. The proposed optimization model considers simplification of mathematical expressions calculated from online variables as the power transfer distribution factor (PTDF) and line outage distribution factor (LODF). Thus, a first algorithm that calculates economic re-dispatch for online operation to avoid overloads during the normal operation and a second algorithm that calculates online emergency economic re-dispatch when overloads occur due to line and generator contingencies are proposed in this paper. The results show that the proposed algorithms avoid overload before and after contingencies, improving power system security, and at the same time reducing operational costs. This scheme allows a reduction of power generation units in the electricity market during online operation that considers line overloads in the power system.

Keywords: electricity markets; economic dispatch; optimization; maximum capacity; online operation; power system; unit commitment

---

1. Introduction

Power system security is an important subject for electricity companies that operate transmission networks, which depend on the system operating condition and contingencies and can lead to interruption of customer service [1–3]. Thus, some methodologies seek to identify the safe operation of the power system [2]. Although these procedures allow a safer operation, the large number of exogenous variables, such as maintenance that exceeds the planned time, emergency outage of transmission equipment, and volatility of renewable sources, lead the operation to a risk zone and new emergency actions.

A classic economic dispatch is commonly used to identify the generation costs of each unit, which considers constraints as Kirchhoff's first and second laws with AC or DC models [4]. The security-constrained economic dispatch (SCED) could include physical limits of equipment, operative limits, and contingencies [5], and the solution is found with a security-constrained optimal power flow (SCOPF), in which the objective function can be the minimization of losses, minimization of power demand rationing, minimization of the cost of operation, and others. Security constraints imply the modeling of the power grid, which leads to an important high computational cost [6]; therefore, economic dispatch models with security constraints are usually implemented by means of metaheuristic methods that imply lengthy execution time, or by means of exact models such as the interior point algorithm, which sometimes causes difficulties with convergence [7].

With this, the economic re-dispatch is made to avoid overloads at the lowest cost, as implemented in [8,9]. In this case, particle swarm optimization (PSO) is used to optimize as a stable algorithm with respect to the convergence of the non-linear modeling of the power flow equations.

Other authors have implemented a similar economic dispatch model with classic constraints by using learning machines [10]. In addition, an interesting algorithm was presented in [11], which selects the generation to participate in an economic dispatch using a direct acyclic graph (DAG). This model has been proposed as an alternative for large networks and various operational areas. In [12], a model that considers mixed integer linear programming (MILP) to minimize switch opening as a solution to reduce overloads is implemented. Further, in [13] the overloads are reduced through fuzzy logic, whose model tries to recreate the actions of the network operators; however, this model does not take into account the generation cost.

In [14], the online economic dispatch is implemented using the metaheuristics of Fuzzy Logic and Genetic Algorithm, which avoids the modeling of the complete AC system and the problems of non-convergence of exact solution methods. On the other hand, an exact solution method based on Primal Dual interior Point (PDIP) can be used [15]; however, in order to execute the final solution algorithm, different simplification stages of the dispatch scenarios to be executed by clustering can be used, so that the appropriate selection of the scenarios will have a direct impact on the consolidation of the global optimum and compliance with the system constraints.

Middle- and long-term solutions seek to guarantee security and reliability within the network. On the one hand, middle-term solutions use contingency analysis through simulations to find network expansion that mitigate the constraints found in the analysis. On the other hand, short-term solutions and online operation are based on controlling constraints where the operator or intelligent algorithms monitor and control system variables. However, during online operation there is no time to perform network expansion; here, techniques consider variation of the network topology, new economic dispatches, stress control, load shedding, and others.

Security-constrained economic dispatch (SCED) uses algorithms that are of high computational cost, so they are not normally used in the online operation. Therefore, despite previous plans for the operation through middle- and short-term analysis to carry out the secure operation of power systems, deviations of the programmed resources and the projected variables are evidenced in the operation. Therefore, this paper proposes an online economic re-dispatch to mitigate overloads of transmission elements after $N-1$ contingencies to reduce the risk of collapse [16]. Because of the number of variables that contain the problem, the mathematical formulation can be simplified by distribution factors (DFs) such as the power transfer distribution factor (PTDF), the line outage distribution factor (LODF) [17], and the outage transfer distribution factors (OTDF) [4,5,18–20]. This technique has an advantage that allows the linearization of the power flow's equations around an operation point. Thus, the contributions of the paper are defined as follows:

- a mathematical model is formulated for the economic re-dispatch, together with the security constraints to be used for the online operation;
- the system models are simplified to find quick solutions to the problem of contingencies, and are used in the formulation of the economic re-dispatch with constraints;
- the input of the economic re-dispatch program is referred to the topology changes in the power grid, the availability of power generation, and the generation costs of the units;
- and an on-line re-dispatch is presented to reduce the operational risk of generation outage and changes of power from renewable energies.

The rest of the paper is organized in three more sections: Section 2 describes the main mathematical model of the online economic re-dispatch for the normal operation and the model of the online economic re-dispatch proposed for overloads; the sensitivity factors, which are applicable to the power flow and the online operation are included in the formulation. Section 3 presents the power system test case and the results obtained by applying the formulation. Finally, Section 4 includes the conclusions, recommendations and the future work in this research.

2. Methodology

In this section the mathematical formulation used for the application of the economic re-dispatch schemes, and the simplification of the formulation to apply the electrical constraints of the system are presented. In addition, we present the algorithms used for the economic re-dispatch in normal operation and the economic re-dispatch after contingencies.

2.1. Economic Re-Dispatch with Power Demand Rationing Cost

The online economic dispatch model to reduce overloads after N−1contingencies can be formulated as in Equation (1), where ΔPG_i is the power generation change in bus i, CG_i is the operational generation cost in bus i, ΔPR_i power load shedding change in bus i, CR_i is the load shedding cost in bus i, and $i = 1, 2, 3 \ldots m$ buses. This equation corresponds to the objective function, which is used to minimize the operating cost of the system. The expression considers the sum of the generation cost and the cost of power demand rationing. It should be noted that the generation cost (CG_i) is assumed as a constant value that is expressed in terms of the cost per power supplied ($/MWh), as well as the cost of power demand rationing. However, the latter is much greater than the generation cost and it can be considered as a constant value of great magnitude, because it is only required to help the convergence of the model:

$$Minimize\ F = \sum_{i=1}^{m}(\Delta PG_i * CG_i + \Delta PR_i * CR_i). \tag{1}$$

The model starts with an initial condition, dispatching the central generation plants (PG_i^0), the power flowing through transmission elements (P_i^0) and the power demanded in the load buses of the power system (PD_i). In the present work, it is assumed that all these initial parameters are obtained online by measurement units as the supervisory control and data acquisition (SCADA), because this information can be managed from this type of application.

When there is no feasible safe dispatch and the power demand rationing can be applied to bus i, the model would reflect that situation, obtaining a value of ΔPR_i greater than zero. In this case, the final power demanded in bus i (PD_i') would be reduced as expressed in Equation (2):

$$PD_i' = PD_i - \Delta PR_i. \tag{2}$$

Therefore, the decision variable for power rationing (ΔPR_i) is modeled as a generator in each bus with power limits on the demand at each bus, as expressed in Equation (3):

$$0 \leq \Delta PR_i \leq PD_i,\ i = 1, 2 \ldots, m. \tag{3}$$

The active generation or generation synchronized to the system is the only one that should be considered in the optimization model, because it is the only one that can be managed in real time. This generation could consider renewable generation sources such as solar, wind, and even battery banks (BESS).

Starting from the initial operation (PG_i^0, P_i^0, and PD_i), the model calculates the deltas of generation (ΔPG_i), which are required to maintain a preventive condition where the N−1contingencies do not generate overloads. Once the model converges, the new power for each generator in bus i ($\Delta PG_i\prime$) will have a value calculated by Equation (4):

$$\Delta PG_i\prime = PG_i^0 + \Delta PG_i. \tag{4}$$

Equation (5) is used to prevent the total generation deviation from exceeding the allowed level of automatic generation control (AGC) for the online re-dispatch algorithm, where $ResAGC$ represents the power reserves used for generation re-dispatch. This value of AGC can be managed by the $ResAGC$ parameter, which could even be 0 MW. Equation (6) is used to avoid the operating limits of each generator are violated, where $PGmax_i$ and $PGmin_i$ are the maximum and minimum power generation in bus i, respectively:

$$- ResAGC \leq \sum_{k=1}^{m} \Delta PG_i \leq ResAGC, \, i = 1, 2 \ldots, m. \tag{5}$$

$$PGmin_i \leq PG_i^0 + \Delta PG_i \leq PGmax_i, i = 1, 2 \ldots, m. \tag{6}$$

The power that flows for each transmission element (P_i) is calculated from the initial state P_i^0, which corresponds to the power measured online at transmission lines and transformers. To this value, the calculated delta of the power flow (ΔPG_k) and delta power rationing (ΔPR_k) are added, multiplied by the power transfer distribution factors PTDF that relate to the contribution of the power injection in the bus with respect to the transmission element (λ_{ik}), as shown in Equation (7):

$$P_i = P_i^0 + \sum_{k=1}^{m} \lambda_{ik} * \Delta PG_k + \sum_{k=1}^{m} \lambda_{ik} * \Delta PR_k, i = 1, 2 \ldots, n. \tag{7}$$

Equation (8) corresponds to the security constraints that prevent the transmission elements from being overloaded above their allowed values, after generation changes and without contingencies; where $Pnom_i$ is the rated power flow of each transmission element i without contingencies:

$$- Pnom_i \leq P_i \leq Pnom_i, i = 1, 2 \ldots, n. \tag{8}$$

Finally, the constraints in Equation (9), ensure that no N$-$1contingency leads to the overload of other elements, by including the sensitivity factor LODF (γ_{ij}) between two lines. Herein, the term $Pmax$ refers to the maximum power flow for transmission element i under single contingencies:

$$-Pmax_i \leq P_i + P_j * \gamma_{ij} \leq Pmax_i$$
$$i = 1, 2 \ldots, n., \, j = 1, 2 \ldots, n, \, i \neq j. \tag{9}$$

2.2. Economic Re-Dispatch without Power Demand Rationing Cost

If an economic dispatch solution is required without power rationing, the objective function is defined as presented in Equation (10). This function considers minimizing the operating cost of the system:

$$Minimize \, F = \sum_{i=1}^{m} (\Delta PG_i * CG_i) + \sum_{i=1}^{n} \alpha * PAd_i. \tag{10}$$

The constraints defined above are maintained, and the power flowing through each transmission element is calculated from the initial power of the system, as shown in Equation (11):

$$P_i = P_i^0 + \sum_{k=1}^{m} \lambda_{ik} * \Delta PG_k, i = 1, 2 \ldots, n. \tag{11}$$

In addition, Equation (2) is delete and Equation (3) is modified with the expression shown in Equation (12):

$$-(Pmax_i + PAd_i) \leq P_i + P_j * \gamma_{ij} \leq Pmax_i + PAd_i$$
$$i = 1, 2 \ldots, n., \, j = 1, 2 \ldots, n, \, i \neq j. \tag{12}$$

With this modification the model that considers electricity rationing is eliminated, and now the capacity of the transmission elements after contingencies is considered in the formulation. The variable PAd_i allows the convergence of the model, even when the system is unsecure with critical contingencies, allowing security to be maximized. The critical contingencies correspond to all those contingencies that generate overloads and that said overload cannot be alleviated by any feasible online economic dispatch. Therefore, the result of this model corresponds to the most secure economic dispatch possible without considering the rationing and improving the capacity of the transmission elements to contingency.

Because the process is iterative according to the total number of critical contingencies, this implies that the model must be linearized around the point of operation whenever the topology of the system is altered from the evaluation of each critical contingency. The above implies that the matrix of PTDF sensitivity factors must be updated for each topology. However, there is another possibility to update the sensitivity matrix PTDF, by replacing this matrix with the calculation of the sensitivities known as OTDF [21].

According to this, the power flow for transmission elements that was considered in Equation (7) is modified and the new expression is shown in Equation (13), where:

$$P_i = P_i^0 + \sum_{k=1}^{m} \psi_{ik} * \Delta PG_k + \sum_{k=1}^{m} \psi_{ik} * \Delta PR_k \, , i = 1, 2 \ldots, n. \tag{13}$$

where ψ_{ik} represents the sensitivity factor OTDF, that models the power flow distribution in a bus k, to the monitored element i, considering the outage of an element j (for our case the critical contingency evaluated). Therefore, the calculation of the sensitivities for the modelling variations in the power flow after various topological changes becomes simpler, if it is formulated from the OTDF sensitivities matrix instead of PTDF.

2.3. Algorithm for the Online Economic Re-Dispatch

The general model of the online economic re-dispatch algorithm used in the research is presented in Figure 1. The interaction between the power system information SCADA, the AC simulator, and the linear optimization model is observed.

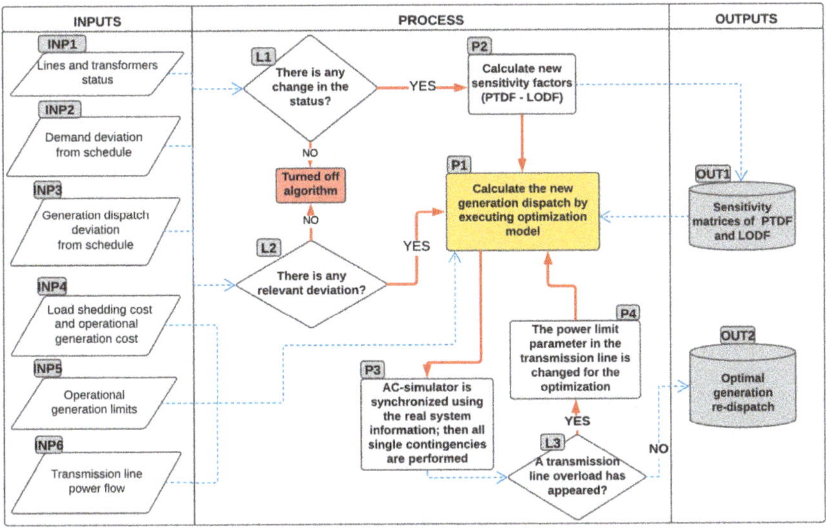

Figure 1. First algorithm for the online economic dispatch.

The model is divided into three main stages. The first one (inputs) corresponds to the capture of information corresponding to the physical system. The second one (process) corresponds to the evaluation of this information and integration of the optimization models and AC simulator. Finally, the third one (outputs) corresponds to the economic re-dispatch solution in each operating state which could be implemented in an automated system or through operational actions of operators.

In the first stage of the complete methodology, the inputs of the system are considered as the required variables for the procedure. The input E1 corresponds to the status of the transmission lines and transformers, representing the elements that are in service (ON = 1) or out of service (OFF = 0). The input E2 defines the difference between the forecasted demand and the actual demand, representing the power deviation of the system. The input E3 corresponds to the difference between the scheduled power dispatch and the current power dispatch presented in the online transaction; for this, each generator must be monitored to calculate the deviation. The input E4 represents the price of each generator for the online operation period. The input E5 refers to the minimum and maximum limits of the generation plants, which, in the online operation, are considered variables, mainly in thermal plants and will depend on factors such as generator temperature, generator's configuration (cycle center combined), start ramps, etc. The input E6 is the linearized model obtained from the current power system condition, and refers to the values of real power that flows through the various transmission elements that are monitored; it is not necessary to monitor or model the complete power system for use in the economic re-dispatch with overloads and after contingencies.

The second stage corresponds to the logic and subroutines of the general process. For example, the process L1 identifies the state of a line or transformer defined as in service (ON = 1) and out of service (OFF = 0). This logic can be implemented from measures of each element or switch positions. The process L2 corresponds to the logic in which the presence of a deviation of generation (E1) or demand (E2) is evaluated, so that the algorithm recognizes the presence of a deviation, it must be fulfilled; thus, if a deviation is greater than a value ArP, then the logic L1 can be expressed as the inequality $\Delta Pi > ArrP$.

Now, P1 refers to the economic re-dispatch optimization model with constraints and $N-1$ contingencies. Herein, the optimization model does not consider the calculation of power demand rationing. Therefore, the optimization model defined in P1 uses the objective function of Equation (10) and the constraints of Equations (11), (12), and (13). P1 also corresponds to the main subprocess of the algorithm, because the solution of the economic re-dispatch is obtained from it, which is refined through the iterations with the AC simulator. Then, the process P2 includes the sensitivity factors PTDF and LODF, and the optimization model P1 is linearized from a current operating point. This process must be carried out whenever the topology of the network is modified, due to the deviations in the distribution of the power flows.

In the process P3 the operating state of the AC simulator must be coordinated, which implies that the state of lines and transformers in the simulator must be the same as in the real power system, as well as the power demand in the bus of the power system. However, the power dispatch to be implemented in the simulator must be that obtained through the process optimization model P1. Once the simulator is coordinated, $N-1$ contingencies are executed with AC load flow. From the $N-1$ contingency results obtained from process P3, the process L3 verifies that no overloads are presented in the elements. If an overload is detected in this process, then the process P4 is executed, correcting the maximum power overload with this change in the parameter of the optimization model; the process P1 is executed again thus obtaining an economic re-dispatch result, starting from the new iteration. If no overloads are detected in the process L3, then the final value is obtained.

2.4. Online Emergency Economic Re-Dispatch Algorithm

The model presented in Figure 1 should not generate power demand rationing as a solution to the problem. Instead, an adaptable optimization model is necessary for the calculation of the secure dispatch and lowest cost without sacrificing the power demand and minimizing the overload of transmission elements after N−1 contingencies. However, if critical contingencies are detected when applying the algorithm presented in Figure 1, then the complementary algorithm presented in Figure 2 is executed. This new processes perform the calculation of a new dispatch and emergency load shedding, for each one of the critical contingencies found in the P1 process. That is, in this second phase of the algorithm, a new dispatch is calculated with possible load shedding for each critical contingency. This complementary sub-algorithm for critical contingencies corresponds to an optional algorithm, and is useful in cases in which the network is not fully covered before all the possible N−1 contingencies, or is useful in cases in which the only solution to avoid collapse is the load shedding.

According to the above, it is assumed that an emergency load shedding is considered because of critical contingencies. The previous algorithm could have an operation similar to that of a systemic protection supplementary scheme (SPS), but considering the minimization of the generation cost and minimization of load shedding, depending on the various operation points and topologies presented in online operation. The power demand rationing should always be the minimum possible. For this the complementary algorithm of Figure 2 has been proposed, which, given the initial dispatch calculated in process P1, obtains the value of the new dispatch and emergency load shedding for each critical contingency.

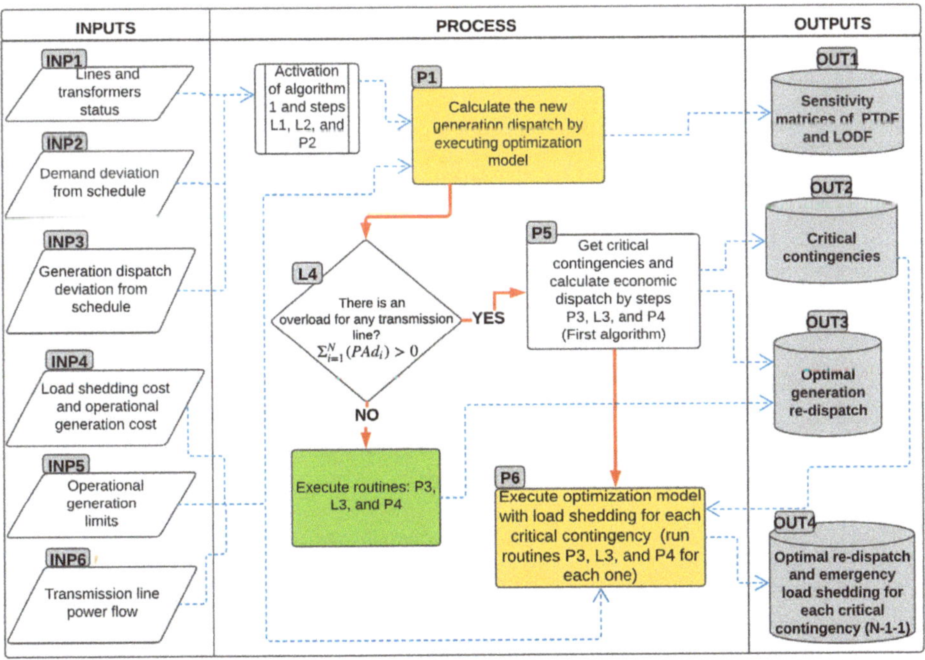

Figure 2. Second algorithm for the online dispatch with load shedding after critical contingencies.

In the process L4, the presence of critical contingencies is detected through the evaluation of the variable (PAd_i). If there is no risk due to critical contingencies, the algorithm behaves as shown in Figure 1; otherwise, the complementary algorithm shown in Figure 2, composed of P5 and P6, is executed. With the process P5, a most secure generation dispatch is obtained, despite the existence of critical contingencies. Additionally, the process L3 and processes P3 and P4—through which the economic re-dispatch is obtained from P1—is refined by iteration with the AC simulator.

Then, the process P6 considers an adapted optimization model, this time, with the objective of providing a solution of an economic re-dispatch with the possibility of load shedding for each of the critical contingencies. Therefore, the objective function formulated by Equation (1) is used, considering the constraints of the model presented from Equations (2) to (6). Additionally, the constraint in Equation (7) is not considered in the first part of the algorithm, because the calculation of the new economic re-dispatch and load shedding is performed after detecting critical contingencies. Thus, the optimization model is applied to the degraded network after critical contingency; therefore, from each critical contingency, a result of the new economic dispatch and different load shedding is obtained. Finally, the constraint presented in Equation (10) is not required for process P6.

3. Results

3.1. IEEE 39–Bus Power System Test Case

Figure 3 shows the IEEE 39–bus power system test case, which is a simplified model of the New England power grid. This system has ten generators; generator one is an equivalent representation of the rest of the power grid. This power system topology allows one to assess multiple power demand and generation dispatch scenarios.

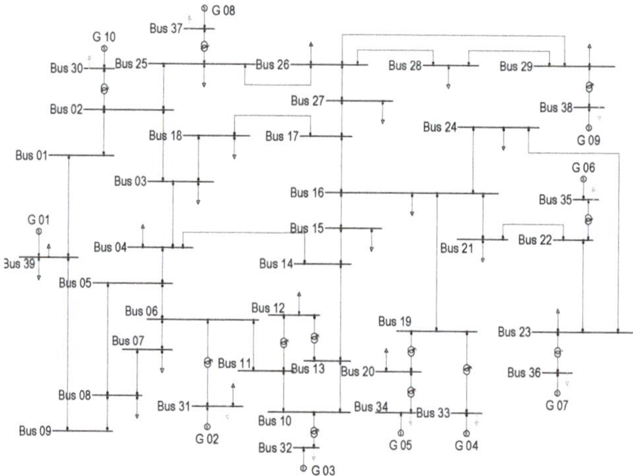

Figure 3. IEEE 39–bus power system test case.

3.2. Input Parameters

Table 1 presents the demand value and initial economic dispatch considered in the research. This dispatch has been calculated using the optimization model with the aim of starting from a secure point, which will be altered before topology changes and generator outages. This is also done to obtain the economic re-dispatch using the proposed online economic dispatch algorithm.

Table 1 shows that the power demand rationing cost tends to an infinite value, and the optimization must search to minimize this value modifying each bus of the system. For example, in Colombia the power demand rationing cost oscillates between 130 and 2608 USD/MWh [22]. From the generation dispatch and demand presented in Table 1, the initial operating state of the power system is obtained as shown in Figure 4, in which each line and transformer obtains a color according to the level of overload presented between 0% and 100% of its maximum loadability. In this case the load flow is simulated, and the power system is represented under normal operating conditions and considering all the elements in service.

By representing the maximum loadability of transmission elements, considering N−1contingencies, and using the same method of coloring elements used in Figure 4, the heat diagram of Figure 5 is obtained, in which branches increased overloading risks when N−1contingencies occurred.

Additionally, when comparing Figures 4 and 5, loadability has increased with N−1contingencies and that affects power system security. For example, line 23–24, under normal operating conditions has an overload of less than 50% and the power flow can increase to approximately twice the current value. However, for the same line after N−1contingencies, loadability reaches values higher than 90%, implying a critical state.

Table 1. Initial operation of the power system.

Bus name	Initial dispatch [MW]	Initial demand [MW]	Minimum generation [MW]	Maximum generation [MW]	Generation cost [USD/MWh]	Rationing Cost [USD/MWh]
Bus 03	0	327	0	0	0.00	99,999,999
Bus 04	0	336	0	0	0.00	99,999,999
Bus 07	0	233	0	0	0.00	99,999,999
Bus 08	0	499	0	0	0.00	99,999,999
Bus 12	0	7	0	0	0.00	99,999,999
Bus 15	0	320	0	0	0.00	99,999,999
Bus 16	0	333	0	0	0.00	99,999,999
Bus 18	0	160	0	0	0.00	99,999,999
Bus 20	0	620	0	0	0.00	99,999,999
Bus 21	0	275	0	0	0.00	99,999,999
Bus 23	0	248	0	0	0.00	99,999,999
Bus 24	0	313	0	0	0.00	99,999,999
Bus 26	0	143	0	0	0.00	99,999,999
Bus 27	0	285	0	0	0.00	99,999,999
Bus 28	0	211	0	0	0.00	99,999,999
Bus 29	0	291	0	0	0.00	99,999,999
Bus 30 G10	570	0	0	850	83.34	99,999,999
Bus 31 G2	571	9	0	595	100.00	99,999,999
Bus 32 G3	254	0	0	680	150.00	99,999,999
Bus 33 G4	680	0	0	680	83.34	99,999,999
Bus 34 G5	510	0	0	510	53.34	99,999,999
Bus 35 G6	221	0	0	680	200.00	99,999,999
Bus 36 G7	595	0	0	595	116.67	99,999,999
Bus 37 G8	249	0	0	595	216.67	99,999,999
Bus 38 G9	652	0	0	850	186.67	99,999,999

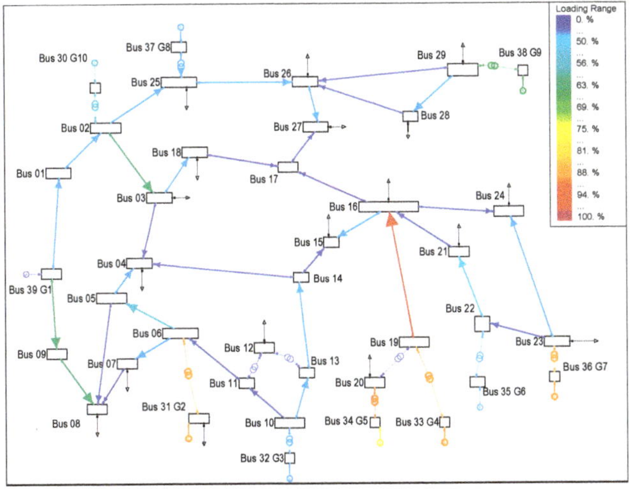

Figure 4. Maximum loadability without N−1contingencies.

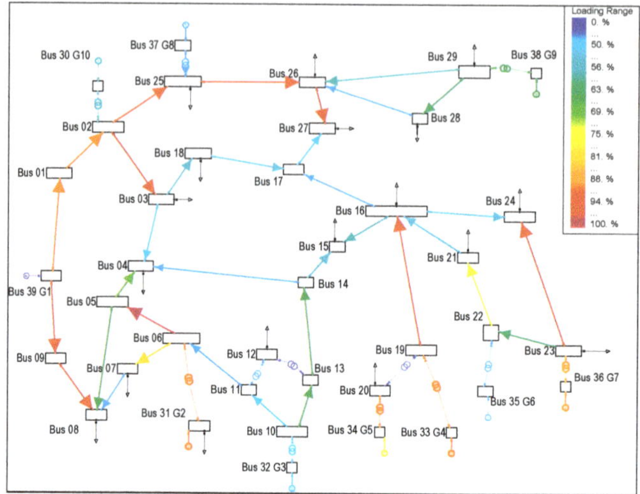

Figure 5. Maximum loadability with N−1contingencies.

As discussed above, the optimization model uses security constraints based on sensitivity factors, such as: PTDF, LODF and OTDF. The main advantages of using these sensitivity factors are: first, they simplify the security constraint model for both power flow and contingencies calculations; and second, they include small errors in the optimization result when the system works with AC model. In this method, the constraint calculation is made based on the initial and real states of the power system, which means that a linearized model is achieved around the operating point and a simple model is obtained to solve the difficult problem of the economic dispatch with online security constraints. For example, Figure 6 shows a comparison between the AC and DC power flow for a 230 kV line with 260 km, obtaining that the largest error of the calculation is given by the angular opening and the voltage along the line. Therefore, the error when using sensitivity factors is small, because the problem is linearized around the real state of the power system.

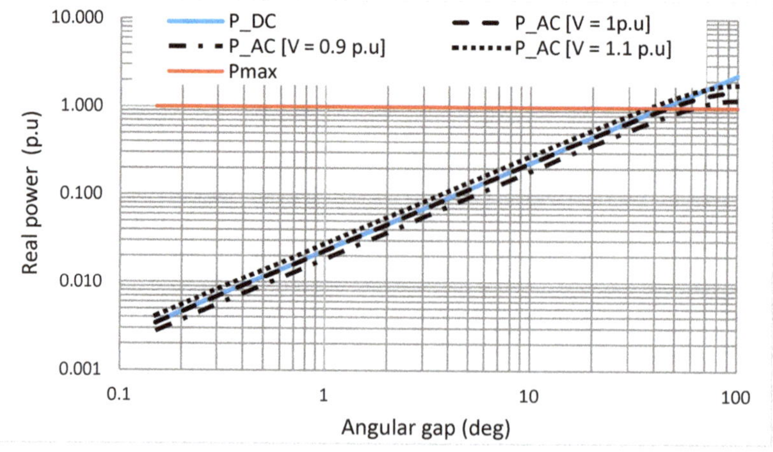

Figure 6. Comparison between the DC power flow and the AC power flow with sensitivity factors with three different voltage magnitudes.

3.3. Line Outages

The algorithm must again perform a generation dispatch calculation based on the changes presented in the power system (L1 and L2 of Figure 1). In the case of the unexpected line outage, logic L1 will activate the algorithm, which will obtain a new economic re-dispatch. In accordance with the above, permanent line outage events that do not represent critical contingencies activate the first economic dispatch algorithm presented in Figure 1 and do not activate the second economic dispatch algorithm presented in Figure 2. For example, in the IEEE 39–bus power system the lines outages are: (a) Line 3–18, (b) Line 16–21, and (c) Line 17–18, and those allow calculating an economic re-dispatch.

For each event a, b and c, the algorithm calculates the economic dispatch and the results are shown in Table 2. The economic dispatch before the outage of line 16–21 is approximately equal to the dispatch of the base case. Therefore, it is not necessary to implement a generation re-dispatch for this case. On the other hand, the output of lines 3–18 and 17–18 originate a more relevant change in the generation dispatch. Additionally, generation plants G4, G5, G6, and G7 remain almost invariant before the outage of these three lines. As generators G4, G5, and G6 are the most economic units, then those generators are not dispatched again because of security constraints related to overloads.

Table 2. Results of the economic re-dispatch after the outage of lines 03–18, 16–21, and 17–18.

Generator	Base case [MW]	Outage of line 03–18 [MW]	Outage of line 16–21 [MW]	Outage of line 17–18 [MW]
G 02	571.216	595.000	572.959	593.792
G 03	253.564	252.937	250.292	203.264
G 04	680.000	680.000	680.000	680.000
G 05	510.000	510.000	510.000	510.000
G 06	221.431	199.869	222.199	221.431
G 07	595.000	595.000	595.000	595.000
G 08	248.729	47.946	248.729	26.604
G 09	651.835	850.000	651.835	850.000
G 10	569.838	566.004	571.287	619.125

None of the three outages of lines 3–16, 16 21, and 17–18 create a risk situation after $N-1$contingencies. However, the resulting economic re-dispatch corresponds to an opportunity to save generation costs, because each outage of each of these lines creates a different topology, so that power system security changes. Thus, the algorithm performs online monitoring of the cost reductions, avoiding the overload risk after $N-1$contingencies, and contributing to the improvement of power system security and operating costs. By considering hour-operation periods, the outage of line 17–18 saves around USD 12,316; the outage of line 16–21 saves only USD 42.15; and the outage of line 03–18 saves USD 8859.73. According to the previous results, we can say that before the unexpected outage of lines 03–18 and 17–18, it is not necessary to carry out a re-dispatch if only the security constraints after $N-1$contingencies are required. However, the resulting economic re-dispatch corresponds to an opportunity to save the operating cost.

Finally, the verification of the deviation value of each of the new economic re-dispatch is important because the power delta lost or gained after each dispatch could affect the operation of the secondary frequency control or AGC. In addition, the generation delta should not be very large, since the AGC must have available a power reserve to respond to frequency events, and in this case the permissive availability of AGC of + −10 MW has been assumed. For the outage of line 03–18 there is a power generation deviation of −4.857 MW, for the outage of line 16–21 there is a deviation of 0.688 MW, and for the outage of line 17–18 there is a deviation of −2,397 MW.

3.4. Critical Contingencies

To involve the operation of the algorithm presented in Figure 2, two lines have been selected, which correspond to the outages of lines 05–08 and 13–14, and which cause critical contingencies $N-1$. Therefore, the algorithm will present two operation stages: the first is presented in Figure 1, where the algorithm must calculate the secure operation and most economical re-dispatch without considering load shedding, and without minimizing the overload after $N-1$contingency. In the second stage presented in Figure 2, the algorithm must calculate the new redistribution and load shedding corresponding to each of the critical contingencies that remain unprotected in the first stage.

In accordance with the above, Figure 7 shows the results of the economic re-dispatch of the first stage, which allows the lowest possible overload after $N-1$contingencies at the lowest generation cost. However, as stated above, the evaluated line outages create critical contingencies $N-1$. Thus, before the outage of any line (05–08 and 13–14), there is no totally secure economic re-dispatch; that is, under this condition there will be at least one $N-1$contingency that causes an overload above the maximum limit of the overloaded element. In Table 3, we present in detail the values of the economic re-dispatch to consider, before the outage of lines 05–08 or 13–14.

Table 3. Results of the economic re-dispatch after outage of lines 05-08 and 13–14.

Generator	Base case [MW]	Outage of lines 05–08 [MW]	Outage of lines 13–14 [MW]
G 02	571.216	595.000	586.299
G 03	253.564	416.762	0.000
G 04	680.000	680.000	680.000
G 05	510.000	510.000	510.000
G 06	221.431	97.853	377.327
G 07	595.000	595.000	595.000
G 08	248.729	47.575	115.680
G 09	651.835	850.000	850.000
G 10	569.838	499.423	577.307

In relation to the deviations of the generation cost, when the outage of line 13–14 is presented, the cost increases by USD 3271.93; therefore, in this case, the algorithm detects the need to increase the generation of more expensive security to reduce the risk of contingencies $N-1$. On the other hand, the result of the cost of generation before the exit of the line 05–08 leads to a reduction of the same in USD 9818.76, which represents an operating benefit, and at the same time leads to a reduction of risk due to overload between $N-1$contingencies.

When evaluating $N-1$contingencies before the outage of line 05–08 considering the economic re-dispatch presented in Table 3, it is observed that lines 09–39, 08–09 and 06–07 present an overload on their maximum values, as seen in Figure 7.

After the outage of line 05–08, the algorithm calculates a fictitious increase in the capacity of lines 09–39, 08–09 and 06–07, resulting in 42.3 MW, 28.2 MW and 142.5 MW, respectively. The fictitious increase in the capacity of transmission lines is calculated by means of the variable PAd_i, of Equations (8) and (10). From the process P5 of Figure 2, critical contingencies are obtained when considering the outage of line 05–08. For this case, the critical contingencies correspond to the lines: 08–09, 06–07, 09–39 and 01–39, as shown in Figure 8.

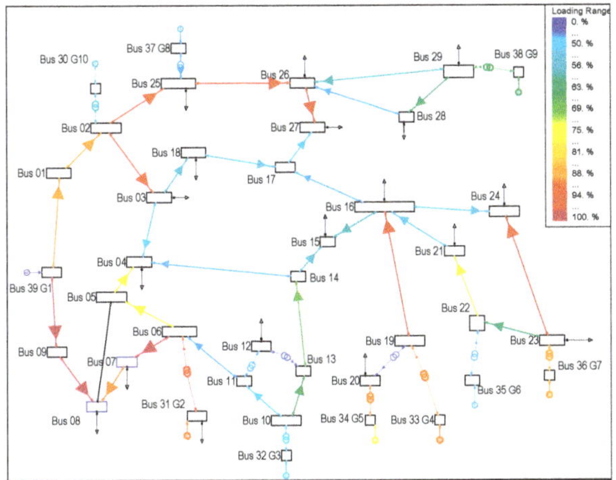

Figure 7. Simulation of maximum loadability of elements when considering N−1contingencies after the outage of line 05-08.

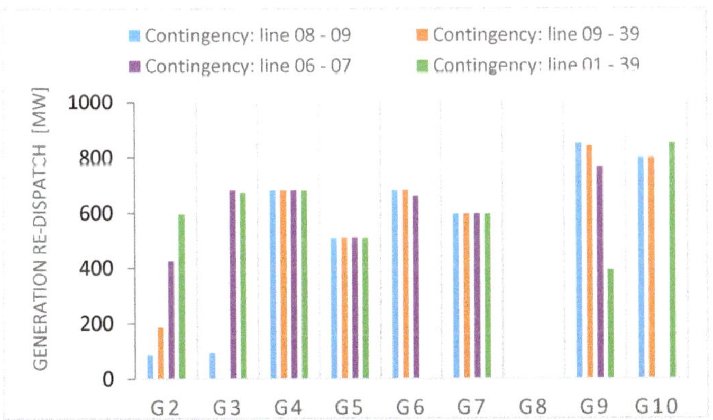

Figure 8. Emergency dispatch after a contingency that considers the outage of line 05–08.

Table 4 shows the results obtained by simulating critical N−1contingencies, considering the outage of line 05–08. The loadability of each element of the power system is presented. For each critical contingency, it is required to carry out the dispatch presented in Figure 9, as well as the corresponding load shedding as shown in Table 4.

Table 4. Results after a contingency that considers the outage of line 05–08.

Results	Line 08–09	Line 09–39	Line 06–07	Line 01–39
Overloaded elements	Line 06–07	Line 06–07	Line 08–09; Line 09–39	Line 09–39
Load shedding	136.7 MW in Bus 07	142.8 MW in Bus 07	25.9 MW in Bus 08	0 MW
Cost of power redistribution and load shedding [USD]	$377,699.28	$388,001.39	$123,821.14	−$58,200.51
Generation deviation [MW]	10.0	10.0	−10.0	7.6

As shown in Table 4, the contingency of line 01–39, would be the only critical contingency that would not increase the operation cost. The other contingencies cause a considerable increase in the operation cost, due to the fact that they require load shedding.

3.5. Economic Re-Dispatch Algorithm Operating after Generation Tripping

For this case, a generation tripping event is considered, which activates the online economic re-dispatch algorithm using the L2 logic, presented in Figure 1, for this case a complete outage of generator G8 in bus 37. Figure 9 shows the loadability status of the transmission elements after the outage of G8 and with N−1contingencies. High loadability are presented in lines 01–39, 01–02, 02–25, 09–39, 08–09, because the slack bus, which represents the generator with AGC control of the power system, responds to the loss of the G8 generation with a real power increase and generates power congestion on those transmission lines. With an event that creates generation-demand imbalance, as the outage of G8, the redistribution of the AGC generators is required, greater than the value considered in the *ResAGC* parameter of Equation (5). Because after the outage of a non-radial transmission line, there would be no generation-demand imbalance; therefore, the adjustment value of the AGC (*ResAGC*) may be lower than after the event that generates a generation-demand imbalance.

For the event of the outage of G8, the *ResAGC* value has been modified by 248.729 MW, which corresponds to the value of the G8 generator dispatch after the outage of G8. Figure 10 shows the variation of the economic re-dispatch calculated from the generation tripping (G8), with which the system will be safe again with N−1contingency. Table 5 shows the values of the economic re-dispatch calculated after the outage of G8, a generation deviation of −19.58 MW is presented, which means that an increase of the *ResAGC* parameter is required, given that the most economical solution is the generators G2 and G3, generating an unbalance greater than 10 MW.

It should be noted that for the simplicity of this model, a cost of 0.0 USD/MWh has been assumed for the AGC generator, because the generator G1 represents not only the AGC, but also the rest of the equivalent system; therefore, the AGC is assumed to be part of the equivalent system and for simplicity the cost of this equivalent generator is assumed to be zero. Deviation of the total generation is −19.58 MW and the generation cost is USD −4945.

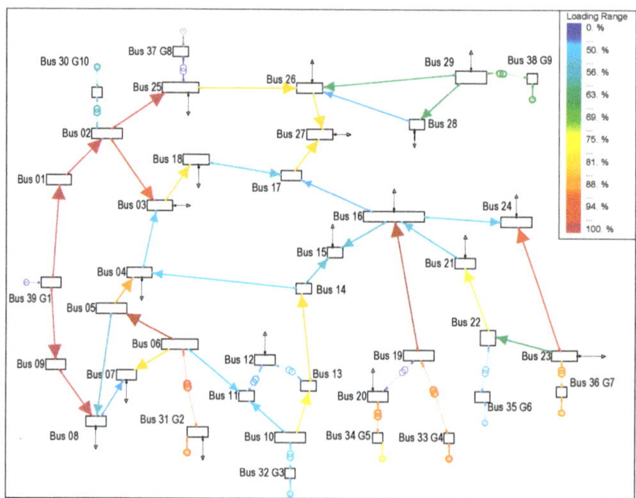

Figure 9. Maximum loadability of elements after outage of G8.

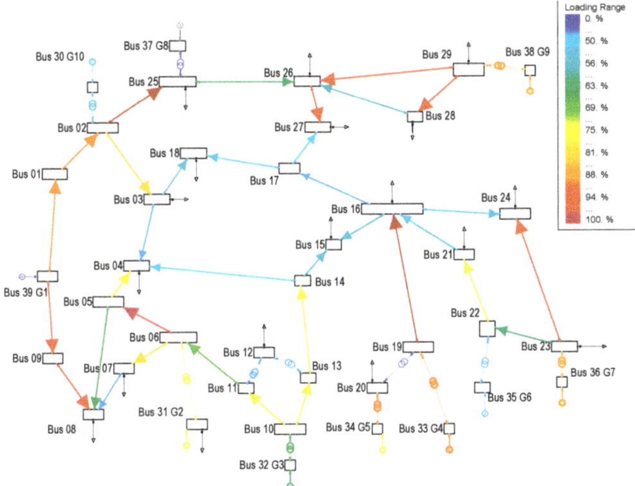

Figure 10. Maximum loadability of elements after outage of G8, with the new economic re-dispatch algorithm.

Table 5. Economic re-dispatch values after the outage of generator G8.

Generator	Base Case [MW]	Outage of G8 [MW]
G 02	571.216	450.93
G 03	253.564	424.237
G 04	680.000	680.000
G 05	510.000	510.000
G 06	221.431	221.431
G 07	595.000	595.000
G 08	248.729	0.000
G 09	651.835	850.000
G 10	569.838	550.435

4. Conclusions

This paper presented an online generation dispatch scheme to minimize generation costs and the loadability of elements in the power system. The proposed method incorporated the power sensitivity factors (LODF and PTDF) in an optimization model with variables obtained from online operation, considering the generation cost, and with overload constraints. The results showed that the online economic re-dispatch identifies the operating cost reduction and reduces the overload risk created by $N-1$ contingencies, contributing to a safety operation, and at the same time reducing the operating costs of the power system. In addition, it is a useful proposal when there are highly variable power plants such as power systems with solar and wind power.

This operating scheme could also be very useful in power systems in which there are usually appreciable differences in demand, dispatch or topology, between the scheduled operation and the actual operation. The complementary algorithm to deal with critical $N-1$ contingencies, corresponds to a useful proposal in cases in which the network is not fully covered before all the possible $N-1$ contingencies. It is also considered a necessary operation scheme, in cases in which the only solution to avoid collapse is load shedding. Therefore, instead of executing a programmed power demand rationing, the model uses a prediction of the required load shedding and emergency economic re-dispatch, which would be implemented only in cases of critical contingency.

The model considered sensitivity factors in order to simplify calculations, and it provided an advantage when considering large-scale systems divided into many operational areas. The online economic re-dispatch presented in the present work does not require metaheuristic techniques, so the minimization of the generation cost is done by an exact and simple method, which means it is suited to an online operation. Thanks to the model based on the objective function with the variable PAD_i, it is possible to achieve the convergence of the model, even when the system is unsafe with critical contingencies. Therefore, the result of this model corresponds to the security constraint generation economic re-dispatch without the need to calculate power demand rationing.

Author Contributions: Conceptualization, methodology, software, validation, formal analysis, investigation, and writing original draft were performed by O.A.-C. and J.E.C.-B. Formal analysis, writing, review, and editing was performed by F.E.H.V.

Funding: This research received no external funding.

Acknowledgments: This work was supported by the Universidad Nacional de Colombia, Sede Medellín. We would like to thank to the Department of Electrical Engineering and Automation and the Grupo de Investigación en Tecnologías Aplicadas GITA for the continuous support to our research.

Conflicts of Interest: The authors declare no conflict of interest.

References

1. Morison, K.; Wang, L.; Kundur, P. Power system security assessment. *IEEE Power Energy Mag.* **2004**, *2*, 30–39. [CrossRef]
2. Kirschen, D.S. Power system security. *Power Eng. J.* **2002**, *16*, 241–248. [CrossRef]
3. Balu, N.; Bertram, T.; Bose, A.; Brandwajn, V.; Cauley, G.; Curtice, D.; Fouad, A.; Fink, L.; Lauby, M.G.; Wollenberg, B.F.; et al. On-line power system security analysis. *Proc. IEEE* **1992**, *80*, 262–282. [CrossRef]
4. Ronellenfitsch, H.; Timme, M.; Witthaut, D. A Dual Method for Computing Power Transfer Distribution Factors. *IEEE Trans. Power Syst.* **2016**, *32*, 1007–1015. [CrossRef]
5. Zhu, J. *Optimization of Power System Operation*; John Wiley & Sons, Inc.: Hoboken, NJ, USA, 2009; ISBN 9780470466971.
6. Men, K.; Chung, C.Y.; Lu, C.; Zhang, J.; Tu, L. Online re-dispatching of power systems based on modal sensitivity identification. *IET Gener. Transm. Distrib.* **2015**, *9*, 1352–1360. [CrossRef]
7. del Valle, Y.; Venayagamoorthy, G.K.; Mohagheghi, S.; Hernandez, J.-C.; Harley, R.G. Particle Swarm Optimization: Basic Concepts, Variants and Applications in Power Systems. *IEEE Trans. Evol. Comput.* **2008**, *12*, 171–195. [CrossRef]
8. Sen, S.; Chanda, S.; Sengupta, S.; Chakrabarti, A.; De, A. Alleviation of line congestion using Multiobjective Particle Swarm Optimization. In Proceedings of the 2011 International Conference on Electrical Engineering and Informatics, Bandung, Indonesia, 17–19 July 2011; pp. 1–5.
9. Kuruseelan, S. A Novel Method for Generation Rescheduling to Alleviate Line Overloads. *Int. J. Electr. Energy* **2014**, *2*, 167–171. [CrossRef]
10. Verma, S.; Saha, S.; Mukherjee, V. Optimal rescheduling of real power generation for congestion management using teaching-learning-based optimization algorithm. *J. Electr. Syst. Inf. Technol.* **2018**, *5*, 889–907. [CrossRef]
11. Maharana, M.K.; Swarup, K.S. Graph theory based corrective control strategy during single line contingency. In Proceedings of the 2009 International Conference on Power Systems, Kharagpur, India, 27–29 December 2009; pp. 1–6.
12. Gupta, A.K.; Kiran, D.; Abhyankar, A.R. Flexibility in transmission switching for congestion management. In Proceedings of the 2016 National Power Systems Conference (NPSC), Bhubaneswar, India, 19–21 December 2016; pp. 1–5.
13. Pandiarajan, K.; Babulal, C.K. Overload alleviation in electric power system using fuzzy logic. In Proceedings of the 2011 International Conference on Computer, Communication and Electrical Technology (ICCCET), Tamilnadu, India, 18–19 March 2011; pp. 417–423.
14. Chahar, R.K.; Chuahan, A. Devising a New Method for Economic Dispatch Solution and Making Use of Soft Computing Techniques to Calculate Loss Function. In *Software Engineering. Advances in Intelligent Systems and Computing*; Springer: Singapore, 2019; pp. 413–419.

15. Bie, P.; Chiang, H.-D.; Zhang, B.; Zhou, N. Online Multiperiod Power Dispatch With Renewable Uncertainty and Storage: A Two-Parameter Homotopy-Enhanced Methodology. *IEEE Trans. Power Syst.* **2018**, *33*, 6321–6331. [CrossRef]
16. Kirschen, D.S. Do Investments Prevent Blackouts? In Proceedings of the 2007 IEEE Power Engineering Society General Meeting, Tampa, FL, USA, 24–28 June 2007; pp. 1–5.
17. Tejada-Arango, D.A.; Sanchez-Martin, P.; Ramos, A. Security Constrained Unit Commitment Using Line Outage Distribution Factors. *IEEE Trans. Power Syst.* **2018**, *33*, 329–337. [CrossRef]
18. Eastern Interconnection Reliability Assessment Group (ERAG). *Study Procedure Manual*; ReliabilityFirst–NPCC (RN): Cleveland, OH, USA; Midwest Reliability Organization–ReliabilityFirst–SERC–Southwest Power Pool (MRSS): Cleveland, OH, USA; ReliabilityFirst Corporation: Cleveland, OH, USA, 2015; pp. 1–38.
19. Arenas-Crespo, O.; Candelo-Becerra, J.E. A power constraint index to rank and group critical contingencies based on sensitivity factors. *Arch. Electr. Eng.* **2018**, *67*, 247–261. [CrossRef]
20. Chen, Y.C.; Dominguez-Garcia, A.D.; Sauer, P.W. Generalized injection shift factors and application to estimation of power flow transients. In Proceedings of the 2014 North American Power Symposium (NAPS), Pullman, WA, USA, 7–9 September 2014; pp. 1–5.
21. Chen, Y.C.; Dominguez-Garcia, A.D.; Sauer, P.W. Measurement-Based Estimation of Linear Sensitivity Distribution Factors and Applications. *IEEE Trans. Power Syst.* **2014**, *29*, 1372–1382. [CrossRef]
22. Unidad de Planeación Minero Energética—UPME Costo Incremental Operativo de Racionamiento de Energía. Available online: http://www.upme.gov.co/CostosEnergia.asp (accessed on 1 March 2019).

Article

An Expeditious Methodology to Assess the Effects of Intermittent Generation on Power Systems

Gracita Batista Rosas [1], Elizete Maria Lourenço [2,*], Djalma Mosqueira Falcão [3] and Thelma Solange Piazza Fernandes [2]

[1] Electrical Engineering Department, Brazilian Energy Utility Companhia Paranaense de Energia (COPEL), Curitiba 81200-240, Brazil; gracita.rosas@copel.com
[2] Electrical Engineering Department, Federal University of Paraná, Curitiba 81531-980, Brazil; thelma@eletrica.ufpr.br
[3] Electrical Engineering Department, Federal University of Rio de Janeiro/COPPE, Rio de Janeiro 21941-972, Brazil; falcao@nacad.ufrj.br
*   Correspondence: elizete@eletrica.ufpr.br; Tel.:+55-41-3361-3688

Received: 2 March 2019; Accepted: 21 March 2019; Published: 23 March 2019

Abstract: This paper proposes an expeditious methodology that provides hourly assessments of the effect of intermittent wind and solar power generation on the electrical quantities characterizing power systems. Currents are measured via circuit breakers to confirm the correct sizing of devices based on their rated currents. Nodal voltage magnitudes are assessed for compliance with limits imposed by regulatory authorities, whereas the active power produced by hydroelectrical generators is assessed for reserve energy. The proposed methodology leverages a fuzzy extended deterministic optimal power flow that uses in power balance equations the average hourly values of active power generated by wind and solar sources as well as hourly energy load. The power grid is modeled at the substation level to directly obtain power flow through circuit breakers. Uncertainties in power system electrical quantities are assessed for an optimal solution using a Taylor series associated with deviations from the average values of the active power produced by the wind and solar sources. These deviations are represented using a fuzzy triangular model reflecting the approximations of the probability density functions of these powers. The methodology takes into account a subjective investigation that focuses on the qualitative characteristic of these energy sources' behaviors.

Keywords: circuit breaker; network modeling; optimal power flow; wind and solar energies

---

1. Introduction

The challenges of the expansion and operation of Electrical Power Systems (EPS) have intensified over time owing to the incessant increase in the complexity of these systems. Nowadays, additional factors have been incorporated in these challenges, including uncertainties in the generation and consumption of electric energy as dictated by the growing participation of intermittent sources of energy, domestic electric vehicles, and energy storage systems. Because of these uncertainties, the technical specifications of the EPS components must be assessed, especially those associated with substations' internal devices, as is the case with circuit breakers. Therefore, it is of fundamental importance that the tools in development that seek to model these uncertainties are also capable of assessing the loading of these devices in their analysis processes.

Optimal power flow (OPF) is recognized as a powerful tool for planning an operation and has gained importance in analysis involving the incorporation of distributed generation [1–3]. OPF is able to indicate the ideal size of these generation units to avoid compromising voltage stability [4,5]. The uncertainties of wind and solar energy sources have been widely explored in OPFs, whether these were coordinated with conventional generation or in association with energy storage systems [6–8].

In this context, some works have explored the characteristics of the OPF to analyze the impact of the expansion of distributed generation on transmission line loading [9], on the presence of domestic electric vehicles in power grids [10,11], on high-voltage direct current connections [12,13], and in frequency control actions [14].

The uncertainties of energy sources are often represented in an OPF by probabilistic approaches, which are unable to incorporate qualitative information residing in human knowledge. In this case, fuzzy logic has been used to represent qualitative, vague, or incomplete knowledge in mathematical models in several areas of knowledge, including the OPF [15]. The literature presents the use of fuzzy logic combined with OPF as promising for EPS analysis [16], mainly in the definition of functions with conflicting objectives [17,18], in operational limit constraints [19,20], and in solutions that define the date of need, location, and sizing of new energy resources [21]. Some techniques incorporate fuzzy logic in the OPF solution through a linear relation [22], or through sensitivity analyses [23]. This allows for the assessment of the influence of uncertainties in energy generation in the electrical quantities of EPS.

Another promising application of the OPF has been the inclusion of switch and circuit breaker representation, hereinafter referred to as switchable branches, in power grid modeling. Through continuous functions, this representation has been used successfully in assessments of system reconfiguration [24] and in restoring the maximum loads in distribution systems [25]. The representation of these devices by integer variables is proposed in [26] to obtain the optimum dispatch and topology of an electric network to meet static loads. In some applications, the concept of switching transforms contingencies (criterion $n - 1$) into inequality constraints as a way of guaranteeing safety in the electrical system [27]. An OPF with a switching technique has also been used to alleviate transmission line congestion [28], in addition to improving the hosting of distributed generation systems [29] and minimizing losses [30] through network reconfiguration.

The conventional bus-branch representation of the power grid appears in the vast majority of OPF applications. However, developments related to the explicit representation of switches and circuit breakers in state estimation studies [31] have been incorporated into the power flow problem formulation [32], allowing for the processing of power grids at the substation level. In this type of modeling, the power flows through switchable branches are incorporated into the vectors of state variables together with the nodal complex voltages. This incorporation avoids possible numerical problems in the search for an optimal solution, motivated by the representation of impedances with very high or very low values. This is necessary to indicate the open and closed positions of these devices, respectively. As a consequence, this modeling allows for the direct determination of the power flow distribution through the components of substations, which is necessary for the analysis of electrical device loading, especially in circuit breakers. Preliminary studies demonstrated the feasibility of modeling at the substation level in the formulation of the OPF problem [33].

This work is inserted in this context and proposes an expeditious methodology capable of determining, for each hour of the day, the uncertainties in EPS electrical quantities owing to the uncertainties of wind and solar energy sources. The proposed approach consists of a Fuzzy Extended Deterministic OPF (FED-OPF) composed by two stages, which aims to explore the subjectivity of qualitative knowledge associated with these energy sources' behaviors. The first stage solves a deterministic OPF with extended formulation, referred to as ED-OPF, that is capable of processing modeled power grids at the substation level. In this first stage, the average values of the active power generated by wind and solar energy sources are used as input data together with the data referring to hourly load curves. In the second stage, the uncertainties of the EPS electric quantities are determined by the sum of the deterministic variables (determined in the first stage) with the fuzzy variables. The fuzzy variables are determined through qualitative sensitivity analysis applied to the postprocessing of the ED-OPF, considering maximum and minimum values for the active power of wind and solar energy sources. This procedure allows that the qualitative behavior of wind and solar energies sources be incorporated to the solution.

The proposed methodology allows for the assessment, in an expeditious way, of the influence of the uncertainties of the active power produced by wind and solar energy sources: (a) in the active power produced by conventional generators for energy reserve analysis, (b) in the nodal voltage magnitudes in order to comply with the limits imposed by regulatory authorities, and, mainly, (c) in the currents through circuit breakers. These results are essential to ascertain the correct sizing of the rated current of these devices.

2. Fuzzy Extended Deterministic Optimal Power Flow

Originating from an extension of the economic dispatch [34], the OPF allows for the determination of the optimum point of operation in relation to a predefined objective, while it complies with the operational limits imposed by power grids. Over the years, mainly owing to uncertainties in the generation and consumption of electric energy, the OPF was marked by a progression in the numerical techniques of solutions and in problem formulation. Among these techniques of solution, it is worth mentioning the Primal Dual Interior Points Method, initially proposed for the optimal dispatch of reactive power, which proved to be robust in solutions of large systems [35].

Based on these assumptions and considering the subjectivity of qualitative knowledge about the wind and solar energy sources behavior, this paper proposes a fuzzy extended deterministic optimal power flow (FED-OPF) capable of determining, for each hour of the day, the uncertainties of EPS electrical quantities owing to the uncertainties of these energy sources. These uncertainties are determined mainly for the currents through circuit breakers, to verify the correct sizing of these devices by the rated current, for the active power produced by conventional generators for energy reserve analysis and for the nodal voltage magnitudes in order to comply with the limits imposed by the regulatory authorities. The hourly analysis via the proposed FED-OPF is composed by two stages, as described in the following sections.

2.1. Extended Deterministic Optimal Power Flow: First Stage

In this work, the modeling of a power grid at the substation level is incorporated in the ED-OPF formulation. This modeling allows for the explicit representation of the switches and circuit breakers that make up the substation arrangements. This is different from bus-branch conventional modeling, which considers these arrangements as a single bus. To illustrate substation-level modeling, Figure 1 shows a simple five-bus system (a) where bus three started to rely on this modeling, resulting in an eleven-bus system (b).

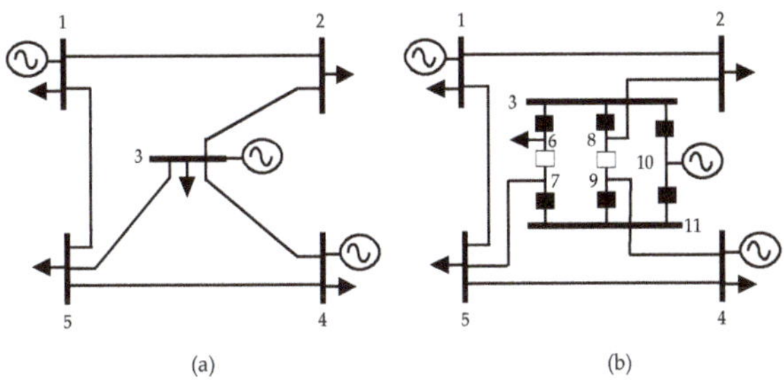

Figure 1. (a) Bus-branch modeling and (b) substation level modeling.

With this new modeling, it is possible to obtain, in a direct way, information of electrical quantities through the substation components. This includes the power flows through circuit breakers, thus creating subsidies to confirm some technical specifications of these devices. As a consequence, this modeling excludes the need to use complementary analysis tools to obtain this information, as occurs with bus-branch conventional modeling.

It should be emphasized that the modeling at the substation level is applied only to previously selected substations, which are the substations through which it is desired to determine the power flow distribution and the influence of wind and solar sources. This means that only a small number, usually the substations connected to solar and wind sources, will be modeled at this level of detail.

In the following formulation, the details of the modeling of a power grid at the substation level and of the wind and solar power sources in the ED-OPF are presented:

$$\min C(\mathbf{Pg}) = \mathbf{Pg}^t \, \mathbf{Q} \, \mathbf{Pg} + \mathbf{b}^t \, \mathbf{Pg} + \mathbf{co} \tag{1}$$

subject to:

$$\mathbf{Pg} + \overline{\mathbf{Pg}}_{win} + \overline{\mathbf{Pg}}_{sol} - \mathbf{Pd} = \mathrm{real}\{\mathrm{diag}(\mathbf{V}).(\mathbf{YV})^*\} + \mathbf{tfl} \tag{2}$$

$$\mathbf{Qg} + \mathbf{Bs} - \mathbf{Qd} = \mathrm{imag}\{\mathrm{diag}(\mathbf{V}).(\mathbf{YV})^*\} + \mathbf{ufl} \tag{3}$$

$$f\left(\theta_{ij}, \mathbf{t}_{ij}\right)^{P\theta} = 0 \tag{4}$$

$$f\left(\mathbf{V}_{ij}, \mathbf{u}_{ij}\right)^{QV} = 0 \tag{5}$$

$$\mathbf{Pg}_{min} \leq \mathbf{Pg} \leq \mathbf{Pg}_{max} \tag{6}$$

$$\mathbf{Qg}_{min} \leq \mathbf{Qg} \leq \mathbf{Qg}_{max} \tag{7}$$

$$\mathbf{V}_{min} \leq \mathbf{V} \leq \mathbf{V}_{max} \tag{8}$$

$$\mathbf{a}_{min} \leq \mathbf{a} \leq \mathbf{a}_{max} \tag{9}$$

$$\mathbf{b}_{min} \leq \mathbf{Bs} \leq \mathbf{b}_{max} \tag{10}$$

$$\boldsymbol{\varphi}_{min} \leq \boldsymbol{\varphi} \leq \boldsymbol{\varphi}_{max} \tag{11}$$

$$\mathbf{fl}_{min} \leq \mathbf{fl} \leq \mathbf{fl}_{max} \tag{12}$$

Equation (1) represents the conventional objective function that minimizes the costs of active power generation **Pg** by conventional generators, where **Q** is the diagonal matrix with quadratic cost coefficients, **b** is the vector of linear cost coefficients and **co** is the constant cost vector. Equations (6) and (7) represent the minimum and maximum limit constraints of the active and reactive powers **Pg** and **Qg** produced by these generators, respectively. Equations (8) to (12) represent these limit constraints for the nodal voltage magnitudes **V**, taps of the transformers **a**, powers injected by static compensators **Bs**, angles of the shift transformers φ, and the active power flows on transmission lines **fl**, respectively.

The necessary adaptations in the ED-OPF formulation for the processing of power grids at the substation level are performed by extending the state vector so as to include the active (**t**) and reactive (**u**) power flow through each switchable branch along with the conventional nodal complex voltages. This procedure avoids the use of atypical values to represent the open and closed position of switchable branches in the problem formulation, eliminating the numeric problem such values would cause in the optimal solution search.

Equations (2) and (3) represents, respectively, the active and reactive bus power balance (or bus injected power), that is, the difference between the generated power and the demand at each bus. Please notice that, as a consequence of the current Kirchhoff law, the injected power at bus "k" is equal to the sum of the power flow through all adjacent branches of bus "k". When the substation level modeling is applied to the network, as proposed in this article, the sum of the power flow through adjacent branches must consider the set of adjacent conventional branches (transmission lines and transformers) as well as the set of adjacent switchable branches. Power flow through conventional branches are determined as usually, that is, as a function of the complex voltage. However, power flow through switchable branches are written directly as a function of the new state variables (**t** and **u**).

To incorporate the power flows through switchable branches in power balance equations (2) and (3) of the ED-OPF, the bus-switchable branches incidence matrix $\mathbf{A}_{bb}$ is defined. With the number of rows equal to the number of buses of the system and the number of columns equal to the number of switchable branches, the element i-j of $\mathbf{A}_{bb}$ is defined as

- -1 if the switchable branch j is connected to bus i and the flow is oriented to enter in bus i.
- 1 if the switchable branch j is connected to bus i and the flow is oriented outward from bus i.
- 0 if switchable branch j is not connected to bus i.

In (2), **tfl** represents the vector of the sum of the switchable branches active power flow in the buses. It can be written using the incidence matrix $\mathbf{A}_{bb}$ as indicated in (13):

$$\mathbf{tfl} = \mathbf{A}_{bb} \cdot \mathbf{t} \tag{13}$$

Similarly, in (3), **ufl** represents the contribution, that is, the sum of the reactive power flows through switchable branches to the system buses, and can be written as:

$$\mathbf{ufl} = \mathbf{A}_{bb} \cdot \mathbf{u} \tag{14}$$

where **t** and **u** are the new state variable vectors referring to the active and reactive power flows in switchable branches, respectively.

Equation (4) represents the active operational constraints that model the closed and open positions of the switches and circuit breakers. These constraints can be reformulated in function of specific incidence matrices for this purpose, as presented in (15):

$$f(\theta, t)^{P\theta} = \begin{bmatrix} \mathbf{A}_c & 0 \\ 0 & \mathbf{A}_o \end{bmatrix} \cdot \begin{bmatrix} 0 \\ t \end{bmatrix} = \begin{bmatrix} \mathbf{A}_c \cdot \theta \\ \mathbf{A}_o \cdot t \end{bmatrix} \tag{15}$$

The incidence matrix $\mathbf{A}_c$ has a number of lines equal to the number of closed circuit breakers and a number of columns equal to the number of total buses of the system. This matrix represents the operational constraints of closed circuit breakers of the type $\theta_i - \theta_j = 0$, where θ is the angle of the complex voltages. Thus, the values of this matrix are equal to 1 in column "i" and equal to -1 in column "j" of the line corresponding to the closed circuit breaker i-j.

The incidence matrix $\mathbf{A}_o$ has a number of lines equal to the number of open circuit breakers and a number of columns equal to the total number of circuit breakers. This matrix represents the operational constraints of open circuit breakers of type $t_{ij} = 0$. The values of this matrix will only be different from zero and are necessarily equal to 1 in the line corresponding to open circuit breaker i-j and the column corresponding to the state variable (active power flow) associated with this circuit breaker.

A similar reformulation can be performed on the reactive operating restrictions of (5), as shown in (16):

$$f(\mathbf{V}, \mathbf{u})^{QV} = \begin{bmatrix} \mathbf{A}_c & 0 \\ 0 & \mathbf{A}_o \end{bmatrix} \cdot \begin{bmatrix} V \\ u \end{bmatrix} = \begin{bmatrix} \mathbf{A}_c \cdot V \\ \mathbf{A}_o \cdot u \end{bmatrix} \tag{16}$$

From (16), it is possible to obtain the operational restrictions of the closed circuit breakers of type $V_i - V_j = 0$, in addition to the operational restrictions of the open circuit breakers of type $u_{ij} = 0$.

Regarding the representation of alternative sources of energy, (2) considers in the balance of the active power the average values of the active power produced by wind and solar energy sources $\overline{Pg}_{win}$ and $\overline{Pg}_{sol}$, respectively. The average values can be determined from real historic data, as discussed in the Section 3.

The solution of the proposed extended deterministic optimal power flow is achieved by the Primal Dual Interior Points Method [35] as associated with the Lagrange technique, the conditions of Karush Kuhn Tucker, and the Newton Method.

This solution defines the deterministic variables of EPS electrical quantities as function of average values of the active power produced by wind and solar energy sources and demands the traditional expensive computational time of solutions involving OPFs.

2.2. Electrical Quantity Uncertainties of EPS: Second Stage

The information regarding the uncertainties of generation and consumption of electric energy in the OPF is usually represented by probabilistic approaches through probability density functions [4,36,37]. However, these approaches based on quantitative information are not capable of representing the qualitative information that resides in human knowledge. With the fuzzy technique, the incomplete, vague, and qualitative knowledge present in human beings began to be represented by mathematical models in several areas of knowledge, including in OPFs [15].

In addition to not considering qualitative information, usually the processing of probabilistic data in the OPF demands a high computational effort. However, in several practical situations, expeditious and even approximate responses of the electric quantities of the power systems are enough for the analysts of these systems. An example of such situations is the investigation of the necessity of replacing circuit breakers with equipment that has a greater rated current capacity, usually common in expansions and changes in EPS.

Based on the above discussion, this work proposes an hourly evaluation of the electrical quantities of the electric system in relation to the intermittent behavior of wind and solar energy sources, based on the association illustrated in the Figure 2. In this illustration the area delimited by the mean and $+/-3$ standard deviations of the Probability Density Function (PDF) normal distribution of the active power produced by wind and solar energy sources is associated with the area delimited by the membership function of a fuzzy triangular model. This approach allows composing, through fuzzy variables and a real brief historical data about these energy sources, an hourly qualitative knowledge about the average, minimum, and maximum values of powers produced by these sources. Section 3 exemplifies the composition of this qualitative knowledge.

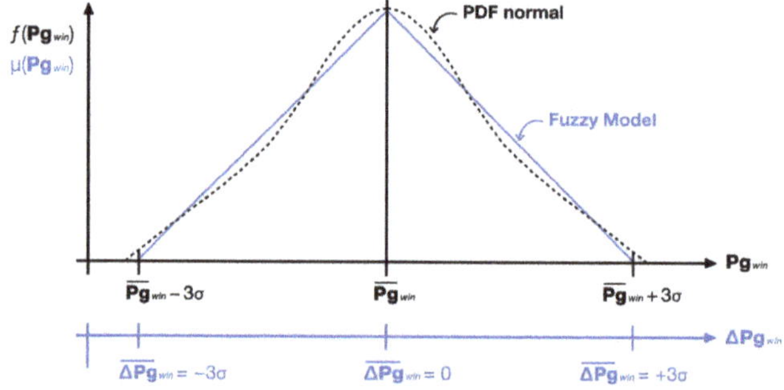

Figure 2. Probability Density Function (PDF) and fuzzy triangular model for wind and solar energy sources.

The Figure 2 shows the Probability Density Function (PDF) (in black), where: $f(\mathbf{Pg}_{\text{win}})$: probability density function of the active power produced by wind energy sources; $\overline{\mathbf{Pg}}_{\text{win}}$: hourly average values of the active power produced by wind energy sources; σ: standard deviation of the active power produced by wind energy sources.

For the fuzzy triangular model (in blue), the Figure 2 shows: $\mu(\mathbf{Pg}_{\text{win}})$: membership function of the active power produced by wind energy sources; $\overline{\Delta\mathbf{Pg}}_{\text{win}}$: hourly variations in relation to the average values of the active power produced by wind energy sources, here called fuzzy variables. For the solar energy sources, we can use the same definition, changing the subscript "win" to "sol".

As the qualitative average values of the active power produced by these sources $\overline{\mathbf{Pg}}_{\text{win}}$ and $\overline{\mathbf{Pg}}_{\text{sol}}$ were considered in the optimal solution of the ED-OPF of the first stage (Equation (2)), it is possible to define the fuzzy variables of the other system electrical quantities as a function of the fuzzy variables $\overline{\Delta Pg}_{win}$ and $\overline{\Delta Pg}_{sol}$. For this purpose, let us consider the Lagrangian function of the problem formulation (1)–(12). The first derivative (or gradient) of this function in the optimal solution must be null, in order to comply with the optimality conditions of Karush Kuhn Tucker. This gradient is represented by (17):

$$\rho\left(\mathbf{Pg},\ \overline{\mathbf{Pg}}_{win},\ \overline{\mathbf{Pg}}_{sol},\mathbf{V},\mathbf{t},\mathbf{y}\right) = 0 \tag{17}$$

where, $\mathbf{Pg}$ is the vector of the active power produced by conventional generators; $\overline{\mathbf{Pg}}_{win}$ and $\overline{\mathbf{Pg}}_{sol}$ are the vectors of the qualitative average values of the active power produced by wind and solar energy sources, respectively; $\mathbf{V}$ is the vector of the nodal voltage magnitudes; $\mathbf{t}$ is the vector of the active power flows through circuit breakers; and $\mathbf{y}$ is the vector of the other electrical quantities of the system, including the dual variables.

Based on this, the expansion of (17) in Taylor series, in the directions of fuzzy variables $\overline{\Delta Pg}_{win}$ and $\overline{\Delta Pg}_{sol}$, still in the optimal solution, can be formulated by:

$$\begin{aligned}
\rho&\left(\mathbf{Pg} + \Delta\mathbf{Pg},\ \overline{\mathbf{Pg}}_{win} + \overline{\Delta\mathbf{Pg}}_{win},\ \overline{\mathbf{Pg}}_{sol} + \overline{\Delta\mathbf{Pg}}_{sol},\mathbf{V} + \Delta\mathbf{V},\mathbf{t} + \Delta\mathbf{t},\ \mathbf{y} + \Delta\mathbf{y}\right)\\
&= \rho\left(\mathbf{Pg},\ \overline{\mathbf{Pg}}_{win},\overline{\mathbf{Pg}}_{sol},\mathbf{V},\mathbf{t},\mathbf{y}\right) + \frac{\partial\mathbf{p}}{\partial\mathbf{Pg}}\Delta\mathbf{Pg} + \frac{\partial\mathbf{p}}{\partial\overline{\mathbf{Pg}}_{win}}\overline{\Delta\mathbf{Pg}}_{win} + \frac{\partial\mathbf{p}}{\partial\overline{\mathbf{Pg}}_{sol}}\overline{\Delta\mathbf{Pg}}_{sol} + \frac{\partial\mathbf{p}}{\partial\mathbf{V}}\Delta\mathbf{V} + \frac{\partial\mathbf{p}}{\partial\mathbf{t}}\Delta\mathbf{t} + \frac{\partial\mathbf{p}}{\partial\mathbf{y}}\Delta\mathbf{y}
\end{aligned} \tag{18}$$

This relationship implies that:

$$\begin{bmatrix} \dfrac{\partial\mathbf{p}}{\partial\mathbf{Pg}} & \dfrac{\partial\mathbf{p}}{\partial\mathbf{V}} & \dfrac{\partial\mathbf{p}}{\partial\mathbf{t}} & \dfrac{\partial\mathbf{p}}{\partial\mathbf{y}} \end{bmatrix}\begin{bmatrix} \Delta\mathbf{Pg} \\ \Delta\mathbf{V} \\ \Delta\mathbf{t} \\ \Delta\mathbf{y} \end{bmatrix} = -\frac{\partial\mathbf{p}}{\partial\overline{\mathbf{Pg}}_{win}}\overline{\Delta\mathbf{Pg}}_{win} - \frac{\partial\mathbf{p}}{\partial\overline{\mathbf{Pg}}_{sol}}\overline{\Delta\mathbf{Pg}}_{sol} \tag{19}$$

Or that:

$$\mathbf{W}'\begin{bmatrix} \Delta\mathbf{Pg} \\ \Delta\mathbf{V} \\ \Delta\mathbf{t} \\ \Delta\mathbf{y} \end{bmatrix} = -\frac{\partial\mathbf{p}}{\partial\overline{\mathbf{Pg}}_{win}}\overline{\Delta\mathbf{Pg}}_{win} - \frac{\partial\mathbf{p}}{\partial\overline{\mathbf{Pg}}_{sol}}\overline{\Delta\mathbf{Pg}}_{sol} \tag{20}$$

Defining $\mathbf{W}$ as a Hessian matrix, the equation can be rewritten as:

$$\begin{bmatrix} \Delta\mathbf{Pg} \\ \Delta\mathbf{V} \\ \Delta\mathbf{t} \\ \Delta\mathbf{y} \end{bmatrix} = \mathbf{W}^{-1}\left(-\frac{\partial\mathbf{p}}{\partial\overline{\mathbf{Pg}}_{win}}\overline{\Delta\mathbf{Pg}}_{win} - \frac{\partial\mathbf{p}}{\partial\overline{\mathbf{Pg}}_{sol}}\overline{\Delta\mathbf{Pg}}_{sol}\right) \tag{21}$$

Equation (21) defines the fuzzy variables of the active power produced by conventional generators $\Delta\mathbf{Pg}$, of the nodal voltage magnitudes $\Delta\mathbf{V}$, of the active power flows through switches and circuit breakers $\Delta\mathbf{t}$, and of the other system quantities $\Delta\mathbf{y}$. With these fuzzy variables, it is possible to determine the uncertainties of the electrical quantities of the EPS. These uncertainties result from the sum of the deterministic variables calculated in the optimal ED-OPF solution as determined in the first stage, with the fuzzy variables determined in this stage, as shown below:

$$\mathbf{Pg}_{uncertain} = \mathbf{Pg} + \Delta\mathbf{Pg} \tag{22}$$

$$\mathbf{V}_{uncertain} = \mathbf{V} + \Delta\mathbf{V} \tag{23}$$

$$\mathbf{t}_{uncertain} = \mathbf{t} + \Delta\mathbf{t} \tag{24}$$

$$\mathbf{y}_{uncertain} = \mathbf{y} + \Delta\mathbf{y} \tag{25}$$

where $\mathbf{Pg}_{uncertain}$ are the uncertainties regarding the active power produced by conventional generators, necessary for energy reserve analysis; $\mathbf{V}_{uncertain}$ are the uncertainties regarding the nodal voltage magnitudes, necessary to analyze the accomplishment of the limits imposed by the regulatory authorities; $\mathbf{t}_{uncertain}$ are the uncertainties regarding the active power flows through circuit breakers, necessary to verify the correct sizing of these devices by the rated current; and $\mathbf{y}_{uncertain}$ are the uncertainties regarding the other electrical quantities that can be used according to necessity.

In summary, the uncertainties of EPS electrical quantities are determined, in the second stage, by the sum of deterministic variables (determined in the first stage) and fuzzy variables. The fuzzy variables are determined through qualitative sensitivity analysis (Equations (17)–(21)) applied to the postprocessing of the ED-OPF optimal solution, as function of variation (maximum and minimum values) of the active power of wind and solar energy sources, always considering an hourly based analysis.

The fuzzy variables participate only in the postprocessing stage of the OPF and avoid the incorporation of these uncertainties into the iterative process of the OPF solution. The result is a significant reduction in the computational time when compared with the first stage. The second stage provides rapid and authentic responses (as discussed in next section) for a general analysis of the electrical quantities of the system as a function of the variation of wind and solar energy sources.

3. Test System Data

To implement the proposed methodology, a real power network composed of 139 buses that comprises the northeast Brazilian coastal region was used. The system contains buses at different voltage levels ranging from 13.8 kV to 500 kV. In addition, the available data includes real hourly load curves, static compensators, capacitor banks, bus and line reactors, previously selected substations modeled at the substation level, and large hydroelectric power plants: Sobradinho (1.00 GW), Xingó (3.6 GW), Luiz Gonzaga (1.5 GW), and Paulo Afonso Complex (2.4 GW). The power system has approximately 400 MW of thermoelectric power plants.

Additionally, the power system has 19 wind farms that represent approximately 2 GW of installed power. Figure 3 illustrates the daily variation of active power produced by these wind farms in the period between June 2016 and July 2017. All of this information corresponds to daily real data properly processed and grouped to subsidize the analysis of the proposed tool in the most realistic way possible.

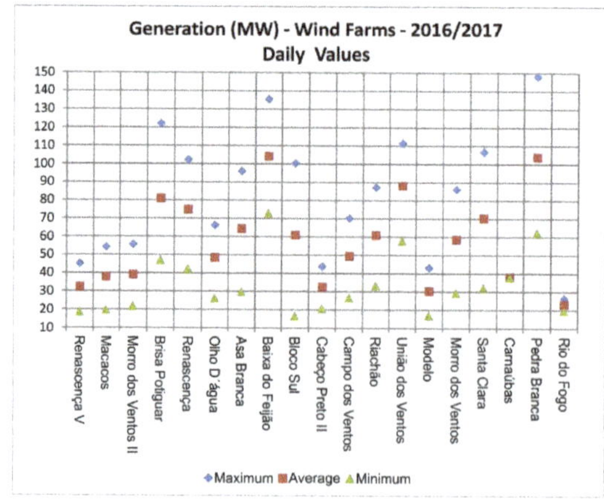

Figure 3. Maximum, average, and minimum daily values of active power (MW) wind farms.

To illustrate the proposed qualitative definition of maximum, average, and minimum of active power of the renewable energy sources, the Brisa Potiguar wind farm (BP) were chosen. Figure 4 presents the active power average values $\overline{Pg}_{win}$ of BP for each hour of the day, which was determined from the available real historic data.

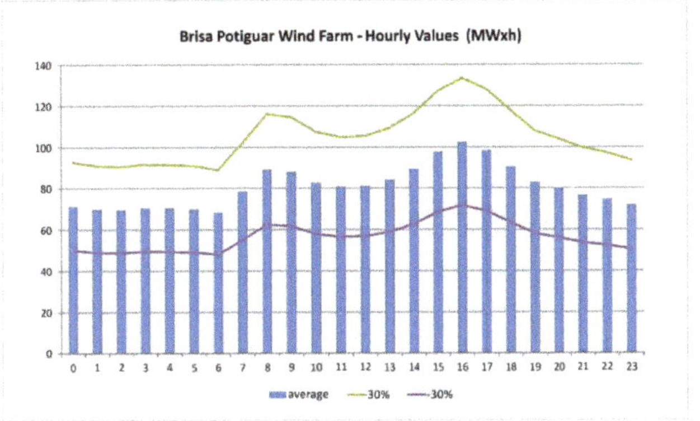

Figure 4. Brisa Potiguar wind farm daily behavior.

These values were used in the first stage (Section 2.1—Equation (2)) of the proposal methodology, in order to define the deterministic variables of EPS electrical quantities.

According to real historical data, the maximum and minimum values of active power of BP are defined for each hour of the day. These values correspond to the variation of $+/-30\%$ of the average values of these powers, which are represented the fuzzy variables $\overline{\Delta Pg}_{win}$ (see Figure 2). These values are used in the second stage (Section 2.2—Equations (17)–(21)) of the proposal methodology, in order to determine the fuzzy variables of EPS electrical quantities. The uncertainties of EPS electrical quantities are then computed by the sum of deterministic variables, determined in the first stage, and fuzzy variables, determined in the second stage, as discussed in Section 2.2—Equations (22–25).

In summary, the average values of active power of wind and solar energy sources are determined through the brief real historic data. The maximum and minimum values of these active powers are defined by the hourly variation of $+/-30\%$ of the average value. It should be emphasized that the whole proposed strategy assumes an hourly based analysis, such that fuzzy triangular model (based in the normal PDF) is used to model the hourly variation of the wind and solar generators in the second stage of the proposed methodology.

The same methodology is applied to active power produced by solar energy sources, which are randomly distributed in six of the 69 kV buses. The solar energy participates in the formulation of the problem only in the period of solar incidence between 6–18 h, and the average value of these powers $\overline{Pg}_{sol}$ is dimensioned in 10% of the active power demanded in these buses. The fuzzy variables $\overline{\Delta Pg}_{sol}$, related to the active power produced by solar energy sources are also defined by values equivalent to $+/-30\%$ of the average values.

4. Simulations and Results

The developments presented in this work were carried out using the MATLAB computational tool. The first and most important result consists of authenticating the proposed methodology. In order to validate the proposed approach, exhaustive solutions of conventional OPF considering each increase/decrease variation for wind and solar sources, were used.

Figure 5 illustrates this validation for the active power of the hydroelectric plant of Xingó.

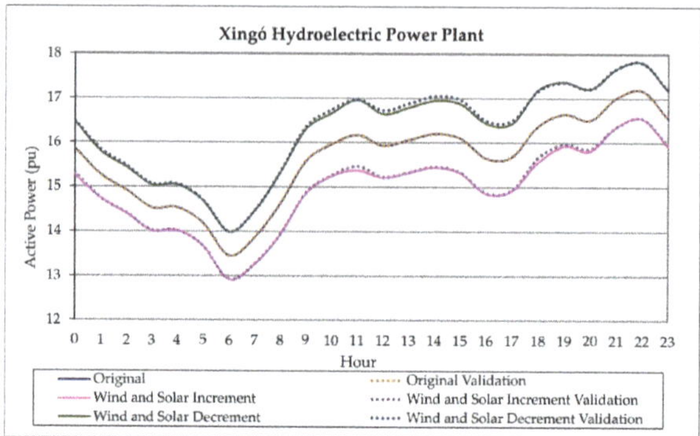

Figure 5. Authentication—active power of Xingó hydroelectric power plant.

The "Original" trace of Figure 5 represents the deterministic values of the active power of the hydroelectric plant of Xingó, obtained with the ED-OPF solution, as described in Section 2.1. In this case, the qualitative average values of the active power of the wind and solar energy sources were used in the power balance equations. The "Wind and Solar Increment" trace represents the uncertainties of the active power of the hydroelectric plant of Xingó. These uncertainties are determined by the sum of determinist values ("Original" trace) with fuzzy values. In this case, according to the technique described in Section 2.2, the fuzzy values were determined through qualitative sensitivity analysis applied to the postprocessing of the ED-OPF, as function of 30% (maximum values) increase in the active power of wind and solar energy sources. The "Wind and Solar Decrement" trace also represents the uncertainties of the active power of the hydroelectric plant of Xingó, but now considering a decrease of 30% (minimum values) in the active power of these energy sources.

Simulations were performed in a conventional OPF to authenticate the proposed methodology. The "Original Validation" trace represents the values obtained in the solution of a conventional OPF that considers in the power balance equations the qualitative average values of the active power of the wind and solar energy sources. The "Wind and Solar Increment Validation" trace represents the values obtained in the conventional OPF that considers in the active power balance equations the qualitative average values of the active power of these energy sources increased by 30%, while the "Wind and Solar Decrement Validation" trace represents the values obtained in the conventional OPF when in these equations the qualitative average values of the active power of these energy sources is reduced in 30%. The adherence of the results indicates the effectiveness of the proposed methodology in obtaining the desired results in an expeditious way, as proposed.

This authentication is also successfully achieved for the other electrical quantities of the system. The following sections present the definition of the uncertainties of the power flows through circuit breakers, of the active power produced by conventional generators, and of the nodal voltage magnitudes as a function of the uncertainties of the active power produced by wind and solar energy sources. These analyses are academic and have no intention of interfering with the operation or expansion of the actual system.

Similar proposals, that reach solution with reduced computational time as proposed in our paper, and that simultaneously contemplates in the OPF, renewable energies, switching (substation model level), and qualitative information, were not found in the literature. These characteristics make us confident that the proposed approach presents innovative ideas and presents itself as a promising tool for analyzing the impact of intermittent energy sources in the electrical quantities of the power systems.

4.1. Power Flows through Circuit Breakers

The verification of the correct sizing of the rated current of circuit breakers, often motivated by expansions or alterations in power systems, has attracted the attention of analysts of these systems in different parts of the world. In these analyses, the representation of the substation of interest at the substation level, as proposed in this paper, allows for the assessment in a direct way of the power flows through these devices. As most analytical tools consider bus-branch modeling, this analysis requires the adoption of additional procedures that overload the area professionals and hamper this task. The incorporation of network modeling at the substation level in the OPF eliminates this step, thus facilitating and more effectively subsidizing the analysis work by these professionals.

To illustrate the effectiveness of this attribute of the proposed methodology, the Sobradinho 230 kV substation was chosen for modeling at the substation level owing to its connections with wind and solar power sources, as well as sheltering an important hydroelectric plant of the same name with 1.05 GW of installed power. Figure 6 illustrates the configuration of this substation, including the most common directions of power flows through the circuit breakers of this installation.

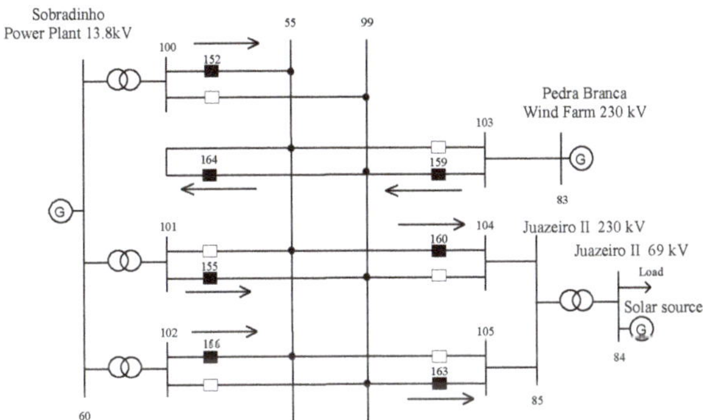

Figure 6. Power flow distribution at Sobradinho 230 kV substation.

Table 1 presents the most relevant results related to this analysis involving the Sobradinho 230 kV substation. Column 1 indicates the hour at which the variation was assessed (corresponding to the solar incidence period), while the second to fifth columns show the variations in the active power flows through the indicated circuit breaker when the active power produced by wind and solar sources varied from $+/-30\%$ (always in relation to the average values according to Section 2.1).

Table 1. Active Power Flow Variations (MW) Through Circuit Breakers of Sobradinho 230 kV Substation.

Hour	Circuit Breaker (CB) 164				CB 159		CB 160	
	Wind Variation (MW)		Solar Variation (MW)		Wind Variation (MW)		Solar Variation (MW)	
	30%	−30%	30%	−30%	30%	−30%	30%	−30%
06	26.5	−26.5	0.67	−0.66	39.8	−39.8	2.0	−2.0
07	26.4	−26.4	0.70	−0.69	39.6	−39.6	2.1	−2.1
08	25.3	−25.3	0.74	−0.76	38.0	−38.0	2.2	−2.3
09	22.8	−22.8	0.79	−0.80	34.2	−34.2	2.4	−2.4
10	19.1	−19.1	0.82	−0.81	28.7	−28.7	2.4	−2.4
11	15.8	−15.8	0.82	−0.81	23.8	−23.8	2.4	−2.5
12	13.6	−13.6	0.81	−0.80	20.4	−20.4	2.4	−2.4
13	11.9	−11.9	0.82	−0.82	17.9	−17.9	2.5	−2.5
14	11.3	−11.3	0.83	−0.83	17.0	−17.0	2.5	−2.5
15	11.4	−11.4	0.83	−0.84	17.2	−17.2	2.5	−2.5
16	12.3	−12.3	0.81	−0.81	18.4	−18.4	2.4	−2.4
17	13.0	−13.0	0.79	−0.80	19.5	−19.5	2.4	−2.4
18	14.7	−14.7	0.84	−0.82	22.1	−22.1	2.5	−2.5

The results indicate that the active power flows through the circuit breakers of this substation were influenced by the simultaneous generation of wind and solar energy sources. The exception was with the active power flows through circuit breaker 159, which were influenced only by the wind energy source since this device was connected to the Pedra Branca 230 kV wind farm. Similar analyses can be performed for the active power flows through circuit breakers 160 and 163, which connect the solar source, and therefore were influenced only by this source of energy.

Once the power flows through the circuit breakers of the Sobradinho 230 kV substation were determined, the maximum currents through these devices were calculated as a function of the apparent powers. From this calculation, it was observed that the maximum ratio between these currents and the rated current was 22.2% for circuit breaker 159, ensuring the correct dimensioning of these devices by the rated current.

The João Câmara II 138 kV bus was also modeled at the substation level owing to the connection of approximately 700 MW of installed power by wind farms. Figure 7 shows the bus configuration and indicates the most common orientation of power flows through the circuit breakers of this substation.

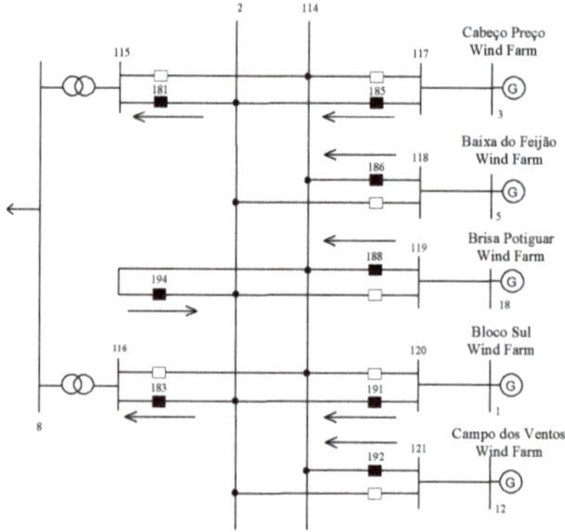

Figure 7. Power flow distribution at João Câmara II 138 kV substation.

Table 2 lists the maximum current values through the main circuit breakers of this substation, as well as the proportion of these values in relation to their rated values. These analyses are performed for the original condition, that is, without the variation of the average values of the active power of the wind sources, and with a variation of +/−30% in relation to these values.

Table 2. Maximum Currents (A) Through Circuit Breakers of João Câmara II 138 kV Substation.

System	Circuit Breakers				
	181	186	188	191	194
Original (A)	789.6	491.2	442.9	300.3	1134.9
Wind Energy Increment (A)	1012.1	606.5	562.7	390.8	1431.7
Wind Energy Decrement (A)	579.2	396.9	331.0	209.9	866.3
Circuit Breaker Rated Current (A)	2000	2000	2000	2000	2000
Original Current/Rated current (%)	39.5	24.6	22.1	15.0	56.7
(Maximum Current with Wind and/or Solar) /Rated Current (%)	50.6	30.3	28.1	19.5	71.6

The results indicate that the effect of the wind power sources was very significant, taking the maximum current values through these devices near the rated values. For example, the maximum current through circuit breaker 194 reached 71.6% of its rated current value when exposed to the uncertainties of the active power produced by the wind farms connected to the bus under study. The effect of the solar energy sources in this substation was negligible.

It should be noted that this case considers a normal operating condition. Logically, under conditions of contingency (criterion n − 1) or expansions in the system considered, the currents passing through these circuit breakers could be closer to the nominal values and may even exceed them. This type of analysis can be carried out by the proposed methodology and allows for the assessment of the necessity of replacing these devices with equipment that has a greater rated current capacity.

4.2. Active Power: Conventional Generators

Another relevant issue associated with the uncertainties of the active power generated by wind and solar energy sources is related to the analysis of their impact on the established powers for conventional generators. Although it is not advisable to ignore the importance of intermittent sources in reducing the rotating reserve, a careful analysis is also needed to avoid compromising the energy supply to consumers in conditions of unexpected reductions of the share of these energy sources in the energy matrix.

Table 3 lists, for some hours of the day, the variation in megawatts from the average values of the active power produced by the Paulo Afonso IV hydroelectric power plant. These values were obtained as a function of the variation of +/−30% from the average values of the active power generated by wind and solar energy sources.

Table 3. Variations (MW) at the Paulo Afonso IV Hydroelectric Power Plant.

Hour	Wind Variation (MW)		Solar Variation (MW)	
	+30%	−30%	+30%	−30%
10	−90.3	90.3	54.2	−54.2
11	−99.2	99.2	2.5	−2.5
12	−35.6	35.6	−0.8	0.8
13	−97.1	97.1	59.1	−59.1
14	−102.8	102.8	63.3	−63.3
15	−100.1	100.1	59.9	−59.9
16	−37.6	37.6	−0.6	0.6
17	−36.5	36.5	−0.7	0.7
18	−106.0	106.0	−13.8	13.8

These and other results obtained (omitted because of lack of space) show that all conventional generators of the system were sensitized by both the solar and wind energy uncertainties except for the generators of the thermoelectric plants that were not dispatched owing to high cost.

In this context, Table 4 shows the influence of the uncertainties of wind and solar energy on the energy reserve of these conventional generators.

Table 4. Available Powers: Conventional Generators.

Conventional Generators	Active Power			Reactive Power		
	Limits (MW)	Original (%)	Impact Wind and Solar (%)	Limits (MVar)	Original (%)	Impact Wind and Solar (%)
Sobradinho	1005	79	73	−272~388	75	74
Luiz Gonzaga	1480	4	0	−210~210	45	10
Paulo Afonso	2400	79	77	−920~820	67	66
Xingó	3600	46	44	−860~519	79	76

The results presented show a slight reduction in the available powers of generators, which was necessary to increase energy production to supply the reduction in the share of wind and solar sources. The Luiz Gonzaga hydroelectric power plant reached its active power limit with a reduction of 30% in the share of these energy sources. The system under analysis contemplates an installed capacity of approximately 20 GW distributed between the hydroelectric and thermoelectric plants and equivalent generators, in addition to the hourly demand of around 7 GW. Considering energy generation planning based on an operational reserve equivalent to 5% of the demand, the variations of +/−30% of the active power of the wind and solar energy sources guarantee an operational reserve of 350 MW.

4.3. Nodal Voltage Magnitudes

Voltage control is one of the main concerns when there are expansions and changes in energy systems, as it must comply with the limits imposed by regulatory authorities. Thus, for the test system of this work, the uncertainties of the nodal voltage magnitudes were determined as a function of the uncertainty of +/−30% in relation to the average value of the active power of the wind and solar energy sources. Figure 8 illustrates the behavior throughout the day of the nodal voltage magnitudes in the 500 kV Pau de Ferro substation, an important transmission trunk of the system, owing to the combination of the uncertainties of wind and solar energy sources.

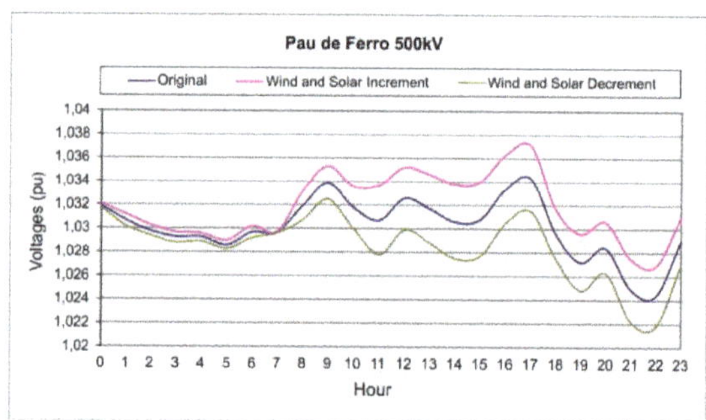

Figure 8. Uncertainties in nodal voltage magnitudes of 500 kV bus of Pau de Ferro substation.

The values obtained for the nodal voltage magnitudes of all buses of the test system are within the limits established by the Brazilian regulatory authority, according to Table 5 [38]. These results show once again the relevance of the proposed methodology that allows for the inference of questions of this order without the need to carry out studies and exhaustive simulations involving the intermittent sources of energy.

Table 5. Voltage magnitude admissible limits in the Brazilian electrical system.

Rated Operating Voltage	Normal Operating Condition	Emergency Operating Condition
(kV)	(pu)	(pu)
≤138	0.95 to 1.05	0.90 to 1.05
230	0.95 to 1.05	0.90 to 1.05
345	0.95 to 1.05	0.90 to 1.05
440	0.95 to 1.046	0.90 to 1.046
500	1.00 to 1.10	0.95 to 1.10
525	0.95 to 1.05	0.90 to 1.05
765	0.90 to 1.046	0.90 to 1.046

4.4. Computational Time

Table 6 presents the computational times demanded in the execution of first and second stages of the proposed methodology.

Table 6. Computational time (seconds).

Hour	First Stage	Second Stage	Hour	First Stage	Second Stage	Hour	First Stage	Second Stage
0	27.6	0.94	8	20.8	0.90	16	19.3	0.98
1	23.3	1.01	9	21.3	0.94	17	22.6	0.99
2	28.8	0.98	10	26.3	0.84	18	24.2	0.94
3	24.7	0.95	11	25.2	1.01	19	56.0	0.90
4	22.8	0.90	12	22.9	0.92	20	25.2	0.95
5	22.7	0.87	13	22.4	0.81	21	24.6	0.87
6	40.8	1.03	14	22.5	0.95	22	26.6	0.90
7	20.4	0.92	15	27.4	0.90	23	20.8	0.92

According to the results presented in Table 6 it is possible to verify the significant computational time reduction in the second stage.

Based on the subjectivity of the qualitative knowledge about the wind and solar energy sources behavior, the second stage of the proposal methodology provides for the analysts fast answers and with good authenticity about electrical quantities of these systems, as indicated in Figure 5.

5. Conclusions

The combination of fuzzy logic with a Taylor series in the postprocessing of the Extended Deterministic Optimum Power Flow allows for the assessment of the impact of the active power uncertainties produced by the wind and solar energy sources in the electrical quantities of energy systems, for each hour of the day. In an expeditious way, based in a subjective investigation that focuses on the qualitative character of these energy sources' behaviors, the proposed methodology allows for the assessment of the impact of these uncertainties on the active power produced by conventional generators. This makes it possible to conduct energy reserve analyses. These impacts are also assessed on the nodal voltage magnitudes. Thus, it is possible to verify if the limits reached comply with the limits of the voltages imposed by the regulatory authorities.

In addition, a representation of the power grid at the substation level made it possible to identify the impact of the variations of the wind and solar energy sources on the distribution of power flows and consequently of the currents through the circuit breakers of substations. This model allows for the evaluation of the technical specifications of such devices, such as rated currents. In face of this new reality of intermittent energy generation systems, this topic has attracted the attention of systems analysts from various parts of the world owing to the need for expansions and changes in EPS.

The results presented highlight the importance of energy reserve analyses and of correct technical specifications of circuit breakers in addition of voltages control, as a function of the forecasted growth of wind and solar energy sources in energy matrices. These results reinforce the relevance of analytical tools that provide professionals the ability to perform these tasks, in an expeditious way, as in the case of the methodology proposed in this work. Future studies are under investigation to assess the impact of such approach in contingencies analysis, including wind farm groups contingencies.

6. Article Contribution

The main contribution of this article is to offer power systems analysts a tool that allows the rapid assessment of the hourly behavior of the electrical quantities of these systems as a function of the variation of wind and solar energy sources. It is not a tool that involves data accuracy, but rather a tool that provides enough and rapid responses to decision making by those analysts. The proposed tool is original and suitable mainly for analysis of currents through circuit breakers, a theme that has attracted the attention of analysts from various parts of the world, due to the need for expansions and topology changes in power systems. The variations in the power injected by these energy sources can increase the value of the currents through the circuit breakers, resulting in an alert to the analysts about the need for replacement with equipment with higher nominal current value.

The assessment of the impact of the variation of the wind and solar energy sources not only on the currents through the circuit breakers, but also on the reserve energy and the nodal voltage magnitudes is reached through the solution of fuzzy extended deterministic optimal power flow, composed of two stages.

As discussed before, in the first stage, the extended deterministic OPF (ED-OPF) solution provides the deterministic variables of the power system electrical quantities. For this stage, we would like to emphasize that the formulation of ED-OPF contemplates an original modeling at the substation level, which enables the direct assessment of the currents through the circuit breakers. In addition, this formulation contemplates the hourly average values of the active powers of wind and solar energy sources in the active power balance equations. The first stage solution demands the traditional expensive computational time of solutions involving OPFs.

In the second stage, the uncertainties of the power system electrical quantities are obtained considering an hourly based analysis. These uncertainties are determined by the sum of determinist variables (first stage) with fuzzy variables. The fuzzy variables are obtained when the maximum and minimum values of these energy sources (corresponding to $+/-30$ % of average values of these energy sources) are applied to the ED-OPF solution, through a qualitative sensitivity analysis. The second stage presents lower computational time when compared to the first stage.

With the proposed tool, power system analysts can run ED-OPF once, which demands longer computational time and work more frequently with the second stage, which demands lower computational time. The second stage provides rapid and authentic responses for a general analysis of the electrical quantities of the system as a function of the variation of wind and solar energy sources.

Author Contributions: All the authors contributed to the realization, analysis of the data, as well as the writing of the paper.

Funding: This research received no external funding.

Acknowledgments: The authors are grateful for the support of the Brazilian electric utilities Companhia Paranaense de Energia (COPEL) (Paraná, Brazil) and Companhia Hidroelétrica do São Francisco (CHESF) (Pernambuco, Brazil).

Conflicts of Interest: The authors declare no conflict of interest.

References

1. Keane, A.; Ochoa, L.F.; Borges, C.L.T.; Ault, G.W.; Alarcon-Rodriguez, A.D.; Currie, R.A.F.; Pilo, F.; Dent, C.; Harrison, G.P. State-of-the-Art Techniques and Challenges Ahead for Distributed Generation Planning and Optimization. *IEEE Trans. Power Syst.* **2013**, *28*, 1493–1502. [CrossRef]
2. Miranda, V. Successful Large-scale Renewables Integration in Portugal: Technology and Intelligent Tools. *CSEE J. Power Energy Syst.* **2017**, *3*, 7–16. [CrossRef]
3. Khaled, U.; Eltamaly, A.M.; Beroual, A. Optimal Power Flow Using Particle Swarm Optimization of Renewable Hybrid Distributed Generation. *Energies* **2017**, *10*, 1013. [CrossRef]
4. Al Abri, R.S.; El-Saadany, E.F.; Atwa, Y.M. Optimal Placement and Sizing Method to Improve the Voltage Stability Margin in a Distribution System Using Distributed Generation. *IEEE Trans. Power Syst.* **2013**, *28*, 326–334. [CrossRef]
5. Warid, W.; Hizam, H.; Mariun, N.; Addul-Wahad, N.I. Optimal Power Flow Using the Jaya Algorithm. *Energies* **2016**, *9*, 678. [CrossRef]
6. Mégel, O.; Mathieu, J.L.; Andersson, G. Hybrid Stochastic-Deterministic Multiperiod DC Optimal Power Flow. *IEEE Trans. Power Syst.* **2017**, *32*, 3394–3945. [CrossRef]
7. Atwa, Y.M.; El- Saadany, E.F. Optimal Allocation of ESS in Distribution Systems with a High Penetration of Wind Energy. *IEEE Trans. Power Syst.* **2010**, *25*, 1815–1822. [CrossRef]
8. Wienholt, L.; Müller, U.P.; Bartels, J. Optimal Sizing and Spatial Allocation of Storage Units in a High-Resolution Power System Model. *Energies* **2018**, *11*, 3365. [CrossRef]

9. Liang, J.; Molina, D.D.; Venayagamoorthy, G.K.; Harley, R.G. Two-Level Dynamic Stochastic Optimal Power Flow Control for Power Systems with Intermittent Renewable Generation. *IEEE Trans. Power Syst.* **2013**, *28*, 2670–2678. [CrossRef]

10. Derakhshandeh, S.Y.; Masoum, A.S.; Deilami, S.; Masoum, M.A.S.; Golshan, M.E. Coordination of Generation Scheduling with PEVs Charging in Industrial Microgrids. *IEEE Trans. Power Syst.* **2013**, *28*, 3451–3461. [CrossRef]

11. An, K.; Song, K.B.; Hur, K. Incorporating Charging/Discharging Strategy of Electric Vehicles into Security-Constrained Optimal Power Flow to Support High Renewable Penetration. *Energies* **2017**, *10*, 729. [CrossRef]

12. Roald, L.; Misra, S.; Krause, T.; Andersson, G. Corrective Control to Handle Forecast Uncertainty: A Chance Constrained Optimal Power Flow. *IEEE Trans. Power Syst.* **2017**, *32*, 1626–1637.

13. Kim, H.Y.; Kim, M.K.; Kim, S. Multi-Objective Scheduling Optimization Based on a Modified Non-Dominated Sorting Genetic Algorithm-II in Voltage Source Converter—Multi-Terminal High Voltage DC Grid-Connected Offshore Wind Farms with Battery Energy Storage Systems. *Energies* **2017**, *10*, 986. [CrossRef]

14. Liu, Y.; Qu, Z.; Xin, H.; Gan, D. Distributed Real-Time Optimal Power Flow Control in Smart Grid. *IEEE Trans. Power Syst.* **2017**, *32*, 3403–3414. [CrossRef]

15. Miranda, V.; Saraiva, J.T. Fuzzy Modeling of Power System Optimal Load Flow. *IEEE Trans. Power Syst.* **1992**, *7*, 843–849. [CrossRef]

16. Zhang, W.; Li, F.; Tolbert, L.M. Review of Reactive Power Planning: Objectives, Constraints, and Algorithms. *IEEE Trans. Power Syst.* **2007**, *22*, 2177–2186. [CrossRef]

17. Abdul-Rahman, K.H.; Shahidehpour, S.M. Static Security in Power System Operation with Fuzzy Real Load Conditions. *IEEE Trans. Power Syst.* **1995**, *10*, 77 87. [CrossRef]

18. He, X.; Wang, W.; Jiang, J.; Xu, L. An Improved Artificial Bee Colony Algorithm and Its Application to Multi-Objective Optimal Power Flow. *Energies* **2015**, *8*, 2412–2437. [CrossRef]

19. Edwin Liu, W.H.; Guan, X. Fuzzy Constraint Enforcement and Control Action Curtailment in an Optimal Power Flow. *IEEE Trans. Power Syst.* **1996**, *11*, 639–645.

20. Assis, T.M.L.; Falcão, D.M.; Taranto, G.N. Dynamic Transmission Capability Calculation Using Integrated Analysis Tools and Intelligent Systems. *IEEE Trans. Power Syst.* **2007**, *22*, 1760–1770. [CrossRef]

21. Lami, B.; Bhattacharya, K. Clustering Technique Applied to Nodal Reliability Indices for Optimal Planning of Energy Resources. *IEEE Trans. Power Syst.* **2016**, *31*, 4679–4690. [CrossRef]

22. Arneja, I.S.; Venkatesh, B. Probabilistic OPF Using Linear Fuzzy Relation. In Proceedings of the 10th International Power & Energy Conference IPEC, Ho Chi Minh City, Vietnam, 11–12 December 2012.

23. Mohapatra, A.; Bijw, P.R.E.; Panigrahi, B.K. Optimal Power Flow with Multiple Data Uncertainties. *Electr. Power Syst. Res.* **2013**, *95*, 160–167. [CrossRef]

24. Gomes, F.V.; Carneiro, S.; Pereira, J.L.R.; Vinagre, M.P.; Garcia, P.A.N.; Araújo, L.R. A New Distribution System Reconfiguration Approach Using Optimum Power Flow and Sensitivity Analysis for Loss Reduction. *IEEE Trans. Power Syst.* **2006**, *21*, 1616–1623. [CrossRef]

25. Borges, T.T.; Carneiro, S.; Garcia, P.A.N.; Pereira, J.L.R. A New OPF Based Distribution System Restoration Method. *Int. J. Electr. Power Energy Syst.* **2006**, *80*, 297–305. [CrossRef]

26. Fisher, E.B.; O'Neill, R.P.; Ferris, M.C. Optimal transmission switching. *IEEE Trans. Power Syst.* **2008**, *23*, 1346–1355. [CrossRef]

27. Schnyder, G.; Glavitsch, H. Integrated Security Control using an Optimal Power Flow and Switching Concepts. *IEEE Trans. Power Syst.* **2008**, *3*, 782–790. [CrossRef]

28. Khanabadi, M.; Ghasemi, H.; Doostizadeh, M. Optimal Transmission Switching Considering Voltage Security and N-1 Contingency Analysis. *IEEE Trans. Power Syst.* **2013**, *28*, 542–550. [CrossRef]

29. Capitanescu, F.; Ochoa, L.F.; Margossian, H.; Hatziargyriou, N.D. Assessing the Potential of Network Reconfiguration to Improve Distributed Generation Hosting Capacity in Active Distribution Systems. *IEEE Trans. Power Syst.* **2015**, *30*, 346–356. [CrossRef]

30. Peng, Q.; Tang, Y.; Low, S.H. Feeder Reconfiguration in Distribution Networks Based on Convex Relaxation of OPF. *IEEE Trans. Power Syst.* **2015**, *30*, 1793–1804. [CrossRef]

31. Monticelli, A.; Garcia, A. Modeling Zero Impedance Branches in Power System State Estimation. *IEEE Trans. Power Syst.* **1991**, *6*, 1561–1570. [CrossRef]

32. Lourenço, E.M.; Simões Costa, A.; Ribeiro, R. Steady-State Solution for Power Networks Modeled at Bus Section Level. *IEEE Trans. Power Syst.* **2010**, *25*, 10–20. [CrossRef]
33. Rosas, G.B.; Lourenço, E.M.; Fernandes, T.S.P. Network Model at Substation Level for Optimal Power Flow Studies. *Controle Automação* **2012**, *23*, 766–781. (In Portuguese) [CrossRef]
34. Huneault, M.; Galiana, F.D. A Survey of the Optimal Power Flow Literature. *IEEE Trans. Power Syst.* **1991**, *6*, 762–770. [CrossRef]
35. Granville, S. Optimal Reactive Dispatch through Interior Point Methods. *IEEE Trans. Power Syst.* **1994**, *9*, 136–146. [CrossRef]
36. Chayakulkheeree, K. Probabilistic Optimal Power Flow: An Alternative Solution for Emerging High Uncertain Power Systems. In Proceedings of the International Electrical Engineering Congress, Pattaya, Thailand, 19–21 March 2014.
37. Lyu, J.K.; Heo, J.H.; Park, J.K.; Kang, Y.C. Probabilistic Approach to Optimizing Active and Reactive Power Flow in Wind Farms Considering Wake Effects. *Energies* **2013**, *6*, 5717–5737. [CrossRef]
38. National System Operator (Brazil). Available online: http://ons.org.br/paginas/sobre-o-ons/procedimentos-de-rede/vigentes-Módulo23-Critériosparaestudos-Submódulo23.3-Diretrizesecritériosparaestudoselétricos (accessed on 19 March 2019).

Article

Optimal Placement of UHF Sensors for Accurate Localization of Partial Discharge Source in GIS

Rui Liang [1,2,*], Shenglei Wu [3], Peng Chi [1], Nan Peng [1] and Yi Li [1]

[1]   School of Electrical and Power Engineering, China University of Mining and Technology, Xuzhou 221116, China; p5411578@163.com (P.C.); pncumt@163.com (N.P.); sgcccq@163.com (Y.L.)

[2]   Jiangsu Laboratory of Coal Mine Electrical and Automation Engineering, China University of Mining and Technology, Xuzhou 221116, China

[3]   State Grid Chongqing Economic Research Institute, Chongqing 401121, China; wushenglei1994@163.com

*   Correspondence: liangrui@cumt.edu.cn; Tel.: +86-516-83592000

Received: 12 February 2019; Accepted: 21 March 2019; Published: 26 March 2019

Abstract: This paper proposes an optimal placement model of ultra-high frequency (UHF) sensors for accurate location of partial discharge (PD) in gas-insulated switchgear (GIS). The model is based on 0-1 program in consideration of the attenuation influence on the propagation of electromagnetic (EM) waves generated by PD in GIS. the optimal placement plan improves the economy, observability, and accuracy of PD locating. After synchronously acquiring the time of the initial EM waves reaching each UHF sensor, PD occurring time can be obtained. Then, initial locating results can be acquired by using the Euclidean distance measuring method and the extended time difference of arriving (TDOA) location method. With the information of all UHF sensors and the inherent topological structure of GIS, the locating accuracy can be further improved. The method is verified by experiment, showing that the method can avoid the influence of false information and obtain higher locating accuracy by revising initial locating results.

Keywords: gas insulated switchgear; partial discharges; electromagnetic wave propagation; optimization; UHF measurement

---

1. Introduction

With the wide use of gas-insulated switchgear (GIS), various failures occur with the growth of service time, and the insulation defects in GIS may lead to partial discharge (PD). PD location can be realized via ultrasonic waves and ultra-high frequency (UHF) electromagnetic (EM) waves [1–3]. The UHF method is an efficient PD detection technology that detects the EM wave generated by PD based on UHF sensors [4–8]. The UHF method has the advantages of high sensitivity and high signal-to-noise ratio. The position of PD can be located by analyzing and processing the EM signal received by the sensors [9–11]. UHF location methods are mainly based on time difference of arriving, time of arriving, angle of arriving, and received signal strength [12,13].

The spatial placement of UHF sensors is the key to PD locating. Studies have been undertaken to improve sensor economy and observability. Though some exploratory research works have been carried out in these fields, most of them focused on transformer or conventional switchgear [14–18]. For instance, a spatial array of UHF sensors is used for transformer PD location. However, a transformer is a multisurface system of which the internal structure is totally different from GIS. The spatial array aims to overcome the reflection and diffraction of electromagnetic waves. Therefore, the methods for transformers and switchgear are not suitable for GIS. The topological structures of GIS and the attenuation of the EM wave especially propagating through the insulators and L-type and T-type structures should be considered more. A few studies have investigated the topological structures of GIS but these do not consider the optimal placement of sensors [19–23]. The arriving time difference of EM waves was used to locate the PD in GIS based on UHF sensors, but its accuracy should be improved [24–26]. This paper is devoted to solving these problems in existing research.

In this paper, the GIS space structure is abstracted into an undirected graph and a 0-1 programming model is established based on the optimization of the optimal economy and the maximum observability. Then, an arrangement plan of UHF sensors can be obtained. With the optimal placement plan, the accurate location method of PD based on extended time difference of arriving (TDOA) is studied. The experimental result indicates that the proposed method can improve the location accuracy and possesses fine fault tolerance.

2. Principle of Method

2.1. Partial Discharge in GIS

GIS, which includes a circuit breaker, disconnector, grounding switch, transformer, and bus, has been widely used due to its low maintenance, high reliability, and compact size. PD refers to the discharge phenomenon occurring in the partial area of the insulator which does not penetrate the conductor under voltage. Under a strong electric field, the insulation defects of GIS will lead to PD, accompanied by the generation of electricity, light, heat, sound, ozone, and nitric oxide, which will corrode the insulating materials. In addition, the charged particles produced by PD will impact the insulating materials. PD is a sign of insulation degradation in GIS, which endangers the safety of equipment and even the power system.

The breakdown field strength of insulators in power equipment is very high. When PD occurs in a small range, the breakdown process is very fast, which produces steep pulse current. Its rising time is less than 1 ns, and ultra-high frequency (UHF) electromagnetic waves are excited. The principle of the UHF method for PD detection is to detect the UHF electromagnetic waves (300 MHz-3 GHz) generated by PD in power equipment by UHF sensor. Then, the information of PD can be obtained. Built-in UHF sensors and external UHF sensors are usually used for UHF methods.

Because corona interference is mainly concentrated below the 300 MHz frequency band, the UHF method can effectively avoid corona interference. It has high sensitivity and anti-interference ability. Naturally, the optimal placement of UHF sensors and location method become the key to partial discharge locating.

2.2. Optimal Placement of UHF Sensors

The optimal placement of UHF sensors should satisfy the economy of scale and provide maximum observability. A necessary requirement for successful locating is that at least two sensors detect an effective discharge EM wave, and the EM wave must pass through both ends of the discharge section. If the EM wave only passes through the head or the end of the detecting section, the locating will fail. 1 represents that the sensor needs to be installed and 0 represents the sensor doesn't need to be installed. Therefore, the placement of UHF sensors can be abstracted as a 0-1 programming model.

The EM wave will be attenuated when passing through the insulators and L-type and T-type structures in GIS. The vertical branches of L-type structures and T-type structures significantly attenuate the EM wave, and the attenuation caused by the horizontal branches of T-type structures and insulators comes after. In order to improve the detection sensitivity and locating reliability, the L-type structures and the T-type structures in GIS should be taken into consideration.

In topology, the number of branches connected to the node is called the degree (d) of the node. Obviously, the degree of an L-type node (node 4 in Figure 1) is 2 and the degree of a T-type node (node 4 in Figure 2) is 3. For general GIS, the node with the largest degree is the cross node, of which the degree is 4, and the node with the smallest degree is the terminal node, of which the degree is 1. The cross node can be considered as the combination of two L-type nodes. To effectively detect the EM wave of PD, the EM wave should not pass through more than one L-type node, one cross node, or one

T-type node. These three nodes are collectively called the nonterminal node, of which the degree is greater than or equal to 2.

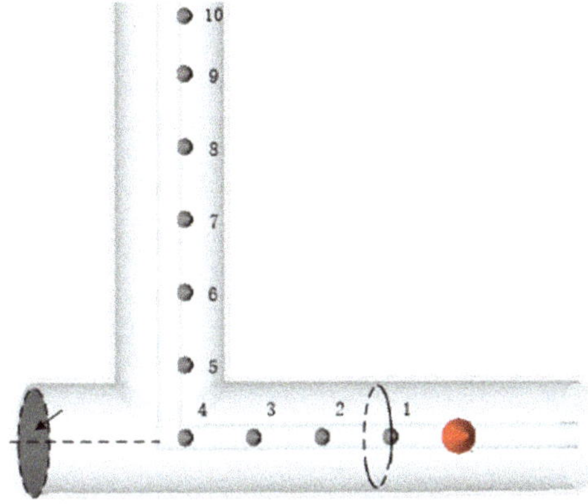

Figure 1. L-type structure of gas-insulated switchgear (GIS).

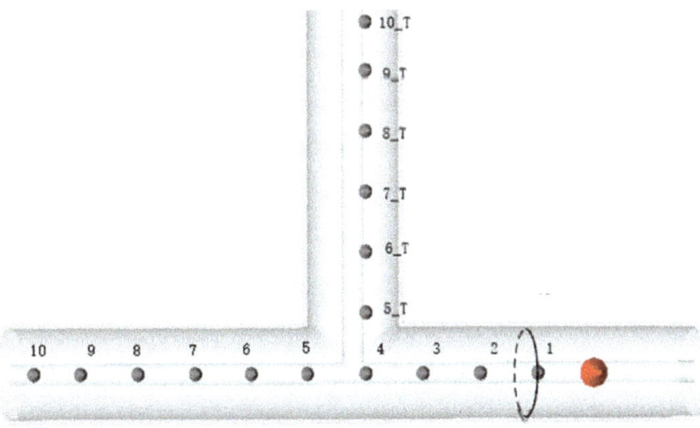

Figure 2. T-type structure of GIS.

The number of branches included in the shortest path of node i to node j is defined as the shortest path length P_{ij}. Under the condition of installing sensors at the terminal node (subject to $x_t = 1$), a suitable nonterminal node should be selected to install a sensor so that the EM wave passes through, at most, one nonterminal node before reaching the sensors. Therefore, the maximum number of nonconfigurable nodes between any adjacent configurable nodes is 1, and P_{ij} should be no more than 2 (subject to $P_{ij} = 2$). An optimization model can be obtained:

$$\min \sum_{k=1}^{N} x_k$$
$$\text{s.t. } P_{ij} \leq 2 \tag{1}$$
$$x_t = 1$$

where t is terminal node, i and j are adjacent nodes, and x_k is the node in which the sensor should be installed. Because the configurable node is unknown, the inequality constraint of the shortest path which contains variable x_k cannot be obtained directly.

The condition above can be described as follows: at least one sensor should be installed on those nodes of which the shortest path length to node k is 1 or 2.

For a node k, an adjacent node matrix V^k is proposed. V^k is $M \times N$ matrix, where N is the number of nodes and M is the number of the nodes, of which the shortest path length to the node k is 2. The nodes of which the shortest path length is 1 don't need to be considered because the shortest path from these M nodes to the node k includes them. For a row vector v of V^k and j ($1 \leq j \leq N$), when $P_{kj} = 1$ or $P_{kj} = 2$, $v_j = 1$ and other elements are 0. Therefore, it forms the constraint that at least one sensor should be installed on the path from the node k to the node of which the shortest path length is 2. $V^{1\sim N}$ is arranged in columns to form a new matrix $V_{M \times N}$, and it can be found as follows:

$$V_{M \times N} X_{N \times 1} \geq I_{M \times 1} \tag{2}$$

where $X^{\mathrm{T}} = [x_1, x_2, \ldots, x_N]$ is the vector to be solved and $I_{M \times 1}$ is $[1, 1, \ldots, 1]^{\mathrm{T}}$.

Reference [27] has proposed the critical point of EM wave propagation. When the values of the critical points are only 0 or 1, the maximum observability can be achieved. The critical points divide the whole system into several sections. R detectable sections are produced; then, the propagation path of the EM wave can be obtained by the topological structure and section length of the GIS. Therefore, a constraint can be found to be

$$G_{2R \times N} X_{N \times 1} \geq I_{2R \times 1} \tag{3}$$

where G is the propagation path matrix of EM wave of each section and the constraint shows that at least one path of the EM wave to the node where the sensor is installed passes through both ends of the section and $I_{2R \times 1} = [1, 1, \ldots, 1]^{\mathrm{T}}$.

By Equations (2) and (3), the 0-1 optimization model can be obtained:

$$\begin{aligned} \min \ &W_{1 \times N}^{T} X_{N \times 1} \\ \text{s.t. } &V_{M \times N} X_{N \times 1} \geq I_{M \times 1} \\ &G_{2R \times N} X_{N \times 1} \geq I_{2R \times 1} \\ &B_{2N \times N} X_{N \times 1} = b_{2N \times 1} \end{aligned} \tag{4}$$

where $W^{\mathrm{T}} = [w_1, w_2, \ldots, w_N]$ is a weight vector and represents the tendency of each node to install a sensor. The elements values of W^{T} are between 0 and 1. The equation constrains whether if a sensor should be installed at a certain node. If it should be installed at the node k, $B(k, k) = 1$ and $b(k) = 1$, otherwise $B(k + N, k) = 1$. The remaining elements in B and b are 0. The model conforms to the standard form of the 0-1 programming model and it can be solved by using *brintprog* tool in MATLAB.

2.3. Initial PD Location Method

The time difference of arriving location method for GIS is based on the arriving time of the EM signals detected by UHF sensors installed at both ends of the PD section. Considering the outer radius of the cavity of GIS is much smaller than its total length, it can be assumed that the EM wave will travel along the path as the pattern in Figure 3 after PD occurs.

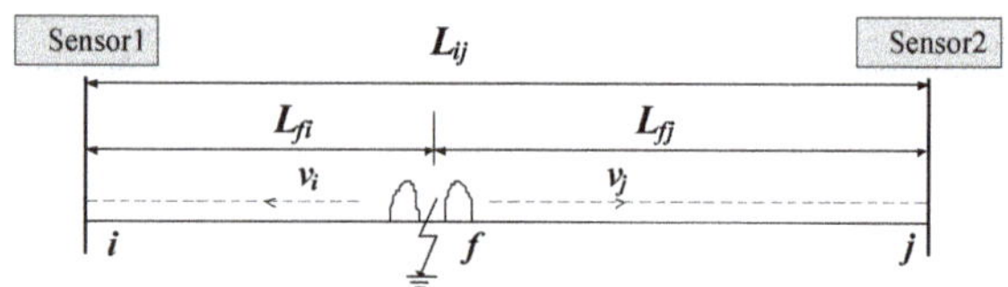

Figure 3. Location method based on time difference of arriving (TDOA).

For Figure 3, $v_i = v_j = c = 0.3$ m/ns is the propagation velocity of the EM wave, and the fault distance from the PD position to sensor 1 is given by

$$L_{fi} = \frac{1}{2}[L_{ij} + (t_i - t_j)c]$$ (5)

$$L_{fj} = L_{ij} - L_{fi}$$ (6)

where L_{fi} and L_{fj} are the distance from the PD position to I and to j, t_i and t_j are the time when sensor I and sensor j detect the signals. Equation (5) is a conventional PD location method, and the PD signals can propagate to other sensors through I and j. In view of the integrity of GIS, the propagation velocity of EM wave can be considered as remaining constant. This method lays a foundation for accurate PD locating using multiple sensors in GIS. Figure 3 can be expanded to Figure 4 below, and the EM waves travel through I and j to adjacent sensor x and y. Replacing I and j with x and y in Equations (5) and (6), the distance of extended TDOA can be expressed as

$$L_{fx} = \frac{1}{2}[L_{xy} + (t_x - t_y)c]$$ (7)

$$L_{fy} = L_{xy} - L_{fx}$$ (8)

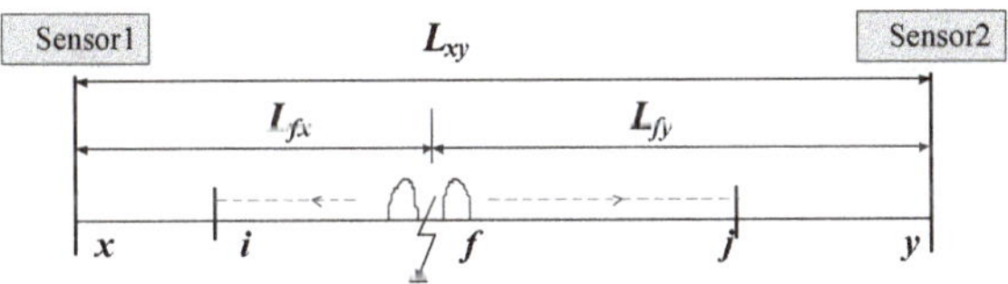

Figure 4. Extended TDOA location method.

Similarly, taking x and j or i and y as the ends of section, the corresponding PD position can be obtained. The extended TDOA method is the basic principle for PD locating of complex GIS. The accurate timing of optical fiber connection and the synchronization of different sampling units to picoseconds are the basis of PD locating among multiple sensors.

By the optimization Equation (4), it is assumed that P UHF sensors are installed at the nodes $S = [S_1 S_2 S_3 \ldots S_P]$ of a GIS. After PD occurs, the arriving time of the EM waves detected by the UHF sensors is $T_M = [T_1 T_2 T_3 \ldots T_P]^T$ and the minimum time in T_M is T_{min}. For a node S_m in S, all adjacent nodes are $S_V = [S_v S_{v2} S_{v3} \ldots S_{vw}]$, where w is the number of adjacent nodes. In Figure 5, S_{vr} can be any node in S_V.

The UHF sensors at S_m and S_p form a time difference location combination when PD occurs between i and j, and the shortest path between the two nodes must pass through an adjacent section L of S_m. The other end of L is the node S_{vr}. For S_m and its adjacent node matrix S_V, w TDOA location combinations can be obtained. For the UHF sensor arrangement matrix S, $P - 1$ combinations can be obtained. Therefore, the approximate solution can be obtained by comparing the similarity between the transmission time of the EM wave from the PD position to the sensors and the arriving time of the EM wave acquired by sensors synchronously.

The Euclidean distance is used to compare the similarity between two vectors. The Euclidean distance can magnify the error caused by sensor communication and data processing. It is suitable for comparing the unknown quantity with the known quantity with possible error and detecting redundancy. The value of the Euclidean distance is inversely proportional to the similarity between the two sets of data. The more similar the two sets of data are, the smaller the value is. Therefore, the initial locating results of PD can be obtained.

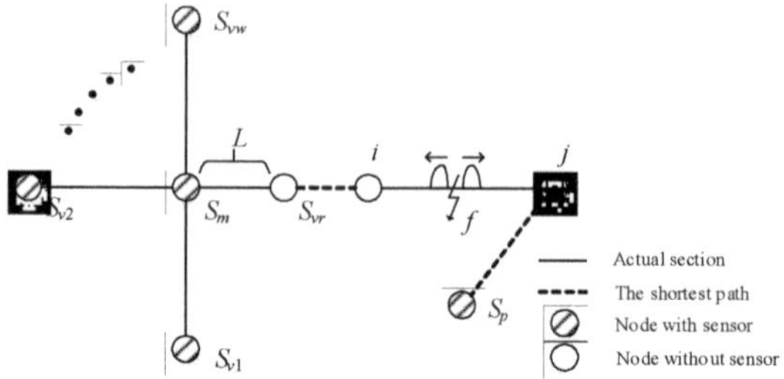

Figure 5. The sensors adjacent to Sm.

By the shortest path algorithm, the time matrix D_f of the EM waves propagating from the $w \times (P-1)$ positions to UHF sensors can be found as

$$D_f = \begin{array}{c} \\ 1 \\ 2 \\ \vdots \\ P-1 \end{array} \begin{array}{cccc} 1 & 2 & \cdots & w \\ \left(\begin{array}{cccc} D_{11} & D_{12} & \cdots & D_{1w} \\ D_{21} & D_{22} & \cdots & D_{2w} \\ \vdots & \vdots & \ddots & \vdots \\ D_{(P-1)1} & D_{(P-1)2} & \cdots & D_{(P-1)w} \end{array} \right) \end{array} \tag{9}$$

where the matrix element $D_{ij} = [D_{1'}\ D_{2'}\ \cdots\ D_P{}']$. Furthermore, the distance measure matrix E can be built as

$$E = \begin{array}{c} 1 \\ 2 \\ \vdots \\ P-1 \end{array} \left(\begin{array}{cccc} E_{11} & E_{12} & \cdots & E_{1w} \\ E_{21} & E_{22} & \cdots & E_{2w} \\ \vdots & \vdots & \ddots & \vdots \\ E_{(P-1)1} & E_{(P-1)2} & \cdots & E_{(P-1)w} \end{array} \right) = \begin{array}{c} 1 \\ 2 \\ \vdots \\ P-1 \end{array} \left(\begin{array}{cccc} \|D_{11}-T_M\| & \|D_{12}-T_M\| & \cdots & \|D_{1w}-T_M\| \\ \|D_{21}-T_M\| & \|D_{22}-T_M\| & \cdots & \|D_{2w}-T_M\| \\ \vdots & \vdots & \ddots & \vdots \\ \|D_{(P-1)1}-T_M\| & \|D_{(P-1)2}-T_M\| & \cdots & \|D_{(P-1)w}-T_M\| \end{array} \right) \tag{10}$$

The initial PD location result is given by the minimum value in the measure matrix E, and the corresponding PD position can be expressed as f'.

2.4. Accurate PD Location Method for GIS

The accuracy of PD locating can be improved by using the information of all UHF sensors installed in GIS. From the initial PD position and the inherent topological structure of GIS, the time of PD occurrence can be calculated. Then, the time of the EM wave propagating to UHF sensors is calculated by the time of PD occurrence and the EM wave velocity to obtain multiple sets of PD location results. With these results, accurate PD locating can be achieved.

The occurring time t_0 of PD can be found to be

$$Tmin - t_0 = \frac{L_{bf'}}{c} \tag{11}$$

where $L_{bf'}$ is the distance from the PD position to the UHF sensor corresponding to T_{min}. By t_0, the distance matrix of the PD signals propagating to each sensor can be calculated to be

$$\begin{aligned} L &= [\ L_{1f}\ \ L_{2f}\ \ \cdots\ \ L_{Pf}\ ]^T \\ &= c(T_M - t_0\eta) \end{aligned} \tag{12}$$

where $\eta = [1\ 1\ \cdots\ 1]^{\mathrm{T}}_{P\times1}$. According to the topological structure of GIS, the PD positions can be found:

$$S = \begin{bmatrix} S_1 & S_2 & \cdots & S_P \end{bmatrix}^{\mathrm{T}} \tag{13}$$

In combination with all the information of UHF sensors, the accurate PD position can be found as follows:

$$S_{act} = \frac{1}{P}\sum_{i=1}^{P} S_i \tag{14}$$

The flow chart of the PD location method for GIS and the steps are as follows (shown in Figure 6):

1) Acquire the EM wave arriving time matrix $T_M = [T_1 T_2 T_3 \ldots T_P]^{\mathrm{T}}$ and the corresponding node matrix $S = [S_1 S_2 S_3 \ldots S_P]$ from UHF sensors after PD occurs.

2) Based on S_m corresponding to T_{min}, $w \times (P-1)$ PD location combinations are composed of w adjacent nodes and $P-1$ UHF sensors.

3) The initial PD position can be calculated by the measure matrix E, and the occurring time t_0 of PD can be found.

4) Combining t_0 with the topology structure of GIS, the PD position can be revised by multiple sensors of the GIS to obtain the accurate PD location results.

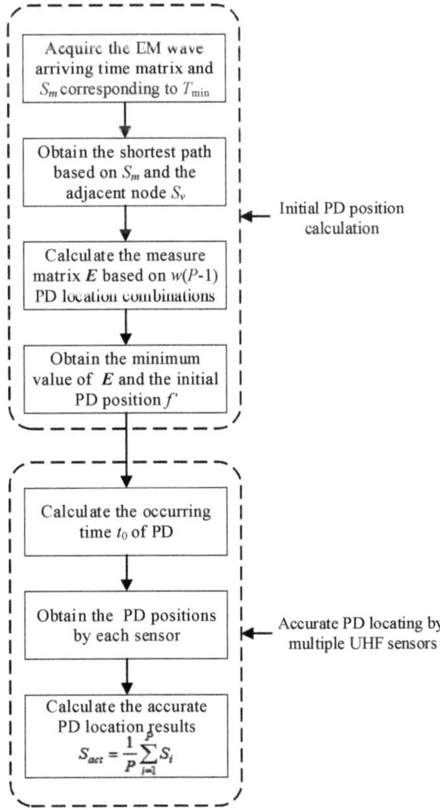

Figure 6. Flow chart of the location method.

2.5. Noise Reduction Method

An improved dual-tree complex wavelet transform (DTCWT) method is proposed to reduce noise. The DTCWT method is used to decompose the noisy signal into a series of detail components which can reflect PD [28]. However, only part of DTCWT components details are closely related to PD, while the others are reflected in noise signals. Therefore, the improved DTCWT method is proposed to select sensitive components according to time-domain kurtosis and envelope spectral kurtosis. According to the principle of maximum kurtosis deconvolution, the sensitive components are deconvoluted to reduce noise by MKD deconvolution. Finally, the denoised components are reconstructed by inverse DTCWT transform to obtain the denoised signal.

The steps of the improved DTCWT method are as follows:

1) Determine the number of decomposition layers of DTCWT. Different layers will affect the accuracy and speed of noise reduction calculation.

2) Decompose the noisy signal by DTCWT. The signal is transformed by n-layer dual-tree complex wavelet transform. The inverse transform of the wavelet coefficients $wi = w_i^{re} + iw_i^{im}$ of layer i is implemented to obtain the wavelet components of layer i. A matrix can be formed by signal components at different scales of signal transformation:

$$Dn = \begin{pmatrix} D1 \\ D2 \\ \vdots \\ Dn \end{pmatrix} \begin{pmatrix} d_1(1) & d_1(2) & \cdots & d_1(k) \\ d_2(1) & d_2(2) & \cdots & d_2(k) \\ \vdots & \vdots & & \vdots \\ d_n(1) & d_n(2) & \cdots & d_n(k) \end{pmatrix} \tag{15}$$

3) Screen Signal Components. The time kurtosis of signal D_i can be found:

$$Kur(d) = \frac{\frac{1}{n}\sum (d_i - \bar{d})^4}{Std(x)^4} \tag{16}$$

The envelope spectral kurtosis can be found:

$$env(t) = \sqrt{x(t)^2 + \{HT[x(t)]\}^2} \tag{17}$$

where HT is Hilbert transform. The higher time kurtosis and envelope spectral kurtosis indicate that there are more partial discharge impulse components in D_i. The product of time domain kurtosis and envelope spectral kurtosis (index TE) of each component is calculated, and the component with large product is selected for reconstruction.

4) The selected DTCWT detail signal component is processed by MKD denoising, and then the inverse DTCWT transform is used to reconstruct the signal. After the inverse transform, the denoised signal of the component can be obtained.

3. Experimental Results and Analysis

3.1. Experimental Platform

The experiment is based on the partial discharge online monitoring system of China Electric Power Research Institute. The precise timing system with optical fiber connection can ensure that the clocks of each sensor can be synchronized to the same nanosecond to realize a precise location system composed of multiple channels. The test system consists of 500 kV GIS model, power supply, PD detector, phase acquisition device, built-in UHF sensor, UHF amplifier, broadband oscilloscope, discharge model, etc (shown in Figures 7 and 8).

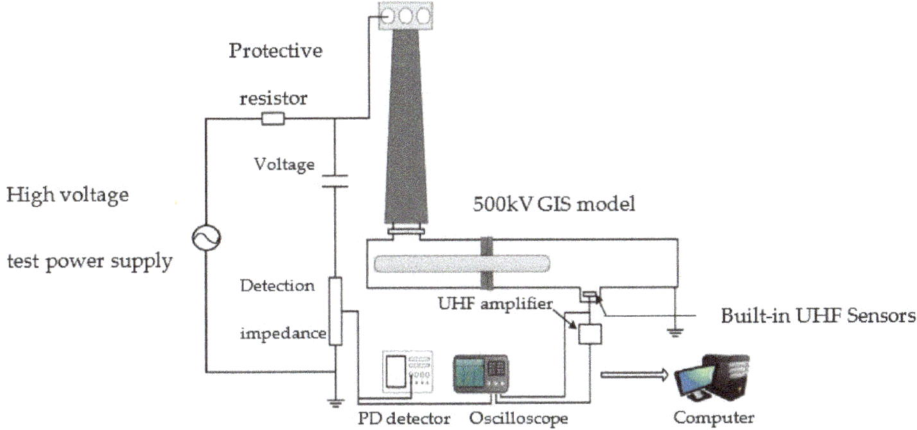

Figure 7. Partial discharge (PD) online monitoring system diagram.

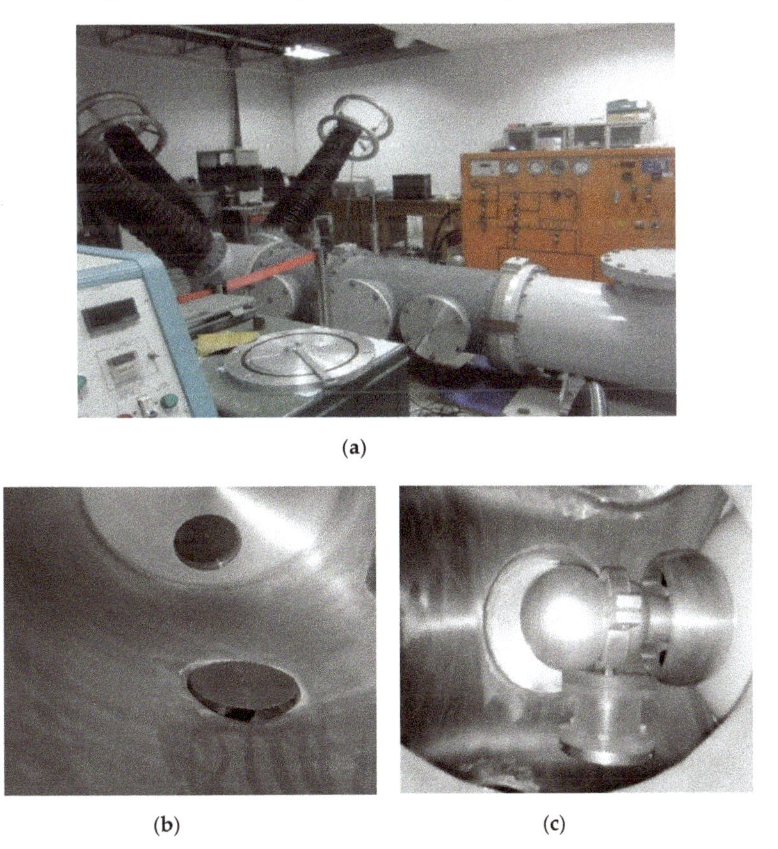

(**a**)

(**b**)   (**c**)

Figure 8. Laboratory for the PD online monitoring system. (**a**) Experimental Platform, (**b**) ultra-high frequency (UHF) sensor in GIS, (**c**) discharge model in GIS.

The whole experimental GIS platform is divided into independent test chambers by pelvic insulators. Operating hand holes are opened on the side of each chamber to facilitate the placement of defect models, such as a pin-to-plate discharge model.

3.2. Noise Reduction

The noise reduction method is validated by a measured partial discharge signal with noise shown in Figure 9.

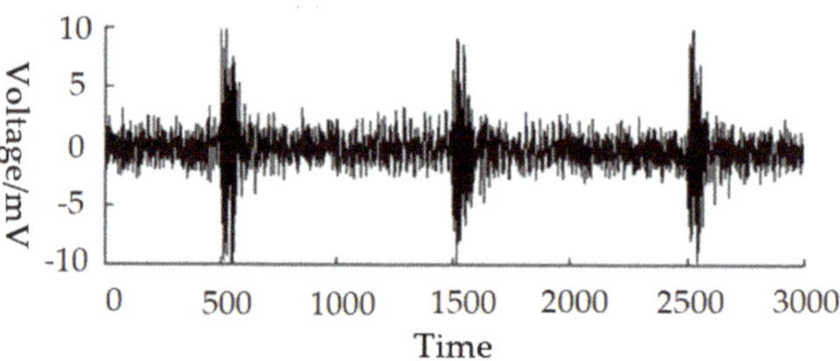

Figure 9. Measured partial discharge signal.

The improved DTCWT method is used to denoise the measured partial discharge signal. The sampling number of UHF signal is 3000. The decomposition level of DTCWT is 5. The signal components of DTCWT, of which the TE values are in the top three, are reconstructed. The noise reduction result is shown in Figure 10.

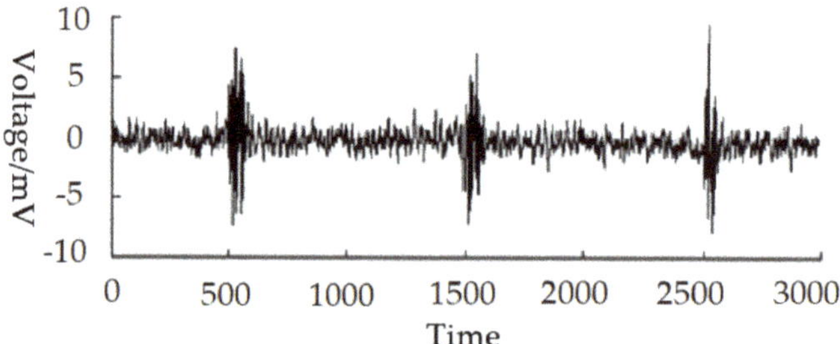

Figure 10. Noise reduction result.

Variation trend parameter (*VTP*) and signal-to-noise ratio (*SNR*) are used to evaluate the denoising effect of UHF PD signals:

$$SNR = 20 \lg \frac{\max\limits_{i=1}^{n} s(i)}{\max\limits_{i=1}^{n} n(i)} \tag{18}$$

$$VTP = \frac{\sum\limits_{i=2}^{n} [f(i) - f(i-1)]}{2 \sum\limits_{i=2}^{n} [s(i) - s(i-1)]} + \frac{\sum\limits_{i=2}^{n} [f(i-1) - f(i)]}{2 \sum\limits_{i=2}^{n} [s(i-1) - s(i)]} \tag{19}$$

The closer the *VTP* value is to 1, the better the waveform fitting effect is. Comparing the DTCWT method with improved DTCWT method, the result is shown in Table 1. The improved DTCWT method has higher *SNR* and the waveform is not obviously distorted, which maintains the characteristics of the original UHF PD signal.

Table 1. The comparison of output variation trend parameter (*VTP*) and signal-to-noise ratio (*SNR*).

	DTCWT	Improved DTCWT
SNR	7.59	20.07
VTP	1.495	1.030

3.3. Optimal Placement

The 500 kV GIS experimental platform was used to verify the effectiveness of the proposed method. The nodes of GIS are numbered and the node number is also the sensor number. The lengths of the sections between the nodes are shown in Table 2.

Table 2. GIS topological parameters.

Section	Length (mm)	Section	Length (mm)
1–2	3826	16–17	1060
2–3	1705	17–18	3413
3–4	1153	16–19	1715
4–5	2494	19–7	3380
5–6	1155	15–20	1035
4–7	1550	20–21	1380
7–8	1133	21–22	2860
8–9	3485	22–23	1380
9–10	2550	23–24	1060
10–11	8210	24–25	1250
11–12	2550	24–26	3460
12–13	3485	26–27	990
13–14	1740	27–28	990
14–15	3380	28–29	3790
15–16	1715	29–14	1740

The undirected graph of the structure of this GIS shown in Figure 11 can be drawn from Table 2. The nodes where UHF sensors should be installed can be obtained by model (4), including nodes 1, 3, 4, 6, 7, 9, 11, 13, 14, 15, 16, 18, 21, 23, 25, 26, and 28. In Figure 11, the system has two tangent ring structures; therefore, non-detection zones may exist. The minimum and maximum values of the EM wave critical points for all sections are 0 and 1 through calculations [27]. Therefore, these sections are totally detectable. In addition, the constraints of model (4) ensure that the sensors can detect the EM waves effectively.

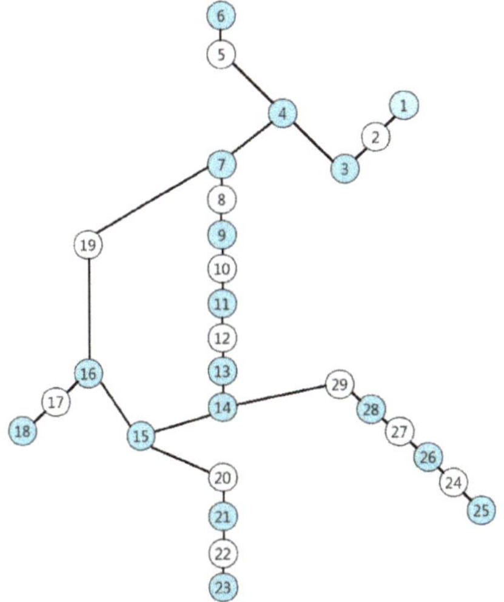

Figure 11. Sensor placement results.

3.4. PD Location Results

Assume a PD occurred at 20 ns in section 8-9, and the distance from the PD position to node 8 is 1000 mm. Each sensor can monitor the signal of each node and 400 MHz high-pass filtering was performed on the EM signal detected by the sensors. The amplitude and waveform of the signal barely change through the filter, so it is easy to obtain the arriving time of the EM wave, as shown in Table 3. From Table 2, $T_{min} = t_7$. Calculating the TDOA of sensor 7 with the other 16 sensors in sequence, the results are shown in Table 4. In Table 4, the minimum value in the Euclidean distance measure matrix corresponds to the result calculated by sensor 7 and 11. Therefore, the PD occurred in section 8-9, 954.3 mm away from node 8. The actual discharge distance is 1000 mm, the absolute error is 45.7 mm, and the relative error is 4.6%. The occurring time t_0 of PD was found to be 20.1 ns by Formula (12). Based on Formulas (13) and (14), the revised PD position can be obtained shown in Table 5. By Formula (15), the PD occurred in section 8-9, 993.5 mm away from node 8. The actual discharge distance is 1000 mm, the absolute error is 6.5 mm, and the relative error is 0.7%.

Table 3. Arriving time of the initial signal.

Sensor Number	Time (ns)	Sensor Number	Time (ns)
1	34.7	15	29.3
3	15.7	16	24.3
4	12.6	18	39.4
6	24.3	21	37.3
7	7.3	23	52.5
9	7.9	25	59.4
11	43.8	26	66.6
13	47.2	28	59.1
14	40.5	–	–

Table 4. Initial results of PD locating.

Sensor Combination	Euclidean Distance	PD Section	PD Position (mm)
(1,7)	8.9	3-4	7.0×10^{-2}
(3,7)	6.8	4-7	1.2×10^{-1}
(4,7)	7.0	4-7	2.4×10^{-1}
(6,7)	8.8	5-6	1.2
(9,7)	1.9	8-9	915.0
(11,7)	0.8	8-9	954.3
(13,7)	2.3	8-9	1113.4
(14,7)	1.9	8-9	903.5
(15,7)	1.7	8-9	954.3
(16,7)	1.8	8-9	915.0
(18,7)	3.9	4-5	1.3×10^{-3}
(21,7)	1.7	8-9	923.5
(23,7)	2.1	8-9	903.5
(25,7)	1.2	8-9	1078.2
(26,7)	1.9	8-9	915.0
(28,7)	1.8	8-9	1093.4

Table 5. Revised results of PD locating.

Sensor Number	PD Position (mm)	Absolute Error (mm)
1	1098.4	98.4
3	1123.9	123.9
4	1087.0	87.0
6	923.5	76.5
7	883.5	116.5
9	1095.8	95.8
11	913.9	86.1
13	913.4	86.5
14	954.3	45.7
15	902.7	97.3
16	906.6	93.4
18	1087.9	87.9
21	1077.2	77.2
23	923.5	76.5
25	893.7	106.3
26	914.2	85.8
28	1092.6	92.6

3.5. Fault Tolerance Analysis

The proposed method has been verified in the example above on PD of 500 kV GIS, but the PD locating is under the premise that there is no error in the detection of the UHF sensors. In practical operation, error data of sensors would be caused by the influence of environmental noise and communication delay.

In order to verify the fault tolerance of the algorithm, the arriving time of the EM wave of the sensor at node 11 was artificially set as 53.8 ns. The minimum arriving time T_{min} remains t_7, and the TDOA calculations of sensor 7 with the other sensors in sequence were carried out. The result shows that the PD distance calculated by the combination of senor 7 and 11 is infinity. Therefore, the information obtained by sensor 11 was judged to be erroneous data. After the erroneous data was eliminated, it can be obtained that the PD occurred on section 8-9, 988.7 mm away from node 8. The absolute error is 11.3 mm and the relative error is 1.1%.

In the same case, the existing methods for PD locating cannot eliminate the impact of erroneous data. The proposed method realizes the identification of the erroneous data through the information of multiple sensors in GIS and ensures the fault tolerance of the algorithm.

3.6. Comparison of Methods

The received signal strength indicator (RSSI) location method is based on the strength of EM wave. PD locating can be achieved by the reconstruction algorithm of RSSI based on pattern matching [29,30]. Twelve experiments were carried out to compare the proposed method and the RSSI method. Location results are shown in Table 6.

Table 6. Method comparison.

	Proposed Method	RSSI
average absolute error (mm)	22.8	107.7
absolute error < 50 mm (%)	100	50
absolute error < 100 mm (%)	100	66.7
absolute error < 200 mm (%)	100	83.3
Maximum absolute error (mm)	42.3	238.5

For Table 6, The average absolute error of the proposed method is 22.8 mm, which is obviously superior to the RSSI method, and its 100% absolute errors are less than 50 mm. The proposed method keeps the ns-level time synchronization between the sensors and the acquisition system. Therefore, the precision and accuracy can be guaranteed.

4. Conclusions

Considering the attenuation of the EM wave when passing through the insulators, L-type and T-type structures in GIS, and the constraints of economy and observability, an optimal placement model of UHF sensors is proposed. The model can meet the requirements and has high reliability, as validated by experimental results.

Based on the optimal placement of the UHF sensors and the arriving time of EM waves, the initial results of PD locating can be calculated by the Euclidean distance and the TDOA method. Then, the accurate results can be calculated by the occurring time of PD and the information of all UHF sensors.

Compared to the existing PD location methods, the proposed method ensures the accuracy and fault tolerance of PD locating. The experimental results show the method has high accuracy and robustness.

Author Contributions: Conceptualization, R.L.; methodology, S.W.; validation, P.C.; formal analysis, N.P. and S.W.; data curation, Y.L.; writing—original draft preparation, S.W.; writing—review and editing, R.L.

Funding: This work was supported by the Fundamental Research Funds for the Central Universities (2017XKQY033).

Conflicts of Interest: The authors declare no conflict of interest.

References

1. Si, W.R.; Li, J.H.; Li, D.J.; Yang, J.G.; Li, Y.M. Investigation of a Comprehensive Identification Method Used in Acoustic Detection System for GIS. *IEEE Trans. Dielectr. Electr. Insul.* **2010**, *17*, 721–732. [CrossRef]
2. OKabe, S.; Yamagiwa, T.; Okubo, H. Detection of Harmful Metallic Particles Inside Gas Insulated Switchgear using UHF Sensor. *IEEE Trans. Dielectr. Electr. Insul.* **2008**, *15*, 701–709. [CrossRef]
3. OKabe, S.; Ueta, G.; Hama, H.; Ito, T.; Hikita, M.; Okubo, H. New aspects of UHF PD diagnostics on gas-insulated systems. *IEEE Trans. Dielectr. Electr. Insul.* **2014**, *21*, 2245–2258. [CrossRef]
4. Mirzaei, H.R.; Akbari, A.; Gockenbach, E.; Miralikhani, K. Advancing new techniques for UHF PD detection and localization in the power transformers in the factory tests. *IEEE Trans. Dielectr. Electr. Insul.* **2015**, *22*, 448–455. [CrossRef]
5. Ishak, A.M.; Ishak, M.T.; Jusoh, M.T.; Dardin Syed, S.F.; Judd, M.D. Design and optimazation of UHF partial discharge sensors using FDTD modeling. *IEEE Sens. J.* **2017**, *17*, 127–133.
6. Wang, Y.; Wang, X.; Li, J. UHF moore fractal antennas for online GIS PD detection. *IEEE Antennas Wirel. Propag. Lett.* **2017**, *16*, 852–855. [CrossRef]
7. Tenbohlen, S.; Denissov, D.; Hoek, S.M.; Markalous, S.M. Partial discharge measurement in the ultra-high frequency (UHF) range. *IEEE Trans. Dielectr. Electr. Insul.* **2008**, *15*, 1544–1552. [CrossRef]
8. Xu, Y.; Liu, W.; Gao, W. Investigation of disc-type sensors using the UHF method to detect partial discharge in GIS. *IEEE Trans. Dielectr. Electr. Insul.* **2015**, *22*, 3019–3027. [CrossRef]
9. Sinaga, H.H.; Phung, B.T.; Blackburn, T.R. Recognition of single and multiple partial discharge sources in transformers based on ultra-high frequency signals. *IET Gener. Transm. Distrib.* **2014**, *8*, 160–169. [CrossRef]
10. Boya, C.; Rojas, M.V.; Ruiz, M.; Robles, G. Location of partial discharges sources by means of blind source separation of UHF signals. *IEEE Trans. Dielectr. Electr. Insul.* **2015**, *22*, 2302–2310. [CrossRef]
11. Gao, W.; Ding, D.; Liu, W. Research on the typical partial discharge using the UHF detection method for GIS. *IEEE Trans. Power Deliv.* **2011**, *26*, 2621–2629. [CrossRef]
12. Zhang, G.B.; Guo, J.B.; Chu, F.H.; Zhang, Y.C. Environmental-adaptive indoor radio path loss model for wireless sensor networks localization. *Int. J. Electron. Commun.* **2011**, *65*, 1023–1031. [CrossRef]
13. Iorkyase, E.T.; Tachtatzis, C.; Atkinson, R.C.; Glover, I.A. Localization of partial discharge sources using radio fingerprinting technique. In Proceedings of the 2015 Loughborough Antennas & Propagation Conference (LAPC), Loughborough, UK, 2–3 November 2015; pp. 1–5.
14. Ito, T.; Kamei, M.; Ueta, G.; Okabe, S. Improving the sensitivity verification method of the UHF PD detection technique for GIS. *IEEE Trans. Dielectr. Electr. Insul.* **2012**, *18*, 1847–1853. [CrossRef]

15. Reid, A.J.; Judd, M.D.; Fouracre, R.A.; Stewart, B.; Hepburn, D.M. Simultaneous measurement of partial discharges using IEC60270 and radio-frequency techniques. *IEEE Trans. Dielectr. Electr. Insul.* **2011**, *18*, 444–455. [CrossRef]
16. Zheng, S.; Li, C.; Tang, Z.; Chang, W.; He, M. Location of PDs inside transformer windings using UHF methods. *IEEE Trans. Dielectr. Electr. Insul.* **2014**, *21*, 386–393. [CrossRef]
17. Hekmati, A.; Hekmati, R. Optimum acoustic sensor placement for partial discharge allocation in transformers. *IET Sci. Meas. Technol.* **2017**, *11*, 581–589. [CrossRef]
18. Dai, D.; Wang, X.; Long, J.; Tian, M.; Zhu, G.; Zhang, J. Feature extraction of GIS partial discharge signal based on S-transform and singular value decomposition. *IET Sci. Meas. Technol.* **2017**, *11*, 186–193. [CrossRef]
19. Gao, W.; Ding, D.; Liu, W.; Huang, X. Propagation attenuation properties of partial discharge in typical in-field GIS structures. *IEEE Trans. Power Deliv.* **2013**, *28*, 2540–2549. [CrossRef]
20. Hikita, M.; Ohtsuka, S.; Wada, J.; Okabe, S.; Hoshino, T.; Maruyama, S. Propagation properties of PD-induced electromagnetic wave in 66 kV GIS model tank with L branch structure. *IEEE Trans. Dielectr. Electr. Insul.* **2011**, *18*, 1678–1685. [CrossRef]
21. Hikita, M.; Ohtsuka, S.; Ueta, G.; Okabe, S.; Hoshino, T.; Maruyama, S. Influence of insulating spacer type on propagation properties of PD-induced electromagnetic wave in GIS. *IEEE Trans. Dielectr. Electr. Insul.* **2010**, *17*, 1642–1648. [CrossRef]
22. Li, T.; Wang, X.; Zheng, C.; Liu, D.; Rong, M. Investigation on the placement effect of UHF sensor and propagation characteristics of PD-induced electromagnetic wave in GIS based on FDTD method. *IEEE Trans. Dielectr. Electr. Insul.* **2014**, *21*, 1015–1025. [CrossRef]
23. Li, X.; Wang, X.; Yang, A.; Xie, D.; Ding, D.; Rong, M. Propogation characteristics of PD-induced UHF signal in 126 kV GIS with three-phase construction based on time-frequency analysis. *IET Sci. Meas. Technol.* **2016**, *10*, 805–812. [CrossRef]
24. Markalous, S.M.; Tenbohlen, S.; Feser, K. Detection and location of partial discharges in power transformers using acoustic and electromagnetic signals. *IEEE Trans. Dielectr. Electr. Insul.* **2008**, *15*, 1576–1583. [CrossRef]
25. Tang, J.; Xie, Y. Partial discharge location based on time difference of energy accumulation curve of multiple signals. *IET Electr. Power Appl.* **2011**, *5*, 175–180. [CrossRef]
26. Zhang, Q.; Li, C.; Zheng, S.; Yin, H.; Kan, Y.; Xiong, J. Remote detecting and locating partial discharge in bushings by using wideband RF antenna array. *IEEE Trans. Dielectr. Electr. Insul.* **2016**, *23*, 3575–3583. [CrossRef]
27. Liang, R.; Xu, C.; Wang, F.; Cheng, Z.; Shen, X. Optimal deployment of fault location devices based on wide area travelling wave information in complex power grid. *Trans. China Electrotech. Soc.* **2016**, *31*, 30–38.
28. Le, Q.; Zhen, Z.; Pei, Y. Image identification of insects based on color histogram and dual tree complex wavelet transform (DTCWT). *Acta Entomol. Sin.* **2010**, *53*, 91–97.
29. Zhang, W.; Bi, K.; Luo, L.; Sheng, G.; Jiang, X. RSSI fingerprinting-based UHF partial discharge localization technology. In Proceedings of the Power & Energy Engineering Conference, Xi'an, China, 25–28 October 2016; pp. 1364–1367.
30. Li, Z.; Luo, L.G.; Sheng, G.H.; Liu, Y.; Jiang, X. UHF partial discharge localisation method in substation based on dimension-reduced RSSI fingerprint. *IET Gener. Transm. Distrib.* **2018**, *12*, 398–405. [CrossRef]

Article

Implementation of User Cuts and Linear Sensitivity Factors to Improve the Computational Performance of the Security-Constrained Unit Commitment Problem

Cristian Camilo Marín-Cano, Juan Esteban Sierra-Aguilar, Jesús M. López-Lezama, Álvaro Jaramillo-Duque * and Walter M. Villa-Acevedo

Research Group in Efficient Energy Management (GIMEL), Departamento de Ingeniería Eléctrica, Universidad de Antioquia, Calle 67 No. 53-108, Medellín 050010, Colombia; cristian1013@gmail.com (C.C.M.-C); juane.sierra@udea.edu.co (J.E.S.-A.); jmaria.lopez@udea.edu.co (J.M.L.-L); walter.villa@udea.edu.co (W.M.V.-A.)
* Correspondence: alvaro.jaramillod@udea.edu.co; Tel.: +57-034-2198597

Received: 20 March 2019; Accepted: 3 April 2019; Published: 11 April 2019

Abstract: Power system operators must schedule the available generation resources required to achieve an economical, reliable, and secure energy production in power systems. This is usually achieved by solving a security-constrained unit commitment (SCUC) problem. Through a SCUC the System Operator determines which generation units must be on and off-line over a time horizon of typically 24 h. The SCUC is a challenging problem that features high computational cost due to the amount and nature of the variables involved. This paper presents an alternative formulation to the SCUC problem aimed at reducing its computational cost using sensitivity factors and user cuts. Power Transfer Distribution Factors (PTDF) and Line Outage Distribution Factors (LODF) sensitivity factors allow a fast computation of power flows (in normal operative conditions and under contingencies), while the implementation of user cuts reduces computational burden by considering only biding *N-1* security constraints. Several tests were performed with the IEEE RTS-96 power system showing the applicability and effectiveness of the proposed modelling approach. It was found that the use of Linear Sensitivity Factors (LSF) together with user cuts as proposed in this paper, reduces the computation time of the SCUC problem up to 97% when compared with its classical formulation. Furthermore, the proposed modelling allows a straightforward identification of the most critical lines in terms of the overloads they produce in other elements after an outage, and the number of times they are overloaded by a fault. Such information is valuable to system planners when deciding future network expansion projects.

Keywords: optimization; power system; Security-Constraint Unit Commitment; sensitivity factors; user cuts

---

1. Introduction

The electric power sector is currently facing rapid changes, mostly related to the integration of renewable energy resources and the new trend of smart grids. In this context, power system operators are continuously seeking novel sources of operational flexibility and improved methods to manage and integrate generation resources, guaranteeing a secure operation. Within the daily operation planning of a power system the unit commitment (UC) is one of the most important decision-making activities. The UC brings up the task of finding an optimal schedule and production level of a set of generation units over a given period of time. It also serves for clearing daily prices in power markets and as a tool used by generation companies for optimizing bidding strategies [1]. In its traditional approach, the UC is run within a typical time horizon of 24 h aiming at minimizing the total operation costs while meeting network demand.

UC is a widely studied problem. Recent works emphasize the inclusion of renewable energy [2–4], energy storage [5,6], uncertainty [7,8] and security constraints [9,10]. This paper focuses on the last issue. Within the daily operation planning of power systems, an important aspect to take into account is monitoring the limits of the electric network not only for normal operating conditions but also under credible contingencies; especially, at periods of peak demand when transmission capacity margins are reduced. Therefore, solving the UC problem considering security constraints becomes a fundamental condition for obtaining an economic, safe and reliable system operation [11]. When transmission limits and security constraints (*N-1* criterion) are included in the UC, system operators find themselves dealing with a much more complex problem, namely the security-constraint unit commitment (SCUC) problem.

The SCUC is a non-linear, large-scale, mixed-integer combinatorial optimization problem that considers numerous operating constraints, such as ramp rates, minimum up/down time, start-up time as well as transmission and generation limits. The SCUC also incorporates contingency analyses that take into account forced outages of network elements. This problem can be represented as a mixed-integer programming (MIP) problem, and from the point of view of its computational complexity is NP-complex [9]. Several optimization methods have been used to approach the SCUC problem. These methods can be broadly classified in three groups: mathematical programming approaches, metaheuristic techniques, and hybrid methods. Classical mathematical programming approaches include Lagrangian Relaxation (LR) [12,13], Benders Decomposition (BD) [14–17] and Branch and Cut (BC) methods [18,19]. Several metaheuristic techniques have also been applied to solve the UC and SCUC problem such as genetic algorithms [20–24], cuckoo search algorithm [25] and swarm intelligence [26]. Finally, hybrid methodologies are presented in [27,28]. The literature regarding the SCUC problem is very extensive. A comparison of transmission constrained unit commitment formulations is presented in [29]. A survey of research work made in the domain of UC and SCUC using various techniques is presented in [30]. Also, literature reviews approaching specific issues such as stochasticity and uncertainty within the SCUC problem are addressed in [31,32], respectively.

The inclusion of security constraints is one of the main complicating factors of the UC problem. That is because its number is directly proportional to the product of the number of transmission lines and the discrete steps of the commitment horizon. Therefore, the SCUC can be greatly simplified if inactive security constraints are identified and eliminated. In [10] , the authors show that inactive constraints can be identified by solving a series of small-scale mixed-integer linear programming (MILP) problems and propose an analytical condition to identify such constraints. They also found that computational requirements and numerical error of the SCUC are greatly reduced if inactive constraints are eliminated. Furthermore, the authors show that the crucial transmission lines affecting the total operating costs are among those associated with the remaining security constraints. In [33], the authors provide a systematic method to construct feasible solutions for the SCUC problem within a LR framework, based on a group of analytical feasibility conditions. In line with the work presented in [10] the authors, in [34] present the concept of umbrella constraints for the security-constraint optimal power flow (SCOPF) problem. This is because few of the constraints actually serve to enclose the feasible region of a typical SCOPF. Hence, the constraints that are not contributing directly to the feasible space can be discarded. The authors establish the necessary and sufficient conditions of the set of feasible solutions by ruling out unnecessary constraints of the SCOPF. Although these conditions allow the identification of inactive and binding constraints, they also involve solving MILP problems, which despite being less complex than the original formulation are not trivial in their solution, especially when dealing with large-scale problems, and therefore BD may be needed [10,34]. In [35] the authors propose an iterative algorithm to speed up the solution of an AC Preventive Security-Constrained Optimal Power Flow (AC-PSCOPF) through contingency filtering techniques; non-dominated *N-1* contingencies are the security constraints bounded to the AC-PSCOPF model. Nevertheless, for performing the *N-1* contingency analysis, AC-power flows, into this algorithm, must be computed.

A fundamental aspect in the modelling of the SCUC problem is the representation of the electrical network. A linear representation of the network (DC model) as in [13,15,16], instead of an AC modelling as in [36–38] allows the use of linear sensitivity factors (LSF) such as power transfer distribution factors (PTDF) and line outage distribution factors (LODF), commonly used in the analysis of electrical systems [39,40]. These factors represent the approximate change in power flows in transmission assets due to a change in power injection (PTDF) or due to a new topology configuration of the network when a transmission element fails (LODF). The use of these factors allows reducing the number of variables and constraints of the SCUC model, increasing computational efficiency. In [41], the authors use PTDF and LODF to calculate generation shifts so that power flows on transmission lines are kept within security limits after single and multiple-line outage events. In [9], the authors present an N-1 security-constraint formulation of the SCUC problem based on LODF combined with an iterative methodology for filtering N-1 congestion constraints. The authors show that such filtering through lazy constraints greatly reduces the computational time of the SCUC problem. A method for estimating these sensitivity factors based on near-real-time measurements and their applications can be found in [42]. Despite advantages of these linear factors, ref [43] proposes a new formulation to reduce the number of variables and constraints to speed up the SCUC solution in large-scale power systems, using BD and the BC in an iterative framework. Nonetheless, post-contingency power flows are modelled using the bus voltage angles.

This paper presents a novel formulation of the SCUC problem that considerably reduces its computing time. This is done through the implementation of user cuts and the use of LSF. On the one hand, the user cuts approach proposed in this paper serves as umbrella constraints which results in a considerable computational burden reduction when considering N-1 security constraints. This is because most of the N-1 contingency constraints are superfluous and do not set up the feasibility space of the SCUC problem [9]. On the other hand, PTDF and LODF sensitivity factors allow a fast computation of power flows both in normal operation and under contingencies. Furthermore, some indexes are proposed to identify the most severe faults as well as the most critical lines. The later information is of paramount importance to system planners since this can be included in expansion plan studies.

This paper is organized as follows: Section 2 describes the mathematical formulation of the SCUC problem. Section 3 presents the proposed approach to solve the SCUC problem. Section 4 shows tests and results. Finally, Section 5 reports conclusions.

2. Problem Statement

2.1. Linear Sensitivity Factors

Distribution factors are linear approximations of the sensitivities of power flows with respect to changes in nodal injections [39]. PTDF represent changes in power flows in lines as regards the transfer of one MW from an injection node to a consumption node. On the other hand, LODF represent the distribution of power flows in the remaining lines after an outage has taken place. The DC model of the transmission network is one of the most used in power system studies. With this model it is possible to compute the power flows using the vector of power injections in each bus $\bar{P}$ and the susceptance matrix $\bar{B}$. Power flows are given in terms of angular differences that can be obtained from Equation (1). Since matrix $\bar{B}$ is non-invertible, the row and column corresponding to the slack bus are eliminated resulting in matrix $\bar{B}'$ as indicated in Equation (2). The computation of the power flow in a transmission line is obtained as the product of the susceptance of the line by the angular difference between the two buses that connect it. This operation can be expressed in matrix form as shown in Equation (3), where $\bar{D}$ is the diagonal matrix of susceptances, $\bar{A}$ is the incidence matrix of the system and $\bar{\theta}$ is the vector of bus voltages angles. Power flows in lines can be expressed as a function of $\bar{P}$, by replacing Equation (2) in Equation (3), as shown in Equation (4). Finally, Equation (5) shows how to compute the matrix of PTDF. It is worth mentioning that when calculating matrix $\bar{A}$, the row corresponding to the slack bus must be removed to perform matrix operations without dimensional problems. This leads to the PTDF matrix not considering the slack bus until this instance of the calculation. Subsequently, to account for the slack bus, a column of zeros in the position of this node is added to the PTDF matrix. This is because there is no power flow change in the transmission lines when the power injection is made in the slack bus and withdrawn from it.

$$\bar{P} = \bar{B} \cdot \bar{\theta} \tag{1}$$

$$\bar{\theta} = \bar{B}'^{-1} \cdot \bar{P} \tag{2}$$

$$\bar{F} = \bar{D}\bar{A}\bar{\theta} \tag{3}$$

$$\bar{F} = \bar{D}\bar{A}\bar{B}'^{-1} \cdot \bar{P} \tag{4}$$

$$\overline{PTDF} = \bar{D}\bar{A}\bar{B}'^{-1} \tag{5}$$

The computation of LODF indices is done by means of an auxiliary matrix $\bar{H}$ as shown in Equation (6). One of the drawbacks of this methodology lies on the fact that there are no defined values of LODF for radial lines. To address this problem, it is suggested in [44] to assign values of 0 (see Equation (7); that is to say, contingencies associated with radial lines are not taken into account. Furthermore, in the main diagonal of the LODF matrix, -1 is assigned, since there is no post-fault power flow for the line that is out of operation. With these established rules, the post-contingency power flow in line l when line k fails $(F_{l,k})$ is given as shown in Equation (8).

$$\bar{H} = \overline{PTDF} \cdot \bar{A}^T \tag{6}$$

$$\overline{LODF} = \begin{cases} \frac{H_{i,j}}{1-H_{j,j}} & \text{if } H_{i,j} \neq 0 \\ 0 & \text{if } H_{j,j} = 1 \\ -1 & \text{if } i = j \end{cases} \tag{7}$$

$$F_{l,k} = F_l + \overline{LODF}_{l,k} \cdot F_k \tag{8}$$

where F_l and F_k are the power flows in normal operation state, that is, prior to the contingency k.

2.2. Modelling Approach: Classic UC formulation

The SCUC model proposed in this paper uses as starting point the classic UC formulation described and modelled in [29,45], respectively. From this initial model, a new one was developed adopting LSF to account for power flow constraints both in normal operative conditions and under contingencies. In this way, the UC model described in [29] is turned into a SCUC problem. The base UC formulation is given by Equations (9)–(29). The objective function, given by Equation (9) consists of minimizing the total generation cost of a set power plants over a given time horizon. In this case, $C_i(t)$ is the cost of power plant i at time t, c^{sh} is the cost of non-attended demand, and L_s^{sh} is the unserved load at bus s at time t. This objective function is subject to a set of constraints as explained below.

$$\text{Minimize} \quad \sum_{t=1}^{T}\sum_{i=1}^{I} C_i(t) + \sum_{t=1}^{T}\sum_{s=1}^{S} c^{sh} \cdot L_s^{sh}(t) \tag{9}$$

Constraints given by Equations (10) and (11) are used to preserve the logic of running, start-up, and shut-down status of generators. Where $y_i(t)$ and $z_i(t)$ are binary variables that are equal to 1 if generator i is started or shut down at the beginning of time interval t, respectively, and 0 otherwise. On the other hand, $x_i(t)$ is a binary variable that is equal to 1 if generator i is committed in time interval t, and 0 otherwise. Equation (10) is used to preserve the logic of running units, while Equation (11) indicates that a given generator cannot be simultaneously on and off.

$$y_i(t) - z_i(t) = x_i(t) - x_i(t-1) \quad \forall t \leq T, i \leq I \tag{10}$$

$$y_i(t) + z_i(t) \leq 1 \quad \forall t \leq T, i \leq I \tag{11}$$

Although there are several technologies of eclectic power plants, for the sake of simplicity only thermal generators were considered. The cost curve of thermal generator units was modelled using B linear segments and is described by Equation (12). In this case, a_i is a fixed production cost that is taken into account only if generator i is committed as indicated by the binary variable $x_i(t)$; $k_{i,b}$ is the slope of the cost curve for segment b of generator i, $g_{i,b}(t)$ is the output of generator i in segment b at time t and $SUC_i(t)$ is the start-up cost of generator i at time t. Equation (13) indicates that the total generation output given by $g_i(t)$ must be equal to the sum of the generation in each segment of the cost curve. Equations (14) and (15) indicate the minimum and maximum output limits for each generation level, respectively.

$$C_i(t) = a_i \cdot x_i(t) + \sum_{b=1}^{B} k_{i,b} \cdot g_{i,b}(t) + SUC_i(t) \quad \forall t \leq T, i \leq I \tag{12}$$

$$g_i(t) = \sum_{b=1}^{B} g_{i,b}(t) \quad \forall t \leq T, i \leq I \tag{13}$$

$$g_i(t) \geq g_i^{min} \cdot x_i(t) \quad \forall t \leq T, i \leq I \tag{14}$$

$$g_{i,b}(t) \leq g_i^{max} \cdot x_i(t) \quad \forall t \leq T, i \leq I, b \leq B \tag{15}$$

Minimum up/down time constraints are expressed by Equations (16)–(18). Where g_i^{on-off} represents the on-off status of generator i at $t = 0$, $L_i^{up,min}$ is the minimum length of time generator i must be on once it is committed, and $L_i^{down,min}$ is the minimum length of time generator i must be off before being committed again.

$$x_i(t) = g_i^{on-off} \quad \forall i \leq I, t \leq L_i^{up,min} + L_i^{down,min} \tag{16}$$

$$\sum_{tt=t-g_i^{up}+1}^{t} y_i(tt) \leq x_i(t) \quad \forall t \geq L_i^{up,min} \tag{17}$$

$$\sum_{tt=t-g_i^{down}+1}^{t} z_i(tt) \leq 1 - x_i(t) \quad \forall t \geq L_i^{down,min} \tag{18}$$

Ramping constraints are given by Equations (19)–(21). These are added to take into consideration how fast a generator can increase or decrease its output. In this case, $ramp_i^{up}$ and $ramp_i^{down}$ are the up and down ramp limits of generator i, $g_i(t)$ is the generator output at time t, and g_i^0 is the output of generator i at $t = 0$.

$$-ramp_i^{down} \leq g_i(t) - g_i(t-1) \quad \forall i \leq I, \quad 2 \leq t \leq T \tag{19}$$

$$ramp_i^{up} \leq g_i(t) - g_i(t-1) \quad \forall i \leq I, \quad 2 \leq t \leq T \tag{20}$$

$$-ramp_i^{down} \leq g_i(t_1) - g_i^0 \quad \forall i \leq I \tag{21}$$

$$ramp_i^{up} \leq g_i(t_1) - g_i^0 \quad \forall i \leq I \tag{22}$$

Generator off counter set-up constraints are given by Equations (23)–(25). These equations are used to take into account the initial conditions of generators. The symbols $|$ and $\wedge$ indicate the logical conditions IF and AND, respectively.

$$w_{i,j}(t) \leq \sum_{tt=SUC_{i,j}^{lim}}^{min\{t-1,SUC_{i,j+1}^{lim}-1\}} z_i(t-j)$$
$$+1|\{j=J-1 \wedge SUC_{i,j}^{lim} \leq g_i^{down,init}+t-1 < SUC_{i,j+1}^{lim}\}$$
$$+1|\{j=J \wedge SUC_{i,j}^{lim} \leq g_i^{down,init}+t-1\} \quad \forall t \leq T, i \leq I, j \leq J \tag{23}$$

$$\sum_{j=1}^{J} w_{i,j}(t) = y_i(t) \quad \forall t \leq T, i \leq I \tag{24}$$

$$SUC_i(t) = \sum_{j=1}^{J} w_{i,j}(t) \quad \forall t \leq T, i \leq I \tag{25}$$

Equations (26)–(29) take into account transmission constraints within a DC modelling of the network. Equation (26) indicates the active power balance constraint. This means that the supply and demand must be equal over all time intervals on each node. In this case, $\theta_s(t)$ is the voltage angle at bus s at time t, and $B_{s,m}$ is the admittance connecting nodes $s-m$. Equation (27) accounts for power flow limits, where $l_{s,m}^{max}$ is the maximum power flow allowed in line connecting nodes $s-m$. Finally, Equations (28) and (29) indicate limits on bus angles and the angular value of the reference bus, respectively.

$$\sum_{i=1}^{I} g_i(t) - \sum_{\{s,m\}\in L|m>s} B_{s,m}(\theta_s(t)-\theta_m(t))+$$
$$\sum_{\{s,m\}\in L|m<s} B_{m,s}(\theta_m(t)-\theta_s(t)) = d_s(t) \quad \forall t \leq T, s \leq S \tag{26}$$

$$-F_{s,m}^{max} \leq B_{s,m}(\theta_s(t)-\theta_m(t)) \leq F_{s,m}^{max} \quad \forall t \leq T, \{s,m\} \in L \tag{27}$$

$$-\pi \leq \theta_s(t) \leq \pi \quad \forall t \leq T, s \leq S \setminus s : reference\ bus \tag{28}$$

$$\theta_s(t) = 0 \quad \forall t \leq T, s : reference\ bus \tag{29}$$

2.3. Improvements and Adaptations to the Classic UC Model

Several modifications were introduced to the aforementioned UC model aiming at considering contingency constraints and improving computational time. The improvements and adaptations performed on the UC model are described below.

- *Net power injection and power balance*

  The first modification introduced in the UC model given by Equations (9)–(29) is the reformulation of the power balance constraint given by Equation (26). We introduce the variable $P_{s,t}^{Net}$ in Equation (30) that indicates the net power injection in bus s at time t. In this case, $A_{s,i}^{g}$ is a matrix that identifies which generator is in each bus. If generator i is located at bus s the corresponding position of $A_{s,i}^{g}$ is equal to one. Otherwise, it is zero; $g_i(t)$ is the output of generator i at time t, $D_{s,t}$ is the forecasted demand at bus s at time t. To guarantee the power balance, the sum of the total generation must be equal to the sum of the total demand for each period of time, as indicated by Equation (31). The formulation of the net power injection provided in Equations (30) and (31) replaces Equation (26). Please note that such formulation does not take into account bus angles, reducing the number of variables and rendering constraints (28) and (29) unnecessary.

$$P_s^{Net}(t) = \sum_i A_{s,i}^g g_i(t) - D_s(t) + L_s^{sh}(t) \quad \forall s \leq S, t \leq T \tag{30}$$

$$\sum_s P_s^{Net}(t) = 0 \quad \forall t \leq T \tag{31}$$

- *Power flows and post-contingency power flows*

Power flows in lines can be obtained as the product of the PTDF matrix and the vector of net power injections. Power flows must be within minimum and maximum limits as indicated in Equation (32). On the other hand, post-contingency power flows can be obtained using the LODF matrix as indicated in Equation (33). In this case, *TCF* stands for *Transmission Capacity Factor*, which is a parameter used to adjust transmission capacity limits. When post-contingency constraints are included in the UC formulation, this one is turned into a SCUC problem. Nevertheless, as it will be explained later, not all security constraints given by Equation (33) need to be incorporated in the model to guarantee network security. This is because most of these constraints may not be binding in the optimal solution. Therefore, only the security constraints that need to be enforced to avoid post-contingency overloads are added as user cuts as explained in the next section.

$$-F_l^{max} \cdot TCF \leq \sum_s \overline{PTDF}_{l,s} \cdot P_s^{Net}(t) \leq F_l^{max} \cdot TCF \quad \forall l \leq L, t \leq T \tag{32}$$

$$-F_l^{max} \cdot TCF \leq F_l(t) + \overline{LODF}_{l,k} \cdot F_k(t) \leq F_l^{max} \cdot TCF \quad \forall l \leq L, k \leq K, t \leq T \tag{33}$$

3. Methodology

The proposed modelling approach to the SCUC problem was inspired by [9], and it has two main features: the use of LSF (described in Section 2.1) for a straightforward computation of power flows and post-contingency power flows (see Equations (32)–(33); and the implementation of user cuts that reduce computational burden by considering only biding *N-1* security constraints. In this section, a detailed description of the so-called user cuts is provided along with its implementation within the SCUC problem.

3.1. User Cuts

For large power systems, most single-element contingencies (*N-1*) do not result in power flow violations of other lines; that is to say, the majority of the *N-1* security constraints may be superfluous and do not set up the feasibility space of the SCUC problem. Therefore, user cuts are implemented in this paper to consider only biding security constraints. In a MILP problem, a user cut is basically an additional linear constraint defined by the user, which is not part of the original model and does not rule out any feasible integer solutions [46]. Suppose the hypothetical two variable pure integer linear programming problem presented in Figure 1. In this case, the black dots indicate the feasible region of the problem and the black polygon indicates the integer hull or IP hull, which is the smallest convex set containing all integer-feasible solutions. The polygon bounded in blue represents the feasible region of the linear relaxation, also known as the linear hull or LP hull. This one is obtained when relaxing integrality constraints while enforcing variable bounds and functional constraints. Please note that the IP hull is always a subset of the LP hull. A user cut is illustrated by a red line, the shaded area shows the portion of the original LP hull that is removed by the user cut. Please note that none of the integer solutions of the original problem are removed by the cut.

In the approach proposed in this paper, the original problem is represented by the classic UC formulation presented in Section 2.2 considering the LSF through Equations (33) to (35). Please note that this first modification of the modelling does not include Equation (33). In consequence, it only deals with normal operative conditions. However with an alternative expression for power flow limits and power flow balance as a function of the PTDF and net power injections.

To turn a classic UC into a SCUC problem it must be guaranteed that N-1 security constraints are properly dealt with. However, in the operation of a power system not every contingency results in overloads. Therefore, it is not necessary or practical to consider the whole set of all security constraints given by Equation (33). Only those single contingencies that result in overloads must be considered. This is done by adding new security constraints as user cuts.

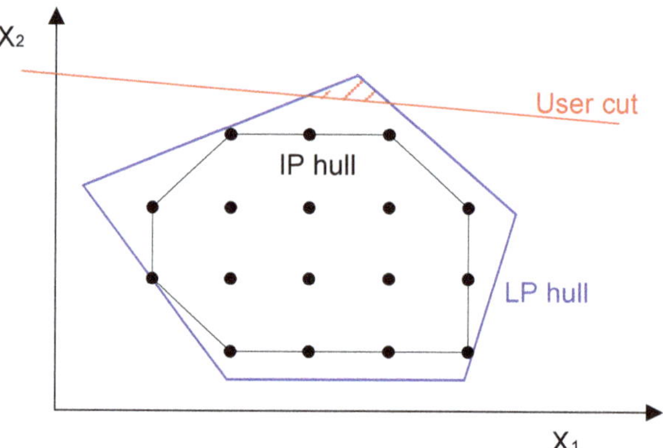

Figure 1. Illustration of user cuts.

3.2. Adding N-1 Security Constraints to the UC Problem

The addition of N-1 security constraints to the UC problem is done by means of user cuts as described below (see Figure 2).

- *Step 0:* read system data.
- *Step 1:* set the Security-Constraint Recorder $SCR_{l,k}(t)$ to zero as indicated in Equation (34). This with the aim of only accounting for the normal operation state and excluding the security constraints (given by Equation (33) in the first iteration of the algorithm.

$$SCR_{l,k}(t) = 0 \quad \forall l \leq L, k \leq K, t \leq T \tag{34}$$

- *Step 2:* solve the model given by Equations (9) to (25) and (30) to (32). Additionally, set the parameter $OP_{l,k}(t)$ to zero, which will store the overloads of every transmission line l, under every contingency k in every period of time t.

- *Step 3:* Estimate post-contingency power flows $\hat{F}_{l,k}(t)$ as indicated in Equation (35), using the optimal power flows $F_l^*(t)$ and $F_k^*(t)$ prior to the contingencies.

$$\hat{F}_{l,k}(t) = F_l^*(t) + \overline{LODF}_{l,k} \cdot F_k^*(t) \quad \forall t \leq T, l \leq L, k \leq K \tag{35}$$

- *Step 4:* verify post-contingency power flow limits in every line l for every period of time t, under every contingency k. If there is an overload in line l, for a contingency k, in the period of time t, assign a value of 1 in the corresponding position of $SCR_{l,k}(t)$, as indicated in Equation (36).

$$\text{If } \left|\hat{F}_{l,k}(t)\right| \geq F_l^{\max} \cdot TCF \rightarrow SCR_{l,k}(t) = 1 \quad \forall t \leq T, l \leq L, k \leq K \tag{36}$$

For those lines presenting overloads, the excess value is stored in the Overload Parameter $OP_{l,k}(t)$. If there is not an overload in line l, under contingency k in the period of time t the corresponding position of $OP_{l,k}(t)$ is set to zero as indicated in Equation (37). The sum over the sets of periods, lines and contingencies yields the total overload (TO) of the system as indicated by Equation (38).

$$OP_{l,k}(t) = \begin{cases} \left|\hat{F}_{l,k}(t)\right| - F_l^{max} \cdot TCF & \text{If } \left|\hat{F}_{l,k}(t)\right| \geq F_l^{max} \cdot TCF \\ \\ 0 & \text{If } \left|\hat{F}_{l,k}(t)\right| < F_l^{max} \cdot TCF \end{cases} \qquad \forall t \leq T, l \leq L, k \leq K \qquad (37)$$

- *Step 5:* compute the total overload (TO) as indicated in Equation (38).

$$TO = \sum_{l,k,t} OP_{l,k}(t) \qquad (38)$$

- *Step 6:* check convergence verifying if TO is lower than a given tolerance (*tol*). If it is true, stop the algorithm and report the solution. Otherwise, return to Step 2 introducing user cuts by adding new security constraints through Equation (33). This is done with the values of l, k, and t for which $\left|\hat{F}_{l,k}(t)\right| \geq F_l^{max}$; which in turn correspond to the positions where $SCR_{l,k}(t) = 1$.

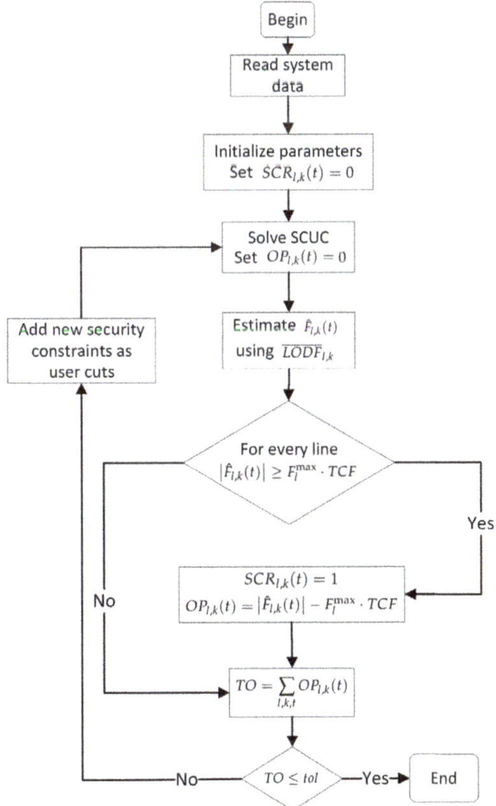

Figure 2. Flowchart of the algorithm for adding security constraints as user cuts.

3.3. Identification of Vulnerable Lines and Critical Contingencies

Using the information stored in $SCR_{l,k}(t)$ it is possible to identify the most vulnerable lines and critical contingencies. For example, adding $SCR_{l,k}(t)$ over t and k as indicated in Equation (39) yields L_l^v; this is a vector that contains the number of times each line is overloaded due to every contingency in each period of time t. Thus, L_l^v provides information regarding the most vulnerable lines of the system.

$$\sum_{t,k} SCR_{l,k}(t) = L_l^v \quad \forall l \leq L \tag{39}$$

The identification of the most critical contingencies (in terms of the number of overloads they cause) is obtained by adding $SCR_{l,k}(t)$ over t and l as indicated in Equation (40). In this case, L_k^s is a vector that contains the number of lines that are overloaded due to a given contingency k over the set t. In consequence, L_k^s identifies the most critical contingencies of the system.

$$\sum_{t,l} SCR_{l,k}(t) = L_k^s \quad \forall k \leq K \tag{40}$$

Finally, it is possible to obtain a mapping of contingencies and overloads adding $SCR_{l,k}(t)$ over t. In this case, $L_{l,k}^c$ is a matrix that maps every contingency with every line overloaded, permitting identification of which contingencies impact which lines. This information is valuable to system planners when deciding over reinforcements on the network or eventual expansion plans.

$$\sum_{t} SCR_{l,k}(t) = L_{l,k}^c \quad \forall k \leq K \quad \forall l \leq L \tag{41}$$

4. Tests and Results

To show the applicability of the proposed modelling approach several tests were performed using the IEEE RTS-96 for a time horizon of 24 h. This power system is made of 73 buses, 120 transmission lines, 96 thermal generators, and 51 loads that add up a maximum demand of 7539 MW. Figure 3 depicts the power system under study. Information regarding initial conditions of generators is necessary to enforce ramping constraints as well as minimum up/down time constraints denoted by Equations (19)–(21) and (16)–(18), respectively. Cost data of generators and their initial conditions are presented in Tables A1 and A2, respectively. The system data and classic UC model used in this paper as starting point can be found in [45]. Furthermore, data and models are available at GitHub [47].

Three different formulations as described in Table 1 were implemented for comparative purposes. Two instances of the problem were considered for all formulations varying line ratings and the gap of relative optimality (inGAP). The inGAP is a user-defined parameter that indicates the grade of optimality (quality of solution) required for the software. The first instance of the problem is solved with an inGAP of 0.001 and a TCF of 1.0 (nominal line ratings); the second one considers an inGAP of 0.005 and a TCF of 0.8 (line rating reduction of 20%). The proposed model was solved using the commercial software of algebraic modelling GAMS version 24.8.5, under CPLEX solver in an 8-core 3.4 GHz Intel Core i7 desktop computer with an 8 GB RAM memory. The results of the simulations are presented in Table 2.

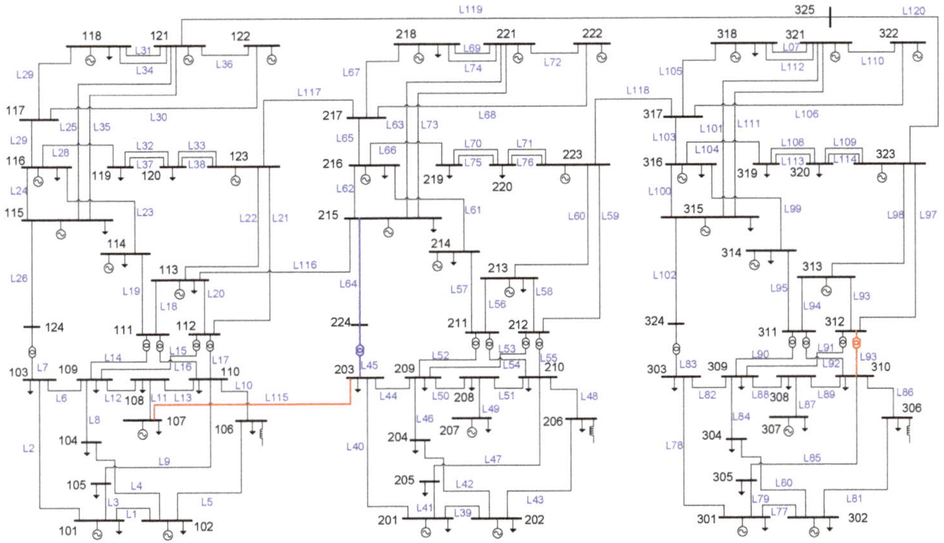

Figure 3. Test system under study (IEEE RTS-96).

Table 1. Formulations of the SCUC problem.

Formulation	Description
A	Conventional network formulation given by Equations (9)–(29) Contingency constraints computed using LODF (Equation (33))
B	Network formulation using PTDF (Equations (26) to (29) are replaced by Equations (30) to (32) Contingency constraints computed using LODF (Equation (33)
C	Network formulation using PTDF (Equations (26) to (29) are replaced by Equations (30) to (32) Contingency constraints computed using LODF (Equation (33) Implementation of user cuts as described in Section 3.1

Table 2. Results obtained for different formulations of the SCUC problem.

Formulation	Objective Function [M\$]		Time [s]		inGAP		outGAP	
	TCF = 1	TCF = 0.8	TCF = 1	TCF = 0.8	TCF = 1	TCF = 0.8	TCF = 1	TCF = 0.8
A	2.6968	2.7024	1220	519	0.0010	0.0050	0.00250	0.00480
B	2.6943	2.7041	1209	345	0.0010	0.0050	0.00109	0.00490
C	2.6940	2.7016	169	321	0.0010	0.0050	0.0010	0.00430
B*	2.6970	2.7024	495	687	0.00254	0.0045	0.00254	0.00437

4.1. Impact of Linear Sensitivity Factors in the Performance of the SCUC Problem

A comparison of results with formulations *A* and *B* is initially presented to illustrate the impact of including LSF within the SCUC formulation. As described in Table 1 formulation *A* considers a classic network modelling and although it considers LODF to account for contingencies, it does not take into account PTDF to model network constraints. It can be seen in Table 2 that formulations *A* and *B* require similar computing time to solve the SCUC problem when using a TCF equal to 1 (formulation *A* takes 1220 seconds and formulation *B* 1209). Although the same inGAP was used for both formulations, the outGAP (which is indicative of the quality of the solution reached by the solver) obtained when using formulation *B* is almost half the one obtained when using formulation *A*. This means that including

LSF to represent network constraints leads to better solutions as can be verified in the first column of Table 2. To illustrate, even more, the superior computational efficiency of formulation B compared to formulation A, an additional simulation with an inGAP of 0.00254, labeled as $B*$ was carried out. Such inGAP is used to basically force the solver to reach a solution with a similar quality as the one already obtained with formulation A. In this case, formulation $B*$ finds such solution in significantly less time (495 seconds). Please note that for obtaining results of similar quality, formulation A takes 1220 seconds while formulation $B*$ takes 495 seconds. This corresponds to a time reduction of nearly 60%. On the other hand, when using a TCF of 0.8 the outGAPs of both formulations are similar (0.00480 for formulation A and 0.00490 for formulation B). However, formulations A takes 519 seconds while formulation B only takes 345 seconds. This indicates a reduction of computational time of 33.5%.

4.2. Impact of User Cuts in the Performance of the SCUC Problem

As described in Table 1 formulation C considers user cuts (described in Sections 3.1 and 3.2) to account for security constraints. As can be observed in Table 2 for a TCF equal to 1, this formulation takes 169 seconds to solve the SCUC problem. This represents an important time reduction (approximately 86%) when compared with formulations A and B that take 1220 and 1209 seconds to solve the same problem, respectively. Furthermore, the outGAP obtained with formulation C is lower than the ones obtained with formulations A and B. This indicates that the proposed formulation not only improves computational time but also quality of solutions. This last aspect can be verified by comparing the data presented in the second column of Table 2. When a TCF of 0.8 is considered, formulation C also outperforms formulations A, B and $B*$ both in terms of computational time and quality of solutions. As regards computational time, formulation C takes 321 seconds to solve the SCUC problem while formulation A takes 519 seconds. On the other hand, to achieve a solution with a similar quality (see outGAP in the last column of Table 2) formulation $B*$ takes 687 seconds. Consequently, formulation C provides in this case a time reduction of 53.2%.

Table 3 presents details of each iteration when using formulation C with an inGAP of 0.005 and TCF = 0.8. The solution to the SCUC problem is found after 6 iterations in 321 seconds. Please note that in the first iteration, normal operation state of the system is solved, that is, security constraints are not considered. However, in the second iteration 68 security constraints are added as user cuts. These correspond to the positions where $SCR_{l,k}(t) = 1$. Subsequently, more user cuts are added in every iteration until all post-contingency power flows under every single contingency over all periods of time are within limits. This is achieved with 144 additional constraints. In contrast, classic modelling approaches (formulations A and B) require 13816 additional constraints to account for security in the UC problem. This corresponds to a reduction of 99% in the number of security constraints required to solve the SCUC problem.

Table 3. Performance of the SCUC formulation (TCF = 0.8 and inGAP = 0.005).

Iteration	Added Constraints N-1	Elapsed Time of Simulation [s]	Objective Function [M$]
1	-	25	2.6947
2	68	66	2.6986
3	21	87	2.6980
4	7	53	2.7021
5	27	57	2.7025
6	21	33	2.7016
total	144	321	-

4.3. Most Vulnerable Lines and Critical Contingencies

As mentioned in Section 3.3, the proposed approach also allows identification of the most vulnerable lines and critical contingencies over the time horizon under consideration and for every N-1 contingency. Figure 4a presents the most vulnerable lines when the SCUC is solved with TFC = 0.8. It can be observed that the lines labeled as L115 and L93 are overloaded 54 and 11 times, respectively due to every *N*-1 contingency over the 24h time horizon. These lines are indicated in red in Figure 3. The most critical contingencies are presented in Figure 4b. Please note that the outage of lines labeled as L45 or L64 impacts 22 other lines. These are indicated in blue in Figure 3. Figure 5 presents a mapping of critical contingencies and vulnerable lines. Please note that it is possible to identify which contingencies cause overloads in which lines. For example, the most critical line already identified in Figure 4a presents 54 overloads. This is line L115 indicated in the first position of the x-axis. In this case, 20 of the overloads of line L115 are due to the outage of line L45, another 20 are due to the outage of line L64, and the remaining 12 are caused by outages of lines L119 and L120, each of them contributing with 6 overloads. This information is valuable to the system planner who may decide over future reinforcements of the corridor where line L115 is located to reduce its criticality. Furthermore, the mapping of critical contingencies vs. Vulnerable lines can be used for the assessment of generation and transmission expansion plans on the security of the system operation.

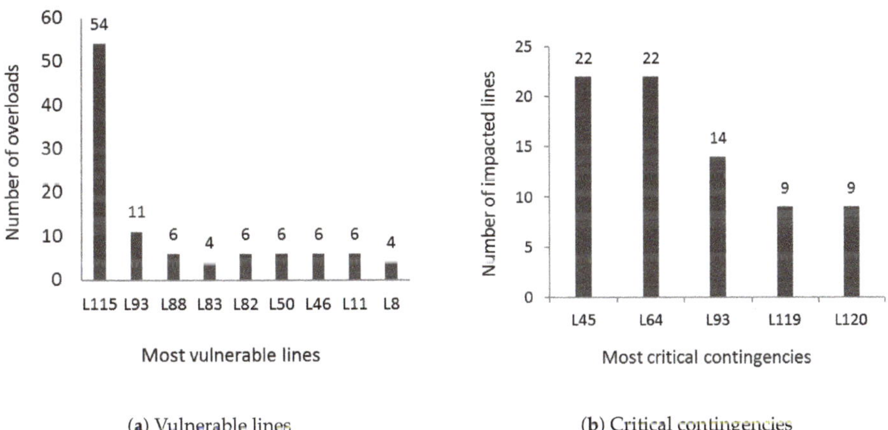

(a) Vulnerable lines   (b) Critical contingencies

Figure 4. Vulnerable lines and critical contingencies considering TCF = 0.8.

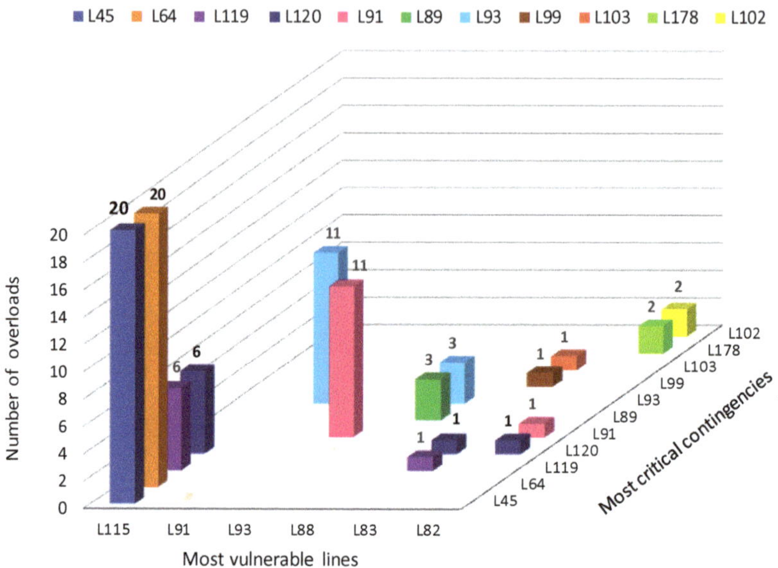

Figure 5. Mapping of vulnerable lines and critical contingencies considering TCF = 0.8.

4.4. Scalability Evaluation

To show the robustness and flexibility of the proposed approach, several tests were performed with a larger power system. The new test system was built duplicating the original IEEE RTS-96 depicted in Figure 3. The new buses were labeled following the same pattern as the original IEEE RTS-96 (prefixing numbers 4, 5, and 6 for each subsystem). The two power systems were connected through two double circuits at 220 kV between buses 221 and 612 and 114 and 613; each with a transmission capacity of 500 MW and a reactance of 0.084 pu. The TCF was modified by a factor of 1.6 to avoid infeasibilities when evaluating contingencies of the circuits connecting the two power systems. The equivalent test system is composed of 146 buses, 244 lines, 192 generators, and 102 demand sides. Table 4 presents the results with formulations B and C for the new test system. Simulations were performed in an Intel ® Xeon E5 @ 2.40 GHz, 44 cores, and 256 GB memory RAM processor.

Table 4. Results obtained for scaled test system.

Formulation	Objective Function [M$]	Time [s]	inGAP	outGAP
B	4.8748	7494.001	0.01	0.00178
C	4.8701	209.687	0.01	0.00100

According to the results presented in Table 4 formulation C outperforms formulation B in both computation time and quality of the solution found. Formulation C takes approximately 209.7 seconds to find a solution, while formulation B takes 7494 seconds. Please note that in this case a time reduction of 97.2% is achieved when using formulation C. For comparative purposes, the same inGAP was used in both formulations. However, the outGAP of formulation C turned out to be lower than the one reached with formulation B. This might be attributable to the fact that formulation C is more compact than formulation B. The quality of solutions can be verified in the second column of Table 4. It is worth mentioning that for this particular power system, formulation B must deal with a set of 1,446,480 constraints associated with the N-1 criterion, while implementing user cuts only requires 16,511 additional N-1 constraints. This represents a reduction of 98.85% of the number of N-1 constraints when using formulation C. Additional simulations were run using a desktop computer which characteristics are previously described in Section 4. In this case, the solution found with formulation C, takes approximately 15 minutes reaching an objective function of 4.8731 MUSD and an outGAP of 0.00158; however, the same simulation could not be performed with formulation B due to lack of RAM memory. Therefore, an additional decomposition algorithm might be necessary to run formulation B in a desktop computer, which is not required when using formulation C. This further emphasizes the advantage of the proposed approach.

5. Conclusions

This paper presented a new modelling approach of the SCUC problem based on PTDF and LODF LSF and user cuts. On the one hand PTDF are used to model power injection balance constraints, avoiding the use of voltage angles and reducing the number of variables and constraints of the UC formulation. On the other hand, LODF are used to evaluate post-contingency power flows and check overloads quickly and reliably. Finally, user cuts are iteratively added to the base problem to account for security constraints. This is because most security constraints of the SCUC problem are not binding in the optimal solution.

The proposed model uses as starting point a classic UC formulation. Subsequently, security constraints are taken into account through an array that registers whether a given contingency in a given time produces overloads on other lines. If this happens, new security constraints are added as user cut to the current problem. This is done iteratively until no overloads are produced by any contingency.

Several tests performed on a benchmark IEEE power system evidenced the robustness and applicability of the proposed model. The main advantage of considering LSF and user cuts to model the SCUC problem lies on the reduction of computing time. Furthermore, better solutions are found with the proposed model. Results show that time reduction ranges from 53% to 97% when compared with a classic formulation set to reach a solution of similar quality (same inGAP parameters). A scalability test showed that such reduction of computing time increases with the size of the test system. Moreover, decomposition techniques as BD or LR are not needed to efficiently address the SCUC problem.

An additional advantage of the proposed methodology lies on its flexibility and ease of implementation. The proposed approach to deal with the SCUC problem was devised to be solver-independent. Therefore, unlike other research works reported in the specialized literature, it does not depend on built-in functions or callbacks.

A value-added element of the methodology is the identification of the most critical lines and most severe contingencies. The mapping of critical contingencies vs. vulnerable lines can be used for the assessment of generation and transmission expansion plans on the security of the system operation. Finally, the tight and compact modelling proposed in this paper can be used in a future work as a basis for other studies such as the integration of renewable energy resources within a SCUC problem.

Author Contributions: Conceptualization, Á.J.-D., C.C.M.-C., J.E.S.-A., and W.M.V.-A.; Data curation, C.C.M.-C., and J.E.S.-A.; Formal analysis, Á.J.-D., C.C.M.-C., J.E.S.-A., and W.M.V.-A.; Funding acquisition, Á.J.-D. and J.M.L.-L.; Investigation, J.E.S.-A., C.C.M.-C., Á.J.-D., J.M.L.-L. and W.M.V.-A.; Methodology, J.E.S.-A., and C.C.M.-C.; Project administration, Á.J.-D.; Software, C.C.M.-C., and J.E.S.-A.; Validation, C.C.M.-C., and J.E.S.-A.; Visualization, C.C.M.-C., and J.E.S.-A.; Writing—original draft, J.M.L.-L.; Writing—review & editing, W.M.V.-A., J.M.L.-L. and Á.J.-D.

Funding: This research was funded by Colciencias. Project code 1115-745-54929; contract 056-2017; proyect name: *"Despacho Económico Multiperiodo Integrando Energías Renovables Intermitentes"*.

Acknowledgments: The authors would like to thank Colciencias, Universidad de Antioquia and the Research Group on Efficient Energy Management (GIMEL).

Conflicts of Interest: The authors declare no conflict of interest.

Abbreviations

The following abbreviations are used in this manuscript:

SCOPF   Security-Constraint Optimal Power Flow
UC   Unit Commitment
SCUC   Security-Constrained Unit Commitment
MILP   Mixed-Integer Linear Programming
PTDF   Power Transfer Distribution Factors
LODF   Line Outage Distribution Factors

Nomenclature

The nomenclature used in this paper is provided here for quick reference:

Indices

b	Index of generating unit cost curve segments, 1 to B
i	Index of generating units, 1 to I
j	Index of generating unit star-up cost, 1 to K
l, k	Index of lines, 1 to L
s, m	Index of buses, 1 to S
t, tt	Index of hours, 1 to T

Parameters

a_i	Fixed production cost of generator ($)
B_{sm}	Admittance of line connecting nodes s-m (S)
$d_s(t)$	Demand at bus s (MW)
g_i^{down}	Minimum down time of generator i (h)
g_i^{up}	Minimum up time of generator i (h)
$g_i^{down,init}$	Time that generator i has been down before $t = 0$ (h)
$g_i^{up,init}$	Time that generator i has been up before $t = 0$ (h)
g_i^0	Output if generator i at $t = 0$ (MW)
g_i^{max}	Rated capacity of generator i (MW)
g_i^{min}	Minimum output of generator i (MW)
$g_{i,b}^{max}$	Capacity of segment b of the cost curve of generator i (MW)
g_i^{on-off}	On-Off status of generator i at $t = 0$ (equal to 1 if $g_i^{up,init} > 0$ and 0 otherwise)
$k_{i,b}$	Slope of the segment b of the cost curve of generator i ($/MW)
c^{sh}	Cost of non-attended demand ($/MW)
$A_{s,i}^g$	Generation map for thermal units
F_{sm}^{max}	Capacity of the line between nodes s and m (MW)
$F_{s,m}^{max}$	Capacity of the line between nodes s and m (MW)
TCF	Transmission capacity factor
$L_i^{down,min}$	Length of time the generator i must be off at the start time of the planning horizon (h)
$L_i^{up,min}$	Length of time the generator i must be on at the start time of the planning horizon (h)
M	Large number used of linearization - larger than the maximum number of hours a unit can be on or off
$ramp_i^{down}$	Ramp-down limit of generator i (MW/h)
$ramp_i^{up}$	Ramp-up limit of generator i (MW/h)
$SUC_{i,j}^{cost}$	Cost steps in start-up cost curve of generator i ($)
$SUC_{i,j}^{lim}$	Time steps in start-up cost curve of generator i (h)
$PTDF_{l,s}$	Matrix of Power transfer distribution factors
$LODF_{l,k}$	Matrix of Line Outage distribution factors
$SCR_{l,k}(t)$	Security-Constraint Recorder
L_l^v	Vector of vulnerable lines
L_k^s	Vector of critical contingencies
$L_{l,k}^c$	Matrix of vulnerable lines vs. critical contingencies

Variables

$C_i(t)$	Operating cost of generator i at time t ($)
$count_i^{down}$	Generator i down time period counter
$g_i(t)$	Generator i output at time t (MW)
$g_{i,b}(t)$	Output of generator i of segment b at time t (MW)
$L_s^{sh}(t)$	Unserved load at bus s at time t (MW)
$SUC_i(t)$	Start-up cost of generator i at time t ($)
$w_{i,j}(t)$	Binary variable equal to 1 if generator i is started at time t after being off for j hours, and 0 otherwise
$x_i(t)$	Binary variable equal to 1 if the generator i is producing at time t, and 0 otherwise
$y_i(t)$	Binary variable equal to 1 if the generator i is started at the beginning of time t, and 0 otherwise
$z_i(t)$	Binary variable equal to 1 if the generator i is shut down at the beginning of time t, and 0 otherwise
$\theta_s(t)$	Voltage angle at bus s (rad)
$P_s^{Net}(t)$	Net power injection in bus s at time t (MW)
$f_l(t)$	Line flow in line l at time t (MW)

Appendix A.

Tables A1 and A2 present generator cost data and initial conditions of generators, respectively.

Table A1. Generator cost data.

Unit Type	Range (MW)	kib ($/MW)	Range (MW)	kib ($/MW)	Range (MW)	kib ($/MW)
1	5.4–7.6	29.453	7.6–9.8	30.120	9.8–12	30.856
2	8–12	28.967	12–16	29.243	16–20	29.703
3	26–34	28.313	34–42	29.256	43–50	30.498
4	40–52	18.423	52–64	19.228	64–76	20.102
5	40–60	17.590	60–80	18.280	80–100	18.966
6	54.24–87.83	23.810	87.83–121.41	24.525	121.41–155	25.240
7	104–135	17.193	135–166	17.708	166–197	18.225
8	140–210	26.213	210–280	26.708	280–350	27.200
9	100–200	6.961	200–300	7.230	300–400	7.499

Table A2. Initial conditions of generators.

G	On/off	Init	G	On/off	Init	G	On/off	Init
1	1	20	33	8	20	65	900	20
2	400	20	34	89	20	66	90	20
3	220	70	35	66	70	67	789	70
4	2	0	36	66	0	68	456	0
5	17	0	37	66	0	69	375	0
6	4	0	38	66	0	70	375	0
7	66	0	39	66	0	71	170	0
8	33	0	40	1	0	72	170	0
9	11	100	41	1	100	73	170	100
10	2	90	42	56	90	74	800	90
11	2	80	43	56	80	75	2500	80
12	2	177	44	56	177	76	2500	177
13	2	155	45	56	155	77	2500	155
14	2	0	46	56	0	78	2500	0
15	6	12	47	56	12	79	1000	12
16	7	12	48	56	12	80	203	12
17	8	12	49	98	12	81	600	12

Table A2. *Cont.*

G	On/off	Init	G	On/off	Init	G	On/off	Init
18	9	0	50	124	0	82	46	0
19	5	0	51	1000	0	83	236	0
20	8	134	52	1000	134	84	236	134
21	8	123	53	1000	123	85	64	123
22	8	0	54	50	0	86	6	0
23	8	377	55	50	377	87	8	377
24	8	47	56	90	47	88	90	47
25	8	48	57	900	48	89	5	48
26	8	0	58	900	0	90	6	0
27	8	0	59	900	0	91	7	0
28	8	50	60	900	50	92	8	50
29	8	50	61	900	50	93	9	50
30	8	150	62	900	150	94	7	150
31	8	0	63	900	0	95	66	0
32	8	0	64	900	0	96	55	0

References

1. Song, M.; Amelin, M. Purchase Bidding Strategy for a Retailer With Flexible Demands in Day-Ahead Electricity Market. *IEEE Trans. Power Syst.* **2017**, *32*, 1839–1850. [CrossRef]
2. Lee, H.; Tekin, C.; van der Schaar, M.; Lee, J. Adaptive Contextual Learning for Unit Commitment in Microgrids with Renewable Energy Sources. *IEEE J. Sel. Top. Signal Process.* **2018**, *12*, 688–702. [CrossRef]
3. Li, N.; Hedman, K.W. Economic Assessment of Energy Storage in Systems With High Levels of Renewable Resources. *IEEE Trans. Sustain. Energy* **2015**, *6*, 1103–1111. [CrossRef]
4. Carrión, M.; Zárate-Miñano, R.; Domínguez, R. A Practical Formulation for Ex-Ante Scheduling of Energy and Reserve in Renewable-Dominated Power Systems: Case Study of the Iberian Peninsula. *Energies* **2018**, *11*, 1939. [CrossRef]
5. Pozo, D.; Contreras, J.; Sauma, E.E. Unit Commitment With Ideal and Generic Energy Storage Units. *IEEE Trans. Power Syst.* **2014**, *29*, 2974–2984. [CrossRef]
6. Lv, M.; Lou, S.; Wu, Y.; Miao, M. Unit Commitment of a Power System Including Battery Swap Stations Under a Low-Carbon Economy. *Energies* **2018**, *11*, 1898. [CrossRef]
7. Pandžić, H.; Dvorkin, Y.; Qiu, T.; Wang, Y.; Kirschen, D.S. Toward Cost-Efficient and Reliable Unit Commitment Under Uncertainty. *IEEE Trans. Power Syst.* **2016**, *31*, 970–982. [CrossRef]
8. Zhao, B.; Conejo, A.J.; Sioshansi, R. Unit Commitment Under Gas-Supply Uncertainty and Gas-Price Variability. *IEEE Trans. Power Syst.* **2017**, *32*, 2394–2405. [CrossRef]
9. Tejada-Arango, D.A.; Sánchez-Martín, P.; Ramos, A. Security Constrained Unit Commitment Using Line Outage Distribution Factors. *IEEE Trans. Power Syst.* **2018**, *33*, 329–337. [CrossRef]
10. Zhai, Q.; Guan, X.; Cheng, J.; Wu, H. Fast Identification of Inactive Security Constraints in SCUC Problems. *IEEE Trans. Power Syst.* **2010**, *25*, 1946–1954. [CrossRef]
11. Fu, Y.; Shahidehpour, M.; Li, Z. Long-term security-constrained unit commitment: hybrid Dantzig-Wolfe decomposition and subgradient approach. *IEEE Trans. Power Syst.* **2005**, *20*, 2093–2106. [CrossRef]
12. Al-Agtash, S. Hydrothermal scheduling by augmented Lagrangian: consideration of transmission constraints and pumped-storage units. *IEEE Trans. Power Syst.* **2001**, *16*, 750–756. [CrossRef]
13. Guan, X.; Guo, S.; Zhai, Q. The conditions for obtaining feasible solutions to security-constrained unit commitment problems. *IEEE Trans. Power Syst.* **2005**, *20*, 1746–1756. [CrossRef]
14. Lu, B.; Shahidehpour, M. Unit commitment with flexible generating units. *IEEE Trans. Power Syst.* **2005**, *20*, 1022–1034. [CrossRef]
15. Martinez-Crespo, J.; Usaola, J.; Fernandez, J.L. Security-constrained optimal generation scheduling in large-scale power systems. *IEEE Trans. Power Syst.* **2006**, *21*, 321–332. [CrossRef]
16. Fu, Y.; Shahidehpour, M. Fast SCUC for Large-Scale Power Systems. *IEEE Trans. Power Syst.* **2007**, *22*, 2144–2151. [CrossRef]

17. Alemany, J.; Magnago, F. Benders decomposition applied to security constrained unit commitment: Initialization of the algorithm. *Int. J. Electr. Power Energy Syst.* **2015**, *66*, 53–66. [CrossRef]
18. Guan, X.; Zhai, Q.; Papalexopoulos, A. Optimization based methods for unit commitment: Lagrangian relaxation versus general mixed integer programming. In Proceedings of the 2003 IEEE Power Engineering Society General Meeting (IEEE Cat. No.03CH37491), Toronto, ON, Canada, 13–17 July 2003; Volume 2, pp. 1095–1100. [CrossRef]
19. Bragin, M.A.; Luh, P.B.; Yan, J.H.; Stern, G.A. Novel exploitation of convex hull invariance for solving unit commitment by using surrogate Lagrangian relaxation and branch-and-cut. In Proceedings of the 2015 IEEE Power Energy Society General Meeting, Denver, Colorado, 26–30 July 2015; pp. 1–5. [CrossRef]
20. Swarup, K.S.; Yamashiro, S. Unit commitment solution methodology using genetic algorithm. *IEEE Trans. Power Syst.* **2002**, *17*, 87–91. [CrossRef]
21. Senjyu, T.; Yamashiro, H.; Shimabukuro, K.; Uezato, K.; Funabashi, T. Fast solution technique for large-scale unit commitment problem using genetic algorithm. *IEE Proc. Gener. Trans. Distrib.* **2003**, *150*, 753–760. [CrossRef]
22. Aghdam, F.H.; Hagh, M.T. Security Constrained Unit Commitment (SCUC) formulation and its solving with Modified Imperialist Competitive Algorithm (MICA). *J. King Saud Univ. Eng. Sci.* **2017**. [CrossRef]
23. Roque, L.A.C.; Fontes, D.B.M.M.; Fontes, F.A.C.C. A Metaheuristic Approach to the Multi-Objective Unit Commitment Problem Combining Economic and Environmental Criteria. *Energies* **2017**, *10*, 2029. [CrossRef]
24. Jo, K.H.; Kim, M.K. Improved Genetic Algorithm-Based Unit Commitment Considering Uncertainty Integration Method. *Energies* **2018**, *11*, 1387.
25. Zhao, J.; Liu, S.; Zhou, M.; Guo, X.; Qi, L. An Improved Binary Cuckoo Search Algorithm for Solving Unit Commitment Problems: Methodological Description. *IEEE Access* **2018**, *6*, 43535–43545. [CrossRef]
26. Singhal, P.K.; Naresh, R.; Sharma, V. Binary fish swarm algorithm for profit-based unit commitment problem in competitive electricity market with ramp rate constraints. *IET Gener. Trans. Distrib.* **2015**, *9*, 1697–1707. [CrossRef]
27. Trivedi, A.; Srinivasan, D.; Pal, K.; Saha, C.; Reindl, T. Enhanced Multiobjective Evolutionary Algorithm Based on Decomposition for Solving the Unit Commitment Problem. *IEEE Trans. Ind. Inform.* **2015**, *11*, 1346–1357. [CrossRef]
28. Arora, V.; Chanana, S. Solution to unit commitment problem using Lagrangian relaxation and Mendel's GA method. In Proceedings of the 2016 International Conference on Emerging Trends in Electrical Electronics Sustainable Energy Systems (ICETEESES), Sultanpur, India, 11–12 March 2016; pp. 126–129. [CrossRef]
29. Pandžić, H.; Qiu, T.; Kirschen, D.S. Comparison of state-of-the-art transmission constrained unit commitment formulations. In Proceedings of the 2013 IEEE Power Energy Society General Meeting, Vancouver, BC, USA, 21–25 July 2013; pp. 1–5. [CrossRef]
30. Bhardwaj, A.; Kamboj, V.K.; Shukla, V.K.; Singh, B.; Khurana, P. Unit commitment in electrical power system-a literature review. In Proceedings of the 2012 IEEE International Power Engineering and Optimization Conference, Melaka, Malaysia, 6–7 June 2012; pp. 275–280. [CrossRef]
31. Zheng, Q.P.; Wang, J.; Liu, A.L. Stochastic Optimization for Unit Commitment—A Review. *IEEE Trans. Power Syst.* **2015**, *30*, 1913–1924. [CrossRef]
32. Jurković, K.; Pandšić, H.; Kuzle, I. Review on unit commitment under uncertainty approaches. In Proceedings of the 2015 38th International Convention on Information and Communication Technology, Electronics and Microelectronics (MIPRO), Opatija, Croatia, 25–29 May 2015; pp. 1093–1097. [CrossRef]
33. Wu, H.; Guan, X.; Zhai, Q.; Ye, H. A Systematic Method for Constructing Feasible Solution to SCUC Problem With Analytical Feasibility Conditions. *IEEE Trans. Power Syst.* **2012**, *27*, 526–534. [CrossRef]
34. Ardakani, A.J.; Bouffard, F. Identification of Umbrella Constraints in DC-Based Security-Constrained Optimal Power Flow. *IEEE Trans. Power Syst.* **2013**, *28*, 3924–3934. [CrossRef]
35. Capitanescu, F.; Glavic, M.; Ernst, D.; Wehenkel, L. Contingency Filtering Techniques for Preventive Security-Constrained Optimal Power Flow. *IEEE Trans. Power Syst.* **2007**, *22*, 1690–1697. [CrossRef]
36. Fu, Y.; Shahidehpour, M.; Li, Z. Security-constrained unit commitment with AC constraints. *IEEE Trans. Power Syst.* **2005**, *20*, 1538–1550. [CrossRef]
37. Fu, Y.; Shahidehpour, M.; Li, Z. AC contingency dispatch based on security-constrained unit commitment. *IEEE Trans. Power Syst.* **2006**, *21*, 897–908. [CrossRef]

38. Bragin, M.A.; Luh, P.B.; Yan, J.H.; Stern, G.A. An efficient approach for Unit Commitment and Economic Dispatch with combined cycle units and AC Power Flow. In Proceedings of the 2016 IEEE Power and Energy Society General Meeting (PESGM), Boston, MA, USA, 17–21 July 2016; pp. 1–5. [CrossRef]

39. Guler, T.; Gross, G.; Liu, M. Generalized Line Outage Distribution Factors. *IEEE Trans. Power Syst.* **2007**, *22*, 879–881. [CrossRef]

40. Leveringhaus, T.; Hofmann, L. Comparison of methods for state prediction: Power Flow Decomposition (PFD), AC Power Transfer Distribution factors (AC-PTDFs), and Power Transfer Distribution factors (PTDFs). In Proceedings of the 2014 IEEE PES Asia-Pacific Power and Energy Engineering Conference (APPEEC), Kowloon, Hong Kong, 7–10 December 2014; pp. 1–6. [CrossRef]

41. Song, C.S.; Park, C.H.; Yoon, M.; Jang, G. Implementation of PTDFs and LODFs for Power System Security. *J. Int. Counc. Electr. Eng.* **2011**, *1*, 49–53. [CrossRef]

42. Chen, Y.C.; Domínguez-García, A.D.; Sauer, P.W. Measurement-Based Estimation of Linear Sensitivity Distribution Factors and Applications. *IEEE Trans. Power Syst.* **2014**, *29*, 1372–1382. [CrossRef]

43. Eslami, M.; Moghadam, H.A.; Zayandehroodi, H.; Ghadimi, N. A New Formulation to Reduce the Number of Variables and Constraints to Expedite SCUC in Bulky Power Systems. *Proc. Natl. Acad. Sci. India Sect. A Phys. Sci.* **2018**. [CrossRef]

44. Wood, A.J.; Wollenberg, B.F.; Sheble, G.B. *Power Generation, Operation, and Control*; Wiley-IEEE Press: Piscataway, NJ, USA, 2013.

45. Pandzic, H.; Dvorkin, Y.; Qiu, T.; Wang, Y.; Kirschen, D. *Unit Commitment under Uncertainty—GAMS Models*; Library of the Renewable Energy Analysis Lab (REAL), University of Washington: Seattle, DC, USA. Available online: https://www2.ee.washington.edu/research/real/gams_code.html (accessed on 27 Feburary 2019).

46. IBM®-IBM Knowledge Center. Differences between User Cuts and Lazy Constraints, 2014. Avalaible online: https://www.ibm.com/support/knowledgecenter/SSSA5P_12.6.1/ilog.odms.cplex.help/CPLEX/UsrMan/topics/progr_adv/usr_cut_lazy_constr/04_diffs.html (accessed on 15 Feburary 2019).

47. IceMerman. Unit Commitment with User Cuts Implementation. 2019. Available online: https://github.com/IceMerman/SCUC-UserCuts (accessed on 1 March 2019).

Article

Optimization of a Power Line Communication System to Manage Electric Vehicle Charging Stations in a Smart Grid

Sara Carcangiu, Alessandra Fanni * and Augusto Montisci

Department of Electrical and Electronic Engineering, University of Cagliari, 09123 Cagliari, Italy;
s.carcangiu@diee.unica.it (S.C.); amontisci@diee.unica.it (A.M.)
* Correspondence: fanni@diee.unica.it; Tel.: +39-320-437-2966

Received: 26 February 2019; Accepted: 1 May 2019; Published: 9 May 2019

Abstract: In this paper, a procedure is proposed to design a power line communication (PLC) system to perform the digital transmission in a distributed energy storage system consisting of fleets of electric cars. PLC uses existing power cables or wires as data communication multicarrier channels. For each vehicle, the information to be transmitted can be, for example: the models of the batteries, the level of the charge state, and the schedule of charging/discharging. Orthogonal frequency division multiplexing modulation (OFDM) is used for the bit loading, whose parameters are optimized to find the best compromise between the communication conflicting objectives of minimizing the signal power, maximizing the bit rate, and minimizing the bit error rate. The off-line design is modeled as a multi-objective optimization problem, whose solution supplies a set of Pareto optimal solutions. At the same time, as many charging stations share part of the transmission line, the optimization problem includes also the assignment of the sub-carriers to the single charging stations. Each connection between the control node and a charging station has its own frequency response and is affected by a noise spectrum. In this paper, a procedure is presented, called Chimera, which allows one to solve the multi-objective optimization problem with respect to a unique frequency response, representing the whole set of lines connecting each charging station with the central node. Among the provided Pareto solutions, the designer will make the final decision based on the control system requirements and/or the hardware constraints.

Keywords: power line communication (PLC); energy storage management; vehicle to grid (V2G); smart grid; multi-objective optimization

---

1. Introduction

Traditional power grids have been coupled with communication networks today, leading to the so-called smart grids. A smart grid enables information flows among various components of the grid, ranging from power plants to distributed energy resources, and from local utilities to residential and commercial customers. The purpose is to better monitor and control power generation and consumption. In smart grids, renewable sources, such as wind and solar energy, vary with weather and daylight conditions, so that an energy storage system is required to accumulate spare energy and to feed it back into the grid when required.

In smart grids, electric vehicles have an impact on energy storage through vehicle-to-grid (V2G) technologies [1], in which the electric vehicles (and even hybrids) can be seen as a distributed network of batteries that can store power at off-peak times and help power on the grid when demand peaks. Hence, V2G is useful to provide energy when demand shifts and to reduce electricity costs, to supply energy to energy markets, and to increase the use of localized renewables.

Each vehicle must have some requirements:

1. A connection to the grid for electrical energy flow;
2. A control or logical connection necessary for communication with the grid operator;
3. Controls and metering onboard the vehicle; and
4. There must be an agreement between the owner of the battery and the grid operator that electricity can be put into or drawn from the battery.

Thus, it is important to have an automated and standardized exchange of information between the vehicles and the grid. In this regard, different protocols for the communication are used [2,3]: ISO/IEC 15118 concerns the communication between an electric vehicle and the charging spot whereas the IEC 61850 is related to the communication between the charging spot and the energy provider (Figure 1).

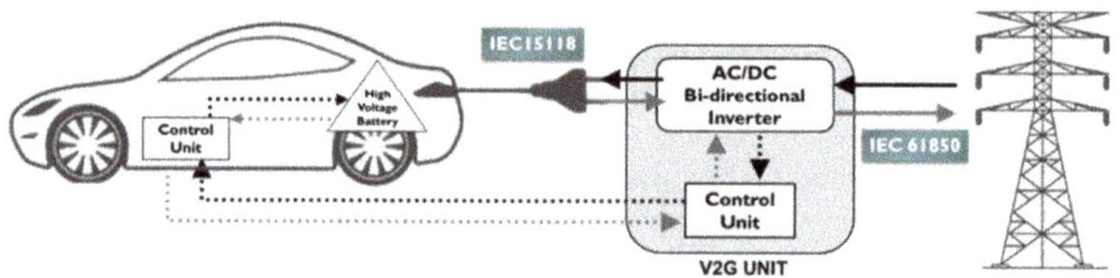

Figure 1. Vehicle-to-grid system.

A heterogeneous set of network technologies can support the smart grid communications ranging from wireless to wired solutions. Among the latter, PLCs have been deployed outdoor for last mile communications and indoor for home area networks [4]. A wide range of PLC technologies are available for different applications. Ultra-narrowband, operating at a very low data rate (100 bps) in the low-frequency band, is used in particular for load control. Narrowband PLC (NB PLC), operating in the 3–500 kHz band to deliver a few hundred kbps, has been used for last mile communications over MV and LV lines. In order to transmit data over a narrowband PLC network, different specifications are used. The most used standards are PRIME, developed by the PRIME Alliance [5], and G3, powered by Maxim company [6]. Both the standards provide source coding techniques to correct as many errors as possible at the receiver side due to the severe channel disturbances. In the range 1.8–86 MHz there is the broadband PLC (BB PLC), which is mainly used for home area networks and can provide several hundred Mbps. The typical examples of broadband PLCs conform to the standards IEEE 1901 [7], HomePlug [8], and ITU-T G.hn [9].

Both the NB-PLC and BB-PLC could be used in the smart grid applications, both in low voltage (LV) and medium voltage (MV) networks, with pros and cons [10,11]. NB-PLC are suitable for smart grid applications where a low-data rate is required, whereas BB-PLC solutions offer higher flexibility and a better trade-off between data rate, latency, robustness and energy efficiency [11]. In [12,13], the authors clarify the role of PLC technology for smart grid applications.

In this paper, to have the maximum flexibility, the BB-PLC technology is used. Specifically, in the V2G application, the most competitive advantage of using PLC is that no installation of an additional communication grid is required since the charging stations must be reached by the existing power grid to be fed. Moreover, the use of dedicated modulation techniques, such as OFDM (orthogonal frequency division multiplexing), allows for a broadband transmission, hence, PLC is competitive in more than in economic terms. Some disadvantages have to be taken into consideration: power lines do not necessarily provide a sure support, due to the presence of different elements; the attenuation of the data could be a problem; the electrical noise on the line limits the speed of the data transmission.

Designing the PLC system for the V2G needs a multi-objective optimization where some conflicting objectives should be considered: the communication capacity, the total transmission power, and the communication error probability. The multi-objective optimization process provides a set of equivalent solutions from which the designer makes the final choice, introducing further fitting criteria [14,15].

The optimization problem must be formalized depending on the modulation technique. Usually, to better exploit the available transmission band, the OFDM [8] system is assumed, which consists of splitting the transmission band into mutually orthogonal sub-carriers. Each one is used as a sub-carrier where the signal can be considered as if it were the only one in the channel. Every sub-carrier is loaded with a bit rate (BR) depending on the specific signal-to-noise ratio (SNR). OFDM is a modulation scheme commonly adopted in several application domains, such as mobile phones, digital TV and radio, and xDSL, each one having specific features and requirements. In particular, in the application presented in the present paper, different transmitters share the same channel, therefore, each sub-carrier must be assigned exclusively to one transmitter.

The optimization of the modulation scheme works by allocating the bits to the sub-carriers, which is referred to as bit-loading [8] problem, and which depends on the properties of the channel and the requirements of the transmission. In the present paper, the optimal design of the modulation system in PLCs is formalized as a multi-objective optimization problem with the following three conflicting objectives: the BR, to be maximized; the bit error rate (BER) and the interference on adjacent equipment, to be minimized [16–18]. The interference mostly depends on the power of the modulated signal and on its evolution in the time.

The problem faced has its own specificity, since the transmissions of the charging stations travel through common branches of the grid. Consequently, they must share the same frequency band. However, each channel has its own frequency response and power spectral density of noise. Hence, the optimal solution of the design problem should consider all the channels simultaneously. Other applicative domains have the same specificity, for example, cellular phones or digital audio broadcasting (DAB) networks.

In the proposed method, the OFDM symbols, to be sent by all the transmitters, are merged into one, which simultaneously travels in all the channels. A fictitious frequency response, called Chimera, is created which allows to perform the multi-channel optimization of the OFDM modulation scheme as if all the transmissions should be sent through a unique channel (named here Chimera channel). To define the Chimera frequency response an integer quadratic optimization problem is formalized whose aim is to assign band resources to the charging stations connected to the same feeder, depending on the number of supplied plugs, maximizing, at the same time, the overall capacity of the transmission.

The rest of the paper is organized as follows. In Section 2, the problem at hand is outlined. In Section 3, the proposed Chimera algorithm is described. Section 4 reports the formalization of the multi-objective problem aiming to off-line design the PLC system. Results of the optimization procedure on a case study are reported in Section 5. Finally, in Section 6 some conclusions are given.

2. V2G Communication System

The present paper aims to optimally design a PLC system that allows to manage a distributed energy storage system integrated with a smart grid. The storage is composed of a number of electric vehicle batteries connected to prefixed charging stations. It is assumed that one modem is installed at the interface between the charging station and the smart grid. Therefore, the lines connecting the charging station to the plugs are not considered in the design problem.

To prove the validity of the method, a simple feeder network is considered, which is a part of a wider network already considered in [19]; this does not invalidate the generality of the results, as the method is applied separately for each single feeder. Moreover, frequency responses and noise spectra are obtained from simulations rather than measurements because the method can be demonstrated regardless of the actual data. Network Simulator 3 (NS-3) [20], which is a free and open source discrete event network simulator, has been used. Figure 2 shows the topology of the case study, with a number of low voltage nodes, 3 kW loads, and four charging stations (checkered squares in Figure 2) supplied by a medium voltage/low voltage substation with 20 kV/400 V, 1MVA transformer. All the lines are three-phase underground commercial cables, whose datasheets are included in the database of the simulator. A phase-neutral cable, among all those available, has been randomly chosen to be the PLC channel. The lengths of the lines are reported in Figure 2. Each charging station supplies a variable number of charging plugs and it is connected to the control center through the smart grid. As each charging station is connected to the control center by a different path, each one is associated to a different frequency response.

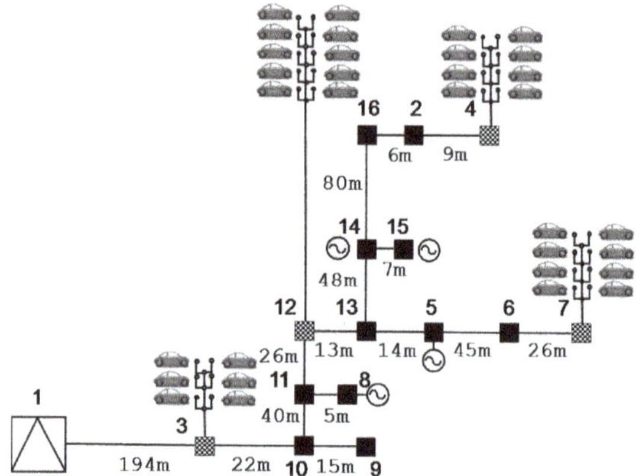

Figure 2. The case study: node 1 is the substation, the charging stations are the checkered nodes 3, 4, 7, and 12, whereas the black nodes are the low voltage 3 kW loads.

As known, the PLC channels are not time invariant due to load connection/disconnection [21]. For this reason, different scenarios have been simulated corresponding to different configurations of loads and active charging plugs and the lower envelop of the frequency responses has been assumed. In Figure 3, the power spectral densities (PSDs) of the signal as a function of the frequency in the nodes corresponding to the charging stations are shown. The frequency responses have been determined by feeding the channel with a white Gaussian noise and by measuring the spectrum of the signal at the receiver side. As a simplifying assumption, only the transmission from the charging station to the control center has been considered, without losing generality. Nonetheless, for each channel, two frequency responses for bi-directional transmission should be assessed, and a separate sub-band should be allocated to each. Thus, transmissions wouldn't need to be synchronized. Since the attenuation affects the performance of the PLC network, increasing with frequency and distance, longer LV lines must use frequencies in lower bands to guarantee a minimum performance.

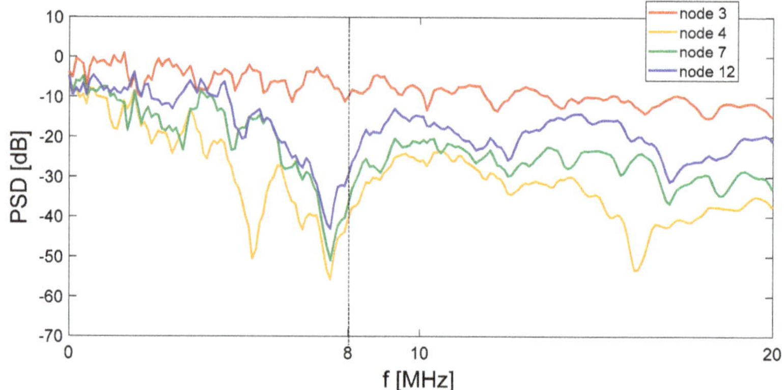

Figure 3. PSD of charging stations #3, #4, #7, and #12 connected to the control node 1.

Depending on the occupation level of the channel, different information can be exchanged between the charging stations and the control system. Note that, because all the charging stations are connected to the same feeder, any transmission reaches all the nodes. However, as each sub-carrier is assigned exclusively to a specific node, each receiver is able to retrieve the transmission addressed to it.

Each node is connected to the control node through a power line, whose frequency response depends on the cable-laying and the length. For this reason, the optimal bit-loading should consider all the different frequency responses and all the possible states of the network. In order to reduce the design optimization complexity, a Chimera frequency response is built starting from the frequency responses of the nodes (the charging stations) connected to the considered feeder.

3. Chimera Algorithm

One of the most significant shortcomings of PLC is the frequency response of the transmission channel, as it has an irregular gain diagram, with deep notches, and an irregular phase diagram. OFDM technology allows mitigation of such shortcomings [22].

As the charging stations served by one feeder share the same physical channel that connects them to the control node, each frequency band has to be assigned exclusively to one charging station. The optimization of the OFDM modulation should take into account the different frequency responses of all the channels. In this paper, an algorithm, called *Chimera*, has been proposed to optimize the allocation of sub-carriers to the nodes served by the same feeder. The output of the algorithm is a single frequency response (Chimera), obtained by combining the entire set of frequency responses of the charging stations. At the same time, the set of transmissions of all nodes is merged into a single bit stream, which is assumed to travel through the Chimera channel.

The bit stream is first subdivided into frames of assigned numbers of bits, which have to be modulated. The frame, in turn, is subdivided into as many strings of bits (words) as the number of sub-carriers, and the length of each word is assigned according to the capacity of the sub-carrier where it will be loaded.

Depending on the single capacity, a constellation, defined in the complex plane, is associated with each sub-carrier. Each point of the constellation is associated to a string of bits, so that each possible word can be transmitted by sending the coordinates of the corresponding point. Therefore, the length of each word, i.e., the number of bits, is equal to the base–2 logarithm of the number of points in the constellation. Having a constellation with many points favors the bit rate, but, at the same time, this reduces the margin among near points, which affects the *BER* of receiving data. In this paper, only a limited number of design parameters are considered, such as the transmission band and its subdivision into sub-carriers, the allocation of bits to each sub-carrier (bit loading) and the maximum power of transmission. Further design parameters, such as zero-padding, cyclic prefix, time guard, and so on [22,23], have been considered here by assuming a proper margin for the duration of the single OFDM symbol.

Let N be the number of nodes of the feeder that share the same transmission band, and K the number of sub-carriers in which the transmission band is subdivided. Each node is associated with a specific frequency response (see Figure 3). It is assumed that each node supplies V_n plugs, whose number can be different for different nodes. Moreover, it is assumed that the nodes communicate only with the central node, therefore, a number of frequency responses equal to the number of nodes have to be taken into account. The same subdivision in sub-bands is applied to all those frequency responses.

The capacity C_{nk} of the *k-th* sub-carrier of the channel n is given by the formula of Shannon-Hartley [24], which depends on the bandwidth B and the signal-to-noise ratio (*SNR*):

$$C_{nk} = B \cdot \log_2(1 + SNR_{nk}) \; [bit/s] \tag{1}$$

The value in Equation (1) is the theoretical upper bound of *BR* in the sub-carrier, but a lower value of *BR* is assumed in order to maintain the *BER* within an acceptable threshold.

The bandwidth is common to all the sub-carriers of all the frequency responses, therefore from here on we consider a normalized value of B. The total capacity of nth channel, corresponding to a specific node, is given by the sum of the capacities of all sub-carriers:

$$C_n = \sum_{k=1}^{K} C_{nk} \tag{2}$$

This expression can be calculated for each frequency response, and it assumes that all the sub-carriers are assigned to each node. An incidence matrix M_{NxK} is used to represent the assignment of sub-carriers to the single nodes. The columns of M correspond to the sub-carriers, whereas the rows correspond to the channels. In each row, each 1 identifies a sub-carrier assigned to the corresponding node. Each column can have at most one element equal to 1 and the others are equal to 0. The columns with all zeros correspond to not assigned sub-carriers. By using matrix M, the capacity $\tilde{C}_n$ assigned to each node can be expressed as:

$$\tilde{C}_n = \sum_{k=1}^{K} m_{nk} C_{nk} = \sum_{k=1}^{K} m_{nk} log_2(1 + SNR_{nk}) \tag{3}$$

where m_{nk} is the element of M corresponding to the frequency response n and the sub-carrier k. The aim of the procedure is to assign to each charging station a capacity proportional to the number of plugs it feeds. This rule can be expressed by the following statement:

$$\frac{1}{V_1} \sum_{k=1}^{K} m_{1k} C_{1k} = \ldots = \frac{1}{V_N} \sum_{k=1}^{K} m_{Nk} C_{Nk} \tag{4}$$

The allocation problem consists in defining the matrix M. As the unknowns are binary, the previous series of equations has not, in general, an exact solution, therefore the following objective function is defined, which has to be minimized:

$$J = \sum_{n=1}^{N} \left(\frac{1}{V_s} \sum_{k=1}^{K} m_{sk} C_{sk} - \frac{1}{V_n} \sum_{k=1}^{K} m_{nk} C_{nk} \right)^2 \tag{5}$$

where s identifies the channel with the lowest total capacity:

$$s = \underset{n \in \{1,\ldots,N\}}{\text{argmin}} \left(\frac{1}{V_n} \sum_{k=1}^{K} C_{nk} \right) \tag{6}$$

The quadratic function (5) allows to obtain the most homogeneous distribution of capacities, but it does not guarantee the total capacity is maximized. For this reason, a further term is added to the objective function, which weighs the capacity assigned to the channel with the lowest total capacity, obtaining:

$$J = \sum_{n=1}^{N} \left(\frac{1}{V_s} \sum_{k=1}^{K} m_{sk} C_{sk} - \frac{1}{V_n} \sum_{k=1}^{K} m_{nk} C_{nk} \right)^2 - \alpha \sum_{k=1}^{K} m_{sk} C_{sk} \tag{7}$$

where α is the weight coefficient; a small value favors the homogeneity among the channels, while a large value tends to increase the capacities assigned to all the channels. Hence, the best α is the one that maximizes the lowest allocated capacity.

In order to avoid both the trivial solution of null matrix M, and the same sub-carrier be assigned to different channels, the following constraint has to be stated:

$$\sum_{n=1}^{N} m_{nk} = 1 \quad \forall k = 1,\ldots,K \tag{8}$$

In case one sub-carrier is not suitable for the assignment, the corresponding column of the matrix M has to be removed.

Summarizing, the allocation problem can be defined as the following quadratic, binary minimization problem:

$$\min J$$
$$\text{s.t.}$$
$$\sum_{n=1}^{N} m_{nk} = 1 \quad \forall k = 1, \ldots, K \tag{9}$$
$$m_{nk} \in \{0; 1\}$$

4. Multi-Objective Optimization

The demand of transmission is maximizing the Bit Rate (BR) (bit/s), which is equal to the number of bits per frame multiplied by the number of OFDM symbols per second. Two other conflicting objectives must be minimized at once. The first is the total power of the signal P_{tot}, which is the sum of the powers of all the sub-carriers. The second conflicting objective is the error bit rate (BER), which depends on the noise spectrum: a probability density function can be associated to each point of the constellation centered in these points and the BER, for a given point, will be equal to the probability that the distance between a received point and its reference position be greater than the semi-distance between two adjacent points. In this paper, Gaussian distribution is assumed, so that there is a direct relationship between BER and the least distance among the points of the constellation. Different assumptions on the statistical distribution of points around the reference position do not change the application of the proposed method.

Note that, by setting the maximum transmission power of each constellation (P_{MAX}), its area is univocally assigned (area of the outer circle in Figure 4). Then, by setting the minimum distance between the points of the constellation (i.e., the BER), also its number of points is determined, hence, also the BR.

Figure 4 refers to the 16–APSK scheme for a generic sub-carrier [25]. In the same figure, the shadow circles centered in the points of the constellation have a radius equal to 3 times the standard deviation σ of the Gaussian distribution. The distance between two adjacent points is greater than 6σ.

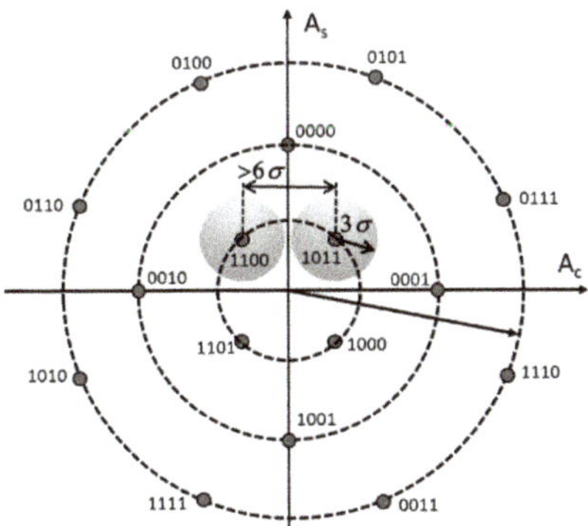

Figure 4. The 16-APSK constellation.

The requirements on the BER are particularly stringent, therefore, the APSK constellations [25] are adopted, which are increasingly considered in 5G mobile communications; for a given power, APSK gives the maximum distance between adjacent points. Without loss of generality, different constellation schemes can be assumed because the inter-dependence among power, BR and BER holds valid.

The water-filling (WF) algorithm [26,27] is used to allocate the bits to the sub-carriers. Note that, WF can be used either to design the constellations or to dynamically allocate the bits to the sub-carriers. In this work, it has been used for the former purpose.

The multi-objective optimization of the bit loading consists in finding the solutions that reconcile the conflicting demands. In particular, a solution is said nondominated, or Pareto optimal, if none of the objective functions can be improved in value without worsening some of the others. The set of such solutions is called Pareto Front [14] and, in the problem at hand, it represents a discrete subspace of the surface of the feasible solutions. In fact, as BR is a discrete variable, except for a discrete set of points in the (P_{tot}, BER) plane, the increase of one or both of them does not correspond to an increase of BR.

The multi-objective optimization aims to find this set of Pareto solutions and then to present them to the designer who will choose among them [14,15]. In the problem at hand, the surface $BR = f(P_{tot}, BER)$ has been densely sampled, and the set of non-dominated points has been selected, which is assumed as the Pareto front.

5. Results

The optimization problem in (9) is solved with a genetic algorithm implemented in the Matlab environment [28]. A transmission band of 1–8 MHz is assumed, because the SNR is unfavorable outside such interval. A width of 100 kHz has been assigned to each sub-carrier, then 70 sub-carriers are available. In practice, the number of actually available sub-carriers could be much lower, because of the great attenuation of the power line and the presence of noise. In this work, for each channel, the power spectrum of the noise has been obtained with the same NS3 simulator, which is in good agreement with experimental results retrieved from the literature [21,29].

The number of variables to be optimized is equal to 280 (70 sub-carriers for four channels), and the initial population has been generated using a uniform random number generator in the range {0;1} After 28,000 iterations the algorithm ends producing the matrix M. This allows us to obtain a single frequency response for all the charging stations by multiplying M with their frequency responses. The values of the genetic algorithm parameters are reported in Table 1. Figure 5 shows the obtained Chimera frequency response assumed to represent all the charging stations in Figure 2. The weight coefficient α has been set equal to 1.

Table 1. Genetic algorithm parameters.

Population Size	Initial Population	Elite Count	Crossover Fraction	Migration Fraction	Constraint Initial Penalty
100	random	5	0.8	0.2	10

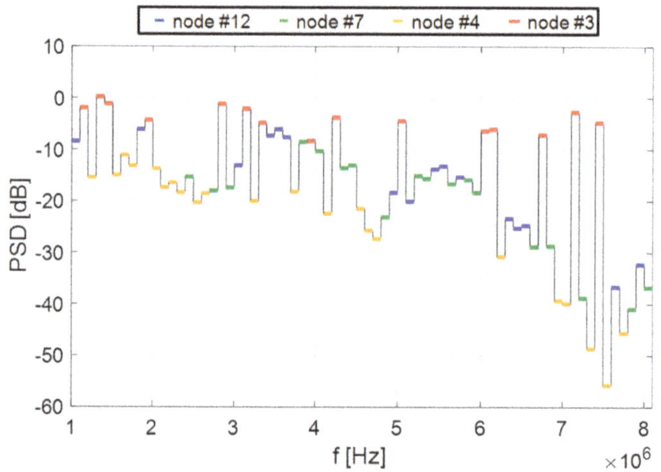

Figure 5. Power spectral density of the Chimera transmission channel: the different colors refer to the allocation of the 4 charging nodes to the 70 sub-carriers.

To perform the offline optimization of BR, BER and P_{tot}, the function $BR = f(P_{tot}, BER)$ is sampled in a regular 100×100 grid, within the interval of 1–10 [W] of P_{tot}, and 10^{-5}–10^{-3} of BER. The upper envelope of the noise PSDs of the four channels has been assumed as unique noise spectrum, as shown in Figure 6. Note that, given the Chimera frequency response and the noise, by applying the water filling algorithm, the power value for each sub-carrier is defined as a function of the total transmitting power P_{tot}. Hence, given the noise, the standard deviation σ is known, then the minimal inter-distance between the constellation points is determined. In this way, the number of points of the constellation is calculated, by obtaining the number of bits associated to each point, then the length of bit frame is obtained, and, finally, the BR.

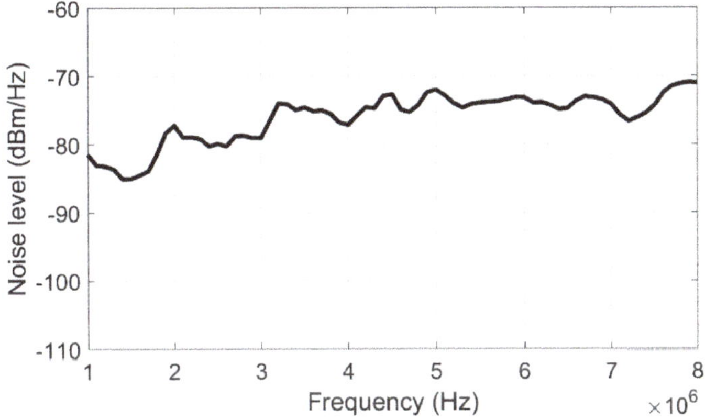

Figure 6. Noise power spectrum.

The sampled surface is shown in Figure 7. The non-dominated points are also represented by dots. The relationship between the three objectives allows us to obtain a BR ranging from 24.3–60 Mb/s.

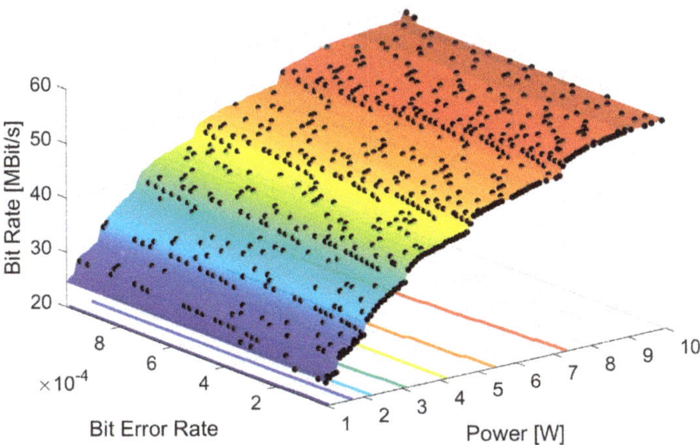

Figure 7. BR as function of BER and P_{tot}, and Pareto front (dots).

Figure 8 gives an equivalent representation of the Pareto front in the plane (P_{tot}, BER). It is a contour plot of the front, and the points represent the Pareto-optimal solutions. As it can be noted, in order to have a minimal BR at least a power of 1 W is needed. Moreover, limiting the available power to about 7 W, a BR of 52.1 Mb/s can be achieved, with a BER of the order of 10^{-5}. If a higher BR (e.g., 60 Mb/s) must be achieved, the BER will strongly affect the transmission.

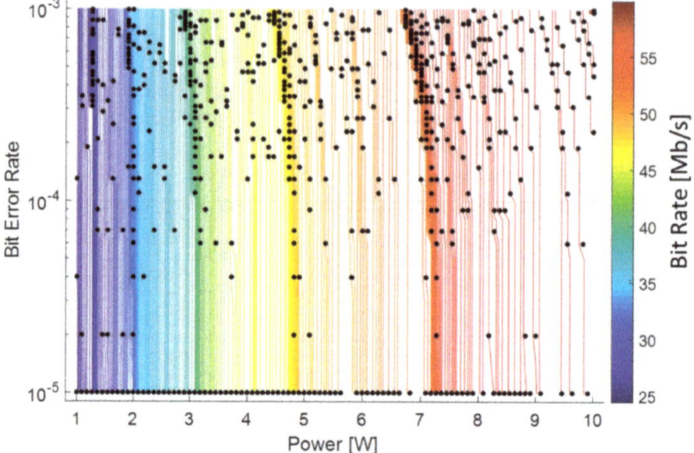

Figure 8. Two-dimensional representation of the Pareto front.

Figure 9 shows the Pareto optimal solutions corresponding to the minimal $BER = 10^{-5}$. Depending on the available total transmitting power, the plot returns the available BR for each plug in kByte/s. As an example, with an available transmitting power of 5 W, the BR is equal to about 190 kByte/s for each of the 32 plugs in the four charging stations.

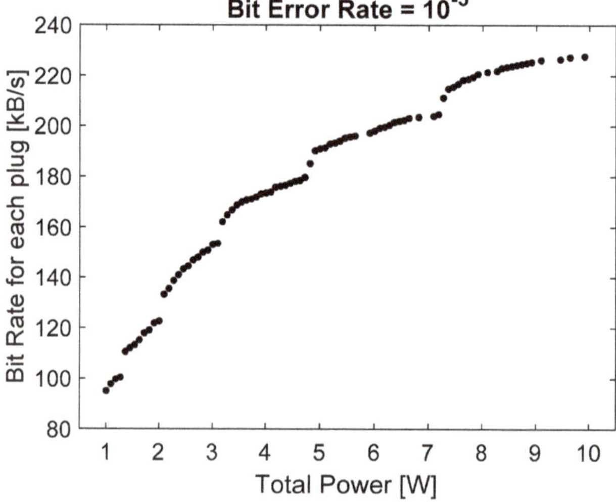

Figure 9. Pareto optimal solutions with a $BER = 10^{-5}$.

Finally, in Figure 10 the 70 APSK constellations corresponding to the Pareto point ($BR = 190$ kB/s, $BER = 10^{-5}$, $P = 5$ W) are shown. The colors refer to the different charging stations. As can be noted, the optimization procedure assigned different constellations to distinct sub-carriers.

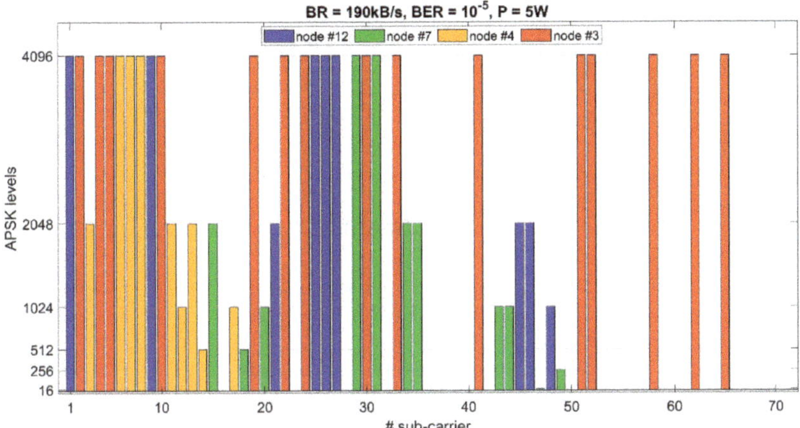

Figure 10. APSK OFDM constellations corresponding to the Pareto point $BR = 190$ kB/s, $BER = 10^{-5}$, $P = 5$ W.

6. Discussion

In a smart grid environment new challenges are posed to the power system operators by electric vehicles. Several studies have been proposed in the literature presenting control and optimization strategies for managing the charging/discharging of the electric vehicles' batteries [30]. However, the communication requirements still remain an open issue. In fact, different charging and discharging management strategies can be adopted ranging from a fully-centralized charge control decision, to distributed (or transactive), and to price control, which has limited communication requirements [31]. In the first case, the decisions are taken at the system level, hence it is generally accepted that it is the best control system in terms of, e.g., security of the power system. However, a more sophisticated communication infrastructure is needed that foresees bidirectional communication flow at the price of higher cost. Power line communication technology stands as a good candidate provided that its design is optimized. The design procedure presented in this paper, allows one to solve the bit-loading problem, finding a compromise among the conflicting objectives of minimal signal power and BER, while maximizing the BR. Among the Pareto-optimal solutions, provided by the optimization process, the designer can perform the definitive choice depending on the hardware and the communication requirements. Once this choice is made, the entire constellation system is designed.

Author Contributions: Conceptualization and analyses were done through the collaboration of all the authors. Software development was done by S.C. Data validation was done by A.M. A.F. wrote and reviewed the paper. All authors have read and approved the final manuscript.

Funding: This research has been funded by "Fondazione di Sardegna" under project "ODIS—Optimization of DIstributed systems in the Smart-city and smart-grid settings", CUP: F72F16003170002.

Conflicts of Interest: The authors declare no conflict of interest. The funders had no role in the design of the study; in the collection, analyses, or interpretation of data; in the writing of the manuscript; or in the decision to publish the results.

References

1. He, Y.; Bhavsar, P.; Chowdhury, M.; Li, Z. Optimizing the performance of vehicle-to-grid (V2G) enabled battery electric vehicles through a smart charge scheduling model. *Int. J. Automot. Technol.* **2015**, *16*, 827–837. [CrossRef]
2. Käbisch, S.; Schmitt, A.; Winter, M.; Heuer, J. Interconnections and communications of electric vehicles and smart grids. In Proceedings of the 1st IEEE International Conference on Smart Grid Communications (SmartGridComm), Gaithersburg, MD, USA, 4–6 October 2010; pp. 161–166.

3. Hansch, K.; Pelzer, A.; Komarnicki, P.; Groning, S.; Schmutzler, J.; Wietfeld, C.; Heuer, J.; Muller, R. An ISO/IEC15118 conformance testing systemarchitecture. In Proceedings of the IEEE-PES General Meeting, Conference & Exposition, National Harbor, MD, USA, 27–31 July 2014; pp. 1–5.

4. Lampe, L.A.; Tonello, M.; Swart, T.G. *Power Line Communications: Principles, Standards and Applications from Multimedia to Smart Grid*, 2nd ed.; John Wiley & Sons: Hoboken, NJ, USA, 2016.

5. Berganza, I.; Sendin, A.; Arriola, J. PRIME: Powerline Intelligent Metering Evolution. In Proceedings of the CIRED Seminar 2008, Frankfurt, Germany, 23–24 June 2008.

6. The G3-PLC Alliance. Available online: http://www.g3-plc.com/ (accessed on 8 January 2019).

7. *Medium Access Control and Physical Layer Specifications*; IEEE Std. 1901–2010; IEEE Standard Association: New York, NY, USA, 30 December 2010.

8. Guerrini, E.; Dell' Amico, G.; Bisaglia, P.; Guerrieri, L. Bit-loading algorithms and SNR estimate for HomePlug AV. In Proceedings of the IEEE International Symposium on Power Line Communications and its Applications, ISPLC 2007, Pisa, Italy, 26–28 March 2007; pp. 77–82.

9. Unified High-Speed Wireline-Based Home Networking Transceivers—System Architecture and Physical Layer Specification. ITU-T Rec. G.9960. 2011. Available online: https://www.itu.int/rec/T-RECG.9960/en (accessed on 8 January 2019).

10. Tonello, A.M.; Versolatto, F.; Pittolo, A. In-home power line communication channel: Statistical characterization. *IEEE Trans. Commun.* **2014**, *62*, 2096–2106. [CrossRef]

11. Tonello, A.M.; Pittolo, A. Considerations on Narrowband and Broadband Power Line Communication for Smart Grids. In Proceedings of the IEEE Smart Grid Communications 2015, Miami, FL, USA, 2–5 November 2015.

12. Ferreira, H.C.; Lampe, L.; Newbury, J.; Swart, T.G. *Power Line Communications: Theory and Applications for Narrowband and Broadband Communications over Power Lines*; Wiley & Sons: Hoboken, NJ, USA, 2010.

13. Galli, S.; Scaglione, A.; Wang, Z. For the Grid and Through the Grid: The Role of Power Line Communications in the Smart Grid. *Proc. IEEE* **2011**, *99*, 998–1027. [CrossRef]

14. Carcangiu, S.; Fanni, A.; Montisci, A. Multiobjective tabu search algorithms for optimal design of electromagnetic devices. *IEEE Trans. Magn.* **2008**, *44*, 970–973. [CrossRef]

15. Carcangiu, S.; Fanni, A.; Montisci, A. Multi objective optimization algorithm based on neural networks inversion. *Lect. Notes Comput. Sci.* **2009**, *5517*, 744–751. [CrossRef]

16. Carcangiu, S.; Montisci, A.; Usai, M. Bit Loading Optimization for Naval PLC Systems. In Proceedings of the IEEE International Symposium on Power Line Communications and its Applications, Udine, Italy, 3–6 April 2011; pp. 84–89.

17. Camplani, M.; Cannas, B.; Carcangiu, S.; Fanni, A.; Montisci, A.; Usai, M. Tabu-Search Procedure for PAPR Reduction in PLC Channels. In Proceedings of the IEEE International Symposium on Industrial Electronics, Bari, Italy, 4–7 July 2010; pp. 2979–2983.

18. Carcangiu, S.; Montisci, A. A Hybrid Analytical-Numerical Approach for PAPR Reduction in PLC Grids. In Proceedings of the 8th Mediterranean Conference on Power Generation, Transmission, Distribution and Energy Conversion, MEDPOWER 2012, Cagliari, Italy, 1–3 October 2012; pp. 1–5.

19. Carcangiu, S.; Celli, G.; Fanni, A.; Garau, M.; Montisci, A.; Pilo, F. Bit loading optimization for smart grid energy storage management. In Proceedings of the IEEE 3rd International Forum on Research and Technologies for Society and Industry, RTSI 2017, Modena, Italy, 11–13 September 2017. 8065946.

20. Aalamifar, F.; Schlögl, A.; Harris, D.; Lampe, L. Modelling power line communication using network simulator-3. In Proceedings of the IEEE Global Communications Conference (GLOBECOM), Atlanta, GA, USA, 9–13 October 2013; pp. 2969–2974.

21. Barmada, S.; Bellanti, L.; Raugi, M.; Tucci, M. Analysis of Power-Line Communication Channels in Ships. *IEEE Trans. Veh. Technol.* **2010**, *59*, 3161–3170. [CrossRef]

22. Ma, Y.H.; So, P.L.; Gunawan, E. Performance analysis of OFDM systems for broadband power line communications under impulsive noise and multipath effects. *IEEE Trans. Power Deliv.* **2005**, *20*, 674–682. [CrossRef]

23. Muquet, B.; Wang, Z.; Giannakis, G.B.; de Courville, M.; Duhamel, P. Cyclic prefixing or zero-padding for wireless multicarrier transmissions? *IEEE Trans. Commun.* **2002**, *50*, 2136–2148. [CrossRef]

24. Carlson, A.B. *Communication Systems*, 3rd ed.; McGraw-Hill: New York, NY, USA, 1986.

25. Liu, Z.; Xie, Q.; Peng, K.; Yang, Z. APSK Constellation with Gray Mapping. *IEEE Commun. Lett.* **2011**, *15*, 1271–1273. [CrossRef]
26. Honda, S.; Umehara, D.; Hayasaki, T.; Denno, S.; Morikura, M. A fast bit loading algorithm synchronized with commercial power supply for in-home PLC systems. In Proceedings of the IEEE International Symposium on Power Line Communications and Its Applications, ISPLC 2008, Jeju City, Jeju Island, Korea, 2–4 April 2008; pp. 336–341.
27. Pérez Palomar, D.; Rodríguez Fonollosa, J. Practical Algorithms for a Family of Waterfilling Solutions. *IEEE Trans. Signal Process.* **2005**, *53*, 686–695. [CrossRef]
28. MathWorks. Available online: https://www.mathworks.com/help/gads/ga.html (accessed on 8 January 2019).
29. Di Bert, L.; Caldera, P.; Schwingshackl, D.; Tonello, A.M. On noise modeling for power line communications. In Proceedings of the IEEE International Symposium on Power Line Communications and Its Applications, ISPLC 2011, Udine, Italy, 3–6 April 2011; pp. 283–288.
30. Samy Faddel, S.; Al-Awami, A.T.; Mohammed, O.A. Charge Control and Operation of Electric Vehicles in Power Grids: A Review. *Energies* **2018**, *11*, 701. [CrossRef]
31. Hu, J.; Morais, H.; Sousa, T.; Lind, M. Electric vehicle fleet management in smart grids: A review of services, optimization and control aspects. *Renew. Sustain. Energy Rev.* **2016**, *56*, 1207–1226. [CrossRef]

Article

A General Intelligent Optimization Algorithm Combination Framework with Application in Economic Load Dispatch Problems

Jinghua Zhang * and Ze Dong

Hebei Engineering Research Center of Simulation Optimized Control for Power Generation, North China Electric Power University, Baoding 071003, China; dongze33@126.com
* Correspondence: 52151053@ncepu.edu.cn; Tel.: +86-139-3086-8002

Received: 17 May 2019; Accepted: 4 June 2019; Published: 6 June 2019

Abstract: Recently, a population-based intelligent optimization algorithm research has been combined with multiple algorithms or algorithm components in order to improve the performance and robustness of an optimization algorithm. This paper introduces the idea into real world application. Different from traditional algorithm research, this paper implements this idea as a general framework. The combination of multiple algorithms or algorithm components is regarded as a complex multi-behavior population, and a unified multi-behavior combination model is proposed. A general agent-based algorithm framework is designed to support the model, and various multi-behavior combination algorithms can be customized under the framework. Then, the paper customizes a multi-behavior combination algorithm and applies the algorithm to solve the economic load dispatch problems. The algorithm has been tested with four test systems. The test results prove that the multi-behavior combination idea is meaningful which also indicates the significance of the framework.

Keywords: population-based intelligent optimization algorithm; multi-behavior combination; algorithm framework; Economic load dispatch (ELD)

---

1. Introduction

Many real-world problems can be modeled as optimization tasks. It is an important solution to solve the optimization problem with the population-based intelligent optimization algorithm such as evolutionary computation, swarm intelligence.

In the early stage, population-based intelligent optimization algorithms tend to be concise, and population intelligence is acquired through the simple behavior of individuals. However, some of the optimization problems in the real world are more complicated, and the existing algorithms may reflect their shortcomings. Economic load dispatch (ELD) is a typical optimization problem in power systems. The goal of ELD is to rationally arrange the power output of each generating unit in a power plant or power system so as to minimize fuel costs on the premise of system load and operational constraints. Due to the valve point effect of the thermal generating unit and the operational constraints, ELD problem presents non-convex, high-dimensional, nonlinear and discontinuous characteristics, which make the problem complicated. As the scale of the power system increases and the model for the problem becomes more sophisticated by considering more conditions, the optimization task becomes more difficult. In recent years, population-based intelligent optimization algorithms have been applied to solve ELD problems. Early research directly adopted classic intelligent optimization algorithms such as the genetic algorithm (GA) [1], particle swarm optimization (PSO) [2,3], or evolution programming (EP) [4]. There are also some studies that improve the classical algorithm for the ELD problem, such as an improvement for GA [5,6], quantum PSO (QPSO) [7], new PSO with local random search

(NPSO-LRS) [8], distributed Sobol PSO and TSA (DSPSO-TSA) [9]. The new algorithms proposed in recent years have also been applied to ELD problems, such as ant colony optimization (ACO) [10], differential evolution algorithm (DE) [11], artificial immune system (AIS) [12], modified artificial bee colony (MABC) [13], improved harmony search (IHS) [14], tournament-based harmony search (THS) [15], biogeography based optimization (BBO) [16], chaotic teaching-learning-based optimization with Lévy flight (CTLBO) [17], grey wolf optimization (GWO) [18]. Additionally, hybrid optimization algorithms have also been proposed by the combination of multiple intelligent optimization algorithms or algorithm components to deal with the ELD problem, such as a hybrid PSO (HPSO) [19], modified TLA (MTLA) [20], DE-PSO method [21], differential harmony search algorithm [22], hybrid differential evolution with biogeography-based optimization (BBO-DE) [23].

The current research on population-based intelligence optimization algorithms presents a new trend. Novel algorithms proposed recently tend to be complicated, such as simulating the more complicated biological population, natural phenomena, or social organization. A population no longer simply adopts one behavior; conversely multiple behaviors are performed during the search process. In [24–27], individuals play different roles in the population and perform different behaviors; in [28–30], the population performs different operations step by step during the process of completing the optimization task; or the search process is divided to different stages and the behavior of the population is different in stages [25]. In conclusion, the research direction of the current new algorithm is to use complex mechanisms to deal with diverse and complex real-world problems.

On the other hand, the researchers try to improve the algorithm through the combination of multiple algorithms or algorithm components with different characteristics. Because the algorithms or components of different characteristics may be suitable for different problems, the robustness of the algorithm may be improved; furthermore, the combination of algorithms with different characteristics is likely to complement each other and get better results. Therefore, an ensemble of multiple algorithms or algorithm components could make the optimization algorithm particularly efficient for complicated optimization problems [31].

Some studies combine multiple search operators in an algorithm. In the literature [24,25,27,28], different search operators are switched with specific probabilities. Some studies execute the operators in the specific order [29]. In [32], several operators are all calculated and the results with good fitness are selected. In the research of differential evolution algorithms, a variety of mutation operators are proposed, and the research on adaptive ensemble of multiple mutation operators has attracted the interest of researchers [33–37]. Some studies have further explored the combination of multiple crossover operators and multiple mutation operators in differential evolutionary algorithms [38]. In addition to differential evolution algorithms, there are also studies on ensembles of multiple search operators based on other algorithms, such as based on PSO [39,40], ABC [41], BBO [42], TLA [20]. There are also studies on ensemble of other algorithm components, such as individual topology [43,44], constraint processing components [45]. In addition to the combination of algorithm components, there are also studies with algorithm-level combinations which directly combine or adaptively select different algorithms [46,47].

From the current research on the improvement of the existing algorithms or new algorithms, it seems that an important trend is to combine multiple search behaviors, and construct high-level population running mechanisms. Therefore, in this paper, we introduce the idea of algorithm combination into application practice. Different from the traditional algorithm design, this paper focuses on the ideas and methods of algorithm combination and implements it as a general framework. Because no free lunch theorem [48] indicated that no matter how the algorithm improves, it is impossible to be better than all other algorithms on all problems. Therefore, an algorithm designed for a problem may be not suitable to another problem. The researchers maybe need to design corresponding optimization algorithms for each type of optimization problem. Thus, a flexible framework may be useful since various algorithms (components) can be combined and different combination modes and strategies can be adopted, which can bring diversity for algorithm design.

We adopt the idea of multi-behavior combination and proposed a unified multi-behavior combination model. Then, an agent-based framework is designed to support the model. The framework has the following three functions: (1) the existing algorithms and techniques can be extracted and reused as multiple behaviors; (2) the performance of the algorithm can be improved through the appropriate multi-behavior combination; and (3) through the framework technology, the corresponding combination algorithm can be customized for specific problems. At last, we customize a multi-behavior combination algorithm for the power system economic load dispatch problems. The test results show the effectiveness of the multi-behavior combination algorithm and the feasibility of the framework.

The remainder of this paper is organized as follows. Section 2 defines an intelligent optimization algorithm combination model based on the idea of multi-behavior combination. Section 3 gives an agent-based general framework for the model. Section 4 defines the ELD problem, and gives our algorithm customized with the framework. Section 5 gives the test with four classical ELD cases. Finally, Section 6 presents the conclusions.

2. A Unified Multi-Behavior Combination Model for Intelligent Optimization Algorithm

The ensemble of multiple algorithms or multiple algorithm components is a research hotspot of intelligent optimization algorithms. In this paper, we try to give a general algorithm combination framework which is hoped to support various algorithm combination mode. Since both algorithm and algorithm component represent specific search behaviors, this kind of research tries to introduce different behaviors in the search process of population. Therefore, this paper analyzes the characteristics of the population-based intelligent optimization algorithm in order to determine the unified multiple-behavior combination model.

2.1. Multi-Behavior Combination

Any population-based intelligent optimization algorithm requires a population with multiple individuals. The algorithm needs to perform iteration in the search process. The individual of the population evolves according to the algorithm mechanism in each iteration. Some algorithms simulate complex multi-population model, and multiple populations coexist and communicate by specific communication mechanisms. Therefore, a population-based intelligent optimization algorithm can be considered as a three-dimensional model, as shown in Figure 1. We consider the population-based optimization algorithm to be a three dimensional structure: x-dimension, y-dimension, and z-dimension; where x represents the individual dimension of the population, y represents the iterative dimension of the loop, and z represents the parallel population dimension. The x-y forms an algorithm plane that can search independently. Multiple algorithm planes are parallel on the z-dimension and influence each other by exchange individuals. A simple algorithm degenerates to a point in the z-dimension, meaning a single population, and has the same behavior in both the x- and y-dimensions, representing that the algorithm has only one behavior. Current algorithm research tends to improve algorithm performance and robustness through a combination of multiple behaviors. From the three-dimensional model, it can be seen that each dimension can be combined with different behaviors, and a combination of behaviors in different dimensions contains a different optimization idea that should bring different effects. This article divides the multi-behavior combination into three levels:

(1) Individual level: the combination of multiple behaviors at individual level refers to that individuals of a population may have different behaviors. Each individual selects a specific behavior according to certain rules. Individual level combination is a combination of ways to bring richer behaviors in a population and can make the entire population more diverse.

(2) Iteration level: the combination of multiple behaviors at iteration level means that different evolutionary generations may use different behaviors during evolutionary process. The multiple behaviors can be related to different stages according to their evolutionary state, or be executed alternately.

(3)   Population level: the combination of multiple behaviors at the population level means that each sub-population may have its own behavior. In this combination, each sub-population is independent, and a sub-population communication mechanism is needed.

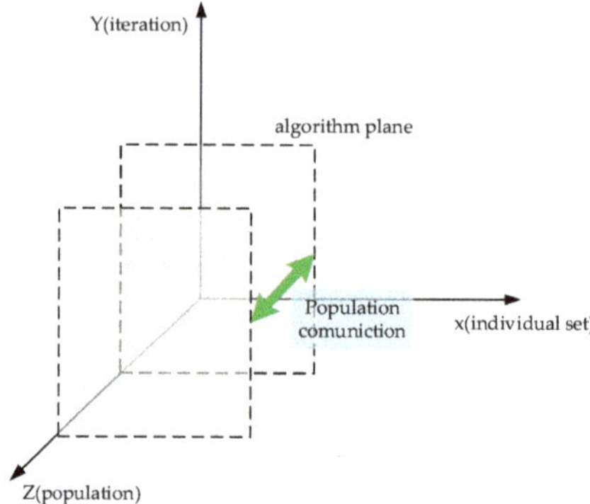

Figure 1. A three-dimensional model for population-based intelligent optimization algorithm.

2.2. Behavior

Different algorithms or algorithm components are used to represent particular behaviors. The combination of multiple behaviors can be a combination of algorithms or algorithm components. Algorithm component is constituent element of the algorithm. For multi-behavior combination, it is necessary to extract the behavior pattern from the algorithms in unit of one iteration which we called the operator of an algorithm. Since the algorithm is a population-based algorithm, the core operator is individual (population) evolution operator, which represents an evolutionary behavior from the parent individuals (population) to the child individuals (population). Some algorithms can be further decomposed to extract the auxiliary operators, which are the certain technological elements of an algorithm such as parameter control strategy, constraint processing technology, etc., and can also be used as the combination elements. This paper categorizes the behaviors into the following six types of operator:

(1)   Individual (population) evolution operator: individual (population) evolution operator is the core operator of various algorithms. Most algorithms are based on individual evolution, and a small number of algorithms are the overall population evolution, such as CMA-ES [49]. The operator directly operates on the individuals or population to generate the offspring.

(2)   Parent individual selection operator: the parent individual is the individual which applies in individual evolution operation; therefore, it is an auxiliary operator for an individual evolution operator, such as random selection, selection based on individual topology, selection based on fitness, selection based on distance, etc.

(3)   Population selection operator: the operator selects the individuals entering the next generation between the parent population and the offspring population. A variety of population selection strategies have been adopted in the algorithms. The PSO algorithm directly uses the offspring population as the next generation population, the DE algorithm adopts the binary selection, the GA adopts the fitness-based selection, and some studies improve the population selection with considering the diversity.

(4)   Parameter control operator: the parameter control operator is used to set the parameter values of the algorithm. The parameter control operator also acts as a behavior because different parameter values can cause different search trajectory. The current research has proposed a variety of parameter control strategies. Some algorithms use fixed parameters, and some use only simple

strategies, such as random values, descending linearly by evolution generation. Some studies proposed adaptive parameter control strategies. Some of these strategies are general and can be extracted as operators.

(5)   Population size control operator: population size is also a parameter of the algorithm, but it has great difference from the operation related parameters. It can also affect the population behavior. Thus, population size control strategy can also be extracted as an operator.

(6)   Constraint processing operator: the population-based intelligent optimization algorithm is suitable for solving single-object unconstrained (boundary constraint) optimization problems. Additional mechanisms are needed for constraint processing when solving constrained optimization problems. The most straightforward approach is to discard the infeasible solution, but this approach may lose performance, because infeasible solutions also carry valid information, and when the search space is not continuous, the intelligent optimization algorithm may fall into local extrema. Therefore, some studies explore constraint processing techniques. Generally, constraint processing occurs at the population selection step, therefore, it can be embedded into the population selection operator to construct the constraint-based population selection operator.

The multi-behavior combination can be an ensemble of multiple individual evolution operators, or multiple auxiliary operators, or an ensemble of multi-type multiple operators.

2.3. Combination Strategy

In addition to three combination modes, a combination of multiple behaviors also needs to establish corresponding combination strategies. For individual level combination, each individual's behavior is needed to be determined; and for the combination at iteration level, each evolutionary generation is needed to determine the behavior for it; and the combination at population levels should determine sub-population behavior, and establish sub-population communicate mechanism. In this paper, we adopt the reverse thinking to take the behavior as the core. For an ensemble of behaviors, if individual level combination is used, individual groups should be determined for each behavior, including group size and grouping methods; if an iteration level combination is used, evolutionary generations should be determined for each behavior; and if population level combination is used, sub-population (is also an individual group) should be determined for each behavior including communication mechanism. Thus, each behavior has its individual group, evolutionary generations, and change strategy. Thereby, the behavior combination can be unified in a general framework. The combination of multiple behaviors can be in a competitive or collaborative manner.

2.3.1. Collaborative Strategy

The collaborative strategy means that the related behaviors have a cooperative relationship, that they will support each other. The collaborative relationship is generally pre-customized, and the goal of collaboration is achieved through different search characteristics of multiple behaviors. Therefore, according to the behavior characteristic, the individual group, or generations, or communicating mechanism can be defined under specific combination mode. The communication mechanism of individual groups can be re-grouping (which is more suitable for individual level combination), or migrating and mixing individuals between different individual groups (which is more suitable for population level combination).

2.3.2. Competitive Strategy

The competition strategy means that multiple behaviors compete with each other and decide the winner. The winner would acquire more resources as the reward. The competitive resource of the individual level combination is the number of individuals, and the resource of the iteration level combination is execution probabilities. The population level combination is generally suitable for the collaborative strategy, but the sub-populations can also compete for the amount of communication resources.

Although the resources to be competed are different, they can be unified. We define the resources for competition as probability, that is, the resource occupancy rate of each behavior. It is defined as a vector $P = \{P_1, P_2,... P_m\}$, m is the number of behaviors, and P_i is the resource occupancy of the ith behavior, $1 \leq i \leq m$. The basis of the competition strategy is the evaluation model of behavior. Define evaluation model data for all behaviors as a vector $soState = \{soState_1, soState_2, \ldots, soState_m\}$, $soState_i$ is the model score for the ith behavior, and $soState_{best}$ is the score for the optimal behavior.

The paper gives two competing strategies, one is a full-competitive strategy and the other is a semi-competitive strategy. The full-competitive strategy maximizes the resources of the excellent behavior set, and the definition of the excellent behavior set is shown in Equation (1), P_i is adjusted according to Equations (2) and (3) [50], *rate* is the maximum adjustment ratio.

$$best = \left\{ i | soState_i > avg(soState) \wedge \frac{soState_i - avg(soState)}{soState_{best} - avg(soState)} > 0.5 i \in [1, m] \right\} \tag{1}$$

$$\Delta_i = rate \cdot \frac{soState_{best} - soState_i}{soState_{best}} i \notin best \tag{2}$$

$$P_i = \begin{cases} P_i + \frac{\sum_{i \notin best} \Delta_i}{|best|} & \text{if } i \in best \\ P_i - \Delta_i & \text{otherwise} \end{cases} \tag{3}$$

The semi-competitive strategy defines a base ratio vector $P_{base} = \{P_{base,1}, P_{base,2},..., P_{base,m}\}$, and $P_{base,i}$ defines the resource ratio that the ith behavior at least occupies. The remaining ratios are proportionally assigned based on the model scores of the behaviors. The adjustment method of P_i is as follows:

$$P_i = P_{base,i} + \left(1 - \sum_{i=1}^{m} P_{base,i}\right) \cdot \frac{soState_i}{\sum_{i=1}^{m} soState_i} \tag{4}$$

2.4. Evaluation Model

Two kinds of evaluation may be needed in the process of behavior combination. One type of evaluation is to evaluate the population in order to understand the evolution state and select the appropriate algorithm. The other type of evaluation is the evaluation of behavior since the optimization performance of each behavior needs to be known for the competitive strategy. It may be necessary to decide which behavior is the winner based on the relative performance of multiple behaviors. Both types of evaluation are judged according to the population. Various evaluation mechanisms are established through the relevant information of the parent population and its offspring population. We divide the evaluation mechanism into two categories:

2.4.1. Fitness Based Evaluation

The evaluation mechanism based on fitness is a more intuitive evaluation mechanism. Obtaining the optimal fitness is the goal of the algorithm. Therefore, the fitness evaluation mechanism is suitable for evaluating the performance of the behavior.

(a) Success rate (*SR*). The *SR* refers to the successful improvement rate for offspring to parent population. It is as follows:

$$SR = \frac{|SI|}{NP} \tag{5}$$

where NP is the individual number of the population, SI is the set of individuals which enter the next generation from offspring population, and $|SI|$ means the individual number of SI.

(b) Successful individual fitness improvement mean (*SFIM*). The *SFIM* is a mean value for all successful individual. If the population selection operator is binary selection as DE used, it can be computed as Equation (7). If the population selection operator is not binary selection, the Equation (8) can be used.

$$\Delta f_k = |f(u_{k,g}) - f(x_{k,g})| \tag{6}$$

$$SFIM = \frac{\sum_{k=1}^{|SI|} \Delta f_k}{|SI|} \tag{7}$$

$$SFIM = \frac{|\sum_{k=1}^{|PI|} f(x_{k,g}) - \sum_{k=1}^{|SI|} f(u_{k,g})|}{|SI|} \tag{8}$$

where Δf_k is the fitness improvement value of the offspring individual $u_{k,g}$ to the parent individual $x_{k,g}$. f is the fitness function (objective function), SI has the same meaning with Equation (5), and PI is the set of parent individuals being replaced.

2.4.2. Evaluation Based on Other Information

Although the evaluation mechanism based on fitness is the most direct and effective method, it lacks other state information of the reference population. Considering the population performing the optimization as a stochastic system, information entropy can be used to evaluate the state of the population. Information entropy can be seen as a measure of system ordering, and the changes in entropy can be used to observe changes in populations. We also adopt an evaluation model based on information entropy measurement in the target space. The information entropy for discrete information is shown as follows:

$$H - -\sum_{i=1}^{n} p_i \log_2 p_i \tag{9}$$

where n represents the number of discrete random variables, and p_i represents the probability of the ith discrete random variable, $p_i \in [0,1]$, and $p_1 + p_2 + ... + p_n = 1$.

Let NP to be the individual number of the population, and the maximum and minimum objective values of the population are f_{min}, f_{max}. Divide $[f_{min}, f_{max}]$ in the target space into NP sub-domains as the NP discrete random variables. Then p_i is the individual percentage of the ith sub-domain.

$$p_i = c_i / NP \tag{10}$$

where c_i represents the number of individuals with objective values in the ith sub-domain. H is the information entropy of the population representing the population diversity.

According to the multi-behavior combination model, a multi-behavior combination algorithm can be defined according to Table 1.

Table 1. Multi-behavior combination algorithm attributes.

Algorithm Attributes	Attributes Value Range
combination method	{individual level, iteration level, population level}
combination strategy	{collaborative strategy, competition strategy}
behavior operator set	{six categories of operators}
evaluation model	{SR, SFIM, Entropy, Compound model, none}

3. Agent-based General Algorithm Combination Framework

In order to customize the multi-behavior combination algorithm conveniently, the model needs to be designed as a framework. In this section we construct an agent-based general framework.

First, an operator component library is built for the operators that represent various candidate behaviors. The core of the framework is to construct the search agent which represents a behavior in runtime. Multiple behavior set are expressed by a search agent list. Therefore, operators that implement a behavior are encapsulated in a search agent. Each search agent is autonomous and can perform the search in a centralized or distributed environment independently. The combination of algorithms or combination of operator components is transformed into a combination of search agents. Multi-behavior combination is modeled by a combination agent which describes the combination scheme and determines the execution logic in order to schedule the search agents. The combination component library includes some related combination modes and strategies. A multi-agent system environment is defined for information exchange among search agents, including the entire population information, general settings of algorithm, information communication region. The framework structure is shown in Figure 2.

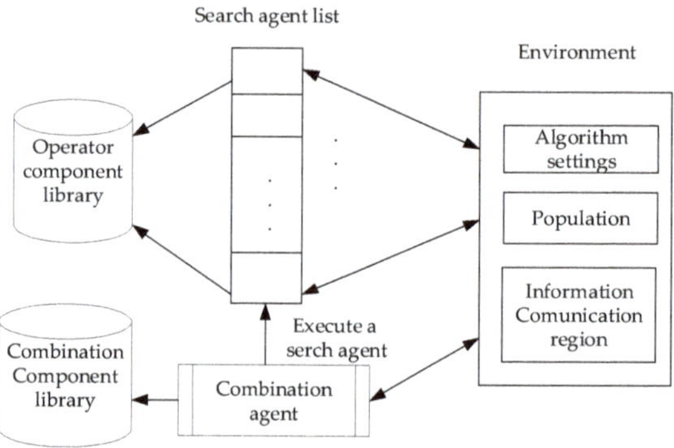

Figure 2. Agent-based framework structure.

3.1. Operator Component Library

The operator components in the operator component library are designed or extracted from existing algorithms. The six types of operators summarized in Section 2.2 can be embedded in the framework. A uniform interface for each type of operator is defined. Each operator takes one iteration as the extraction unit. The data structure corresponding to an operator component should be constructed if some information needs to be maintained between the two iterations of the operator. Most algorithms have the same structure on evolution operation: at first parent individual selection, then individual evolution to generate offspring, finally population selection. These algorithms can be decomposed into the three types of operator components if they are general and can be reused. And these algorithms can be assembled by the corresponding operator components when in runtime. There are also algorithms that cannot be decomposed to general components, thus, the entire evolution operation in one iteration can be extracted into an independent individual evolution operator component.

3.2. Search Agent

Each search agent represents a behavior that can search independently, and therefore, it is an iterative unit of a complete algorithm. It contains the information needed by a complete algorithm, and is defined as a five-tuple:

SAinfo = (AInfo, GroupInfo, RuntimeInfo, OffspringInfo, ModelInfo)

- AInfo is a data structure for algorithm information. The algorithm information describes its corresponding operator combination. The individual evolution operator is the core operator. According to the core operator, the auxiliary operators (parent individual selection operator, the population selection operator) are selected if needed, then the parameters of the algorithm can be determined, and the parameter control operator can be selected.
- GroupInfo is a data structure for individual group information. It means the population of an algorithm, including the group size, individual set, individual type.

- RuntimeInfo is a data structure for runtime information. Runtime information comes from the algorithm running process, because some information should be retained during the iteration. Various operator components may require their specific information; thus, the related runtime information structure is constructed when the operator component is executed for the first time.
- OffspringInfo is the result of search behavior, and it represents the set of offspring individuals.
- ModelInfo is the evaluation data of the behavior. It is a vector including one or more selected evaluation models from {*SR, SFIM, Entropy*}.

A single-step search behavior is defined to enable the search agent to run step by step. The algorithm is identical for all search agents, and the difference is the operators and group. The algorithm is showed in Algorithm 1.

Algorithm 1. Single-step behavior of search agent.

Input: AllInfo, GroupInfo, RuntimeInfoOutput: OffspringInfo, ModelInfo
foreach parameter strategy in AllInfo
Execute the parameter strategy operator component to generate the parameter value set.
end foreach
Execute the evolution operation (including an operator list in AllInfo) and return the offspring set.
Compute the objective of the offspring.
foreach parameter strategy in AllInfo
Execute the parameter strategy component to collect information and construct the parameter strategy model.
end foreach
Collect the algorithm information to generate ModelInfo.

3.3. Combination Agent

The combination agent includes the data structure for describing the combination scheme and the algorithm for executing the combination mechanism. Each search agent represents a specific behavior. The combination scheme defines the combination mode and combination strategy for multiple behaviors (multiple search agents). In the framework, we predefined three basic combination modes as {individual level combination, iteration level combination, population level combination}. Each combination mode needs to define its population grouping strategy and execution strategy, which determines the individual group and execution manner for all search agents in a combination mode. The predefined strategies are shown in the Table 2 and are supported by the combination component library which includes the three combination mode components, grouping components, competitive strategy components, group communication components. The flow chart of three combination mode components in one iteration is shown in Figure 3. It can be seen that the other strategy components are called by the combination mode components. The combination agent executes corresponding combination mode components according to the combination scheme described.

Table 2. Combination Scheme Settings.

Combination Mode	Individual Group			Iteration (Execution Manner)
	Group Size	**Grouping Method**	**Group Communication**	
individual level	fixed full-competition semi-competition	Split the population	Regrouping (Every few generations)	every generation
iteration level	fixed	share the population	none	fixed full-competition semi-competition
population level	fixed	Split the population	migration of individuals (Every few generations)	every generation

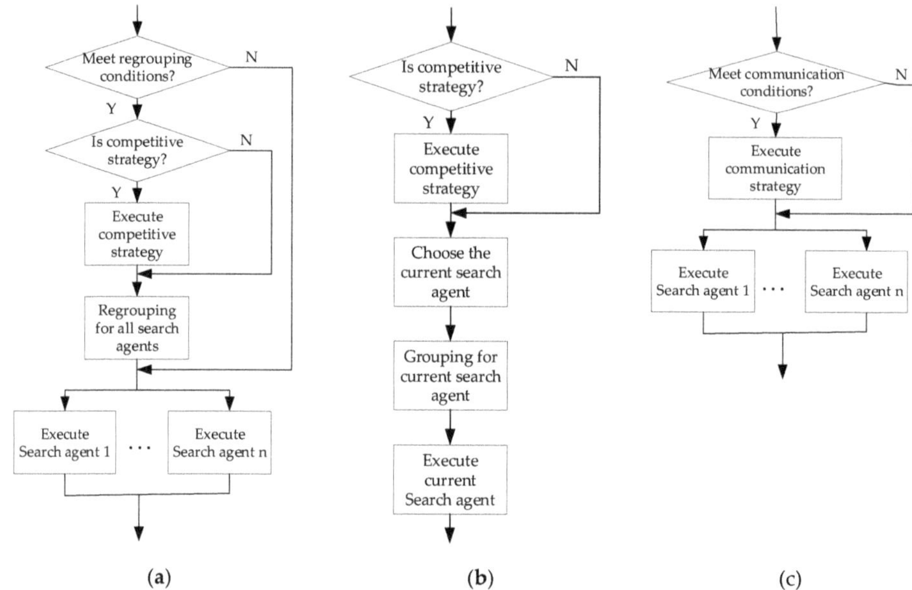

Figure 3. The flow charts of three combination mode components in one iteration: (**a**) Individual level combination; (**b**) Iteration level combination; (**c**) Population level combination.

3.4. Algorithm Reuse and Algorithm Customization under the Framework

Most population-based optimization algorithms can be integrated into the framework. In this paper, an algorithm that adopts the same behavior in three dimensions is called a single-behavior algorithm. An algorithm that adopts different behaviors in any dimension is called a multi-behavior combination algorithm. Most existing algorithms can be classified as single-behavior algorithms or multi-behavior combination algorithms.

Both single-behavior and multi-behavior algorithms can be decomposed to extract the six types of technical unit operators if they have. For the multi-behavior combination algorithms, the combination modes adopted by the algorithms generally belong to the three basic combination modes. However, the combination strategy may be various. We predefined some combination strategies, which may also be extracted as components from the current multi-behavior algorithms and reused by the combination agent.

By analyzing the behaviors and the combination method of the existing algorithms, the operator components and the combination strategy components are extracted, and the existing algorithms can be recombined in the framework. More importantly, the new algorithm can be customized under the framework.

Customizing an algorithm first requires the determination of the optimization idea of multi-behavior combination, selection of the corresponding operators, and the combination of related operators into multiple search agents. Then, the combination scheme is designed, including combination mode, combination strategy, grouping strategy, and some fixed parameters. Then, the algorithm can be executed by the framework. The algorithm of the framework is as Algorithm 2.

Algorithm 2. Algorithm of the framework.

Initialize the algorithm settings (population size, max evaluation number, boundary constraint).
Construct the data structure of combination scheme.
Construct the data structure of each search agent and generate search agent list.
Initialize the population and evaluation by objective function.
while (not satisfied the termination condition)
execute combination agent
acquire the combination mode according to combination scheme data structure.
execute the corresponding combination mode component (Figure 3).
end combination agent
record the best result
end while

To demonstrate the working principle of this framework, we give the examples on introducing DE and its variants jDE [51], SHADE [52], and teaching-learning-based optimization algorithm (TLBO) [29] into the framework:

3.4.1. DE, jDE and SHADE

At first, the individual evolution operator, the parent individual selection operator, the population selection operator, and the parameter control operator can be extracted from the three algorithms as operator components. The DE and jDE has the same individual evolutionary operator: DE/rand/1 mutation strategy and the binomial crossover strategy. The individual evolution operator of the SHADE algorithm adopts the DE/current-to-pBest/1 mutation strategy and binomial cross strategy. The DE and jDE also have the same parent individual selection operator: random selection. The parent individual of SHADE has two types, one is the best parent individual, the other is the ordinary parent individual. The best parent individual selection operator adopts a random selection among the top p% individuals (sorted by fitness), and the ordinary parent individual selection operator adopts a random selection among population and failure individual archive. For the population selection operator, the three algorithms all use binary greedy selection. For the parameter control operator, DE uses the fixed value strategy, jDE and SHADE use adaptive parameter control strategies [51,52], which can be reused and are suitable to be extracted as operator components.

The three algorithms are all single-behavior algorithms. When using one of them, the related operator components can be selected and assembled into a search agent. For example, the SHADE algorithm has four operator components. The identifier (ID) of each component is specified in the AllInfo structure of the search agent. Algorithm 1 automatically calls the corresponding component for an iterative execution. The predefined execution order of operator components is: (1) execute the parameter components for the first call and generate the parameter value set, (2) execute the best parent individual selection component to generate the optimal parent individual set, (3) execute the ordinary parent individual selection component to generate other parent individual set, (4) execute the individual evolution component to generate the offspring, (5) execute the population selection component to generate the population of next generation, (6) execute the second call of the parameter control component to generate the parameter strategy model (needed by adaptive parameter control strategy), and (7) execute the algorithm evaluation component and generate algorithm evaluation model. If any step is not required, the component ID is set to 0, and the step is skipped. The search agent is called by the framework iteratively; thus, the SHADE algorithm can be supported by the framework.

3.4.2. Teaching-Learning-Based Optimization (TLBO)

TLBO algorithm is a multi-behavior combination algorithm. It contains two behaviors in the iteration level, one is teaching behavior and the other is learning behavior. It can be introduced into the framework by four steps: (1) extract operators for each behavior; (2) select the operator components and construct the search agent for each behavior with description in AllInfo of each search agent; and (3) define the combination method of the two behaviors and describe it in the data structure of the combination agent. For the TBLO algorithm, the combination mode is an iteration level combination. For this combination mode, it does not need to divide the population for the search agents. It only needs to assign individuals according to the GroupInfo of the search agent. The group of the two behaviors in the TLBO algorithm is the entire population which can be described by the GroupInfo. The multi-behavior combination strategy is a collaborative strategy, the search agents are executed in a predefined fixed order: one by one. (4) The framework executes the combination agent iteratively, and the combination agent calls the component for iteration level combination. The algorithm flow of the component for iteration level combination is showed in Figure 3b.

Most of the algorithms adopt one combination mode for multiple behavior and can be customized by simple settings under the framework. However, a few studies explore the more complex combination, which can be seen as a further combination of these three basic combination modes. The components of the framework can also be reused to construct complex multi-level combination algorithm; however, the combination agent may need some modification.

4. Economic Load Dispatch Model and Algorithm Customizing

4.1. The Model of Economic Load Dispatch Problem

Economic load dispatch is a typical optimization problem in power systems. The goal is to minimize fuel costs by rationally arranging the active power output of each generating unit in a power plant or power system on the premise of meeting system load demand and operational constraints.

4.1.1. Problem Definition

The objective function of the optimization problem is as follows:

$$\min F = \sum_{i=1}^{Ng} F_i(P_i) \tag{11}$$

where F is total fuel cost, Ng is the total number of online generating units of the system, P_i is the power output (in MW) of ith generator, $F_i(P_i)$ is the cost function of the ith generator as Equation (12), if the valve point effect is considerd, $F_i(P_i)$ can be described as Equation (13).

$$F_i(P_i) = a_i P_i^2 + b_i P_i + c_i \tag{12}$$

$$F_i(P_i) = a_i P_i^2 + b_i P_i + c_i + |e_i \sin(f_i(P_i^{\min} - P_i))| \tag{13}$$

where a_i, b_i, c_i, e_i, f_i is the cost coefficients of the ith generator.

In addition to the objective function, the problem needs to meet some constraints:

4.1.2. Generator Output Constraint

$$P_i^{\min} \leq P_i \leq P_i^{\max} \tag{14}$$

where P_i^{min} and P_i^{max} are lower and upper bounds for power output of the ith generator.

4.1.3. Power Balance Constraints

$$\sum_{i=1}^{Ng} P_i = P_D + P_L \tag{15}$$

where P_D is the total system loads and P_L is the total power loss in all transmission lines. P_L can be obtained by B-coefficient method as Equation (16).

$$P_L = \sum_{i=1}^{Ng} \sum_{j=1}^{Ng} P_i B_{ij} P_j + \sum_{i=1}^{Ng} B_{0i} P_i + B_{00} \tag{16}$$

where B_{ij}, B_{0i}, B_{00} are coefficients of the power loss matrix.

4.1.4. Ramp Rate Limit Constraints

$$P_i - P_i^{t-1} \leq UR_i \,,\, P_i^{t-1} - P_i \leq DR_i \tag{17}$$

$$\max(P_i^{\min}, P_i^{t-1} - DR_i) \leq P_i \leq \min(P_i^{\max}, P_t^{t-1} + UR_i) \tag{18}$$

where P_i^{t-1} is the previous power output, UR_i and DR_i are the up-ramp and down-ramp limit of the ith generator.

4.1.5. Prohibited Operating Zones Constraints

The generating units may have certain zones where operation is restricted. Consequently, the feasible operating zones are discontinuities as follows:

$$\begin{cases} P_i^{\min} \leq P_i \leq P_{i,1}^{pzl} \\ P_{i,j-1}^{pzu} \leq P_i \leq P_{i,j}^{pzl} \\ P_{i,n_i}^{pzu} \leq P_i \leq P_i^{\max} \end{cases} \tag{19}$$

where n_i is the total number of prohibited operating zones, $P_{i,j}^{pzl}$ and $P_{i,j}^{pzu}$ are the lower and upper limits of the jth prohibited zone for the ith generating unit respectively.

4.2. Customizing Multi-Behavior Combination Algorithm

The algorithm we customized for ELD problem is a multi-behavior combination algorithm. The behaviors come from the operator component library. At present, we only give small number of operators including some individual evolution operators (mainly DE algorithm operator and its invariants including mutation step and crossover step), some parent individual selection operators, some parameter control operators, and a population selection operator (just DE adopted) and its two constraint-based version with penalty functions method [45] and ε-constraint handling method [45].

The ε-constraint handling method gives the constraints a relaxation scope, and if the constraint violation value of individuals are less than ε, the individuals are considered feasible. Then in the population selection operator (an operator of one-to-one choice between parent individual and offspring individual), two feasible individuals are compared and selected according to objective value, and in the other case, individuals are selected according to constraint violation degree. The ε is varied as follows [45]:

$$\varepsilon(0) = v(x_\theta)$$
$$\varepsilon(k) = \begin{cases} \varepsilon(0)(1 - \frac{k}{T_c})^{cp} & 0 < k < T_c \\ 0 & k \geq T_c \end{cases} \tag{20}$$

where $\varepsilon(0)$ is the initial ε value, X_θ is the top θth individual and $\theta = (0.05 \times NP)$, v is constraint violation function. cp and T_c are parameters, and T_c is a specific generation in iteration.

For the multi-behavior algorithm, we give three behaviors which come from DE algorithm and its variants jDE [51], SHADE [52], etc. The three behaviors are constructed with the operators in the operator component library. The design for behaviors is showed in Table 3. Behavior1 and behavior2 have the similar algorithms with difference in parameter control operator. Behavior2 and behavior3 have the similar parameter control operator with difference in algorithm operation. Behavior 1 is used for randomness and diversity. Behavior2 is a fitness improvement instructed search because its parameter control strategy tends to the parameter values which bring larger fitness improvement. Behavior 3 is an optimal individual set instructed search, and is used for gathering in the optimal regions. The combination method of the three behaviors is designed as Table 4. Based on the combination method in Table 4, the three behaviors are executed in a fixed order from behavior 1 to 3 for the iteration level, so the search process is a repeated local small loop from diversity exploration to exploitation.

Table 3. The multi-behavior design with operators.

Behavior	Operator Components	Design
1	Individual evolution operator:	DE/rand/1 mutation strategy, Binomial crossover strategy (comes from DE algorithm).
	Parent individual selection operator:	Random selection strategy.
	Population selection operator:	Binary selection (come from DE algorithm) with ε-constraints handling method (cp = 5, T_c = 0.7).
	parameter control operator:	F and CR: jDE algorithm strategy with parameter domain in [0,1].
	other operators:	none
2	Individual evolution operator:	The same as behavior 1
	Parent individual selection operator:	The same as behavior 1
	Population selection operator:	The same as behavior 1
	parameter control operator:	F and CR: SHADE algorithm strategy with parameter domain in [0,1].
	other operators:	none
3	Individual evolution operator:	DE/current-to-pBest/1 mutation strategy with archive, Binomial crossover strategy (come from SHADE algorithm)
	Parent individual selection operator:	Best individual: random selection from %p top individual set with fitness sorting. Other individuals: random selection from population and archive
	Population selection operator:	The same as behavior 1.
	parameter control operator:	p: linear decreasing strategy based on generation from 0.5 to 0F and CR: SHADE parameter control strategy with domain in [0,1]
	other operators:	none

Table 4. Behavior Combination design.

Combination Settings	Design
combination mode	Iteration level combination.
group size	Fixed (NP: population size).
grouping method	Share (The entire population).
group communication	None.
iteration	Collaborative strategy: fixed (executed one by one)

Just like the example in Section 3.4, the three behaviors are organized as search agents. It only needs to describe the operator components designed in Table 3 with AlInfo of search agent, then the search agent can execute these components in order with one iteration. However, the execution manner of multiple search agents is determined by the design in Table 4. The combination agent interprets the combination design and executes the search agents in a designed manner. The combination method is similar to the TLBO algorithm, but we have three behaviors. The behavior combination happens at the iteration level and the combination strategy is collaborative strategy which adopts the fixed execution manner: the three behaviors are executed one by one.

The algorithm of this paper is a general constrained optimization algorithm. The optimization objective is Equation (11). $X_i = \{P_{i,1}, P_{i,2}, \ldots P_{i,Ng}\}$ is an individual vector, and is also a solution of the ELD problem, which means the power outputs of all the online generating units. In the current population-based intelligent optimization algorithm for solving ELD problems, the constraint processing can be divided into two categories. One is to process the individuals that violate the constraints and use the problem-specific knowledge to modify the individuals and satisfy the constraints as much as possible. However, because the constraints are more difficult to fully satisfy, the algorithm still needs to provide constraint processing technology. Another type of algorithm does not perform special processing, but only uses general constraint processing techniques. In this paper, the constraint of Equation (14) is processed as boundary. When the transmission loss is not considered, the equality constraint is simply processed according to Equation (15) by recomputing the power output of the last generating unit, and the other inequality constraints are not processed. When considering the transmission loss, the constraint relationship is complicated, and we do not make special treatment, the constraints are processed only through the ε-constraint handling method. The tolerance of the equality constraint is set as 10–4, which can be updated according to the problem. In order to facilitate the comparison, the algorithm proposed by this paper is named as multi-behavior combination differential evolution algorithm (MBC-DE).

5. Experiment and Discussion

In this section, the proposed algorithm and framework is applied in ELD problems. The ELD problem is a non-convex, nonlinear, discontinuous constrained optimization problem. The equality constraint involved is power balance constraint, inequality constraints are ramp rate limit constraints, and prohibited operating zones constraints. The valve point effect may or may not be considered in the objective function, and transmission losses may or may not be considered in the equality constraint. The transmission losses are calculated by B-coefficient method. We use four test system which varies with different difficulty levels. For each test case, 20 independent runs are performed and the results of the best, worst and mean fuel costs are recorded and compared with other algorithms. Each behavior of the multi-behavior algorithm is tested independently in order to analyze the effects of the multi-behavior combination. The three single-behavior algorithms are named as Behavior1~Behavior3. The population size is set as 50 for all cases, and the max iteration is 2000 generation.

(1) Test Case 1: Test case 1 is a small system. The system is comprising six generators, and the ramp rate limit, prohibited operating zone and transmission losses are all considered without valve point loading effect. The system data is from [2,3]. The load demand is 1263 MW.

Table 5 gives the best output for this case, it can be seen that the equality constraint accuracy is less than 10-4, and all the inequality constraints are satisfied (see the system data in [2]). Table 6 give the comparison between single behavior and multi-behavior algorithm. Among the three behaviors, Behavior3 performs best, and MBC-DE inherits its characteristics and obtains a smaller standard deviation. Table 7 give the comparison between MBC-DE and other algorithms. From the data in the table, the algorithm MBC-DE goes beyond some algorithms, but there are also some algorithms that are better than MBC-DE. In test case 1, MBC-DE falls into a local extremum. The three sub-algorithms of MBC-DE are all differential evolution and the search trajectory is similar. No behavior can help MBC-DE to jump out of this local extremum. This also give us the confidence on algorithm combination research by introducing new algorithm or algorithm components into the framework, to enrich the algorithm behaviors.

Table 5. Best outputs for test case 1 with PD = 1263 MW (6-units system).

Unit	1	2	3	4	5	6
Output power (MW)	446.716	173.145	262.797	143.490	163.918	85.3562
Total Power output (MW)				1275.42190006		
Transmission loss (MW)				12.422		
Fuel Cost ($/h)				15,444.185		

Table 6. Algorithm comparison between multi-behavior and single-behavior for test case 1.

Algorithm	Cost ($/h)			std
	best	mean	worst	
Behavior1	15,444.185	15,451.853	15,489.857	1.4515660×10
Behavior2	15,444.185	15,445.892	15,461.454	4.5020314
Behavior3	15,444.185	15,444.185	15,444.185	1.3139774×10^{-6}
MBC-DE	15,444.185	15,444.185	15,444.185	8.8040610×10^{-7}

Table 7. Algorithm comparison between MBC_DE and other algorithms for test case 1.

Algorithm	Cost ($/h)			std
	best	mean	worst	
GA [6]	15,459.00	15,469.00	15,469.00	41.58
PSO [2]	15,450.00	15,454.00	15,492.00	14.86
NPSO-LRS [8]	15,450.00	15,450.50	15,452.00	NA
AIS [12]	15,448.00	15,459.70	15,472.00	NA
DE [11]	15,449.77	15,449.78	15,449.87	NA
DSPSO-TSA [9]	15,441.57,	15,443.84	15,446.22	1.07
BBO [16]	15,443.096	15,443.096	15,443.096	NA
SOH-PSO [53]	15,446.02	15,497.35	15,609.64	NA
θ-PSO [54]	15,443.1830	15,443.2117	15,443.3548	0.0291
MABC [13]	15,449.8995	15,449.8995	15,449.8995	NA
MP-CJAYA [55]	15,446.17	15,451.68	15,449.23	NA
CTLBO [16]	15,441.697	15,441.9753	15,441.7638	1.94×10^{-2}
MBC-DE	15,444.185	15,444.185	15,444.185	8.8040610×10^{-7}

NA: Not available.

(2) Test Case 2: Test case 2 is a 15 units system. The ramp rate limit and transmission losses are considered. There are 4 generators having prohibited operating zone. The valve point loading effect is neglected. The system data is also from [2,3]. The load demand of the system is 2630MW.

Table 8 gives the best output of MBC-DE for this case, it can be seen that the equality constraint accuracy is also less than 10-4, and all the inequality constraints are satisfied (see the system data in [2]). Table 9 give the comparison between single behavior and multi-behavior algorithm. Among the three behaviors, Behavior3 still performs best, and MBC-DE is better than Behavior3 in mean and worst value. We believe this is due to the combination of multiple behavior, although Behavior2 and 1 is worse than Behavior3. Table 10 shows the comparison between MBC-DE and other algorithms. The data in the table shows that the algorithm we proposed achieved better results than others in case 2.

Table 8. Best outputs for test case 2 with PD = 2630 MW (15-units system).

Unit	Output Power (MW)	Unit	Output Power (MW)	Unit	Output Power (MW)
1	455.000	6	460.000	11	79.9997
2	380.000	7	430.000	12	79.9999
3	130.000	8	690.829	13	25.0001
4	130.000	9	605.011	14	15.0005
5	169.999	10	160.000	15	15.0003
Total Power output (MW)			2659.58290292		
Transmission loss (MW)			29.58300000		
Fuel cost ($/h)			32692.399		

Table 9. Algorithms comparison between multi-behavior and single-behavior for test case 2.

Algorithm	Cost ($/h)			std
	best	mean	worst	
behavior1	32,696.300	32,721.648	32,766.651	1.9540042×10
behavior2	32,692.645	32,697.776	32,741.991	1.0539320×10
behavior3	32,692.399	32,694.789	32,740.250	1.0700474×10
MBC-DE	32,692.399	32,692.509	32,692.815	1.2095983×10^{-1}

Table 10. Algorithms comparison between MBC_DE and other algorithms for test case 2.

Algorithm	Cost ($/h)			std
	best	mean	worst	
GA [8]	33,113.00	33,228.00	33,337.00	49.31
PSO [2]	32,858.00	33,105.00	33,331.00	26.59
SOH-PSO [53]	32,751.39	32,878	32,945	NA
DSPSO-TSA [9]	32,715.06	32,724.63	32,730.39	8.4
IPSO [56]	32,709	32,784.5	NA	NA
SQPSO [57]	32,704.5710	32,707.0765	32,711.6179	1.077
θ-PSO [54]	32,706.6856	32,711.4955	32,744.0306	9.8874
AIS [9]	32,854.00	32,892.00	32,873.25	NA
MDE [58]	32,704.9	32,708.1	32,711.5	NA
MP-CJAYA [55]	32,706.5158	32,708.8736	32,706.7150	NA
MBC-DE	32,692.399	32,692.509	32,692.815	1.2095983×10^{-1}

NA: Not available.

(3) Test Case 3: test case 3 is a 40 units system with valve point effect being considered. The ramp rate limit and transmission losses are neglected. The load demand is 10,500 MW. The system data is from [4].

Table 11 gives the best output of MBC-DE for this case; it can be seen that the equality constraint is satisfied. Table 12 give the comparison between single behavior and multi-behavior algorithm. Among the three behaviors, Behavior1 and Behavior2 all performs well, and MBC-DE is better than them. We think the Behavior3 act as the supporting role, although it works poorly on its own. Table 13 give the comparison between MBC-DE and other algorithms. The data in the table show that the algorithm we proposed achieved better results. For the best value, MBC-DE is better than others, but for the mean value, MDE and DE/BBO are better.

Table 11. Best outputs for test case 3 with PD = 10,500 MW (40-units system).

Unit	Output Power (MW)	Unit	Output Power (MW)	Unit	Output Power (MW)
1	110.800	15	394.279	29	10.000
2	110.800	16	394.279	30	87.800
3	97.400	17	489.279	31	190
4	179.733	18	489.279	32	190
5	87.800	19	511.279	33	190
6	140	20	511.279	34	164.800
7	259.600	21	523.279	35	194.398
8	284.600	22	523.279	36	200
9	284.600	23	523.279	37	110
10	130.000	24	523.279	38	110
11	94.000	25	523.279	39	110
12	94.000	26	523.279	40	511.279
13	214.760	27	10.000		
14	394.279	28	10.000		
Total power output (MW)	10,500				
Fuel cost ($/h)	121,412.5355				

Table 12. Algorithm comparison between multi-behavior and single-behavior for test case 3.

Algorithm	Cost ($/h)			std
	best	mean	worst	
Behavior1	121,420.90	121,453.58	121,506.89	3.0363190×10
Behavior2	121,412.54	121,456.32	121,517.82	3.0129673×10
Behavior3	121,424.15	121,498.80	121,695.91	6.6662083×10
MBC-DE	121,412.54	121,450.32	121,481.33	2.1658087×10

Table 13. Algorithm comparison between MBC_DE and other algorithms for test case 3.

Algorithm	Cost ($/h)			std
	best	mean	worst	
SOH-PSO [53]	121,501.14	121,853.07	122,446.30	NA
SQPSO [57]	121,412.5702	121,455.7003	121,709.5582	NA
BBO [16]	121,426.953	121,508.0325	121,688.6634	NA
DE/BBO [23]	121,420.8948	121,420.8952	121,420.8963	NA
MDE [58]	121,414.79	121,418.44	NA	NA
MP-CJAYA [55]	121,480.10	121,861.08	NA	NA
MBC-DE	121,412.54	121,450.32	121,481.33	2.1658087×10

NA: Not available.

(4) Test Case 4: test case 4 is a large-scale power system with 140 generators. The valve point effect is considered. The ramp rate limit and transmission losses are neglected. The load demand is 49,342 MW. The system data is from [59].

Table 14 gives the best output of MBC-DE for this case; and the equality constraint is satisfied. Table 15 give the comparison between single behavior and multi-behavior algorithm. Among the three behaviors, Behavior3 has the best result and the worst result, this means that the algorithm has large fluctuations and instability. MBC-DE has improved in both the optimal value and stability which represents the effect of algorithm combination. Table 16 gives the comparison between MBC-DE and other algorithms. The data in the table show that the algorithm we proposed achieved better results.

Table 14. Best outputs for test case 4 with PD = 49342 MW (140-units system).

Unit	Power Output (MW)	Unit	Power Output (MW)	Unit	Power Output (MW)	Unit	Power Output (MW)
1	118.667	36	500.000	71	137.693	106	953.998
2	188.994	37	241.000	72	325.495	107	951.999
3	189.997	38	241.000	73	195.039	108	1006.00
4	189.997	39	773.999	74	175.038	109	1013.00
5	168.540	40	768.999	75	175.180	110	1021.00
6	189.790	41	301.105	76	175.627	111	1015.00
7	489.999	42	300.797	77	175.552	112	94.0000
8	489.997	43	245.412	78	330.048	113	94.0006
9	496.000	44	242.746	79	531.000	114	94.0003
10	495.996	45	246.348	80	530.990	115	244.009
11	496.000	46	249.979	81	400.966	116	244.003
12	495.999	47	244.958	82	560.012	117	244.274
13	506.000	48	249.639	83	115.000	118	95.0021
14	509.000	49	249.960	84	115.000	119	95.0003
15	506.000	50	249.282	85	115.000	120	116.000
16	505.000	51	165.348	86	207.007	121	175.000
17	506.000	52	165.000	87	207.000	122	2.00007
18	505.997	53	165.002	88	175.634	123	4.00695
19	505.000	54	165.281	89	175.017	124	15.0000
20	504.998	55	180.002	90	175.255	125	9.00339

Table 14. *Cont.*

Unit	Power Output (MW)	Unit	Power Output (MW)	Unit	Power Output (MW)	Unit	Power Output (MW)
21	505.000	56	180.000	91	175.048	126	12.0054
22	505.000	57	103.628	92	580.000	127	10.0026
23	504.999	58	198.001	93	645.000	128	112.030
24	505.000	59	311.997	94	983.997	129	4.00002
25	537.000	60	282.396	95	978.000	130	5.00497
26	536.999	61	163.000	96	682.000	131	5.00042
27	548.997	62	950.044	97	719.998	132	50.0405
28	549.000	63	160.040	98	717.999	133	5.00001
29	501.000	64	170.063	99	720.000	134	42.0004
30	501.000	65	489.868	100	964.000	135	42.0006
31	506.000	66	198.341	101	958.000	136	41.0030
32	505.999	67	474.705	102	1007.00	137	17.0039
33	505.994	68	489.234	103	1006.00	138	7.06653
34	506.000	69	130.002	104	1013.00	139	7.00002
35	500.000	70	234.746	105	1020.00	140	26.0098
Total power output (MW)					49342		
Fuel Cost ($/h)					1559810.6		

Table 15. Algorithms comparison between multi-behavior and single-behavior for test case 4.

Algorithm	Cost ($/h)			std
	best	mean	worst	
behavior1	1,560,264.0	1,561,026.0	1,562,979.2	6.1392827×10^2
behavior2	1,560,336.3	1,561,104.0	1,562,181.8	5.6010767×10^2
behavior3	1,560,080.0	1,562,559.7	1,565,170.9	1.4819202×10^3
MBC-DE	1,559,810.6	1,560,195.7	1,561,194.1	3.3572187×10^2

Table 16. Algorithms comparison between MBC_DE and other algorithms for test case 4.

Algorithm	Cost ($/h)			std
	best	mean	worst	
SDE [60]	1,560,236.85	NA	NA	NA
GWO [18]	1,559,953.18	1,560,132.93	1,560,228.40	1.024
MBC-DE	1,559,810.6	1,560,195.7	1,561,194.1	3.3572187×10^2

NA: Not available.

Implementing the algorithm in the form of a framework maybe lose some efficiency. The reason is that in order to achieve generality and easy assemblage, various technical units need to be designed as components which can be implemented as functions in Matlab, and thus the overhead of function call is generated. This gap can be reduced through programming technology. In fact, the running time of the algorithm is less than 30s with Matlab R2018a environment on a 1.99 GHz, 16 GB RAM Notebook computer. The time is feasible for some online systems. If the system has strict real-time requirements, the framework can also be used as a debugging tool to determine the algorithm. Then, code refactoring can be performed. The algorithm we proposed is not restricted by the constraint form and has a certain versatility.

6. Conclusions

Through the experiments in this paper, we can see that the performance of the population-based intelligent optimization algorithm can be improved by appropriate multi-behavior combination. A behavior has specific search trajectory, so it is easy to trap into the same local extremum. Multiple behaviors have different search trajectories, and can lead to more diversity, so the combination of multiple behaviors is more likely to be suitable for complex optimization problems. The ELD

Problem has a complex solution space caused by multiple constraints, the algorithms are prone to trap into the local extremum. From the related research, it can be seen that different algorithms are gradually improving the optimization results, which also shows that the previous algorithm has fallen into the local extremum. Therefore, combining the search characteristics of different algorithms makes it possible to obtain good results. The algorithm in this paper is a combination of three DE variants. From the test results, the proposed algorithm with multiple behaviors is better than the single behavior used by the algorithm. However, the sub-algorithms all belong to the DE algorithm, thus, their search trajectories are similar, the algorithm results are improved only a little, and some studies have yielded better values. This makes us more convinced of the significance of implementing a multi-behavior combination framework. With more candidate behaviors, the algorithm may be improved further. Therefore, our follow-up work is to introduce more algorithms into the framework to enrich the candidate behaviors. Another research opportunity involves improving the multi-behavior combination strategy.

Multi-behavior combination is an important research idea of current intelligent optimization algorithms. This article hopes to introduce this idea into practical applications. Different from the traditional research, the focus of this paper is to realize the idea as a general framework. Various algorithm components can be extracted from the all kinds algorithm and introduced into the framework. By performing behavior design and the combination method design, the new algorithm can be assembled. It brings convenience to the customization of new algorithms and to algorithm combination research more generally. Although the algorithm is tailored to solve the ELD problem, it is obvious that the framework can apply to other practical problems with simple customization.

Author Contributions: J.Z. did the programming and writing; Z.D. instructed the idea and writting.

Funding: This research received no external funding.

Conflicts of Interest: The authors declare no conflict of interest.

References

1. Chiang, C.-L. Genetic-based algorithm for power economic load dispatch. *IET Gener. Transm. Distrib.* **2007**, *1*, 261–269. [CrossRef]
2. Gaing, Z.-L. Particle swarm optimization to solving the economic dispatch considering the generator constraints. *IEEE Trans. Power Syst.* **2003**, *18*, 1187–1195. [CrossRef]
3. Gaing, Z.-L. Closure to Discussion of 'Particle swarm optimization to solving the economic dispatch considering the generator constraints. *IEEE Trans. Power Syst.* **2004**, *19*, 2122–2123. [CrossRef]
4. Sinha, N.; Chakrabarti, R.; Chattopadhyay, P.K. Chattopadhyay. Evolutionary Programming Techniques for Economic Load Dispatch. *IEEE Trans. Evol. Comput.* **2003**, *7*, 83–94. [CrossRef]
5. Chiang, C.L. Improved genetic algorithm for power economic dispatch of units with valve-point effects and multiple fuels. *IEEE Trans. Power Syst.* **2005**, *20*, 1690–1699. [CrossRef]
6. He, D.K.; Wang, F.L.; Mao, Z.Z. Hybrid genetic algorithm for economic dispatch with valve-point effect. *Electr. Power Syst. Res.* **2008**, *78*, 626–633. [CrossRef]
7. Mahdi, F.P.; Vasant, P. Quantum particle swarm optimization for economic dispatch problem using cubic function considering power loss constraint. *IEEE Trans. Power Syst.* **2002**, *17*, 108–112.
8. Selvakumar, A.I.; Thanushkodi, K. A new particle swarm optimization solution to nonconvex economic dispatch problems. *IEEE Trans. Power Syst.* **2007**, *22*, 42–51. [CrossRef]
9. Khamsawang, S.; Jiriwibhakorn, S. DSPSO–TSA for economic dispatch problem with nonsmooth and noncontinuous cost functions. *Energy Convers. Manag.* **2010**, *51*, 365–375. [CrossRef]
10. Pothiya, S.; Ngamroo, I.; Kongprawechnon, W. Ant colony optimization for economic dispatch problem with non-smooth cost functions. *Int. J. Electr. Power Energy Syst.* **2010**, *32*, 478–487. [CrossRef]
11. Noman, N.; Iba, H. Differential evolution for economic load dispatch problems. *Electr. Power Syst. Res.* **2008**, *78*, 1322–1331. [CrossRef]
12. Panigrahi, B.K.; Yadav, S.R.; Agrawal, S.; Tiwari, M.K. A clonal algorithm to solve economic load dispatch. *Electr. Power Syst. Res.* **2007**, *77*, 1381–1389. [CrossRef]

13. Secui, D.C. A new modified artificial bee colony algorithm for the economic dispatch problem. *Energy Convers. Manag.* **2015**, *89*, 43–62. [CrossRef]

14. Dos Santos Coelho, L.; Mariani, V.C. An improved harmony search algorithm for power economic load dispatch. *Energy Convers. Manag.* **2009**, *50*, 2522–2526. [CrossRef]

15. Al-Betar, M.A.; Awadallah, M.A.; Khader, A.T.; Bolaji, A.L. Tournament-based harmony search algorithm for non-convex economic load dispatch problem. *Appl. Soft Comput.* **2016**, *47*, 449–459. [CrossRef]

16. Bhattacharya, A.; Chattopadhyay, P.K. Biogeography-based optimization for different economic load dispatch problems. *IEEE Trans. Power Syst.* **2010**, *25*, 1064–1077. [CrossRef]

17. He, X.Z.; Rao, Y.Q.; Huang, J.D. A novel algorithm for economic load dispatch of power systems. *Neurocomputing* **2016**, *171*, 1454–1461. [CrossRef]

18. Pradhan, M.; Roy, P.K.; Pal, T. Grey wolf optimization applied to economic load dispatch problems. *Int. J. Electr. Power Energy Syst.* **2016**, *83*, 325–334. [CrossRef]

19. Lu, H.; Sriyanyong, P.; Song, Y.H.; Dillon, T. Experimental study of a new hybrid PSO with mutation for economic dispatch with non smooth cost function. *Int. J. Electr. Power Energy Syst.* **2010**, *32*, 921–935. [CrossRef]

20. Niknam, T.; Azizipanah-Abarghooee, R.; Aghaei, J. A new modified teaching learning algorithm for reserve constrained dynamic economic dispatch. *IEEE Trans. Power Syst.* **2013**, *28*, 749–763. [CrossRef]

21. Niknam, T.; Mojarrad, H.D.; Meymand, H.Z. A novel hybrid particle swarm optimization for economic dispatch with valve-point loading effects. *Energy Convers. Manag.* **2011**, *52*, 1800–1809. [CrossRef]

22. Wang, L.; Li, L.P. An effective differential harmony search algorithm for the solving non-convex economic load dispatch problems. *Int. J. Electr. Power Energy Syst.* **2013**, *44*, 832–843. [CrossRef]

23. Bhattacharya, A.; Chattopadhyay, P.K. Hybrid differential evolution with biogeography-based optimization for solution of economic load dispatch. *IEEE Trans. Power Syst.* **2010**, *25*, 1955–1964. [CrossRef]

24. Mirjalili, S.; Lewis, A. The Whale Optimization Algorithm. *Adv. Eng. Softw.* **2016**, *95*, 51–67. [CrossRef]

25. Meng, X.B.; Gao, X.Z.; Lu, L.H.; Liu, Y. A new bio-inspired optimisation algorithm: Bird Swarm Algorithm. *J. Exp. Theor. Artif. Intell.* **2016**, *28*, 673–687. [CrossRef]

26. Meng, X.; Liu, Y.; Gao, X.Z.; Zhang, H.Z. A New Bio-inspired Algorithm: Chicken Swarm Optimization. In *International Conference in Swarm Intelligence (ICSI 2014)*; Springer: Cham, Switzerland, 2014; pp. 86–94.

27. Chu, S.C.; Tsai, P.W.; Pan, J.S. Cat Swarm Optimization. In Proceedings of the 9th Pacific Rim International Conference on Artificial Intelligence (PRICAI 2006): Trends in Artificial Intelligence, Guilin, China, 7–11 August 2006.

28. Mirjalili, S.; Mirjalili, S.M.; Hatamlou, A. Multi-Verse Optimizer: A nature-inspired algorithm for global optimization. *Neural Comput. Appl.* **2015**, *27*, 495–513. [CrossRef]

29. Rao, R.V.; Savsani, V.J.; Vakharia, D.P. Teaching-learning-based optimization: A novel method for constrained mechanical design optimization problems. *Comput. Aided Des.* **2011**, *43*, 303–315. [CrossRef]

30. Yang, X. Flower pollination algorithm for global optimization. In *International Conference on Unconventional Computation and Natural Computation (UCNC2012)*; Springer: Berlin/Heidelberg, Germany, 2012; pp. 240–249.

31. Guohua, W.; Rammohan, M.; Nagaratnam, S.P. Ensemble strategies for population-based optimization algorithms—A survey. *Swarm Evol. Comput.* **2018**, *9*, 1–17.

32. Wang, Y.; Cai, Z.X.; Zhang, Q.F. Differential Evolution with Composite Trial Vector Generation Strategies and Control Parameters. *IEEE Trans. Evol. Comput.* **2011**, *15*, 55–66. [CrossRef]

33. Qin, A.K.; Huang, V.L.; Suganthan, P.N. Differential Evolution Algorithm with Strategy Adaptation for Global Numerical Optimization. *IEEE Trans. Evol. Comput.* **2009**, *13*, 398–417. [CrossRef]

34. Li, X.T.; Ma, S.J.; Hu, J.H. Multi-search differential evolution algorithm. *Appl. Intell.* **2017**, *47*, 231–256. [CrossRef]

35. Wang, S.C. Differential evolution optimization with time-frame strategy adaptation. *Soft Comput.* **2017**, *21*, 2991–3012. [CrossRef]

36. Yi, W.C.; Gao, L.; Li, X.Y.; Zhou, Y.Z. A new differential evolution algorithm with a hybrid mutation operator and self-adapting control parameters for global optimization problems. *Appl. Intell.* **2015**, *42*, 642–660. [CrossRef]

37. Elsayed, S.M.; Sarker, R.A.; Essam, D.L. Training and testing a self-adaptive multi-operator evolutionary algorithm for constrained optimization. *Appl. Soft Comput.* **2015**, *26*, 515–522. [CrossRef]

38. Fan, Q.Q.; Zhang, Y.L. Self-adaptive differential evolution algorithm with crossover strategies adaptation and its application in parameter estimation. *Chemom. Intell. Lab. Syst.* **2016**, *151*, 164–171. [CrossRef]

39. Wang, C.; Liu, Y.C.; Chen, Y.; Wei, Y. Self-adapting hybrid strategy particle swarm optimization algorithm. *Soft Comput.* **2016**, *20*, 4933–4963. [CrossRef]
40. Gou, J.; Guo, W.P.; Wang, C.; Luo, W. A multi-strategy improved particle swarm optimization algorithm and its application to identifying uncorrelated multi-source load in the frequency domain. *Neural Comput. Appl.* **2017**, *28*, 1635–1656. [CrossRef]
41. Wang, H.; Wang, W.; Sun, H. Multi-strategy ensemble artificial bee colony algorithm for large-scale production scheduling problem. *Int. J. Innov. Comput. Appl.* **2015**, *6*, 128–136. [CrossRef]
42. Xiong, G.J.; Shi, D.Y.; Duan, X.Z. Multi-strategy ensemble biogeography-based optimization for economic dispatch problems. *Appl. Energy* **2013**, *111*, 801–811. [CrossRef]
43. Cai, Y.Q.; Sun, G.; Wang, T.; Tian, H.; Chen, Y.H.; Wang, J.H. Neighborhood-adaptive differential evolution for global numerical optimization. *Appl. Soft Comput.* **2017**, *59*, 659–706. [CrossRef]
44. Zhang, A.Z.; Sun, G.Y.; Ren, J.C.; Li, X.D.; Wang, Z.J.; Jia, X.P. A dynamic neighborhood learning-based gravitational search algorithm. *IEEE Trans. Cybern.* **2018**, *48*, 436–447. [CrossRef] [PubMed]
45. Wang, Y.; Cai, Z.X. A Dynamic Hybrid Framework for Constrained Evolutionary Optimization. *IEEE Trans. Cybern.* **2012**, *42*, 203–217. [CrossRef] [PubMed]
46. Wu, G.H.; Shen, X.; Li, H.F.; Chen, H.K.; Lin, A.P.; Suganthan, P.N. Ensemble of differential evolution variants. *Inf. Sci.* **2018**, *423*, 172–186. [CrossRef]
47. Zhou, J.J.; Yao, X.F. Multi-population parallel self-adaptive differential artificial bee colony algorithm with application in large-scale service composition for cloud manufacturing. *Appl. Soft Comput.* **2017**, *56*, 379–397. [CrossRef]
48. Wolpert, D.H.; Macready, W.G. No free lunch theorems for optimization. *IEEE Trans. Evol. Comput.* **1997**, *1*, 67–82. [CrossRef]
49. Hansen, N.; Ostermeier, A. Completely Derandomized Self-Adaptation in Evolution Strategies. *Evol. Comput.* **2001**, *9*, 159–195. [CrossRef] [PubMed]
50. LaTorre, A. A Framework for Hybrid Dynamic Evolutionary Algorithms: Multiple Offspring Sampling (MOS). Ph.D. Thesis, Universidad Polite'cnica de Madrid, Madrid, Spain, 2009.
51. Brest, J.; Greiner, S.; Boskovic, B.; Mernik, M.; Zumer, V. Self-adapting control parameters in differential evolution: A comparative study on numerical benchmark problems. *IEEE Trans. Evol. Comput.* **2006**, *10*, 646–657. [CrossRef]
52. Tanabe, R.; Fukunaga, A. Success-history based parameter adaptation for differential evolution. In Proceedings of the IEEE Congress on Evolutionary Computation 2013, Cancún, México, 20–23 June 2013.
53. Chaturvedi, K.T.; Pandit, M.; Srivastava, L. Self organizing hierarchical particle swarm optimization for nonconvex economic dispatch. *IEEE Trans. Power Syst.* **2008**, *23*, 1079–1087. [CrossRef]
54. Hosseinnezhad, V.; Babaei, E. Economic load dispatch using θ-PSO. *Int. J. Electr. Power Energy Syst.* **2013**, *49*, 160–169. [CrossRef]
55. Yu, J.T.; Kim, C.H.; Wadood, A.; Khurshiad, T.; Rhee, S.B. A Novel Multi-Population Based Chaotic JAYA Algorithm with Application in Solving Economic Load Dispatch Problems. *Energies* **2018**, *11*, 1946. [CrossRef]
56. Safari, A.; Shayegui, H. Iteration particle swarm optimization procedure for economic load dispatch with generator constraints. *Expert Syst. Appl.* **2011**, *38*, 6043–6048. [CrossRef]
57. Hosseinnezhad, V.; Rafiee, M.; Ahmadian, M.; Ameli, M.T. Species-based quantum particle swarm optimization for economic load dispatch. *Int. J. Elect. Power Energy Syst.* **2014**, *63*, 311–322. [CrossRef]
58. Amjady, N.; Sharifzadeh, H. Solution of non-convex economic dispatch problem considering valve loading effect by a new modified Differential Evolution algorithm. *Int. J. Elect. Power Energy Syst.* **2010**, *32*, 893–903. [CrossRef]
59. Park, J.B.; Jeong, Y.W.; Shin, J.R.; Lee, K.Y. An improved particle swarm optimization for nonconvex economic dispatch problems. *IEEE Trans. Power Syst.* **2010**, *25*, 156–166. [CrossRef]
60. Reddya, A.S.; Vaisakh, K. Shuffled differential evolution for large scale economic dispatch. *Electr. Power Syst. Res.* **2013**, *96*, 237–245. [CrossRef]

Article

An Improved DA-PSO Optimization Approach for Unit Commitment Problem

Sirote Khunkitti [1], Neville R. Watson [2], Rongrit Chatthaworn [1], Suttichai Premrudeepreechacharn [3] and Apirat Siritaratiwat [1,*]

1   Department of Electrical Engineering, Faculty of Engineering, Khon Kaen University, Khon Kaen 40002, Thailand; sirote_khunkitti@kkumail.com (S.K.); rongch@kku.ac.th (R.C.)
2   Department of Electrical and Computer Engineering, University of Canterbury, Christchurch 8140, New Zealand; neville.watson@canterbury.ac.nz
3   Department of Electrical Engineering, Faculty of Engineering, Chiang Mai University, Chiang Mai 50200, Thailand; suttic@eng.cmu.ac.th
*   Correspondence: apirat@kku.ac.th; Tel.: +66-89-712-2893

Received: 18 May 2019; Accepted: 17 June 2019; Published: 18 June 2019

Abstract: Solving the Unit Commitment problem is an important step in optimally dispatching the available generation and involves two stages—deciding which generators to commit, and then deciding their power output (economic dispatch). The Unit Commitment problem is a mixed-integer combinational optimization problem that traditional optimization techniques struggle to solve, and metaheuristic techniques are better suited. Dragonfly algorithm (DA) and particle swarm optimization (PSO) are two such metaheuristic techniques, and recently a hybrid (DA-PSO), to make use of the best features of both, has been proposed. The original DA-PSO optimization is unable to solve the Unit Commitment problem because this is a mixed-integer optimization problem. However, this paper proposes a new and improved DA-PSO optimization (referred to as iDA-PSO) for solving the unit commitment and economic dispatch problems. The iDA-PSO employs a sigmoid function to find the optimal on/off status of units, which is the mixed-integer part of obtaining the Unit Commitment problem. To verify the effectiveness of the iDA-PSO approach, it was tested on four different-sized systems (5-unit, 6-unit, 10-unit, and 26-unit systems). The unit commitment, generation schedule, total generation cost, and time were compared with those obtained by other algorithms in the literature. The simulation results show iDA-PSO is a promising technique and is superior to many other algorithms in the literature.

Keywords: dragonfly algorithm; metaheuristic; particle swarm optimization; unit commitment

---

1. Introduction

The development of electricity markets has made it even more crucial to determine the optimal generator schedule to minimize costs while meeting load demand. Traditional economic dispatch (ED) does not perform decisions on which generators to commit and assumes all generators must be dispatched within their minimum and maximum generator limits. Unit Commitment (UC) is the optimization problem of determining the optimal set of in-service and out-of-service generating units and their output during the scheduling period to minimize the total production costs while satisfying all the constraints [1]. In the UC problem, two decision processes involved are unit scheduling and ED. The unit scheduling process is to determine the on/off status of generating units in each hour of the planning horizon while considering minimum up- and down-time of the units. ED aims to find the optimal power generation of the in-service generating units to meet the load demand and spinning reserve during each hour while maintaining generating unit limits.

The UC problem has been considered to be a large-scale, non-convex, and mixed-integer non-linear combinatorial optimization problem, which makes the UC problem difficult to be solved. In the past, many methods have been proposed to solve the UC problem [2]. Some of the proposed techniques for solving the UC problem are; integer programming [3,4], branch-and-bound methods [5], dynamic programming (DP) [6–11], mixed-integer programming [12], Lagrangian relaxation methods (LR) [13,14], priority list method [15]. However, each of these methods has some drawbacks when solving the UC problem. For instance, the integer and mixed-integer programming methods, which use linear programming to find an integer part of the solution require too large memory for large systems, and this results in a large computation burden. The computation time of the Branch-and-bound increases exponentially with system size. Although DP is flexible, it sometimes requires a large amount of computation time if various constraints are considered. The disadvantage of LR is the difficulty confronted in providing optimal solutions when solving complex problems. The priority list method is fast and easy to implement, but it cannot confirm the quality of the solution for the same reason as LR.

Apart from these traditional techniques, many metaheuristic algorithms have been applied, such as; genetic algorithms (GA) [16], particle swarm optimization (PSO) combined with the Lagrangian relaxation (PSO-LR) [17], evolutionary programming (EP) [18], new genetic approach (NGA) [19], local convergence averse binary particle swarm optimization (LCA-PSO) [20], improved binary particle swarm optimization (IPSO) [20], mutation-based particle swarm optimization (MPSO) [20], a two-stage genetic-based technique (TSGA) [21], inter-coded genetic algorithm (ICGA) [22], binary-coded genetic algorithm (BCGA) [22], simulated annealing (SA) [23], Seeded Memetic algorithm (SM) [23], a hybrid algorithm comprising of particle swarm optimization and grey wolf optimizer (PSO-GWO) [24] and hybrid particle swarm optimization (HPSO) [25]. These have been successfully applied to solving the UC problem due to their ability to find a near global solution and deal with large-scale non-linear problems. Moreover, several works have previously studied the scheduling of generation units in small to large power systems. For example, fuzzy-based particle swarm optimization (FPSO) has been proposed to minimize the operation cost and emission for ships [26], conditional value-at-risk (CVaR) method has been introduced to maximize the expected profit of a microgrid operator [27], a hybrid PSO and selective PSO method (PSO&SPSO) has been used to solve a proposed a day-ahead operational scheduling framework for reconfigurable microgrids (RMGs) [28], a metaheuristic approach based on PSO has been applied to solve an optimal simultaneous hourly reconfiguration and day-ahead scheduling framework in smart distribution systems [29], a stochastic model for optimal scheduling of security-constrained UC associated with demand response (AC-SUCDR) has been presented in [30], a two-stage stochastic programming model has been developed to minimize the expected cost of microgrid under different time-based rate programs [31], and a Fuzzy Self-Adaptive Particle Swarm Optimization (FSAPSO) has been applied to solve multi-operation management of a typical microgrids and of a renewable microgrid [32,33].

Many metaheuristic optimization algorithms have been proposed to solve other types of complex optimization problems such as in an optimal power-flow (OPF). Examples are; grey wolf optimizer (GWO) [34], dragonfly algorithm (DA) [35], ant colony optimization (ACO) [36] and artificial bee colony (ABC) [37]. However, these algorithms cannot solve a mixed-integer combinational optimization problem in their native form. A hybrid dragonfly algorithm and particle swarm optimization (DA-PSO) is a recent optimization method which has been applied to efficiently solve a complex optimization problem which is a multi-objective optimization problem [38]. Nevertheless, it is unable to solve the mixed-integer combinational optimization problem. Therefore, this paper proposes an improved DA-PSO algorithm (iDA-PSO) that can solve the UC problem. This is achieved by applying a sigmoid function to the DA-PSO to find the optimal on/off status of generating units, which is the mixed-integer part of the UC problem. The algorithm is tested of four test systems of differing sizes. Five-unit, six-unit, ten-unit, and 26-unit generating systems are used to investigate the effectiveness of the proposed approach. The simulation results were compared with other algorithms in the literature.

2. Formulation of the UC Problem

The UC problem aims to find the optimal generation schedule, which is gauged by the value of the objective function while satisfying a set of constraints.

2.1. Objective Function

The objective function is the total production costs over the scheduling horizon, and this must be minimized to obtain the optimal generator schedule. The total production costs consist of fuel cost and start-up cost of the operating units. Therefore, the objective function is:

$$TPC = \sum_{t=1}^{T} \sum_{i=1}^{Ng} [f_{Cost}(P_{gi}^t) + ST_i^t(1 - u_i^{t-1})]u_i^t \tag{1}$$

where TPC is the total production cost (\$), T is the total scheduling period, N_g is the number of generating units, P_{gi}^t is the active power generation of the ith unit at time t, ST_i^t is the start-up cost of the ith unit at time t, u_i^t is the on or off status of the ith unit at time t, and $f_{Cost}(P_{gi}^t)$ is the fuel cost function of the ith unit for the generator power output P_{gi}^t which is calculated as:

$$f_{Cost}(P_{gi}^t) = a_i P_{gi}^2 + b_i P_{gi} + c_i \tag{2}$$

where a_i, b_i, and c_i are the fuel cost coefficients of the ith generator.

The start-up cost is the cost of bringing the off-line unit on-line. It depends on the time that the unit has been off-line before starting up which is presented as follows:

$$ST_i^t = \begin{cases} HSC_i & if \quad MDT_i \leq T_{i,off}^t \leq (MDT_i + CSH_i) \\ CSC_i & if \quad \quad T_{i,off}^t > (MDT_i + CSH_i) \end{cases} \tag{3}$$

where HSC_i is the hot start-up cost of the ith unit, CSC_i is the cold start-up cost of the ith unit, MDT_i is the minimum down-time of the ith unit, $T_{i,off}^t$ is the number of off hours of the ith unit until time t and CSH_i is the cold start hour of the ith unit.

2.2. Constraints

The optimization of the objective function must satisfy constraints imposed by the operational requirements. The set of constraints are as follows:

2.2.1. Power Balance Constraint

$$\sum_{i=1}^{Ng} P_{gi}^t u_i^t = P_D^t \tag{4}$$

where P_D^t is the active power demand at time t.

2.2.2. Spinning Reserve Constraint

$$\sum_{i=1}^{Ng} P_{gi(max)} u_i^t \geq P_D^t + P_R^t \tag{5}$$

where $P_{gi(max)}$ is the maximum active power of the ith unit, and P_R^t is the active power reserve at time t.

2.2.3. Generation Limit Constraints

$$P_{gi(\min)} \leq P^t_{gi} \leq P_{gi(\max)} \tag{6}$$

where $P_{gi(\min)}$ is the minimum active power of the ith unit.

2.2.4. Minimum Up-Time Constraint

$$T^t_{i,on} \geq MUT_i \tag{7}$$

where $T^t_{i,on}$ is the number of on hours of the ith unit until time t, and MUT_i is the minimum up-time of the ith unit.

2.2.5. Minimum Down-time Constraint

$$T^t_{i,off} \geq MDT_i \tag{8}$$

3. Overview of DA-PSO Optimization Algorithm and Related Algorithms

DA-PSO optimization algorithm is a hybrid algorithm which original combined the frameworks of the DA and PSO algorithms. This section aims to describe the formulations and concepts of the related algorithms including DA, PSO, and DA-PSO.

3.1. DA

DA is a metaheuristic method motivated by the flocking behavior of dragonflies in nature [35], and it has been successfully applied to solve complicated optimization problems, such as the OPF problem [39]. There are two main swarming goals of dragonflies, which are hunting (or static swarm), and migrating (or dynamic swarm). These can be related to two main phases of optimization, which are exploitation and exploration phases. The behavior of swarms follows three traditional rules [40]. The first rule is separation, which is to ensure collision avoidance. That is individuals avoid colliding with others in the neighborhood. Secondly, alignment, referring to velocity matching of an individual to that of other individuals in the neighborhood. The other is cohesion meaning the distance away of individuals to the center of mass of the neighborhood. Moreover, since survival is the main propose of any swarm, all the population should be attracted to food sources and repelled by the presence of enemies. Accordingly, the position updating of individuals are imitated from the aforementioned behavior, and can be mathematically formulated as follows:

Separation is formulated as follows:

$$S_i = -\sum_{j=1}^{N} X - X_j \tag{9}$$

where S_i is the separation of the ith individual, N is the number of neighboring individuals, X is the current individual position, X_j is the position of the jth neighboring individual.

Alignment is formulation is:

$$A_i = \frac{\sum_{j=1}^{N} V_j}{N} \tag{10}$$

where A_i is the alignment of the ith individual, Vj is the velocity of the jth neighboring individual.

Cohesion is formulation is:

$$C_i = \frac{\sum_{j=1}^{N} X_j}{N} - X \tag{11}$$

where C_i is the cohesion of the ith individual.

Attraction towards a food source is formulated as:

$$F_i = X^+ - X \tag{12}$$

where F_i is the food source of the ith individual, X^+ is the food source position.

Repulsion from an enemy is formulated as:

$$E_i = X^- + X \tag{13}$$

where E_i is the enemy of the ith individual, X^- is the enemy position.

The velocity of artificial dragonflies can be simulated by considering step vector (ΔX) representing the direction of their movement, which is calculated by the following equation:

$$\Delta X^{t+1} = (sS_i + aA_i + cC_i + fF_i + eE_i) + \omega^t \Delta X^t \tag{14}$$

where ΔX is the step vector of an artificial dragonfly, t is the present iteration, s is the separation weight, a is the alignment weight, c is the cohesion weight, f is the food factor, e is the enemy factor. The inertia weight factor, ω^t, is given by:

$$\omega^t = \omega_{max} - \frac{\omega_{max} - \omega_{min}}{Iter_{max}} \times Iter \tag{15}$$

The position of the artificial dragonflies is another factor to be considered to simulate their movement, which is computed using:

$$X^{t+1} = X^t + \Delta X^{t+1} \tag{16}$$

where X is the position of an artificial dragonfly.

In the case of no neighboring solutions, the artificial dragonflies need to employ a *Levy* flight, which is a random walk to improve the exploration phase. The position of dragonflies in this situation is given by:

$$X^{t+1} = X^t + Levy(d) \times X^t \tag{17}$$

where the following equation is used to calculate the *Levy* flight:

$$Levy(d) = 0.01 \times \frac{r_1 \times \sigma}{|r_2|^{\frac{1}{\beta}}} \tag{18}$$

where r_1, r_2 are two uniformly generated random number in [0,1], β is a constant which is equal to 1.5 in this work. The parameter σ is calculated using the following equation:

$$\sigma = \left(\frac{\Gamma(1+\beta) \times \sin\left(\frac{\pi\beta}{2}\right)}{\Gamma\left(\frac{1+\beta}{2}\right) \times \beta \times 2^{\left(\frac{\beta-1}{2}\right)}} \right)^{1/\beta} \tag{19}$$

where $\Gamma(x) = (x-1)!$

3.2. PSO

PSO is one of the well-known population-based evolutionary and swarm intelligence algorithms, and has been successfully applied to solve many problems in different fields [41–43]. Moreover, PSO has been effectively employed to be hybrid with many other optimization algorithms because of its simplicity and fast convergence speed [24,38,44]. PSO was originally proposed by Eberhart and Kennedy in 1995 by mimicking the concepts of bird flocking and fish schooling behaviors [45]. In PSO, each particle flies around a multi-dimensional search space and represents a possible solution in an optimization problem. Each particle comprises of a position X_i and a velocity V_i. The particles are initialized in the search space with random velocity and position values. In each iteration, the velocity of each particle is updated based on its personal best experience, $X^t_{pbest_i}$, and the best experience among the whole swarm, X^t_{gbest}, found so far. Therefore, the velocity and position of each particle can be mathematically formulated as follows:

$$V^{t+1}_i = \omega^t \times V^t_i + C_1 \times rand_1 \times (X^t_{pbest_i} - X^t_i) + C_2 \times rand_2 \times (X^t_{gbest} - X^t_i) \tag{20}$$

$$X^{t+1}_i = X^t_i + V^{t+1}_i \tag{21}$$

where V_i is the velocity of the ith particle, t is the number of iteration, ω^t is defined as in (15), C_1 and C_2 are acceleration coefficients, $rand_1$ and $rand_2$ are uniformly generated random numbers, X_i is the position of the ith particle, X_{pbest_i} is the personal best position of the ith particle, X_{gbest} is the global best position among the whole swarm.

3.3. DA-PSO

DA-PSO is a recently developed hybrid metaheuristic algorithm motivated by combining the advantages of the DA and PSO algorithms [38]. PSO applies both personal and global best experiences of the particles to find the optimal solution, is consequently good at exploitation, and often converges on the optimal solution quickly. However, PSO is sometimes trapped in the local optima rather than the global because it converges too quickly on an optimal solution. Conversely, DA is good at exploration since it employs the Levy flight to increase the stochastic behavior in the searching process. However, DA takes too long time to converge on the optimal solution. The hybrid DA-PSO algorithm was proposed to overcome these problems by merging the good exploration of DA together with the good exploitation of PSO, and it has been proven to successfully solve a complicated optimization problem such as multi-objective optimal power-flow (MO-OPF) problems, which is evident in [38]. The idea of the DA-PSO algorithm is that in the exploration phase, DA is employed to initially explore the solution space to provide the global solution area, and the best position of DA is provided. In the exploitation phase, the PSO equations are calculated but the velocity equation of PSO, Equation (20), is modified by replacing the global best position by the provided best position found so far by DA. The PSO then finds a better optimal solution from this starting point. Thus, the modified version of PSO equations can be written as:

$$V^{t+1}_i = \omega^t \times V^t_i + C_1 \times rand_1 \times (X^t_{pbest_i} - X^t_i) + C_2 \times rand_2 \times (X^{t+1}_{DA} - X^t_i) \tag{22}$$

$$X^{t+1}_i = X^t_i + V^{t+1}_i \tag{23}$$

4. An Improved DA-PSO Optimization Approach (iDA-PSO) for UC Problem

The iDA-PSO algorithm is proposed to solve the UC problem by improving the traditional DA-PSO algorithm. An approach for the improvement, the related computational formulations, and the application of the approach are explained below.

4.1. An Approach of Improving DA-PSO to Solve a Binary Problem

Although many efficient metaheuristic algorithms have been proposed in recent years, most of them cannot be applied to solve problems involving binary values such as the UC problem, which is the objective of this work. The contribution of this work is including binary values in the optimization thereby developing an efficient metaheuristic algorithm able to solve the UC problem. The hybrid metaheuristic algorithm DA-PSO operates only on real value; however, it was taken as the starting point to develop the improved DA-PSO (iDA-PSO) approach, which is proposed in this paper.

The Binary PSO (BPSO) was proposed by Kennedy and Eberhart by a modification of the traditional PSO to enable solving binary problems [46]. They also showed that the BPSO could successfully solve the test functions from [47]. In the BPSO, a particle is seen to move by flipping the number of bits. Consequently, the velocity of the particle can be represented by the change of probabilities of bit changed per iteration. In other words, a particle moves in a search space by only taking on values of 0 or 1, where each velocity $(V^t_{i,gi})$ represents the probability of a bit of position $(X^t_{i,gi})$ which takes the value 1. Since the position $(X^t_{i,gi})$ and the personal best $(X^t_{pbest,i,gi})$ are integers (0 or 1), and the velocity $(V^t_{i,gi})$, which is a probability, needs to be limited to be in the range [0,1]. A function used to accomplish this is called the sigmoid function and is mathematically formulated as follows:

$$S(V^t_{i,gi}) = \frac{1}{1 + \exp(-V^t_{i,gi})} \tag{24}$$

The sigmoid function limits the velocity within the appropriate range to be used as a probability. The change in position is defined by comparing with the random uniformly generated numbers between 0 and 1 which is formulated as follows:

$$\text{If } rand() < S(V^t_{i,gi}), \text{ then } X^t_{i,gi} = 1, \text{ else } X^t_{i,gi} = 0 \tag{25}$$

In the UC problem, V_{gimax} is set to limit the range of $V_{i,gi}$, so $S(V^t_{i,gi})$ is not too close to 0 or 1. A higher value of V_{gimax} represents a lower frequency of changing the state of a generator.

To improve the DA-PSO algorithm to be able to solve the UC problem, the sigmoid function described above is applied in the process of the DA-PSO algorithm. The equation of updating the position of dragonflies, Equations (16) and (17), are both replaced by the sigmoid function, Equation (25). Similarly, the position equation of PSO, Equation (23), is also replaced by the sigmoid function equation to find the on/off status of each generator.

4.2. Priority List

A unit operating at its maximum power output normally has a lower cost per produced unit than that operating at other power output levels; hence, a unit should be operated at its maximum power output. Priority list, in this case, is based on the average full-load cost (α) of a unit that is defined as the cost per maximum power of a unit as the following:

$$\alpha_i = \frac{f_{Cost}(P_{gimax})}{P_{gimax}} = a_i P_{gimax} + b_i + \frac{c_i}{P_{gimax}} \tag{26}$$

where a unit with the least α_i is prioritized to be dispatched first.

4.3. Spinning Reserve Constraint Satisfaction

The unit scheduling from the heuristic search may not satisfy the spinning reserve constraint. There are two main ways to deal with the unsatisfying-constraint results. The first one is a penalty function, which transforms the constrained problem into an unconstrained one. However, when the problem is highly constrained, it may be hard to find the near global solution because of the reduction of the search space. The other is to repair the violations that have occurred, which approach used in this paper. The implementation of repairing the spinning reserve violation is expressed below:

Step 1. At each hour t, calculate α_i by using (26) for all uncommitted unit at hour t, and sort them in an ascending order.

Step 2. Calculate the spinning reserve requirement at t as in (5)

Step 3. If the result from step 2 satisfies the spinning reserve constraint, go to step 5; otherwise, go to step 4.

Step 4. Commit one uncommitted unit with the least α_i from step 1.

Step 5. If $t < T$, $t = t + 1$ and go to step 1; otherwise, stop this process.

4.4. Minimum Up-Time and Down-Time Constraints Satisfaction

The results obtained for unit scheduling from the previous process may violate the minimum up- and down-time constraints required in the UC problem. To repair the violations of these constraints, the following implementation is employed.

Step 1. At each hour t, calculate the accumulated current on/off hours of the ith unit at hour t, $T^t_{i,cur}$ by referring to the accumulated hours of the previous state, $T^t_{i,prev}$. If $t = 1$, $T^t_{i, prev}$ = initial state; else $T^t_{i,prev}$ = accumulated on/off hours of the previous state, $T^{t-1}_{i,cur}$.

Step 2. At each unit i

    Step 2.1. If $u_i^t = 1$ and $T^t_{i,prev} \geq 1$, $T^t_{i,cur} = T^t_{i,prev} + 1$

    Step 2.2. If $u_i^t = 1$ and $T^t_{i,prev} \leq -MDT_i$, $T^t_{i,cur} = 1$

    Step 2.3. If $u_i^t = 0$ and $T^t_{i,prev} \leq -1$, $T^t_{i,cur} = T^t_{i,prev} - 1$

    Step 2.4. If $u_i^t = 0$ and $T^t_{i,prev} \geq MUT_i$, $T^t_{i,cur} = -1$

    Step 2.5. If $u_i^t = 0$ and $T^t_{i,prev} < MUT_i$, set $u_i^t = 1$ and $T^t_{i,cur} = T^t_{i,prev} + 1$

    Step 2.6. If $u_i^t = 1$ and $T^t_{i,prev} > -MDT_i$, set $u_i^t = 0$ and $T^t_{i,cur} = T^t_{i,prev} - 1$

Step 3. If $i < N_g$, $i = i + 1$ and go to step 2; otherwise, go to step 4.

Step 4. If $t < T$, $t = t + 1$ and go to step 1; otherwise, stop this process.

4.5. Economic Dispatch

Repairing the minimum up- and/or down-time constraints may result in either excessive generation or spinning reserves, which leads to a high generation cost, or insufficient generation, which cannot meet the load demand and spinning reserve. In case of the excessive spinning reserve, the committed units with the minimum priority will be decommitted by simultaneously considering the minimum up- and down-time constraints and spinning reserve constraint until no unit can be decommitted. In other words, the minimum up- and down-time constraints and the spinning reserve constraints must be checked before decommitting a unit. Moreover, after decommitting a unit, the accumulated current on/off time, $T^t_{i,cur}$, must be updated according to the change of a unit. In the case of the insufficient generation, which cannot meet the load demand and spinning reserve, conversely, the uncommitted units with the highest priority will be committed without violating the minimum up- and down-time constraints until the generations from the committed units satisfy the spinning reserve constraints (i.e., Equation (5)). Similarly, after committing a unit, the accumulated current on/off hours, $T^t_{i,cur}$, must be updated according to the change of a unit. After updating the status of the units without any violations of the constraints, to solve the ED problem, the lambda-iteration method [1] is employed to find the optimal values of P^t_{gi} of all committed units to meet the load demand while satisfying the power balance and generation limit constraints. The implementation of these processes can be explained as follows:

Step 1.   At each hour t, check if $\sum_{i=1}^{Ng} P_{gi(\max)} u_i^t \geq P_D^t + P_R^t$, go to step 2; otherwise, go to step 8.

Step 2.   Calculate α_i by using (26) for all committed unit at hour t, sort them in a descending order, and name it descending order list (DOL^t). Name the first unit in the DOL^t to be the lowest priority (LP^t).

Step 3.   Compute the excessive spinning reserve by $ExcessReserve = \sum_{i=1}^{Ng} P_{gi(\max)} u_i^t - P_D^t - P_R^t$.

Step 4.   Check if $ExcessReserve$ is higher than the maximum power output of the LP^t go to step 5; otherwise, go to step 6.

Step 5.   Check if decommitting the LP^t does not violate its minimum up- or down-time constraint, decommit the LP^t, and update the $T^t_{i,cur}$.

Step 6.   Delete the LP^t from the DOL^t.

Step 7.　Check if the DOL^t is not empty, set the new LP^t to be the first unit of the DOL^t and go to step 3; otherwise, go to step 13.

Step 8.　Calculate α_i by using (26) for all uncommitted units at hour t, sort them in an ascending order, and name it ascending order list (AOL^t). Name the first unit in the AOL^t to be the highest priority (HP^t).

Step 9.　Compute the lacking spinning reserve by $LackReserve = \sum\limits_{i=1}^{Ng} P_{gi(\max)} u_i^t - P_D^t - P_R^t$.

Step 10.　Check if the $LackReserve < 0$, go to step 11; otherwise, go to step 13.

Step 11.　Check if committing the HP^t does not violate its minimum up- or down-time constraint, commit the HP^t, and update the $T^t_{i,cur}$.

Step 12.　Let the HP^t be the next unit in the AOL^t, and go to step 9.

Step 13.　Solve the ED problem by a lambda-iteration method, which finds the optimal value of P^t_{gi} of all on-line units to meet the load demand while satisfying the power balance and generation limit constraints.

Step 14.　If $t < T$, $t = t + 1$ and go to step 1; otherwise, stop this process.

4.6. The Application of the iDA-PSO Approach for Solving the UC Problem

The application of the iDA-PSO approach for solving the UC problem is as follows:

Step 1.　Produce the initial population of dragonflies and particles by randomly generating them to be on or off status (1 or 0) over the time horizon T.

Step 2.　Calculate the objective function of each dragonfly, and set the best one to be the first personal best (X_{pbesti}) of PSO.

Step 3.　Compute the coefficients used in DA (s, a, c, f, e and ω).

Step 4.　Update the food source and enemy of DA.

Step 5.　Compute the representative behavior factors of DA, namely S, A, C, F, and E by (9)–(13).

Step 6.　If each dragonfly consists of at least one neighboring, update the step vector (ΔX) of a dragonfly by (14), and check whether any element of each population violates its limit, then move ΔX of that population into its minimum/maximum limit. Then, update the position of dragonfly (X_{DA}) by sigmoid function as in (25), as described in Section 4.1. However, if a dragonfly does not have any neighboring, calculate the *Levy* flight as in (18) and multiply it by X_{DA}, then update X_{DA} by the sigmoid function, (25), and set ΔX to be zero.

Step 7.　Set the best position provided by DA to be the global best position of PSO (X_{gbest}).

Step 8.　Update the velocity of each particle (V) by (22), and check whether any element of each population violates its limit, then move V of that population into its minimum/maximum limit. Then, apply the sigmoid function, Equation (25), to update the position of each particle (X_{PSO}) as described in Section 4.1.

Step 9.　Change the status of units of the newly generated population to satisfy the spinning reserve constraint as presented in Section 4.3.

Step 10.　Repair the newly generated population violating the minimum up- or down-time constraint as explained in Section 4.4.

Step 11.　Solve ED problem as expressed in Section 4.5 to find the optimal P^t_{gi} of all on-line units of the newly generated population.

Step 12.　Calculate start-up costs, which are hot or cold starts, of the units started in each hour by comparing with the status of the previous hour. For the first hour, compare the status with that of the initial status of each unit.

Step 13.　Calculate the objective function of the newly generated population.

Step 14. Test whether any obtained objective function from an individual is better than that of the previous X_{pbesti}, then the newly generated population is set to be a new X_{pbesti}. Likewise, if the best X_{pbesti} is better than X_{gbest}, that X_{pbesti} is set to be new X_{gbest}.

Step 15. If the maximum number of iterations is not reached, go to step 3; otherwise, stop the implementation and the optimal solution of UC problem is the particle with the non-dominated X_{gbest}.

The flowchart of the iDA-PSO approach for solving the UC problem is presented in Figure 1.

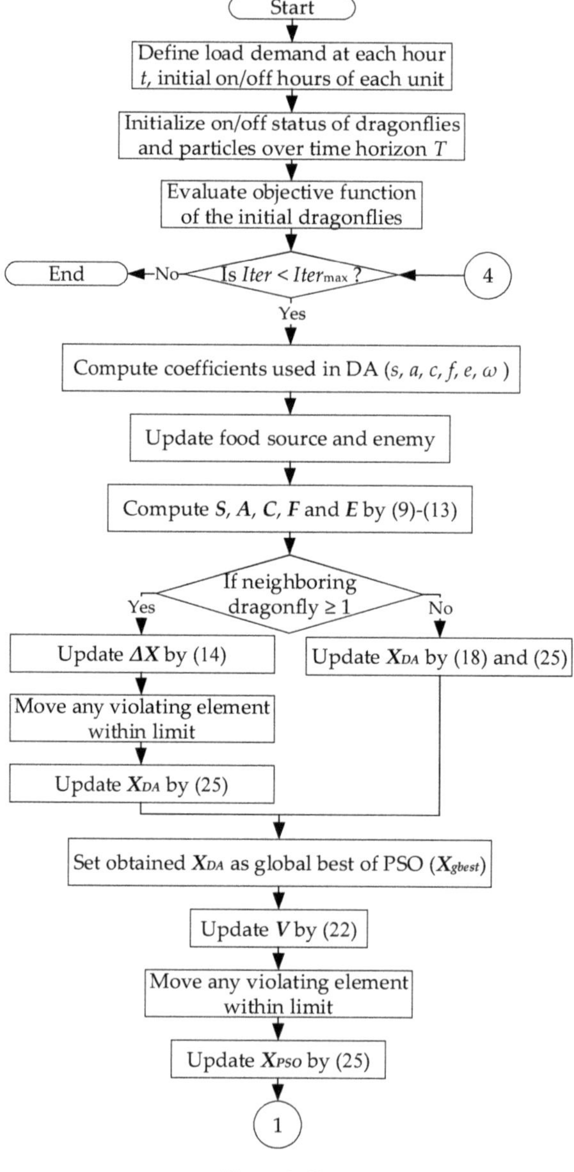

Figure 1. *Cont.*

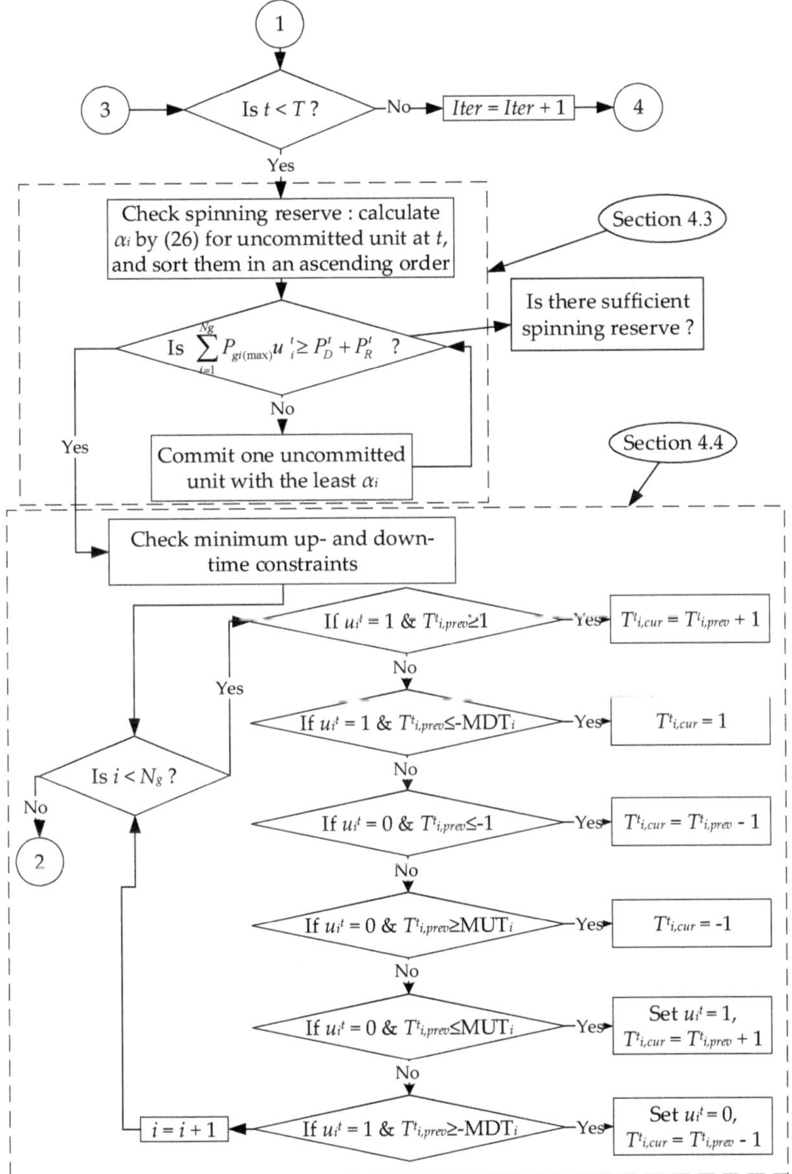

Figure 1. *Cont.*

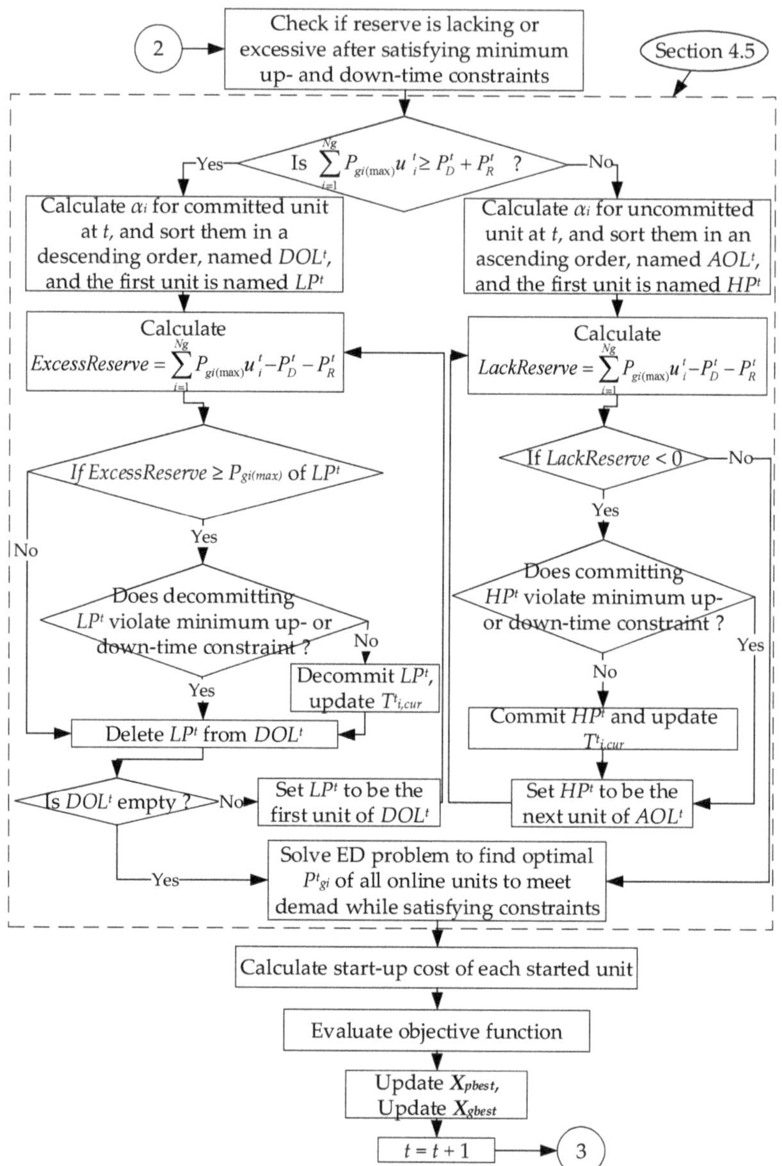

Figure 1. Flowchart of Improved Dragonfly Algorithm-Particle Swarm Optimization (iDA-PSO) approach for Unit Commitment (UC) problem.

5. Numerical Results

The effectiveness of the iDA-PSO algorithm is now examined by solving the UC problem using 24-hour scheduling time horizon for four systems of different sizes. The systems are the 5-unit system [48], 6-unit system [48], 10-unit system [16] and 26-unit system [49]. The spinning reserve requirement is equal to 10% of the total load demand of each hour in the 5-unit, 6-unit, and 10-unit systems. However, the spinning reserve requirement in the 26-unit system is equal to 5% of the total load demand of each hour as in [49]. The data of each system comprising of generator maximum and minimum limits, fuel cost coefficients, minimum up- and down-time limits, hot and cold start costs, cold start hours, and initial status of the units can be found in Tables 1–4. The 24-hour load demand for the 5-unit, 6-unit, 10-unit and 26-unit systems are provided in Tables 5–8, respectively. For each test system, the proposed approach operated for 30 independent runs, and the number of the population and maximum iteration number were set to be 100 and 200, respectively.

Table 1. System data for 5-unit system.

Unit No.	P_{gimax}	P_{gimin}	a ($/MW^2)	b ($/MW)	c ($/h)	MUT_i	MDT_i	HSC_i	CSC_i	CSH_i	IS_i
U1	250	10	0.00315	2	0	1	1	70	176	2	1
U2	140	20	0.0175	1.75	0	2	1	74	187	2	−3
U3	100	15	0.0625	1	0	1	1	50	113	1	−2
U4	120	10	0.00834	3.25	0	2	2	110	267	1	−3
U5	45	10	0.025	3	0	1	1	72	180	1	−2

Table 2. System data for 6-unit system.

Unit No.	P_{gimax}	P_{gimin}	a ($/MW^2)	b ($/MW)	c ($/h)	MUT_i	MDT_i	HSC_i	CSC_i	CSH_i	IS_i
U1	200	50	0.00375	2	0	1	1	70	176	2	1
U2	80	20	0.0175	1.7	0	2	2	74	187	1	−3
U3	50	15	0.0625	1	0	1	1	50	113	1	−2
U4	35	10	0.00834	3.25	0	1	2	110	267	1	−3
U5	30	10	0.025	3	0	2	1	72	180	1	−2
U6	40	12	0.025	3	0	1	1	40	113	1	−2

Table 3. System data for 10-unit system.

Unit No.	P_{gimax}	P_{gimin}	a ($/MW^2)	b ($/MW)	c ($/h)	MUT_i	MDT_i	HSC_i	CSC_i	CSH_i	IS_i
U1	455	150	0.00048	16.19	1000	8	8	4500	9000	5	8
U2	455	150	0.00031	17.26	970	8	8	5000	10000	5	8
U3	130	20	0.002	16.6	700	5	5	550	1100	4	−5
U4	130	20	0.00211	16.5	680	5	5	560	1120	4	−5
U5	162	25	0.0398	19.7	450	6	6	900	1800	4	−6
U6	80	20	0.00712	22.26	370	3	3	170	340	2	−3
U7	85	25	0.00079	27.74	480	3	3	260	520	2	−3
U8	55	10	0.00413	25.92	660	1	1	30	60	0	−1
U9	55	10	0.00222	27.27	665	1	1	30	60	0	−1
U10	55	10	0.00173	27.79	670	1	1	30	60	0	−1

Table 4. System data for 26-unit system.

Unit No.	P_{gimax}	P_{gimin}	a ($/MW^2)	b ($/MW)	c ($/h)	MUT_i	MDT_i	HSC_i	CSC_i	CSH_i	IS_i
U1	400	100	0.0019	7.5031	311.9102	8	5	500	500	10	10
U2	400	100	0.0019	7.4921	310.0021	8	5	500	500	10	10
U3	350	140	0.0015	10.8616	177.0575	8	5	300	200	8	10
U4	197	68.95	0.0026	23.2000	260.1760	5	4	200	200	8	−4
U5	197	68.95	0.0026	23.1000	259.6490	5	4	200	200	8	−4
U6	197	68.95	0.0026	23.0000	259.1310	5	4	200	200	8	−4
U7	155	54.25	0.0049	10.7583	143.5972	5	3	150	150	6	5
U8	155	54.25	0.0048	10.7367	134.3719	5	3	150	150	6	5
U9	155	54.25	0.0047	10.7154	143.0288	5	3	150	150	6	5
U10	155	54.25	0.0046	10.6940	142.7348	5	3	150	150	6	5
U11	100	25	0.0060	18.2000	218.7752	4	2	70	70	4	−3
U12	100	25	0.0061	18.1000	218.3350	4	2	70	70	4	−3
U13	100	25	0.0062	18.0000	217.8952	4	2	70	70	4	−3
U14	76	15.2	0.0093	13.4073	81.6259	3	2	50	50	3	3
U15	76	15.2	0.0091	13.3805	81.4641	3	2	50	50	3	3
U16	76	15.2	0.0089	13.3538	81.2980	3	2	50	50	3	3
U17	76	15.2	0.0088	13.3272	81.1364	3	0	50	50	3	3
U18	20	4	0.0143	37.8896	118.8206	0	0	20	20	2	−1
U19	20	4	0.0136	37.7770	118.4576	0	0	20	20	2	−1
U20	20	4	0.0126	37.6637	118.1083	0	0	20	20	2	−1
U21	20	4	0.0120	37.5510	117.7551	0	0	20	20	2	−1
U22	12	2.4	0.0285	26.0611	24.8882	0	0	0	0	1	−1
U23	12	2.4	0.0284	25.9318	24.7605	0	0	0	0	1	−1
U24	12	2.4	0.0280	25.8027	24.6382	0	0	0	0	1	−1
U25	12	2.4	0.0265	25.6753	24.4110	0	0	0	0	1	−1
U26	12	2.4	0.0253	25.5472	24.3891	0	0	0	0	1	−1

Table 5. 24-hour load demand for 5-unit system.

Hour	1	2	3	4	5	6	7	8	9	10	11	12
Demand	148	173	220	244	259	248	227	202	176	134	100	130
Hour	13	14	15	16	17	18	19	20	21	22	23	24
Demand	157	168	195	225	244	241	230	210	176	157	138	103

Table 6. 24-hour load demand for 6-unit system.

Hour	1	2	3	4	5	6	7	8	9	10	11	12
Demand	166	196	229	267	283.4	272	246	213	192	161	147	160
Hour	13	14	15	16	17	18	19	20	21	22	23	24
Demand	170	185	208	232	246	241	236	225	204	182	161	131

Table 7. 24-hour load demand for 10-unit system.

Hour	1	2	3	4	5	6	7	8	9	10	11	12
Demand	700	750	850	950	1000	1100	1150	1200	1300	1400	1450	1500
Hour	13	14	15	16	17	18	19	20	21	22	23	24
Demand	1400	1300	1200	1050	1000	1100	1200	1400	1300	1100	900	800

Table 8. 24-hour load demand for 26-unit system.

Hour	1	2	3	4	5	6	7	8	9	10	11	12
Dmd.	2223	2052	1938	1881	1824	1825.5	1881	1995	2280	2508	2565	2593.5
Hour	13	14	15	16	17	18	19	20	21	22	23	24
Dmd.	2565	2508	2479.5	2479.5	2593.5	2850	2821.5	2764.5	2679	2662	2479.5	2308.5

The simulation results of the proposed iDA-PSO approach for the 5-unit system are shown in Table 9, and the convergence curve is presented in Figure 2. The unit schedule and generation schedule for the 24-hour duration and the total generation cost are presented in this Table. The total generation cost through the scheduling duration obtained from the proposed iDA-PSO algorithm is equal to $11,830.94. The total generation cost provided by the iDA-PSO solution is better than that obtained by PSO-GWO, which is documented in the literature, for solving this UC problem. PSO-GWO achieved a generation cost of $12,281 [24].

Table 9. Commitment and generation schedule of the 5-unit system by Improve Dragonfly Algorithm-Particle Swarm Optimization (iDA-PSO) approach.

Hour	Unit Schedule					Generation Schedule				
	U1	U2	U3	U4	U5	U1	U2	U3	U4	U5
1	1	0	1	0	0	133	0	15	0	0
2	1	0	0	0	0	173	0	0	0	0
3	1	0	1	0	0	205	0	15	0	0
4	1	0	1	0	0	229	0	15	0	0
5	1	0	1	0	0	244	0	15	0	0
6	1	0	1	0	0	233	0	15	0	0
7	1	0	0	0	0	227	0	0	0	0
8	1	0	0	0	0	202	0	0	0	0
9	1	0	1	0	0	161	0	15	0	0
10	1	0	1	0	0	119	0	15	0	0
11	1	0	0	0	0	100	0	0	0	0
12	1	0	1	0	0	115	0	15	0	0
13	1	0	0	0	0	157	0	0	0	0
14	1	0	0	0	0	168	0	0	0	0
15	1	0	1	0	0	180	0	15	0	0
16	1	0	1	0	0	210	0	15	0	0
17	1	0	1	0	0	229	0	15	0	0
18	1	0	1	0	0	226	0	15	0	0
19	1	0	1	0	0	215	0	15	0	0
20	1	0	0	0	0	210	0	0	0	0
21	1	0	0	0	0	176	0	0	0	0
22	1	0	0	0	0	157	0	0	0	0
23	1	0	0	0	0	138	0	0	0	0
24	1	0	0	0	0	103	0	0	0	0
Total Cost ($) 11,830.94										

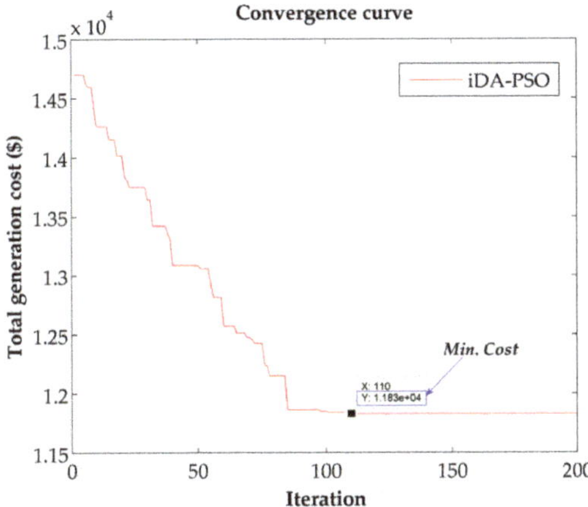

Figure 2. Convergence curve of the iDA-PSO approach for the 5-unit system.

Table 10 presents the unit schedule, generation schedule for the 24-hour duration and the total generation cost obtained by the proposed algorithm for the 6-unit system, and Figure 3 demonstrates the convergence curve of the algorithm for this system. The total generation cost of the proposed approach, which is \$13,292.28, is once again better than that of the PSO-GWO, which is \$13,600 [24].

Table 10. Commitment and generation schedule of the 6-unit system by iDA-PSO.

Hour	Unit Schedule						Generation Schedule					
	U1	U2	U3	U4	U5	U6	U1	U2	U3	U4	U5	U6
1	1	1	1	0	0	0	131	20	15	0	0	0
2	1	1	0	0	0	0	176	20	0	0	0	0
3	1	1	1	0	0	0	194	20	15	0	0	0
4	1	1	1	0	0	0	200	52	15	0	0	0
5	1	1	1	0	0	0	200	68.4	15	0	0	0
6	1	1	1	0	0	0	200	57	15	0	0	0
7	1	1	0	0	0	0	200	46	0	0	0	0
8	1	1	0	0	0	0	193	20	0	0	0	0
9	1	1	0	0	0	0	172	20	0	0	0	0
10	1	0	0	0	0	0	161	0	0	0	0	0
11	1	0	0	0	0	0	147	0	0	0	0	0
12	1	1	0	0	0	0	140	20	0	0	0	0
13	1	1	0	0	0	0	150	20	0	0	0	0
14	1	1	0	0	0	0	165	20	0	0	0	0
15	1	1	0	0	0	0	188	20	0	0	0	0
16	1	1	0	0	0	0	200	32	0	0	0	0
17	1	1	0	0	0	0	200	46	0	0	0	0
18	1	1	0	0	0	0	200	41	0	0	0	0
19	1	1	0	0	0	0	200	36	0	0	0	0
20	1	1	0	0	0	0	200	25	0	0	0	0
21	1	1	0	0	0	0	184	20	0	0	0	0
22	1	1	0	0	0	0	162	20	0	0	0	0
23	1	0	0	0	0	0	161	0	0	0	0	0
24	1	0	0	0	0	0	131	0	0	0	0	0
Total Cost (\$) 13,292.28												

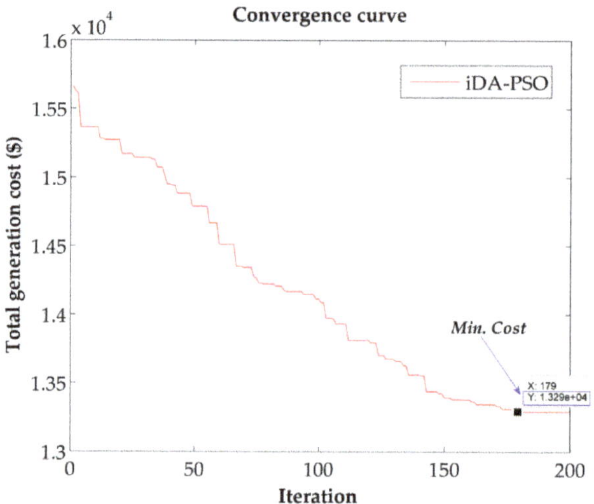

Figure 3. Convergence curve of the iDA-PSO approach for the 6-unit system.

For the 10-unit system, the simulation results including unit and generation schedule for the 24-hour duration and the total generation cost of the iDA-PSO approach are given in Table 11, and its convergence curve is provided in Figure 4. Through the scheduling duration, the total generation cost provided by the iDA-PSO is equal to $565,807.3094, which is slightly worse than those obtained by some algorithms in the literature. However, the total generation cost obtained by the iDA-PSO is significantly better than that of many algorithms presented in the literature. The algorithms GA [16], DP [16], LR [16], PSO-LR [17], EP [18], NGA [19], LCA-PSO [20], IPSO [20], MPSO [20], TSGA [21], ICGA [22], BCGA [22], SA [23], SM [23], PSO-GWO [24], HPSO [25], improve Lagrangian relaxation method (ILR) [25] and greedy randomized adaptive search procedure (GRASP) [50] are compared with the proposed approach as shown in Table 12. The best, average and worst generation costs and the computation times of the proposed iDA-PSO and other algorithms are also presented in Table 12. The computation time of the proposed iDA-PSO is slightly slower than those of some algorithms because of the sequential process of both DA and PSO.

Table 11. Commitment and generation schedule of the 10-unit system by iDA-PSO.

Unit Schedule										Generation Schedule									
U1	U2	U3	U4	U5	U6	U7	U8	U9	U10	U1	U2	U3	U4	U5	U6	U7	U8	U9	U10
1	1	0	0	0	0	0	0	0	0	455	245	0	0	0	0	0	0	0	0
1	1	0	0	0	0	0	0	0	0	455	295	0	0	0	0	0	0	0	0
1	1	0	0	1	0	0	0	0	0	455	370	0	0	25	0	0	0	0	0
1	1	0	0	1	0	0	0	0	0	455	455	0	0	40	0	0	0	0	0
1	1	1	1	1	0	0	0	0	0	455	455	0	65	25	0	0	0	0	0
1	1	1	1	1	0	0	0	0	0	455	455	35	130	25	0	0	0	0	0
1	1	1	1	1	0	0	0	0	0	455	455	85	130	25	0	0	0	0	0
1	1	1	1	1	0	0	0	0	0	455	455	130	130	30	0	0	0	0	0
1	1	1	1	1	1	1	0	0	0	455	455	130	130	85	20	25	0	0	0
1	1	1	1	1	1	1	1	0	0	455	455	130	130	162	33	25	10	0	0
1	1	1	1	1	1	1	1	1	0	455	455	130	130	162	73	25	10	10	0
1	1	1	1	1	1	1	1	1	1	455	455	130	130	162	80	58	10	10	10
1	1	1	1	1	1	1	1	0	0	455	455	130	130	162	33	25	10	0	0
1	1	1	1	1	1	1	0	0	0	455	455	130	130	85	20	25	0	0	0
1	1	0	1	1	1	1	0	0	0	455	455	0	130	115	20	25	0	0	0
1	1	0	1	1	0	0	0	0	0	455	455	0	115	25	0	0	0	0	0
1	1	0	1	1	0	0	0	0	0	455	455	0	65	25	0	0	0	0	0
1	1	0	1	1	0	0	1	0	0	455	455	0	130	50	0	0	10	0	0
1	1	0	1	1	1	1	0	0	0	455	455	0	130	115	20	25	0	0	0
1	1	1	1	1	1	1	1	0	0	455	455	130	130	162	33	25	10	0	0
1	1	1	1	1	1	1	0	0	0	455	455	130	130	85	20	25	0	0	0
1	1	1	0	1	1	0	0	0	0	455	455	130	0	40	20	0	0	0	0
1	1	1	0	0	0	0	0	0	0	455	425	20	0	0	0	0	0	0	0
1	1	1	0	0	0	0	0	0	0	455	325	20	0	0	0	0	0	0	0

Total Cost ($) 565,807.3094

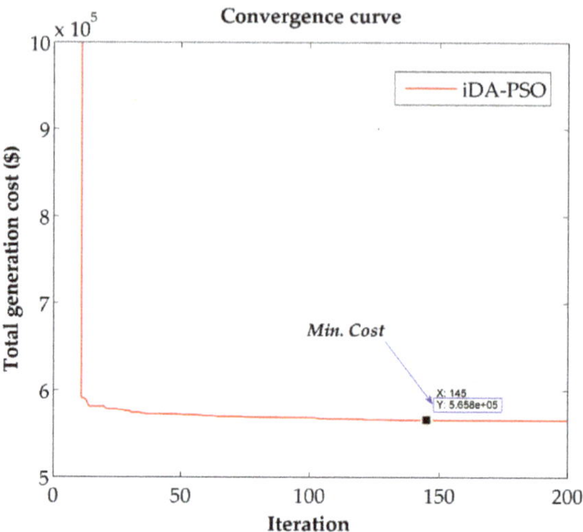

Figure 4. Convergence curve of the iDA-PSO approach for the 10-unit system.

Table 12. Simulation results of the iDA-PSO approach compared with other algorithms in the literature for the 10-generating unit system.

Methods	Total Generation Cost ($)			Time (s)
	Best	**Average**	**Worst**	
GA [16]	565,825	-	570,032	221
DP [16]	565,825	-	-	-
LR [16]	565,825	-	-	257
PSO-LR [17]	565,869	-	-	42
EP [18]	564,551	-	566,231	100
NGA [19]	591,715	-	-	677
LCA-PSO [20]	570,006	-	-	18.34
IPSO [20]	599,782	-	-	14.48
MPSO [20]	574,905	-	-	15.73
TSGA [21]	568,314.56	-	-	-
ICGA [22]	-	566,404	-	7.4
BCGA [22]	567,367	-	-	3.7
SA [23]	565,828	565,988	566,260	3.35
SM [23]	566,686	566,787	567,022	-
PSO-GWO [24]	565,210.2564	-	-	-
HPSO [25]	574,153	-	-	-
ILR [25]	565,823	-	-	-
GRASP [50]	565,825	-	-	17
iDA-PSO	565,807.3094	565,827.0145	565,891.7599	231.31

In the larger 26-unit system, the outcome of the UC for 24-hour duration together with the total generation cost provided by the proposed approach are shown by the non-zero numbers in Table 13, and Figure 5 displays the convergence curve of the proposed approach for this system. The total generation cost obtained by the iDA-PSO is equal to $741,587.7088 and is better than those of other algorithms, including GA [49], discrete binary particle swarm optimization (BPSO) [49], and modified particle swarm optimization (MPSO) [51] in the literature as presented in Table 14.

Table 13. Generation schedule of the 26-unit system by iDA-PSO.

Generation Schedule (Units 1–13)												
1	2	3	4	5	6	7	8	9	10	11	12	13
400	400	350	0	0	0	155	155	155	155	0	35.4	100
400	400	350	0	0	0	155	155	155	155	0	25	29
400	400	350	0	0	0	155	155	155	155	0	25	67
400	400	350	0	0	0	155	155	155	155	0	25	25
400	400	350	0	0	0	155	155	155	155	0	0	0
400	400	350	0	0	0	155	155	155	155	0	0	0
400	400	350	0	0	0	155	155	155	155	0	0	0
400	400	350	0	0	0	155	155	155	155	0	0	0
400	400	350	0	0	101.2	155	155	155	155	0	0	100
400	400	350	0	0	120.4	155	155	155	155	100	100	100
400	400	350	0	68.95	122.05	155	155	155	155	100	100	100
400	400	350	0	68.95	150.55	155	155	155	155	100	100	100
400	400	350	0	68.95	122.05	155	155	155	155	100	100	100
400	400	350	0	120.4	0	155	155	155	155	100	100	100
400	400	350	0	98.3	0	155	155	155	155	100	100	100
400	400	350	0	98.3	0	155	155	155	155	100	100	100
400	400	350	68.95	150.55	0	155	155	155	155	100	100	100
400	400	350	74.8	197	197	155	155	155	155	100	100	100
400	400	350	68.95	181.55	197	155	155	155	155	100	100	100
400	400	350	68.95	124.55	197	155	155	155	155	100	100	100
400	400	350	98.4	0	197	155	155	155	155	100	100	100
400	400	350	83.8	0	197	155	155	155	155	100	100	100
400	400	350	0	0	98.3	155	155	155	155	100	100	100
400	400	350	0	0	0	155	155	155	155	25	97.5	100

Generation Schedule (Units 14–26)												
14	15	16	17	18	19	20	21	22	23	24	25	26
76	76	76	76	0	0	0	4	0	2.4	2.4	2.4	2.4
0	76	76	76	0	0	0	0	0	0	0	0	0
0	0	0	76	0	0	0	0	0	0	0	0	0
0	0	0	61	0	0	0	0	0	0	0	0	0
0	0	15.2	38.8	0	0	0	0	0	0	0	0	0
0	0	15.2	40.3	0	0	0	0	0	0	0	0	0
0	15.2	19.8	76	0	0	0	0	0	0	0	0	0
15.2	53	76	76	0	0	0	0	0	0	0	2.4	2.4
76	76	76	76	0	0	0	0	0	0	0	2.4	2.4
76	76	76	76	0	0	0	4	0	2.4	2.4	2.4	2.4
76	76	76	76	0	0	0	0	0	0	0	0	0
76	76	76	76	0	0	0	0	0	0	0	0	0
76	76	76	76	0	0	0	0	0	0	0	0	0
76	76	76	76	0	0	0	4	0	2.4	2.4	2.4	2.4
76	76	76	76	0	0	0	0	0	0	2.4	2.4	2.4
76	76	76	76	0	0	0	0	0	0	2.4	2.4	2.4
76	76	76	76	0	0	0	0	0	0	0	0	0
76	76	76	76	0	0	0	0	0	0	2.4	2.4	2.4
76	76	76	76	0	0	0	0	0	0	0	0	0
76	76	76	76	0	0	0	0	0	0	0	0	0
76	76	76	76	0	0	0	0	0	2.4	2.4	2.4	2.4
76	76	76	76	0	0	0	0	0	0	2.4	2.4	2.4
76	76	76	76	0	0	0	0	0	0	2.4	2.4	2.4
76	76	76	76	0	0	0	0	2.4	2.4	2.4	2.4	2.4

Total Cost ($) 741,587.7088

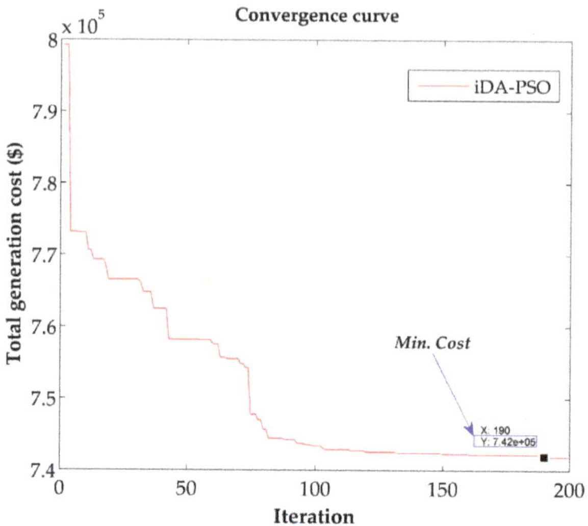

Figure 5. Convergence curve of the iDA-PSO approach for the 26-unit system.

Table 14. Simulation results of the iDA-PSO approach compared with other algorithms for the 26-generating unit system.

Methods	Total Generation Cost ($)			Time (s)
	Best	Average	Worst	
GA [49]	782,373	784,910	786,522	87.33
BPSO [49]	773,191	774,653	776,342	516.57
MPSO [51]	746,600.6	-	-	-
iDA-PSO	741,587.7088	743,176.1415	745,894.2814	327.76

From the generation schedule of each system, it can be noticed that the different units are dispatched in different ways. This is because the different units have different fuel cost coefficients, generation limits, minimum up- and down-time constraints, hot and cold and start-up costs and cold start hours, etc. Therefore, the units which have the cheapest fuel cost coefficient should be prioritized to be firstly dispatched, and the units which have the highest fuel cost coefficient should be dispatched only in the high-demand hour. However, these also depend on the start-up cost of each unit. Another noticeable point is most of the units keep a constant level of production over different time intervals. This is because when any unit has been turned off and turned on again, the start-up cost is added to the total generation cost causing a higher cost. Thus, if the units have low fuel cost coefficients and high maximum power generation, it is unnecessary to turn them off and on again.

According to all simulation results presented in Tables 9–14, the proposed approach can efficiently find the optimal unit schedule during 24-hour time horizon for four different system sizes. The total generation cost obtained by the proposed iDA-PSO approach is better than that of the recently proposed algorithm, PSO-GWO, for the 5- and 6-unit systems. For the 10-unit system, the iDA-PSO could provide considerably better total generation cost than many algorithms in the literature. The iDA-PSO could also produce considerably better total generation cost than several algorithms in the literature for the larger 26-unit system. Thus, adopting the sigmoid function to the recently proposed efficient optimization algorithm, DA-PSO, could make it able to solve the UC problem, which is a mixed-integer combinational optimization problem. The optimal on/off status of generating units, which is the mixed-integer part of the UC problem, could be efficiently provided for all studied systems, and the optimal total generation costs could also be obtained and are significantly better than that of many algorithms in the literature.

6. Conclusions

This paper has presented an improved DA-PSO algorithm that is capable of solving the UC problem in an electrical power system. The DA-PSO is a recent and efficient optimization algorithm, which has been proven to successfully solve a complicated optimization problem, which is a multi-objective, such as the OPF problem. However, DA-PSO cannot solve mixed-integer combination optimization problem such as the UC problem. To overcome this limitation, a new iDA-PSO algorithm has been proposed which employed the sigmoid function to enable finding the optimal on/off status of generation units, while satisfying the system constraints. The four test systems of different sizes (consisting of 5-unit, 6-unit, 10-unit and 26-unit systems) were used to demonstrate the effectiveness of the iDA-PSO algorithm. The proposed approach proved reliable by could successfully finding the optimal results for the generation schedule for a 24-hour duration for the test systems. The total generation costs over the scheduled time horizon obtained by iDA-PSO are less than those of many algorithms reported in the literature. Thus, applying the sigmoid function to the DA-PSO algorithm could enable it to solve the UC problem, which is a mixed-integer combinational problem, and the iDA-PSO also has a superiority over many algorithms reported in the literature. In the future work, the iDA-PSO approach could be improved and tested against other hybrid metaheuristic approaches such as fuzzy adaptive PSO.

Author Contributions: Conceptualization, S.K. and N.R.W.; Methodology, S.K.; Software, S.K.; Validation, S.K., N.R.W., A.S., and R.C.; Formal Analysis, S.K. and N.R.W.; Investigation, S.K.; Resources, N.R.W.; Data Curation, S.R.; Writing Original Draft Preparation, S.R.; Writing-Review and Editing, S.K., N.R.W., A.S., and S.P.; Visualization, S.K.; Supervision, N.R.W., and A.S.; Project Administration, N.R.W., and A.S.; Funding Acquisition, A.S.

Funding: This research was funded by the Thailand Research Fund through the Royal Golden Jubilee Ph.D. Program (Grant no. PHD/0192/2557) to Mr Sirote Khunkitti and Professor Dr Apirat Siritaratiwat.

Conflicts of Interest: The authors declare no conflict of interest.

References

1. Wood, A.J.; Wollenberg, B.F. *Power Generation, Operation, and Control*, 2nd ed.; Wiley: Hoboken, NJ, USA, 1996; ISBN 9780471790556.
2. Padhy, N.P. Unit commitment—A bibliographical survey. *IEEE Trans. Power Syst.* **2004**, *19*, 1196–1205. [CrossRef]
3. Dillon, T.S.; Edwin, K.W.; Kochs, H.-D.; Taud, R.J. Integer Programming Approach to the Problem of Optimal Unit Commitment with Probabilistic Reserve Determination. *IEEE Trans. Power Appar. Syst.* **1978**, *PAS-97*, 2154–2166. [CrossRef]
4. Garver, L.L. Power Generation Scheduling by Integer Programming—Development of Theory. *Trans. Am. Inst. Electr. Eng. Part III Power Appar. Syst.* **1962**, *81*, 730–734. [CrossRef]
5. Cohen, A.; Yoshimura, M. A Branch-and-Bound Algorithm for Unit Commitment. *IEEE Trans. Power Appar. Syst.* **1983**, *PAS-102*, 444–451. [CrossRef]
6. Snyder, W.L.; Powell, H.D.; Rayburn, J.C. Dynamic programming approach to unit commitment. *IEEE Trans. Power Syst.* **1987**, *2*, 339–348. [CrossRef]
7. Lowery, P.G. Generating Unit Commitment by Dynamic Programming. *IEEE Trans. Power Appar. Syst.* **1966**, *PAS-85*, 422–426. [CrossRef]
8. Pang, C.K.; Chen, H.C. Optimal short-term thermal unit commitment. *IEEE Trans. Power Appar. Syst.* **1976**, *95*, 1336–1346. [CrossRef]
9. Pang, C.K.; Sheble, G.B.; Albuyeh, F. Evaluation of dynamic programming based methods and multiple area representation for thermal unit commitments. *IEEE Trans. Power Appar. Syst.* **1981**, *PAS-100*, 1212–1218. [CrossRef]
10. Su, M.-C.; Hsu, Y.-Y. Fuzzy Dynamic Programming: An Application to Unit Commitment. *IEEE Trans. Power Syst.* **1991**, *6*, 1231–1237. [CrossRef]

11. Ouyang, Z.; Shahidehpour, S.M. An intelligent dynamic programming for unit commitment application. *IEEE Trans. Power Syst.* **1991**, *6*, 1203–1209. [CrossRef]

12. Muckstadt, J.A.; Wilson, R.C. An Application of Mixed-Integer Programming Duality to Scheduling Thermal Generating Systems. *IEEE Trans. Power Appar. Syst.* **1968**, *PAS-87*, 1968–1978. [CrossRef]

13. André, M.; Sandrin, P. A New Method for Unit Commitment at Electricité de France. *IEEE Power Eng. Rev.* **1983**, *5*, 38–39.

14. Zhuang, F.; Galiana, F.D. Towards a more rigorous and practical unit commitment by lagrangian relaxation. *IEEE Trans. Power Syst.* **1988**, *3*, 763–773. [CrossRef]

15. Sheble, G.B. Solution of the unit commitment problem by the method of unit periods. *IEEE Trans. Power Syst.* **1990**, *5*, 257–260. [CrossRef]

16. Kazarlis, S.A.; Bakirtzis, A.G.; Petridis, V. A genetic algorithm solution to the unit commitment problem. *IEEE Trans. Power Syst.* **1996**, *11*, 83–92. [CrossRef]

17. Balci, H.; Valenzuela, J. Scheduling electric power generators using particle swarm optimization combined with the Lagrangian relaxation method. *Int. J. Appl. Math. Comput. Sci.* **2004**, *14*, 411–422.

18. Juste, K.A.; Kita, H.; Tanaka, E.; Hasegawa, J. An evolutionary programming solution to the unit commitment problem. *IEEE Trans. Power Syst.* **1999**, *14*, 1452–1459. [CrossRef]

19. Ganguly, D.; Sarkar, V.; Pal, J. A new genetic approach for solving the unit commitment problem. In Proceedings of the 2004 International Conference on Power System Technology, Singapore, 21–24 November 2004; Volume 1, pp. 542–547.

20. Wang, B.; Li, Y.; Watada, J. Re-Scheduling the Unit Commitment Problem in Fuzzy Environment. In Proceedings of the 2011 IEEE International Conference on Fuzzy Systems (FUZZ-IEEE 2011), Taipei, Taiwan, 27–30 June 2011; pp. 1090–1095.

21. Eldin, A.S.; El-sayed, M.A.H.; Youssef, H.K.M. A two-stage genetic based technique for the unit commitment optimization problem. In Proceedings of the 2008 12th International Middle-East Power System Conference, Aswan, Egypt, 12–15 March 2008; pp. 425–430. [CrossRef]

22. Damousis, I.G.; Bakirtzis, A.G.; Dokopoulos, P.S. A solution to the unit-commitment problem using integer-coded genetic algorithm. *IEEE Trans. Power Syst.* **2004**, *19*, 1165–1172. [CrossRef]

23. Simopoulos, D.N.; Kavatza, S.D.; Vournas, C.D. Unit Commitment by an Enhanced Simulated Annealing Algorithm. *IEEE Trans. Power Syst.* **2006**, *21*, 68–76. [CrossRef]

24. Kamboj, V.K. A novel hybrid PSO–GWO approach for unit commitment problem. *Neural Comput. Appl.* **2016**, *27*, 1643–1655. [CrossRef]

25. Sriyanyong, P.; Song, Y.H. Unit commitment using particle swarm optimization combined with Lagrange relaxation. In Proceedings of the 2005 IEEE Power Engineering Society General Meeting, San Francisco, CA, USA, 16 June 2005; Volume 3, pp. 2752–2759. [CrossRef]

26. Kanellos, F.D.; Anvari-Moghaddam, A.; Guerrero, J.M. A cost-effective and emission-aware power management system for ships with integrated full electric propulsion. *Electr. Power Syst. Res.* **2017**, *150*, 63–75. [CrossRef]

27. Vahedipour-Dahraie, M.; Anvari-Moghaddam, A.; Guerrero, J.M. Evaluation of reliability in risk-constrained scheduling of autonomous microgrids with demand response and renewable resources. *IET Renew. Power Gener.* **2018**, *12*, 657–667. [CrossRef]

28. Yaprakdal, F.; Baysal, M.; Anvari-Moghaddam, A. Optimal Operational Scheduling of Reconfigurable Microgrids in Presence of Renewable Energy Sources. *Energies* **2019**, *12*, 1858. [CrossRef]

29. Esmaeili, S.; Anvari-Moghaddam, A.; Jadid, S.; Guerrero, J.M. Optimal simultaneous day-ahead scheduling and hourly reconfiguration of distribution systems considering responsive loads. *Int. J. Electr. Power Energy Syst.* **2019**, *104*, 537–548. [CrossRef]

30. Vahedipour-Dahraei, M.; Najafi, H.R.; Anvari-Moghaddam, A.; Guerrero, J.M. Security-constrained unit commitment in AC microgrids considering stochastic price-based demand response and renewable generation. *Int. Trans. Electr. Energy Syst.* **2018**, *28*, e2596. [CrossRef]

31. Vahedipour-Dahraie, M.; Najafi, H.; Anvari-Moghaddam, A.; Guerrero, J. Study of the Effect of Time-Based Rate Demand Response Programs on Stochastic Day-Ahead Energy and Reserve Scheduling in Islanded Residential Microgrids. *Appl. Sci.* **2017**, *7*, 378. [CrossRef]

32. Moghaddam, A.A.; Seifi, A.; Niknam, T. Multi-operation management of a typical micro-grids using Particle Swarm Optimization: A comparative study. *Renew. Sustain. Energy Rev.* **2012**, *16*, 1268–1281. [CrossRef]

33. Moghaddam, A.A.; Seifi, A.; Niknam, T.; Alizadeh Pahlavani, M.R. Multi-objective operation management of a renewable MG (micro-grid) with back-up micro-turbine/fuel cell/battery hybrid power source. *Energy* **2011**, *36*, 6490–6507. [CrossRef]

34. El-Fergany, A.A.; Hasanien, H.M. Single and Multi-objective Optimal Power Flow Using Grey Wolf Optimizer and Differential Evolution Algorithms. *Electr. Power Compon. Syst.* **2015**, *43*, 1548–1559. [CrossRef]

35. Mirjalili, S. Dragonfly algorithm: A new meta-heuristic optimization technique for solving single-objective, discrete, and multi-objective problems. *Neural Comput. Appl.* **2016**, *27*, 1053–1073. [CrossRef]

36. Sliman, L.; Bouktir, T. Economic Power Dispatch of Power System with Pollution Control using Multiobjective Ant Colony Optimization. *Int. J. Comput. Intell. Res.* **2007**, *3*, 145–153. [CrossRef]

37. Rezaei Adaryani, M.; Karami, A. Artificial bee colony algorithm for solving multi-objective optimal power flow problem. *Int. J. Electr. Power Energy Syst.* **2013**, *53*, 219–230. [CrossRef]

38. Khunkitti, S.; Siritaratiwat, A.; Premrudeepreechacharn, S.; Chatthaworn, R.; Watson, N. A Hybrid DA-PSO Optimization Algorithm for Multiobjective Optimal Power Flow Problems. *Energies* **2018**, *11*, 2270. [CrossRef]

39. Bashishtha, T.K. Nature Inspired Meta-heuristic dragonfly Algorithms for Solving Optimal Power Flow Problem. *Int. J. Electron. Electr. Comput. Syst.* **2016**, *5*, 111–120.

40. Reynolds, C.W. Flocks, herds and schools: A distributed behavioral model. *ACM SIGGRAPH Comput. Graph.* **1987**, *21*, 25–34. [CrossRef]

41. Hu, X.; Shi, Y.; Eberhart, R. Recent advances in particle swarm. In Proceedings of the 2004 Congress on Evolutionary Computation, Portland, OR, USA, 19–23 June 2004; Volume 1, pp. 90–97. [CrossRef]

42. Banks, A.; Vincent, J.; Anyakoha, C. A review of particle swarm optimization. Part I: Background and development. *Nat. Comput.* **2007**, *6*, 467–484. [CrossRef]

43. Niknam, T.; Narimani, M.R.; Aghaei, J.; Azizipanah-Abarghooee, R. Improved particle swarm optimisation for multi-objective optimal power flow considering the cost, loss, emission and voltage stability index. *IET Gener. Transm. Distrib.* **2012**, *6*, 515. [CrossRef]

44. Narimani, M.R.; Azizipanah-Abarghooee, R.; Zoghdar-Moghadam-Shahrekohne, B.; Gholami, K. A novel approach to multi-objective optimal power flow by a new hybrid optimization algorithm considering generator constraints and multi-fuel type. *Energy* **2013**, *49*, 119–136. [CrossRef]

45. Eberhart, R.; Kennedy, J. A New Optimizer Using Particle Swarm Theory. In Proceedings of the Sixth International Symposium on Micro Machine and Human Science, Nagoya, Japan, 4–6 October 1995; pp. 39–43.

46. Kennedy, J.; Eberhart, R.C. A discrete binary version of the particle swarm algorithm. In Proceedings of the 1997 IEEE International Conference on Systems, Man, and Cybernetics. Computational Cybernetics and Simulation, Orlando, FL, USA, 12–15 October 1997; Volume 5, pp. 4104–4108. [CrossRef]

47. De Jong, K.A. *Analysis of the Behavior of a Class of Genetic Adaptive Systems*; University of Michigan: Ann Arbor, MI, USA, 1975.

48. Anita, J.M.; Raglend, I.J. Solution of Unit Commitment Problem Using Shuffled Frog Leaping Algorithm. In Proceedings of the 2012 International Conference on Computing, Electronics and Electrical Technologies (ICCEET), Kumaracoil, India, 21–22 March 2012; pp. 109–115.

49. Gaing, Z.-L. Discrete particle swarm optimization algorithm for unit commitment. In Proceedings of the 2003 IEEE Power Engineering Society General Meeting, Toronto, ON, Canada, 13–17 July 2003; Volume 1, pp. 418–424. [CrossRef]

50. Viana, A.; De Sousa, J.P.; Matos, M. Using GRASP to Solve the Unit Commitment Problem. *Ann. Oper. Res.* **2003**, *120*, 117–132. [CrossRef]

51. Moussa, A.M.; Gammal, M.E.; Attia, A.I. An Improved Particle Swarm Optimization Technique for Solving the Unit Commitment Problem. *Online J. Power Energy Eng.* **2011**, *2*, 217–222.

Article

Research on Multi-Time Scale Optimization Strategy of Cold-Thermal-Electric Integrated Energy System Considering Feasible Interval of System Load Rate

Bin Ouyang [1,2,*], **Zhichang Yuan** [2], **Chao Lu** [2], **Lu Qu** [2] **and Dongdong Li** [1]

[1] Electric Power Engineering, Shanghai University of Electric Power, Shanghai 200090, China
[2] Department of Electrical Engineer, Tsinghua University, Beijing 100084, China
* Correspondence: Oyang2014@163.com; Tel.: +86-188-1722-2035

Received: 15 July 2019; Accepted: 19 August 2019; Published: 22 August 2019

Abstract: The integrated energy system coupling multi-type energy production terminal to realize multi-energy complementarity and energy ladder utilization is of great significance to alleviate the existing energy production crisis and reduce environmental pollution. In this paper, the topology of the cold-thermal-electricity integrated energy system is built, and the decoupling method is adopted to analyze the feasible interval of load rate under the strong coupling condition, so as to ensure the "source-load" power balance of the system. Establishing a multi-objective optimization function with the lowest system economic operation and pollution gas emission, considering the attribute differences and energy scheduling characteristics of different energy sources of cold, heat and electricity, and adopting different time scales to optimize the operation of the three energy sources of cold, heat and electricity, wherein the operation periods of electric energy, heat energy and cold energy are respectively 15 min, 30 min and 1 h; The multi-objective problem is solved by standard linear weighting method. Finally, the mixed integer nonlinear programming model is calculated by LINGO solver. In the numerical simulation, the hotel summer front load parameters of Zhangjiakou, China are selected for simulation and compared with a single time scale system. The simulation results show that the multi-time scale system reduces the economic operation cost by 15.6% and the pollution gas emission by 22.3% compared with the single time scale system, it also has a wider feasible range of load rate, flexible time allocation, and complementary energy selection.

Keywords: cold-thermal-electricity integrated energy system; multi-energy complementation; feasible interval of load rate; multi-objective; multi time scales; optimize operation

---

1. Introduction

As the driving force for social development and the indispensable whole of daily life, energy plays an increasingly important role in modern life. As for the problems that the traditional fossil energy is exhausted, the energy utilization efficiency is low and the environmental pollution is serious which people have to seek an efficient, energy-saving, environment-friendly and renewable energy production and utilization mode. At the same time, due to the differences and limitations in the development of different energy systems, the planning, design, operation and control of each energy system are often separated and lack coordination with each other [1]. As a result, the overall energy utilization rate of the system is low, the energy supply of the system is safe, and the self-healing ability is not strong [2]. Therefore, Integrated Energy Systems (IES) [3–6] that couple multiple energy production and consumption modes, absorb a large amount of renewable energy, realize multi-energy complementation, and improve the overall energy utilization rate should emerge.

In recent years, many scholars have paid close attention to the research and application of integrated energy system. In Europe, the University of Manchester in the United Kingdom took the lead in developing a local integrated energy system, which integrates electricity/gas/thermal energy systems and a user interaction platform. The platform realizes three major functions: energy consumption mode, energy conservation strategy and demand-side response [2]; Denmark is the first country in the world to set a goal of completely separating from fossil fuels. It is expected to use 100% renewable energy by 2050, with emphasis on integrating different energy systems [7]; The University of Aachen and the German Federal Ministry of Economy and Environment launched the E-Energy Project [8] through demand-side response, achieving a high degree of integration of energy, information and capital, promoting the automation of energy service management, and successfully landing in Langenfel, Germany. The EU has set a target of carbon pollution from electricity production by 2050 [9], planned a new route for the EU power grid plan, and is committed to building a trans-European high-efficiency energy system by integrating the energy systems of various countries. In 2001, the U.S. Department of Energy put forward a comprehensive energy system development plan [10] aimed at improving the supply and utilization of clean and renewable energy and further improving the economy and reliability of the energy supply system. Japan established the Japan-Wireless City Alliance in 2010 [11] and is committed to the research of wireless city technology and the demonstration of the national integrated energy system. In China, a number of integrated energy system demonstration projects have been launched. Guangzhou Mingzhu Industrial Park, in combination with the future development direction and technical requirements of the city's power grid [12] that actively building an intelligent industrial park demonstration park for large-scale local consumption of renewable energy through cold/heat/electricity/gas multi-system coupling. Zhangbei wind/Light/Heat/Storage/Transmission Multi-energy Complementary Demonstration Project [13] in Zhangjiakou, Hebei province sets a precedent for large-scale multi-energy complementary power generation by comprehensively utilizing various energy storage and photo-thermal power generation technologies in order to build a strong smart grid. At the same time, the integrated energy system is also facing many difficulties and challenges in planning, modeling and optimizing operation.

In terms of optimal operation, document [14] establishes an economic optimal operation model of distributed energy system compatible with demand-side regulation, fully considers the cold/heat/electricity load in the distributed energy system which proposes quantum fireworks algorithm to solve the model. Literature [15] takes the cold/heat demand of 433 buildings as a model, compares the adaptability of two operation modes of heat determination and electricity determination in buildings, and finally points out that heat determination is not economical due to the high investment cost in the early stage, and the optimization of primary energy saving rate can improve the economic benefits of the factory. However, with the method of heat determination by electricity, when the demand for heat energy is strong, auxiliary equipment is needed to ensure the supply of heat energy. Excess heat is easily dissipated in the environment and wasted when the heat supply is excessive, so it is not ideal. In view of the irrationality of the methods of determining electricity by heat and determining heat by electricity, document [16] proposes energy scheduling algorithms aiming at minimizing operating cost, primary energy consumption (PEC) and carbon dioxide emission (CDE), respectively, points out that the three optimization algorithms do not have a common trend under common circumstances, and points out that the installation and operation of the system should be considered as long as PEC and CDE meet the standard requirements. However, using this method, different buildings need different analysis and evaluation to determine the optimal operating conditions. Literature [17] takes the minimum total operating cost as the optimization objective to maximize the utilization rate of resources. But, the problems mentioned above are all aimed at the single time scale operation of the system, and there are few researches on the multi-time scale characteristics of the integrated energy system. Documents [18] and [19] aim at the defects existing in the operation of the integrated energy system before the day, and use the two time scales of day-ahead load state estimation and real-time rolling correction to realize the optimization of real-time scheduling. Literature [20] considers the dynamic response characteristics of multi-time scale and multi-energy coupling systems, and starts with the modeling direction, constructs a network model for the combined calculation of electric/gas/heat/cold multi-energy flows from both steady and dynamic aspects. In the process of solving the dynamic model, a hybrid stepping time domain simulation method for electromechanical transient simulation of power system and a medium and long-term transient simulation for non-power system are proposed. Finally, the electric heating coupling network is constructed for simulation verification. Document [21] proposes a multi-time scale multi-objective optimal joint scheduling model, which takes into account the different time scale characteristics of wind/water resources, electric/thermal load characteristics, peak load regulation capability, thermoelectric coupling characteristics of different thermal units, transmission capacity and system rotation standby requirements, and proposes a multi-time scale scheduling method and wind power prediction technology. In the short-term prediction technology, a day-to-day plan has been established to optimize the unit output of thermal generators, minimize operating costs and maximize the capacity to accommodate wind energy. Establish a time-of-day plan to optimize the power generation plans of all thermal, hydro and wind power plants on the basis of ultra-short-term wind power forecasting technology. However, in these studies, the attribute differences and energy dispatching characteristics of different energy sources are ignored, for example, the regulation of electric energy is more sensitive and its control mode is more flexible [22,23]; Thermal energy shows natural flexibility in regulation, with long response time and slow dynamic process [24]; However, natural gas needs to be added with conversion links during its use. On the premise that natural gas can be used in large quantities, it is inevitable to add large-capacity energy storage devices [25]; the cold energy change process is slow and requires a long dynamic time, which is not conducive to long-term storage [26]. At the same time, due to the strong coupling relationship of integrated energy systems, various energy systems influence each other, especially the non-linear devices of multiple coupled systems force the system to change frequently in operating load rate under the demand response, resulting in system "source-load" power imbalance. Therefore, based on the feasible interval conditions of system load rate, this paper studies the multi-time scale optimal operation of the cold heat-electricity integrated energy system with the economic operation and the lowest pollutant gas emissions as multi-objective functions.

In the following content arrangement, Chapter 2 describes the topology and working principle of the cold-thermal-electricity integrated energy system. Chapter 3 analyzes the feasible range of load rate of integrated energy system. In Chapter 4, a multi-time scale optimization model is established with economic operation and minimum pollutant gas emissions as objective functions. In Chapter 5, the cold/heat/electricity front load of Jianguo hotel in Zhangjiakou, China is selected for simulation analysis and compared with a single time scale system. Chapter 6 summarizes this article.

2. Topological Structure and Working Principle of Cold-Thermal-Electricity Integrated Energy System

The integrated energy system has various structures, different forms and complicated coupling relationships, there are various energy production and conversion devices in the system. In this paper, the topology structure adopted for the research on the optimal operation of the cold-thermal-electricity integrated energy system is shown in Figure 1.

The cold-thermal-electricity integrated energy system is micro-energy network level. The equipment in the system mainly includes: gas internal combustion engine (GE), flue gas absorption heat pump (AP), cylinder liner water heat exchanger (JW), absorption refrigerator (AC), electric boiler (EB) and electric refrigerator (EC), and two kinds of energy storage equipment including electricity storage (ST) and heat storage (HS) are added. At the same time, in order to make full use of local reliable solar energy resources, a photovoltaic generator set (PV) is added and connected to the power network to ensure the balance of power supply and demand in the system. The whole system takes a gas internal combustion engine and a flue gas absorption heat pump as the core, the gas internal combustion engine directly supplies part of the electric load by consuming natural gas and generating electric energy, and the high-temperature steam generated during working is converted into hot water to supply the heat load through a cylinder liner water heat exchanger; The flue gas generated during the combustion of natural gas can be absorbed and utilized by most of the flue gas absorption heat pumps and converted into heat energy and cold energy for direct supply to users; The absorption refrigerator converts part of the heat energy on the heat bus into cold energy to supply the cold load for use; When the system needs more heat energy or cold energy, some of the heat energy and cold energy deficiency can be made up through the work of the electric boiler or electric refrigerator. The system is also connected with two energy storage devices of heat storage and electricity storage to ensure that the system has sufficient margin of electric/thermal power capacity and increase the stability of the system. In addition, the active connection of photovoltaic generator sets not only effectively utilizes solar energy resources, but also increases the environmental protection and economic benefits of the system. When the demand for electric energy is large, the system can interact with the power grid. At the same time, in order to reduce the construction cost and coordination cost of the information channel and physical channel between the system and the power grid, the system adopts the principle of "connecting to the grid without power output " to purchase electric energy from the power grid, so as to make up for the shortage of electric energy in the system and ensure the stable operation of the system.

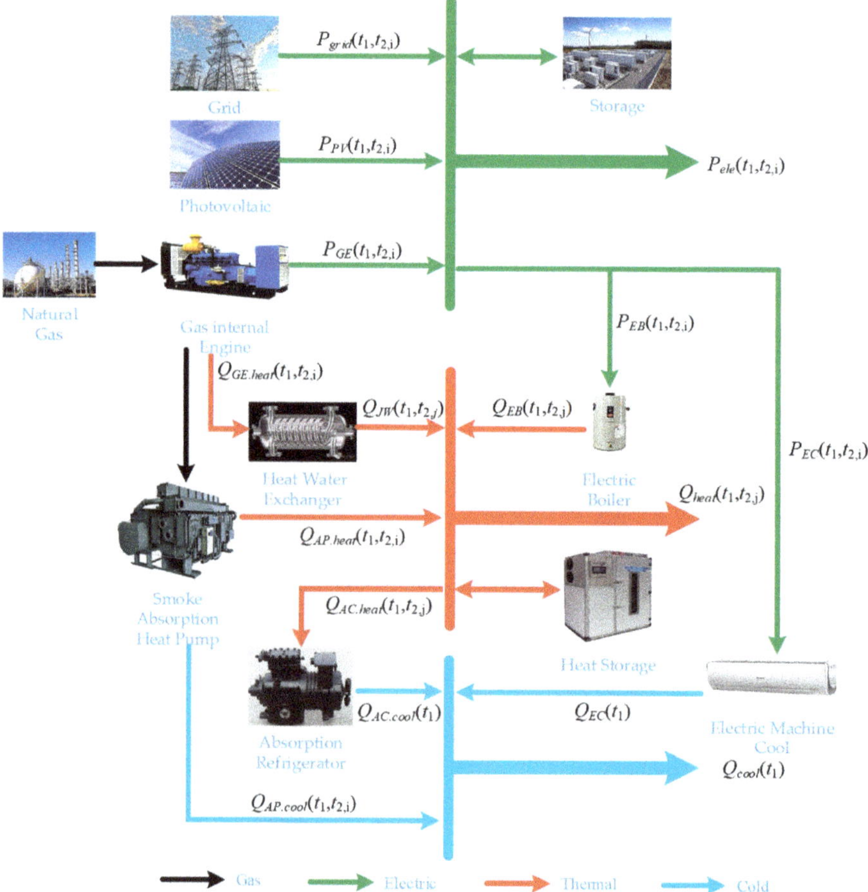

Figure 1. Topology diagram of cold-thermal-electricity integrated energy system.

3. The Feasible Range of System Load Rate

The integrated energy system realizes the coupling relation of multiple types of heterogeneous energy flows, and the system has strong coupling, nonlinearity, multi-energy complementation and mutual influence characteristics. For the cold-thermal-electricity integrated energy system, even if the operating characteristics and operating condition change parameters of each equipment in the system are clearly defined, there are many uncertainties in the energy output of the system under the condition that various energy input terminals are determined. Especially in the research process, it is found that when the cold-thermal-electricity integrated energy system is directly connected to the load side, the output power of the system does not match the load and the load rate of the system cannot find a reliable operation interval, so it cannot operate stably. Therefore, it is necessary to analyze the feasible range of system load rate.

Because of the close coupling relationship between the cold-thermal-electricity integrated energy system, it is difficult to directly analyze the load rate of the system. Therefore, it is possible to decouple the system and analyze the operational load rate characteristics of each decoupling subsystem, thus ensuring the "source-load" power balance of the system.

The decoupling method in this paper is as follows:

1. Build a system model according to the system topology, equipment and power constraints (see Chapter 4 for detailed models);
2. Change the system load factor A input from 0.1, 0.2 ... 1.0;
3. When the external power grid input is 0 KW, calculate the lowest lower limit of the cold, heat and electricity output power of each decoupling system;
4. When the external power grid input is 500 KW, calculate the maximum upper limit of the cooling, heating and electric output power of each decoupling system;
5. Power load ratio curves of each decoupling subsystem are obtained.

Therefore, the feasible range of the load rate of each decoupling subsystem of the cold-thermal-electricity integrated energy system is shown in the following Figure 2:

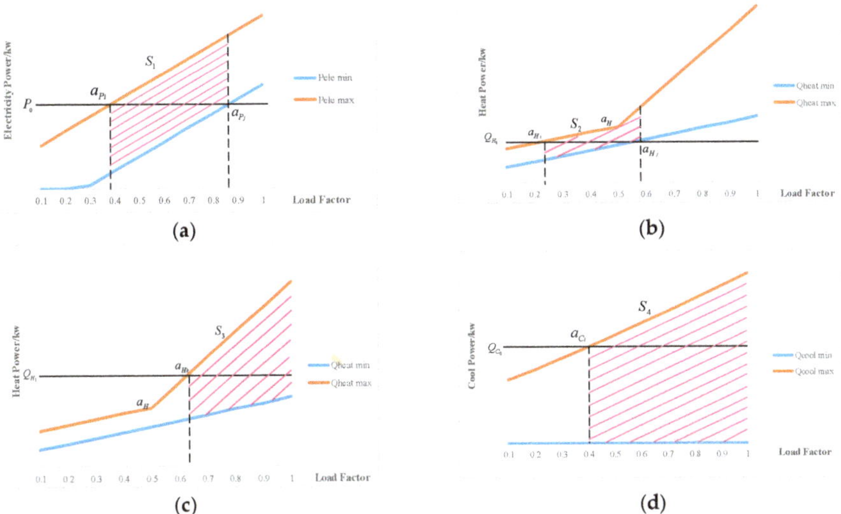

Figure 2. Load rate feasible interval of decoupling subsystem. (**a**) Decoupling power subsystem; (**b**) Decoupling the low load characteristics of the thermal energy subsystem; (**c**) Decoupling the high load characteristics of the thermal energy subsystem; (**d**) Decoupling the cold energy subsystem.

Since the loads in the cold-thermal-electricity integrated energy system are independent of each other and do not interfere with each other. For the analysis of the feasible range of system load rate, as shown in Figure 2a, when the electric power load is P_0, it intersects with the upper and lower limits of electric power at two points of load rate a_{Pi} and a_{Pj}, then the feasible load rate of the power subsystem will be between (a_{Pi}, a_{Pj}), and the electric power output power is shown by shadow S_1 in the figure. There is a critical jump Q_H in the upper limit of the power output of the decoupled thermal energy subsystem, where the system load rate is a_H. When the thermal energy load received by the system is lower than Q_H, it is called the low load characteristic of the thermal energy subsystem, and when the thermal energy load received by the system is higher than Q_H, it is called the high load characteristic of the thermal energy subsystem. As shown in Figure 2b, under the condition of low load characteristics, the upper and lower limits of the thermal energy load Q_{H0} of the system and the power output of the thermal energy decoupling subsystem intersect at two points of the load rates a_{Hi} and a_{Hj}, so that the system load rate is limited between (a_{Hi}, a_{Hj}), and the thermal energy output range of the thermal energy subsystem is shaded S_2 in Figure 2b; When the thermal energy load Q_{H1} is higher than Q_H, the thermal energy subsystem exhibits a high load characteristic, as shown in Figure 2c, the load rate of the system will be higher than a_{Hk}, and the thermal energy output range is shown in shadow S_3. For the cold energy subsystem, as shown in Figure 2d, the lower limit of the cold energy output power is stable, while the upper limit of the output power varies approximately linearly, so that the cold energy system has a wider load rate selection range when outputting power outward. When the system cold energy load is Q_{C0}, the system load rate can fall between (a_{Ci}, 1), and the cold energy output is shown in shadow S_4. To sum up, the analysis of the operating characteristics of the load rate of each decoupling subsystem shows that the loads of inter-cold, heat and electricity in the system are independent and uncertain. In order to balance the "source-load" power of the system, the system load rate A should be in the same interval as the operational load rate of each decoupling subsystem, namely:

$$a \in (a_{Pi}, a_{Pj}) \cap (a_{Hi}, a_{Hj}) \cap (a_{Ci}, 1)$$
$$or\ a \in (a_{Pi}, a_{Pj}) \cap (a_{Hk},\ 1) \cap (a_{Ci}, 1) \tag{1}$$

However, when the load rate intervals of each decoupling subsystem have no intersection, the "source-load" power of the system is unbalanced and cannot operate stably. Considering that the system load fluctuates within a stable range, it is possible to change the operating conditions of the system by adjusting the system equipment parameters and widen the intersection interval of load rates so that the system has a stable feasible interval. This paper will also study the optimal operation of the cold-thermal-electricity integrated energy system on the basis of the feasible load rate range of the system.

4. Optimize Operation

The operation process of the cold-thermal-electricity integrated energy system is complex, with many parameters and various energy outputs. This paper optimizes the operation of the cold-thermal-electricity integrated energy system with multi-objective and multi-time scales. In terms of multi-objective function, the system has the lowest operating cost and the least pollutant gas emission. In terms of multi-time scale, considering the difference of attributes of different energy sources and the different flexible characteristics of operation, for example, short power dispatching period, flexible regulation and fast power release and absorption, therefore, power is operated at $\Delta t_{2,i} = 15$ min as one dispatching time; However, heat energy is easier to adjust and store than cold energy. Therefore, heat energy is operated at $\Delta t_{2,j} = 30$ min as a scheduling time, while cold energy is operated at $\Delta t_1 = 1$ h as a scheduling time. In this way, the operation periods of cold, heat and electricity are multiple of each other, which is beneficial for multiple energy sources to supplement each other and ensures the consistency of multi-time scale coordination. At the same time, in order to reduce system losses and adjustment costs, the load rate of the system is kept unchanged for one hour, i.e., 4 periods of power adjustment.

4.1. Objective Function

The objective function is to construct a multi-objective function with the lowest total operating cost of the system and the lowest pollutant gas emissions, as follows:

$$\min \begin{cases} F_{run} = \sum\limits_{t_1=1}^{24} \sum\limits_{i=1}^{4} \left\{ F_{grid}(t_1,t_{2,i}) + F_{gas}(t_1,t_{2,i}) + F_{main}(t_1) \right\} \\ F_{poll} = \sum\limits_{t_1=1}^{24} \sum\limits_{i=1}^{4} \left\{ \alpha_{sour} \cdot P_{grid}(t_1,t_{2,i}) + \alpha_{tran} \cdot P_{grid}(t_1,t_{2,i}) + \alpha_{PGE} \cdot P_{GE}(t_1,t_{2,i}) \right\} \end{cases} \tag{2}$$

In the total operating cost, the electricity purchase cost, natural gas purchase cost and system equipment maintenance cost of the system and power grid are considered; Pollution gas emissions from the power supply side of the power grid, power loss from the power transmission lines of the power grid, and pollution gas emissions from the operation of the gas internal combustion engine are considered in the pollution gas emissions.

Where F_{run} is the total operating cost of the system; $F_{grid}(t_1,t_{2,i})$ is the electricity purchase cost of the system and the power grid; $F_{gas}(t_1,t_{2,i})$ purchase natural gas for the system; $F_{main}(t_1)$ is the maintenance cost of system equipment; F_{poll} is the amount of pollutant gas discharged; α_{sour} is the pollutant gas emission coefficient on the power supply side of the power grid; α_{tran} is the pollutant gas emission coefficient delivered by power grid lines; $P_{grid}(t_1,t_{2,i})$ purchases power for the system and power grid; α_{PGE} is the pollutant emission coefficient of gas internal combustion engine. $P_{GE}(t_1,t_{2,i})$ is the output electric power of the gas internal combustion engine; t_1 is expressed as the number of hours in a day. t_2 is expressed as the number of minutes in an hour. Since electric energy runs at $\Delta t_{2,i} = 15$ min as a scheduling time, electric energy is scheduled 4 times in an hour, $i = 1, 2, 3$ and 4. Similarly, heat energy is dispatched twice in one hour, with $j = 1$ and 2.

Among them, the system electricity purchase cost is specifically expressed as follows:

$$F_{grid}(t_1,t_{2,i}) = P_{grid}(t_1,t_{2,i}) \cdot \Delta t_{2,i} \cdot f_{grid}(t_1,t_{2,i}) \tag{3}$$

where, $f_{grid}(t_1,t_{2,i})$ is the real-time electricity price of the power grid.

The system purchase cost of natural gas is specifically expressed as follows:

$$F_{gas}(t_1,t_{2,i}) = V_{gas}(t_1,t_{2,i}) \cdot \Delta t_{2,i} \cdot f_{gas}(t_1,t_{2,i}) \tag{4}$$

where $V_{gas}(t_1,t_{2,i})$ is the volume of natural gas consumed by the system; $f_{gas}(t_1,t_{2,i})$ is the price of natural gas.

The system equipment maintenance costs are specified as follows:

$$F_{main}(t_1) = \sum\limits_{i=1}^{4} \sum\limits_{j=1}^{2} \left\{ \begin{array}{l} k_{GE}[P_{GE}(t_1,t_{2,i})] \cdot \Delta t_{2,i} \cdot P_{GE}(t_1,t_{2,i}) + k_{PV}[P_{PV}(t_1,t_{2,i})] \cdot \Delta t_{2,i} \cdot P_{PV}(t_1,t_{2,i}) + k_{AP.cool}[Q_{AP.cool}(t_1,t_{2,i})] \cdot \Delta t_{2,i} \cdot Q_{AP.cool}(t_1,t_{2,i}) + \\ k_{AP.heat}[Q_{AP.heat}(t_1,t_{2,i})] \cdot \Delta t_{2,i} \cdot Q_{AP.heat}(t_1,t_{2,i}) + k_{AC.heat}[Q_{AC.heat}(t_1,t_{2,j})] \cdot \Delta t_{2,j} \cdot Q_{AC.heat}(t_1,t_{2,j}) + \\ k_{batt.dis/cha}[P_{batt.dis/cha}(t_1,t_{2,i})] \cdot \Delta t_{2,i} \cdot P_{batt.dis/cha}(t_1,t_{2,i}) + k_{stor.dis/cha}[Q_{stor.dis/cha}(t_1,t_{2,j})] \cdot \Delta t_{2,j} \cdot Q_{stor.dis/cha}(t_1,t_{2,j}) \end{array} \right\} \tag{5}$$

where $k_{GE}[P_{GE}(t_1,t_{2,i})]$ is the maintenance coefficient of the gas internal combustion engine at different output powers; $k_{PV}[P_{PV}(t_1,t_{2,i})]$ is the maintenance coefficient of photovoltaic generator set; $P_{PV}(t_1,t_{2,i})$ is the generating power of photovoltaic generator set; $k_{AP.cool}[Q_{AP.cool}(t_1,t_{2,i})]$ is the cold power maintenance coefficient of the flue gas absorption heat pump; $Q_{AP.heat}(t_1,t_{2,i})$ is the cold power output by the flue gas absorption heat pump; $k_{AC.heat}[Q_{AC.heat}(t_1,t_{2,i})]$ is the thermal power maintenance coefficient of the flue gas absorption heat pump; $Q_{AP.heat}(t_1,t_{2,i})$ is the output heat power of the flue gas absorption heat pump; $k_{AC.heat}[Q_{AC.heat}(t_1,t_{2,j})]$ is the maintenance coefficient of absorption chiller; $Q_{AC.heat}(t_1,t_{2,j})$ is the heat power absorbed by the absorption refrigerator; $k_{batt.dis/cha}[P_{batt.dis/cha}(t_1,t_{2,i})]$, $k_{stor.dis/cha}[Q_{stor.dis/cha}(t_1,t_{2,j})]$ are the power maintenance coefficients of power storage equipment and heat storage equipment respectively; $P_{batt.dis/cha}(t_1,t_{2,i})$ and $Q_{stor.dis/cha}(t_1,t_{2,j})$ are the interactive power of electric storage equipment and heat storage equipment.

4.2. Constraints

The cold-thermal-electric integrated energy system constraints include: equipment model constraints and power balance constraints.

4.2.1. Equipment Model Constraints

Among them, the equipment model includes a gas internal combustion engine model [27,28], a flue gas absorption heat pump model [29–31], an absorption refrigerator model [32], a cylinder liner water heat exchanger model, an electric boiler model, an electric refrigerator model, a photovoltaic generator set model [33], a heat and electricity storage model [34–36].

- Gas engine model.

Among them, gas-fired internal combustion engines have good electrical energy and thermal energy output characteristics which can be divided into power generation, heat generation and consumption of natural gas.

Power generation of gas internal combustion engine, the maximum real-time output power of an internal combustion engine is limited to $P_{\max}$:

$$\begin{cases} \eta_{GE.elec}(t_1,t_{2,i}) = a_3\left(\frac{P_{GE}(t_1,t_{2,i})}{P_{\max}}\right)^3 + a_2\left(\frac{P_{GE}(t_1,t_{2,i})}{P_{\max}}\right)^2 + a_1\left(\frac{P_{GE}(t_1,t_{2,i})}{P_{\max}}\right) + a_0 \\ P_{GE}(t_1,t_{2,1}) = P_{GE}(t_1,t_{2,2}) = P_{GE}(t_1,t_{2,3}) = P_{GE}(t_1,t_{2,4}) \\ \left| P_{GE}(t_1,t_{2,i}) - P_{GE}(t_1,t_{2,i}-1) \right| \leq P_{GE.\max} \end{cases} \tag{6}$$

Heating power of gas internal combustion engine:

$$\begin{cases} Q_{GE.heat}(t_1,t_{2,i}) = \frac{P_{GE}(t_1,t_{2,i})}{\eta_{GE.elec}(t_1,t_{2,i})}\left(1 - \eta_{GE.elec}(t_1,t_{2,i}) - \eta_L\right) \\ Q_{GE.heat}(t_1,t_{2,1}) = Q_{GE.heat}(t_1,t_{2,2}) = Q_{GE.heat}(t_1,t_{2,3}) = Q_{GE.heat}(t_1,t_{2,4}) \end{cases} \tag{7}$$

Natural gas consumed by gas internal combustion engines:

$$V_{gas}(t_1,t_{2,i}) = \frac{P_{GE}(t_1,t_{2,i}) \cdot \Delta t_2}{\eta_{gas} \cdot \eta_{GE.elec}(t_1,t_{2,i}) \cdot LHV} \tag{8}$$

In the formula, the output electric power $P_{GE}(t_1,t_{2,i})$ of the gas internal combustion engine remains unchanged for one hour, i.e., four periods of electric energy output; $\eta_{GE.elec}(t_1,t_{2,i})$ is the power generation efficiency of the gas internal combustion engine; $P_{\max}$ is the rated power of gas internal combustion engine. The thermal power $Q_{GE.heat}(t_1,t_{2,i})$ output by the gas internal combustion engine remains constant within one hour. η_L is the inherent loss rate of gas internal combustion engine; $P_{GE.\max}$ is the output gradient constraint of gas internal combustion engine. LHV is the low calorific value of natural gas; η_{gas} is the natural gas utilization rate of gas internal combustion engines; a_3, a_2, a_1 and a_0 are fitting constants respectively.

- Smoke absorption heat pump model.

Smoke absorption heat pump model can be divided into three parts: absorbing smoke, outputting heat energy and cold energy.

Smoke absorption heat pump absorbs smoke, absorption heat pump absorbed the flue gas temperature are directly affected parameter COP_{AP}:

$$\begin{cases} T(t_1,t_{2,i}) = b_1 \cdot \left(\frac{P_{GE}(t_1,t_{2,i})}{P_{\max}}\right) + b_0 \\ C_w(t_1,t_{2,i}) = b_3 \cdot T(t_1,t_{2,i}) + b_2 \\ COP_{AP}(t_1,t_{2,i}) = b_5 \cdot \left(\frac{P_{GE}(t_1,t_{2,i})}{P_{\max}}\right)^{b_4} \\ \lambda_{heat}(t_1,t_{2,i}) + \lambda_{cool}(t_1,t_{2,i}) = 1 \end{cases} \tag{9}$$

At the same time, the real-time power of flue gas absorption heat pump has linear constraint values $Q_{AP.heat.\max}$ and $Q_{AP.cool.\max}$.

Smoke absorbs heat output from heat pump:

$$\begin{cases} Q_{AP.heat}(t_1,t_{2,i}) = \lambda_{heat}(t_1,t_{2,i}) \cdot C_w(t_1,t_{2,i}) \cdot (T(t_1,t_{2,i}) - T_{heat}) \cdot COP_{AP}(t_1,t_{2,i}) \cdot L_{heat}(t_1,t_{2,i}) \cdot \eta_{AP.heat} \\ Q_{AP.heat}(t_1,t_{2,1}) = Q_{AP.heat}(t_1,t_{2,2}) = Q_{AP.heat}(t_1,t_{2,3}) = Q_{AP.heat}(t_1,t_{2,4}) \\ \left| Q_{AP.heat}(t_1,t_{2,i}) - Q_{AP.heat}(t_1,t_{2,i}-1) \right| \leq Q_{AP.heat.\max} \\ 0 \leq L_{heat}(t_1,t_{2,i}) \leq L_{heat.\max} \end{cases} \tag{10}$$

Smoke absorption heat pump outputs cold energy:

$$\begin{cases} Q_{AP.cool}(t_1,t_{2,i}) = \lambda_{cool}(t_1,t_{2,i}) \cdot C_w(t_1,t_{2,i}) \cdot (T(t_1,t_{2,i}) - T_{cool}) \cdot COP_{AP}(t_1,t_{2,i}) \cdot L_{cool}(t_1,t_{2,i}) \cdot \eta_{AP.cool} \\ Q_{AP.cool}(t_1,t_{2,1}) = Q_{AP.cool}(t_1,t_{2,2}) = Q_{AP.cool}(t_1,t_{2,3}) = Q_{AP.cool}(t_1,t_{2,4}) \\ \left| Q_{AP.cool}(t_1,t_{2,i}) - Q_{AP.cool}(t_1,t_{2,i}-1) \right| \leq Q_{AP.cool.\max} \\ 0 \leq L_{cool}(t_1,t_{2,i}) \leq L_{cool.\max} \end{cases} \tag{11}$$

where in $T(t_1,t_{2,i})$ is the inlet temperature of the flue gas absorption heat pump; $C_W(t_1,t_{2,i})$ is the specific heat capacity of hot water at different temperatures; $COP_{AP}(t_1,t_{2,i})$ is the energy efficiency coefficient of flue gas absorption heat pump; The heating power $Q_{AP.heat}(t_1,t_{2,i})$) and the cooling power $Q_{AP.cool}(t_1,t_{2,i})$ of the flue gas absorption heat pump remain unchanged within one hour. $\lambda_{heat}(t_1,t_{2,i})$ and $\lambda_{cool}(t_1,t_{2,i})$ are the heating ratio and cooling ratio of flue gas of the flue gas absorption heat pump respectively; T_{heat} and T_{cool} are hot water outlet temperature and cold water outlet temperature respectively. $L_{heat}(t_1,t_{2,i})$ and $L_{cool}(t_1,t_{2,i})$ are the hot water and cold water flows of the flue gas absorption heat pump respectively; $L_{heat.\max}$ and $L_{cool.\max}$ are the maximum heating and cooling flows respectively; $\eta_{AP.heat}$ and $\eta_{AP.cool}$ are the heating and cooling efficiency of flue gas absorption heat pump respectively. $Q_{AP.heat.\max}$ is the gradient constraint of heating power output of flue gas absorption heat pump; $Q_{AP.cool.\max}$ is the gradient constraint of cooling power output of flue gas absorption heat pump; b_5, b_4, b_3, b_2, b_1 and b_0 are fitting constants respectively.

- Absorption refrigerator model.

$$\begin{cases} Q_{AC.cool}(t_1) = COP_{AC} \cdot \sum_{j=1}^{2} Q_{AC.heat}(t_1,t_{2,j}) \\ Q_{AC.heat.\min} \leq Q_{AC.heat}(t_1,t_{2,j}) \leq Q_{AC.heat.\max} \\ \left| Q_{AC.cool}(t_1) - Q_{AC.cool}(t_1-1) \right| \leq Q_{AC.cool.\max} \end{cases} \tag{12}$$

where $Q_{AC.cool}(t_1)$ is the cold power output by the absorption refrigerator; COP_{AC} is the energy efficiency coefficient of absorption refrigerator; $Q_{AC.heat.\min}$ and $Q_{AC.heat.\max}$ are the minimum and maximum thermal power absorbed by the absorption chiller respectively. $Q_{AC.cool.\max}$ is the output gradient constraint of absorption chiller.

In order to stabilize the output power of absorption refrigerator, the absolute value of the difference between the next output power and the previous one is $Q_{AC.cool.\max}$.

- Cylinder liner water heat exchanger model.

$$Q_{JW}(t_1,t_{2,j}) = \eta_{JW} \cdot \sum_{i=2j-1}^{2j} Q_{GE}(t_1,t_{2,i}) \tag{13}$$

where $Q_{JW}(t_1,t_{2,j})$ is the output thermal power of the cylinder liner water heat exchanger; η_{JW} is the heat transfer efficiency of cylinder liner water heat exchanger.

- Electric boiler model.

$$\begin{cases} Q_{EB}(t_1, t_{2,j}) = COP_{EB} \cdot \sum_{i=2j-1}^{2j} P_{EB}(t_1, t_{2,i}) \\ P_{EB.min} \leq P_{EB}(t_1, t_{2,i}) \leq P_{EB.max} \\ \left| Q_{EB}(t_1, t_{2,j}) - Q_{EB}(t_1, t_{2,j} - 1) \right| \leq Q_{EB.max} \end{cases} \tag{14}$$

In the formula, $P_{EB}(t_1,t_{2,i})$ is the input electric power of the electric boiler; $Q_{EB}(t_1,t_{2,j})$ is the output thermal power of the electric boiler; COP_{EB} is the energy production coefficient of electric boilers; $P_{EB.min}$ and $P_{EB.max}$ are respectively the minimum and maximum electric power of electric boilers; $Q_{EB.max}$ is the output gradient constraint of the electric boiler.

- Electric refrigerator model.

$$\begin{cases} Q_{EC}(t_1) = COP_{EC} \cdot \sum_{i=1}^{4} P_{EC}(t_1, t_{2,i}) \\ P_{EC.min} \leq P_{EC}(t_1, t_{2,i}) \leq P_{EC.max} \\ \left| Q_{EC}(t_1) - Q_{EC}(t_1 - 1) \right| \leq Q_{EC.max} \end{cases} \tag{15}$$

where $P_{EC}(t_1,t_{2,i})$ is the input electric power of the electric refrigerator; $Q_{EC}(t_1)$ is the output cold power of the electric refrigerator; COP_{EC} is the energy efficiency coefficient of electric refrigerator. $P_{EC.min}$ and $P_{EC.max}$ are the minimum and maximum electric power of electric refrigerator respectively. $Q_{EC.max}$ is the output gradient constraint of the electric refrigerator.

- Photovoltaic generator set model.

$$P_{PV}(t_1, t_{2,i}) = P_{STC} \frac{G_{ING}(t_1, t_{2,i})}{G_{STC}} [1 - k(T_{out}(t_1, t_{2,i}) - T_s)] \tag{16}$$

where, P_{STC} is the rated output of photovoltaic generator set; $G_{ING}(t_1,t_{2,i})$ is the real-time irradiation intensity; G_{STC} is the rated irradiation intensity of photovoltaic generator set; K is the power generation coefficient of photovoltaic generator set; $T_{out}(t_1,t_{2,i})$ is the external temperature; T_s is the reference temperature of the generator set.

- Electricity storage model.

In the model of electricity storage model, the power balance and charge and discharge constraints are considered.

$$\begin{cases} E_{batt}(t_1, t_{2,i}) = (1 - k_L) \cdot E_{batt}(t_1, t_{2,i} - 1) + \left[\eta_{batt.cha} \cdot P_{batt.cha}(t_1, t_{2,i}) - \frac{P_{batt.dis}(t_1, t_{2,i})}{\eta_{batt.dis}} \right] \cdot \Delta t_{2,i} \\ P_{batt.cha.min} \leq P_{batt.cha}(t_1, t_{2,i}) \leq P_{batt.cha.max} \\ P_{batt.dis.min} \leq P_{batt.dis}(t_1, t_{2,i}) \leq P_{batt.dis.max} \\ E_{batt.min} \leq E_{batt}(t_1, t_{2,i}) \leq E_{batt.max} \end{cases} \tag{17}$$

where in $E_{batt}(t_1,t_{2,i})$ is that real-time capacity of the pow storage equipment; k_L is the self-loss coefficient of power storage equipment; $\eta_{batt.cha}$ and $\eta_{batt.dis}$ are the charging and discharging efficiency of power storage equipment respectively. $P_{batt.cha}(t_1,t_{2,i})$ and $P_{batt.dis}(t_1,t_{2,i})$ are the charging and discharging power of the power storage equipment respectively. $P_{batt.dis.max}$ and $P_{batt.dis.min}$ are the maximum and minimum discharge powers of power storage equipment respectively. $P_{batt.cha.max}$ and $P_{batt.cha.min}$ are respectively the maximum and minimum charging power of power storage equipment. $E_{batt.max}$ and $E_{batt.min}$ are the maximum and minimum storage capacities of power storage equipment respectively.

- Thermal storage model.

Thermal storage model have similar model constraints to electricity storage model.

$$
\left\{
\begin{array}{c}
B_{stor}(t_1, t_{2,j}) = (1 - k_s) \cdot B_{stor}(t_1, t_{2,j} - 1) + \left[\eta_{stor.cha} \cdot Q_{stor.cha}(t_1, t_{2,j}) - \dfrac{Q_{stor.dis}(t_1, t_{2,j})}{\eta_{stor.dis}}\right] \cdot \Delta t_{2,j} \\
Q_{stor.cha.\min} \leq Q_{stor.cha}(t_1, t_{2,j}) \leq Q_{stor.cha.\max} \\
Q_{stor.dis.\min} \leq Q_{stor.dis}(t_1, t_{2,j}) \leq Q_{stor.dis.\max} \\
B_{stor.\min} \leq B_{stor}(t_1, t_{2,j}) \leq B_{stor.\max}
\end{array}
\right. \tag{18}
$$

where $B_{stor}(t_1,t_{2,j})$ is the real-time capacity of heat storage equipment; k_s is the self-loss coefficient of heat storage equipment; $\eta_{stor.cha}$ and $\eta_{stor.dis}$ are the heat absorption and heat release efficiency of heat storage equipment respectively. $Q_{stor.cha}(t_1,t_{2,j})$ and $Q_{stor.dis}(t_1,t_{2,j})$ are the heat absorption and heat release power of heat storage equipment respectively. $Q_{stor.cha.\max}$ and $Q_{stor.cha.\min}$ are the maximum and minimum heat absorption powers of heat storage equipment respectively. $Q_{stor.dis.\max}$ and $Q_{stor.dis.\min}$ are the maximum and minimum heat release power of heat storage equipment respectively. $B_{stor.\max}$ and $B_{stor.\min}$ are the maximum and minimum capacity constraints of heat storage equipment respectively.

4.2.2. Power Balance Constraints

The power balance constraints of cold, heat and electricity are satisfied in the system.

- Electric power balance constraint.

$$
\begin{array}{l}
P_{grid}(t_1, t_{2,i}) + P_{PV}(t_1, t_{2,i}) \quad + P_{GE}(t_1, t_{2,i}) + P_{batt.dis}(t_1, t_{2,i}) \cdot D_{batt.dis}(t_1, t_{2,i}) = \\
\quad P_{batt.cha}(t_1, t_{2,i}) \cdot D_{batt.cha}(t_1, t_{2,i}) + P_{ele}(t_1, t_{2,i}) + P_{EB}(t_1, t_{2,i}) + P_{EC}(t_1, t_{2,i})
\end{array} \tag{19}
$$

In the formula, $D_{batt.dis}(t_1,t_{2,i})$ and $D_{batt.cha}(t_1,t_{2,i})$ are respectively discharge and charge variables of power storage equipment; $P_{ele}(t_1,t_{2,i})$ is the power load.

- Thermal power balance constraint.

$$
\begin{array}{l}
Q_{JW}(t_1, t_{2,j}) + \displaystyle\sum_{i=2j-1}^{2j} Q_{AP.heat}(t_1, t_{2,i}) \quad + Q_{EB}(t_1, t_{2,j}) + Q_{stor.dis}(t_1, t_{2,j}) \cdot D_{stor.dis}(t_1, t_{2,j}) = \\
\quad Q_{stor.cha}(t_1, t_{2,j}) \cdot D_{stor.cha}(t_1, t_{2,j}) + Q_{heat}(t_1, t_{2,j}) + Q_{AC.heat}(t_1, t_{2,j})
\end{array} \tag{20}
$$

In the formula, $D_{stor.dis}(t_1,t_{2,j})$ and $D_{stor.cha}(t_1,t_{2,j})$ are the heat release and heat absorption variables of the heat storage equipment respectively; $Q_{heat}(t_1,t_{2,j})$ is thermal load.

- Cold power balance constraint.

$$
Q_{AC.cool}(t_1) + Q_{EC}(t_1) + \sum_{i=1}^{4} [Q_{AP.cool}(t_1, t_{2,i})] = Q_{cool}(t_1) \tag{21}
$$

where $Q_{cool}(t_1)$ is the cooling load.

4.3. Solution

4.3.1. Multi-Objective Solution Method

This model is a multi-objective mixed integer nonlinear programming model. The multi-objective problem is transformed into a single-objective problem that is easy to solve by adopting a scalar linear weighting method. The results under different weight conditions are compared by changing the weight coefficient to obtain the optimal result. The scalar process is as follows:

Firstly, the optimal values of economic operation and pollutant gas emission under single-objective conditions are solved respectively, and $F_{run.min}$ and $F_{poll.min}$ are obtained.

$$\min \ F_{run} = \sum_{t_1=1}^{24} \sum_{i=1}^{4} \left\{ F_{grid}(t_1, t_{2,i}) + F_{gas}(t_1, t_{2,i}) + F_{main}(t_1) \right\} \tag{22}$$

$$\min \ F_{poll} = \sum_{t_1=1}^{24} \sum_{i=1}^{4} \left\{ \alpha_{sour} \cdot P_{grid}(t_1, t_{2,i}) + \alpha_{tran} \cdot P_{grid}(t_1, t_{2,i}) + \alpha_{PGE} \cdot P_{GE}(t_1, t_{2,i}) \right\} \tag{23}$$

Secondly, the multi-objective optimization problem is transformed into a single-objective problem solving calculation through a linear weighting method. The solving process is as follows:

$$\min F = k_{run} \cdot \frac{F_{run}}{F_{run.min}} + k_{poll} \cdot \frac{F_{poll}}{F_{poll.min}} \ , \ k_{run} + k_{poll} = 1 \tag{24}$$

where F is the mixed objective function value; k_{run} is the economic operation weight coefficient; k_{poll} is the weight coefficient of pollutant gas emission.

By changing the values of the weight coefficients k_{run} and k_{poll} to adjustment and optimization results of the objective function are compared, the sum of weight coefficients is always 1. In this paper, 0.1 is selected as the discrimination degree of two parameters. Therefore, the value of the weight coefficient $(k_{run}, k_{poll}) = (0.1, 0.9), (0.2, 0.8) \ldots (0.9, 0.1)$ changes.

Finally, by changing different weight coefficients, the optimal operating conditions under different weights are calculated, and the result analysis is obtained.

4.3.2. Model Optimization Process

This model is programmed by LINGO software and solved by GLOBAL algorithm. The model optimization process is shown in Figure 3 below.

First of all, initialization the system, system variables and parameters are defined, input cold, thermal, electric load parameters and k_{run}/k_{poll} weight parameters, then, objective function, equipment model constraint and power balance constraint are input.

After that, the LINGO software is used to calculate the optimization model. If there is no feasible solution to the optimization model, then there is no solution to the original optimization problem; otherwise, continue to optimize the process. The results of feasible solutions are compared. If the result of comparison is the minimum of feasible solutions, the global optimal solution is obtained. Instead, continue to compare possible solutions.

Finally, output the optimal result and end the optimization process.

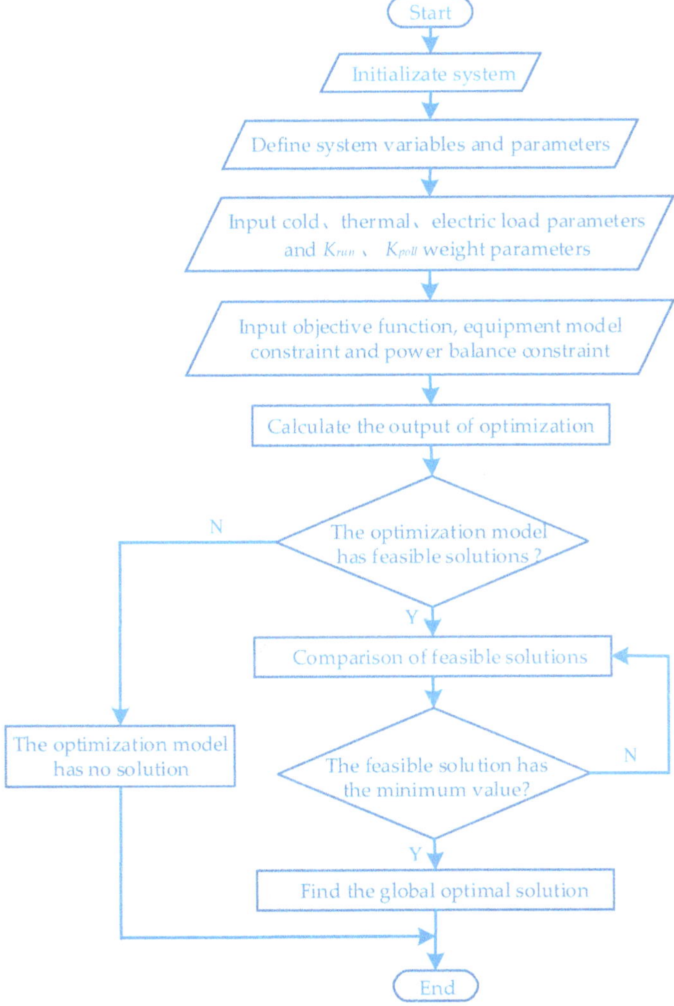

Figure 3. Model optimization process.

5. Numerical Simulation and Operation Load Rate Analysis

5.1. Numerical Simulation and Operation Results Analysis

This paper selects Jianguo hotel in Zhangjiakou, northern China, to analyze the system's cold, heat and electricity load before summer. At the same time, considering the feasible range of system load rate, the system operating conditions are adjusted by changing the system equipment parameters, thus expanding the system's operational load rate and balancing the system's "source-load" power.

The parameters of electricity, heat and cold load in summer are shown in the following Figures 4–6:

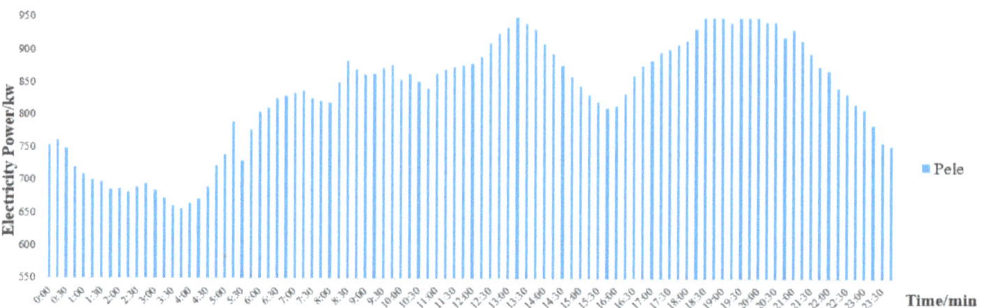

Figure 4. Summer electricity power load chart.

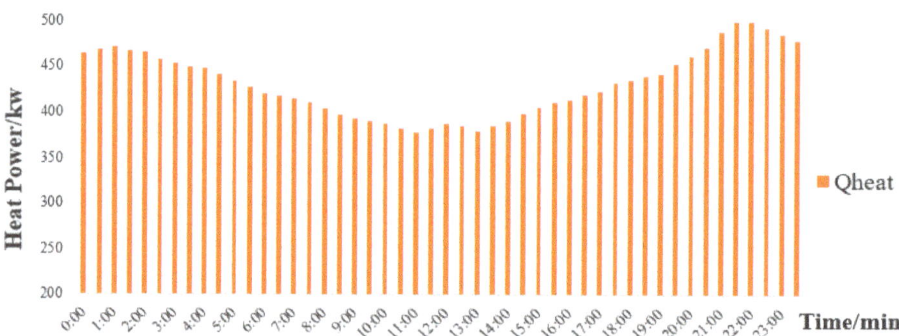

Figure 5. Summer heat load chart.

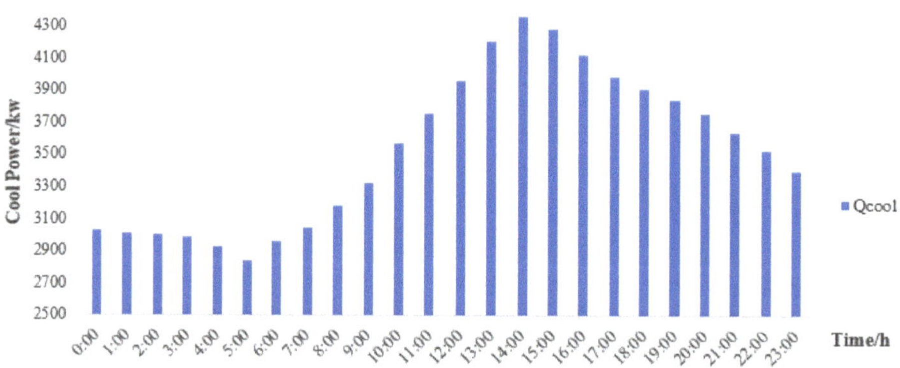

Figure 6. Summer cooling load chart.

The electricity power load before summer presents obvious "peak and valley" characteristics. The power demand is large during peak hours and low during valley hours. Therefore, it is very suitable for implementing time-of-use electricity price. The specific time-of-use electricity price is shown in Table 1 below.

Table 1. Time-of-use tariff.

Name	Time/h	Price/CNY
peak period	8~10, 18~19	0.866
peacetime period	11~17, 20~22	0.559
valley period	0~7, 23	0.223

In summer, the system has a low demand for heat energy, and the load fluctuates. The temperature rises from 10: 00 to 14: 00 noon, and the heat energy load is low. The system has a high demand for hot water and a high heat energy load from 21: 00 to 23: 00 at night.

In summer, the system has a strong demand for cool energy. In the early morning (0:00~5:00), the cool energy load is relatively low, then increases with the temperature rising, reaches the maximum at 14:00, and then continues to decline.

This optimization problem is a mixed integer nonlinear problem. Standard linear method is adopted to solve the multi-objective problem. By changing the weight coefficients of k_{run} and k_{poll}, different optimization results are compared, and the global optimal solution is obtained. In terms of optimization model calculation, this model has a total of 1923 variables and 1824 constraints, including 1128 linear variables and 766 linear constraints. In order to ensure that the output result is the global optimal solution, multiple feasible solutions need to be compared, so the calculation time is also relatively long. The calculation time for each group of different weight coefficients needs at least 3–4 h. The optimal results are obtained by changing the k_{run} and k_{poll} weight coefficients of multi-objective functions as shown in Figure 7 below:

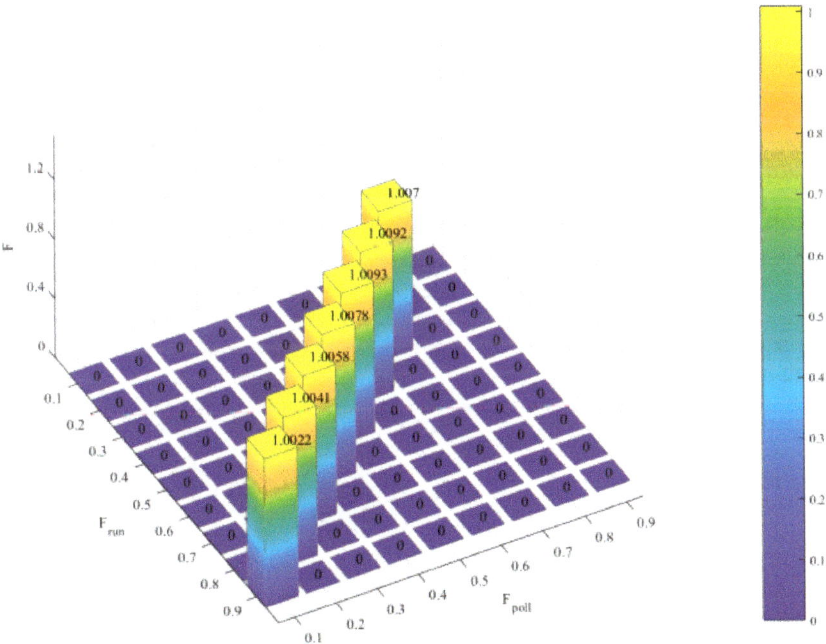

Figure 7. Optimal operation results of different weights in summer.

According to the optimization results of different weights in summer in Figure 7, when the weights (k_{run}, k_{poll}) = (0.1,0.9) and (0.2,0.8), the system cannot operate, and the weights of these two points should be discarded. When the weights k_{run} = 0.9 and k_{poll} = 0.1, the mixed objective function value F is the minimum of 1.0022. At this time, the electric/heat/cool output power of each equipment before summer is shown in the following Figures 8–10:

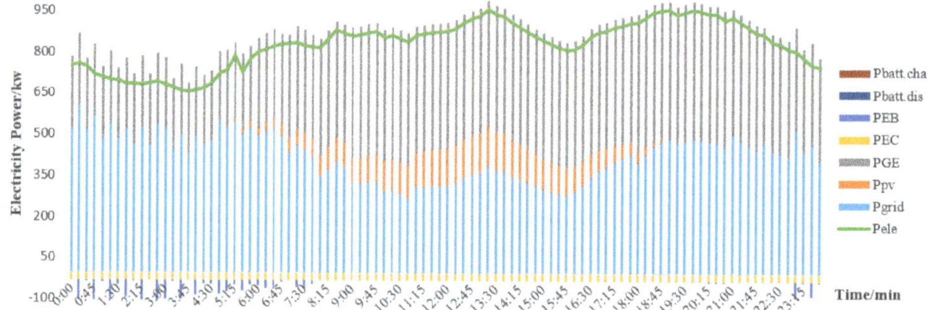

Figure 8. Power output diagram of various equipment before multi-time scale in summer.

As can be seen from Figure 8, the gas internal combustion engine has maintained good electric energy output characteristics. In the 0th to 7th and 23rd hour, the electricity price of the power grid is low, which is suitable for the system to receive a large amount of electric energy of the power grid and meets the requirements of system economy. The gas internal combustion engine is also put into operation, with the starting load rate kept above 55% and the output power stable. In the 8th to 17th hour, the photovoltaic unit is actively connected to the system, the system fully absorbs renewable energy, the power of the power grid is greatly reduced, and the gas internal combustion engine runs stably. At this time, the load rate is kept above 85%, and the electric energy output is stable. During the 18th to 22nd hour of the peak power consumption at night, although the photovoltaic unit will no longer output power, the gas internal combustion engine will remain in a state close to full load output and the power output of the power grid will be limited to a stable range. Electric boilers and electric refrigerators, as units of electric energy consumption, are converted into heat energy and cold energy to supply loads for use. At the same time, the charging and discharging power of the power storage equipment is less throughout the day, which indicates that the system can realize self-absorption, which is beneficial to prolonging the service life of the power storage equipment and the stable operation of the system power.

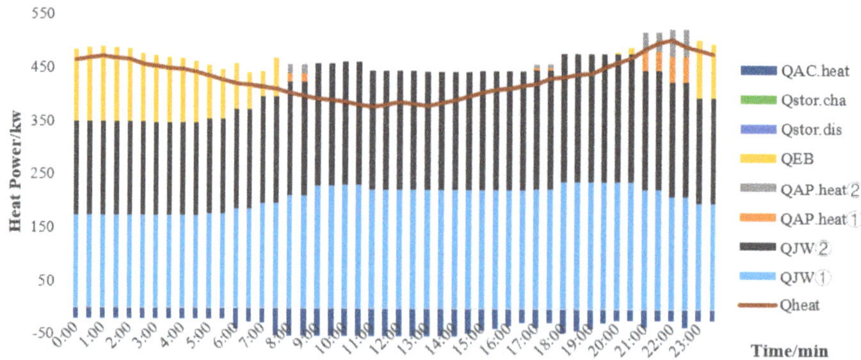

Figure 9. Heat output diagram of various equipment before multi-time scale in summer.

In summer, the demand for heat energy load is relatively low. The heat energy supplied by converting the cylinder liner water directly connected with the gas internal combustion engine into hot water basically bears the base load part of the heat energy load in summer. The electric energy operates twice in one heat energy operation cycle, while the gas internal combustion engine outputs stably in one hour. Therefore, the heat power output by the cylinder liner water is relatively stable in one heat energy operation cycle. The electric boiler works in the electricity valley period and supplements the heat energy load with the flue gas absorption heat pump. The absorption refrigerator absorbs the surplus heat energy from the cylinder liner water during the summer heat energy valley and converts it into cold energy. The heat storage equipment absorbs and releases less heat energy during one day's operation, which indicates that the system has balanced heat power and stable operation.

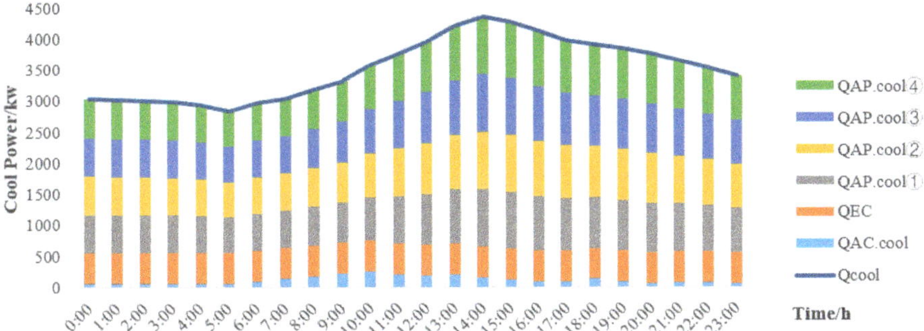

Figure 10. Cooling output diagram of various equipment before multi-time scale in summer.

In summer, the demand for cooling energy load is large, and the electric refrigerator and absorption refrigerator provide only a small part of the cooling energy load all day long. As the flue gas absorption heat pump directly absorbs the flue gas heat energy of the gas internal combustion engine, it can stably output four times of cooling energy in one cooling energy scheduling cycle. Moreover, the flue gas absorption heat pump is sensitive to load changes, has good load following characteristics, and meets the cooling power load demand of the system.

5.2. Analysis of Operation Load Rate Results

In terms of system operation load rate, the loads of cold, heat and electricity are independent of each other. Since the system load rate is affected by the loads of electricity, heat and cold at the same time, analyzing the changes of the three to the load rate will directly determine the stable operation of the system. The operation of electric/heat/cold load and load rate is shown in Figures 11–13.

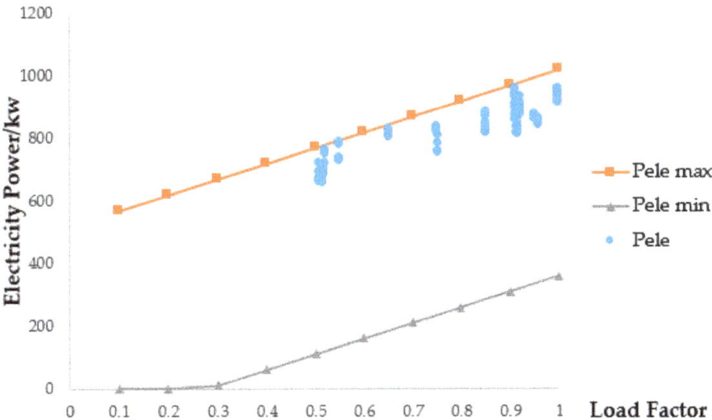

Figure 11. Multi-time scale power load-load rate operation.

As shown in Figure 11, the multi-time scale power load and load rate operating points are mostly close to the upper power limit of the decoupled power system, while the load rate operating points close to the upper power limit are close to full load operation, and the system is stable at this time. The closer the operating point is to the upper limit of power, the more thorough the system is running, the greater the utilization rate of power and the higher the economy of the system.

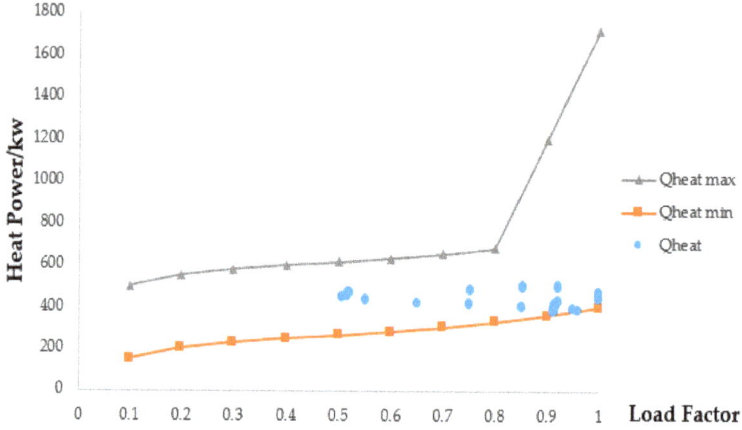

Figure 12. Multi-time scale heat load-load rate operation.

The multi-time scale heat load and load rate operation is shown in Figure 12. The summer heat load demand is low and the load fluctuation range is small, so the heat load and load rate operation points are distributed in a scatter line within the thermal energy output range. At the higher load rate, the operating point is close to the lower limit of the heat energy output boundary, but the heat storage equipment has not been put into operation at this time, indicating that the "source-charge" power of the heat energy subsystem is still balanced.

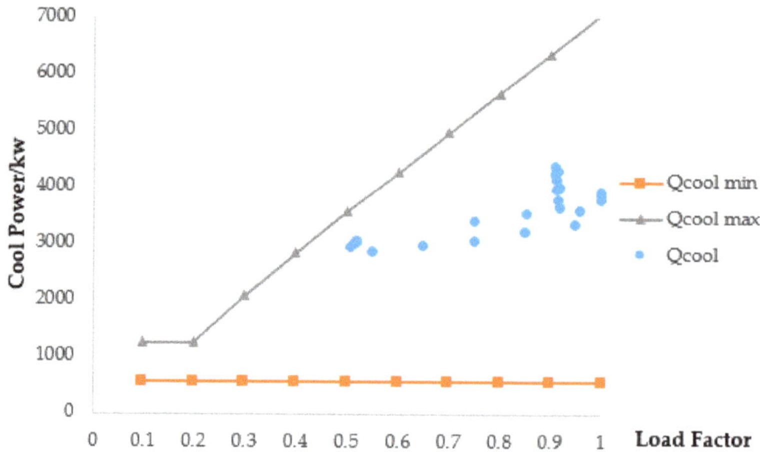

Figure 13. Multi-time scale cooling load-load rate operation.

For summer with large demand for cold energy, due to the strong compatibility and output characteristics of the cold energy system, there is still a wide range of load rate operation range selection under the condition of large fluctuation of cold load throughout the day, so that the load rate operation point can be located inside the output range and the cold energy subsystem can operate stably.

To sum up, under the condition that the cold/heat/electric loads are independent of each other and do not interfere with each other before summer, according to the analysis of the load rate operation results of each decoupling subsystem, the "source-load" power of the system is balanced and stable in operation.

5.3. Comparison between Multi-Time Scale and Single-Time Scale Systems

In order to analyze the difference of time complementarity between systems, this paper compares multi-time scale systems with single time scale systems. In a single-time scale system, according to the principle of minimum adaptability of load scheduling time, i.e., taking the longest scheduling period as the scheduling time, the cold/heat/electric energy systems will be unified into the same time scale scheduling ($\Delta t_1 = 1$ h), the multi-time scale electric energy systems will be unified into one hour by 15 min scheduling, the heat energy systems will be unified into one hour by 30 min scheduling, and the original load parameters will be added into one hour to calculate. As the peak power load of a single-time scale system increases nearly 4 times as much as that of a multi-time scale system, and the gas internal combustion engines cannot match the applicable models due to capacity, power and other reasons, it is necessary to increase the number of gas internal combustion engines to 4 in a single-time scale system and keep the load rates of each gas internal combustion engine consistent so as to facilitate systematic analysis and comparison. Then the single time scale system operation results are shown in the following Figures 14–16:

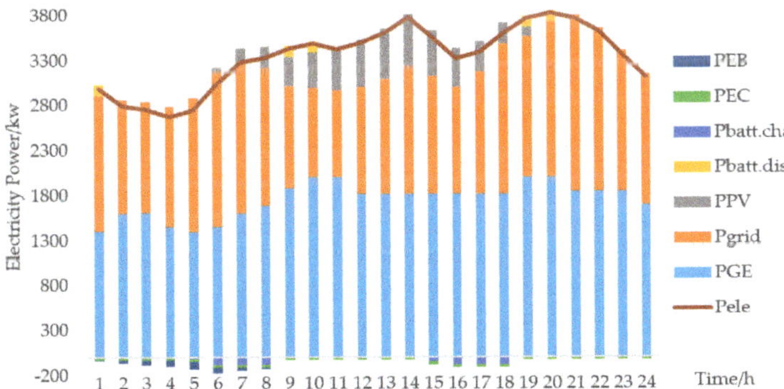

Figure 14. Power output diagram of various equipment before single time scale in summer.

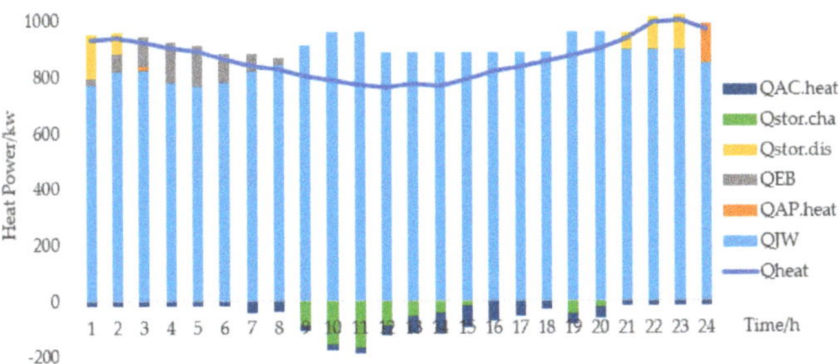

Figure 15. Heat output diagram of various equipment before single time scale in summer.

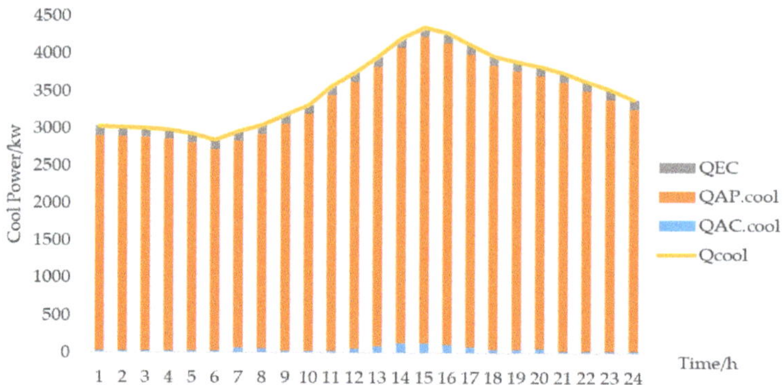

Figure 16. Cooling output diagram of various equipment before single time scale in summer.

The power output of each equipment before the single time scale in summer is shown in Figure 14. The gas internal combustion engine operates at a load rate of more than 80% throughout the day, taking up half of the power load throughout the day. The system has completely absorbed the photovoltaic power and greatly reduced the dependence on the grid power when the system is running at noon (the 10th to 15th hour). Electric boilers and electric refrigerators absorb only a small part of electric energy throughout the day. However, power storage equipment is put into operation at certain times to maintain stable operation of the power system.

The heat energy output of each equipment before the single time scale in summer is shown in Figure 15, and the cylinder liner water heat exchanger outputs sufficient thermal energy. Electric boiler and flue gas absorption heat pump output less; The absorption refrigerator only absorbs a small amount of heat energy and converts it into cold energy. During the first 2, 9, 15 and 19, 23 h of system operation, the heat storage equipment absorbs and releases power to maintain the stability of the thermal system.

The cooling energy output of each equipment before the single time scale in summer is shown in Figure 16. The flue gas absorption heat pump provides almost all-day cooling energy load, while the electric refrigerator and absorption refrigerator are only used as auxiliary equipment, thus the cooling energy power of the system is balanced.

In terms of single-time scale system load rate operation, the relationship between system load and operation load rate is shown in the following Figures 17–19:

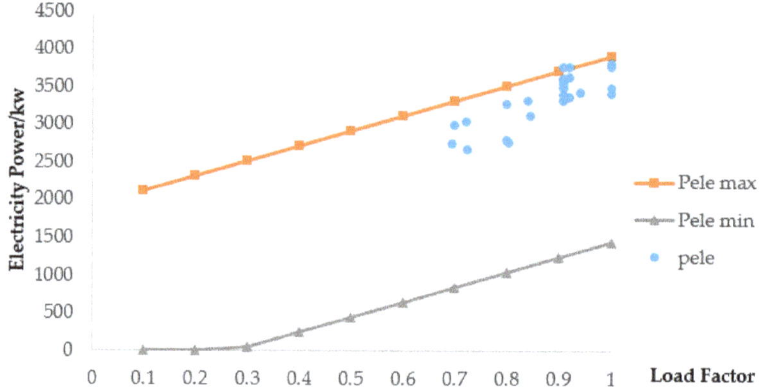

Figure 17. Single time scale power load-load rate operation.

Single-time scale power load-load ratio operation is shown in Figure 17. Although most power load ratio operation points are close to the upper limit of power output of the decoupled power system and the power system operates efficiently, some power load ratio operation points exceed the upper limit of power output boundary. At this time, the power storage equipment is put into operation to absorb some power, otherwise the system cannot operate stably, and at the same time, the energy storage burden of the system is increased.

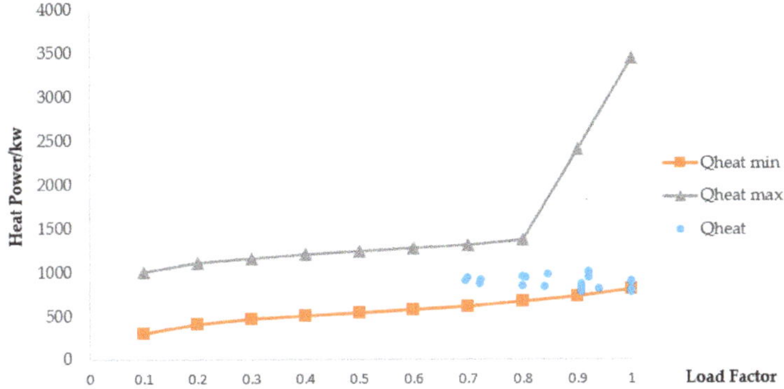

Figure 18. Single time scale heat load-load rate operation.

Single-time scale heat load-load ratio operation is shown in Figure 18. summer heat load is basically linearly distributed. Due to single-time scale adjustment, the heat load of the system is almost twice that of multiple time scales within one hour. At some operating points with high load rate, the lower limit of the power output boundary has been dropped. At this time, the heat storage equipment must be operated to ensure the stability of the system output heat energy.

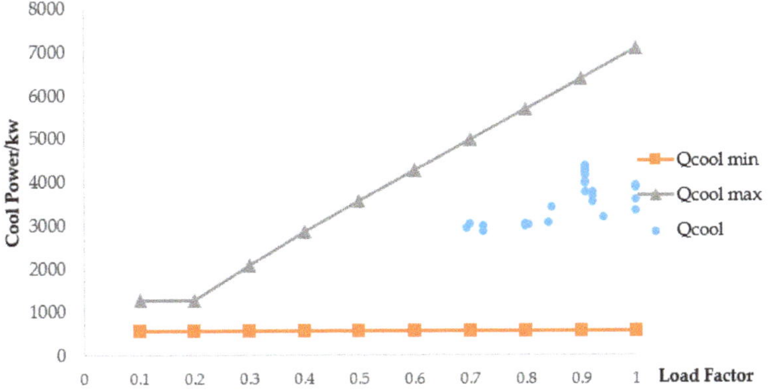

Figure 19. Single time scale cooling load-load rate operation.

The operation of single-time scale cooling load-load ratio is shown in Figure 19. Although the cooling energy scheduling period of single-time scale system is the same as that of multi-time scale system, the system load ratio is determined by the loads of cold, heat and electricity, which makes the overall load ratio of single-time scale system higher and the system is inferior to multi-time scale system in load ratio selectivity.

Comparing the load rates of single-time scale system and multi-time scale system, it is found that the load rates of single-time scale system are kept above 80% throughout the day, while the load rates of multi-time scale system are scattered, ranging from 55% to 100%. On the surface, the single-time scale system load rate is more stable and efficient, but in fact, the single-time scale system maintains a high load rate and the system operates at high load, which will greatly increase the system equipment loss and maintenance cost, which is very unfavorable to the long-term stable operation of the system and also increases the risk of system failure. The load rate of multi-time scale system is stable at 55~70% in low load period, but it can reach full load operation in high load period. The feasible range of system load rate is wider and the system has wider operation selection space.

A comparison of system operation results on multiple time scales and on a single time scale is shown in Table 2 below.

Table 2. Comparison table of results for multi-time scale and single-time scale systems.

	F_{run} (CNY)	F_{poll} (m^3)
multiple time scales	46,771.92	121.84
single time scale	54,068.34	149.01

From Table 2, it can be seen that the results of the multi-time scale system are better than those of the single time scale system in terms of total operating costs and pollutant gas emissions, with a reduction of 15.6% in economic operating costs and 22.3% in pollutant gas emissions. Under the condition of multi-time scale, the system equipment has more flexible time collocation and energy complementary selection, highlighting the multi-energy complementary characteristics of the cold-heat-electricity integrated energy system. However, under the condition of a single time scale, the flexible multi-time scale adjustment mode of the power system and the thermal system is abandoned, and the load burden of the system is increased in the same time dimension, so that the system has to expand the upper limit of equipment capacity or increase the equipment investment when selecting equipment, which is extremely dependent on the energy storage equipment to maintain the "source-load" power balance of the system, thus greatly increasing the investment cost of the system, the difficulty in equipment selection and the maintenance cost of the equipment, and the multi-time scale system has a wider operating load rate range.

6. Summary

The cold-thermal-electricity integrated energy system takes coupling multiple energy production modes to realize multi-energy complementation, energy step utilization and overall energy utilization rate improvement as the core, thus achieving the purposes of high efficiency, energy conservation, environmental friendliness and compatibility. Because it is close to the energy consumption side, the energy loss in the energy transmission process is greatly reduced. The mutual supplement of various energy sources also reduces the losses between energy production at all levels. At the same time, there are still many difficulties in optimizing the integrated energy system. Based on the multi-objective and multi-time scale optimization problem of the cold-thermal-electricity integrated energy system under the condition of the feasible load rate interval, this paper studies the cold-thermal-electricity integrated energy system topology structure, takes the gas internal combustion engine and the flue gas absorption heat pump as the core, considers the feasible load rate interval of the system under the condition of strong coupling, and adopts the method of decoupling analysis of the system to analyze the cold, heat and electrolysis coupling subsystems. The analysis results show that the selection of the system load rate will directly determine the "source-load" power balance of the system. In order to ensure the stable operation of the system under the condition of independent and uncertain loads, the equipment parameters should be adjusted according to the fluctuation range of each load to ensure the operating conditions of the system, so that the system load rate can fall within the stable load rate range of the cold, heat and electrolytic coupling subsystems. On the issue of multi-objective and multi-time scale optimization, the multi-objective function takes into account the lowest system economic operation cost and the lowest pollutant gas emission. At the same time, considering the differences of different energy attributes and energy scheduling characteristics, different time scales are selected to model the equipment model and the power model. The power control is sensitive and the scheduling is fast, with 15 min as an operation cycle. The flexibility of heat energy is obvious, taking 30 min as an operation cycle; Cold energy changes slowly, taking 1 hour as an operation cycle. In the aspect of multi-objective problem solving, the unitary linear weighting method is adopted to convert the multi-objective problem into a single-objective problem, and different weighting coefficients are set to optimize the solution, and the above mixed integer nonlinear problem is calculated by LINGO solver. This paper selects a hotel in Zhangjiakou, northern China, for simulation analysis of summer front cold, heat and electricity loads, and compares it with a single time scale system. The results show that the multi-time scale system reduces the economic operation cost by 15.6% and the pollution gas emission by 22.3% compared with the single time scale system. Under the multi-time scale condition, the system equipment has a wider load rate operation range, flexible time allocation and complementary energy selection, and greatly reduces the equipment capacity, investment cost, equipment selection difficulty and equipment maintenance cost of the single time scale system, which is more conducive to the stable operation of the system.

This study uses a static model. When considering multi-time scale optimization, the model parameters change dynamically. By the way, the optimization time of cold energy is long, so it can be considered to reduce the optimized time scale of cold energy appropriately. Although the optimal weight coefficient (k_{run}, k_{poll}) is selected after the system optimization results, the results show that when the weight coefficient (k_{run}, k_{poll}) is (0.9, 0.1), it performs best, and the target has little impact on the pollution emission control, which is one of the directions that this method can improve. In terms of future work, the system model can be optimized to improve the dynamic parameter changes of equipment, so as to build a real-time coordination and optimization system and improve the stability of comprehensive energy system.

Author Contributions: Methodology, B.O., Z.Y. and L.Q.; Writing—Original Draft Preparation B.O.; Guidance and Editing, Z.Y., C.L., L.Q. and D.L.

Funding: This work supported by "National Key Research and Development Program of China 2018YFB0905105".

Acknowledgments: The research gratefully acknowledge the LINGO software to solving optimization.

Conflicts of Interest: The authors declare no conflict of interest.

Nomenclature

Abbreviations

AC	absorption refrigerator
AP	smoke absorption heat pump
EB	electric boiler
EC	electric refrigerator
GE	gas engine
HS	heat storage equipment
IES	integrated energy system
JW	cylinder liner water heat exchanger
PV	photovoltaic generator set
ST	electricity storage equipment

Parameters and Variables

a	the system load factor
a_3, a_2, a_1, a_0	fitting constant of gas internal combustion engine
$b_5, b_4, b_3, b_2, b_1, b_0$	fitting constant of flue gas absorption heat pump
$B_{stor}(t_1, t_{2,j})$	real-time capacity of heat storage equipment
$B_{stor.max}$	maximum capacity constraint of heat storage equipment, 720 KW
$B_{stor.min}$	minimum capacity constraint for heat storage equipment, 160 KW
COP_{AC}	energy efficiency coefficient of absorption refrigerator, 1.69

$COP_{AP}(t_1,t_{2,i})$	energy efficiency coefficient of flue gas absorption heat pump
COP_{EB}	energy-producing coefficient of electric boiler, 1.8
COP_{EC}	energy efficiency coefficient of electric refrigerator, 4.1
$C_W(t_1,t_{2,i})$	specific heat capacity of hot water at different temperatures
$D_{batt.cha}(t_1,t_{2,i})$	charging variables of power storage equipment
$D_{batt.dis}(t_1,t_{2,i})$	discharge variables of electric storage equipment
$D_{stor.cha}(t_1,t_{2,j})$	heat absorption variables of heat storage equipment
$D_{stor.dis}(t_1,t_{2,j})$	heat release variables of heat storage equipment
$E_{batt}(t_1,t_{2,i})$	real-time capacity of power storage equipment
$E_{batt.max}$	maximum storage capacity of power storage equipment, 400 KW
$E_{batt.min}$	minimum storage capacity of power storage equipment, 100 KW
$f_{gas}(t_1,t_{2,i})$	natural gas price, 1.5 (CNY/m3)
$f_{grid}(t_1,t_{2,i})$	real-time electricity price of power grid
F	the mixed objective function value
$F_{gas}(t_1,t_{2,i})$	cost of system purchase of natural gas
$F_{grid}(t_1,t_{2,i})$	electricity Purchase Expenses for System and Power Grid
F_{run}	the total operating cost of the system
$F_{run.min}$	the minimum optimal values of economic operation
$F_{main}(t_1)$	maintenance cost of system equipment
F_{poll}	emissions of polluting gases
$F_{poll.min}$	the minimum optimal values of economic operation pollutant gas emission
G_{STC}	rated irradiation intensity of photovoltaic generator sets
$G_{ING}(t_1,t_{2,i})$	real-time irradiation intensity
k	generation coefficient of photovoltaic generator set, -0.0047%
k_L	self-loss coefficient of power storage equipment, 0.04
k_s	self-loss coefficient of heat storage equipment, 0.02
k_{run}	the economic operation weight coefficient
k_{poll}	the weight coefficient of pollutant gas emission
$k_{AC.heat}[Q_{AC.heat}(t_1,t_{2,j})]$	maintenance coefficient of absorption refrigerator, 0.02
$k_{AP.cool}[Q_{AP.cool}(t_1,t_{2,i})]$	cold power maintenance coefficient of flue gas absorption heat pump, 0.01
$k_{AP.heat}[Q_{AP.heat}(t_1,t_{2,i})]$	thermal power maintenance coefficient of flue gas absorption heat pump, 0.01
$k_{batt.dis/cha}[P_{batt.dis/cha}(t_1,t_{2,i})]$	power maintenance coefficient of power storage equipment, 0.02
$k_{GE}[P_{GE}(t_1,t_{2,i})]$	maintenance coefficient of gas internal combustion engine under different output power
$k_{PV}[P_{PV}(t_1,t_{2,i})]$	maintenance coefficient of photovoltaic generator set, 0.01
$k_{stor.dis/cha}[Q_{stor.dis/cha}(t_1,t_{2,j})]$	power maintenance coefficient of heat storage equipment, 0.015
$L_{cool}(t_1,t_{2,i})$	cold water flow rate of flue gas absorption heat pump
$L_{cool.max}$	maximum cooling flow rate of flue gas absorption heat pump, 8 L/h
$L_{heat}(t_1,t_{2,i})$	hot water flow rate of flue gas absorption heat pump
$L_{heat.max}$	maximum heating flow of flue gas absorption heat pump, 10 L/h
LHV	low calorific value of natural gas, 9.7 KW/m^3
$P_{batt.cha}(t_1,t_{2,i})$	charging power of power storage equipment
$P_{batt.cha.max}$	maximum charging power of power storage equipment, 120 KW
$P_{batt.cha.min}$	minimum charging power of power storage equipment, 0 KW
$P_{batt.dis}(t_1,t_{2,i})$	discharge power of power storage equipment
$P_{batt.dis.max}$	maximum discharge power of power storage equipment, 90 KW
$P_{batt.dis.min}$	minimum discharge power of power storage equipment, 0 KW
$P_{batt.dis/cha}(t_1,t_{2,i})$	interactive power of power storage equipment
$P_{ele}(t_1,t_{2,i})$	electrical load
$P_{EB}(t_1,t_{2,i})$	electric boiler input power
$P_{EB.max}$	maximum electric power of electric boiler, 80 KW
$P_{EB.min}$	minimum electric power of electric boiler, 0 KW
$P_{EC}(t_1,t_{2,i})$	input electric power of electric refrigerator

$P_{EC.max}$	maximum electric power of electric refrigerator, 60 KW
$P_{EC.min}$	minimum electric power of electric refrigerator, 30 KW
$P_{grid}(t_1,t_{2,i})$	system and Power Grid Power Purchase
$P_{GE}(t_1,t_{2,i})$	output electric power of gas internal combustion engine
$P_{GE.max}$	the output gradient of gas internal combustion engine is limited to 50 KW
P_{max}	the rated power of gas internal combustion engine is 500 KW.
$P_{PV}(t_1,t_{2,i})$	power generation of photovoltaic generator sets
P_{STC}	rated output of photovoltaic generator Sets
$Q_{AC.cool}(t_1)$	cold power output by absorption refrigerator
$Q_{AC.cool.max}$	output gradient constraint of absorption refrigerator, 40 KW
$Q_{AC.heat}(t_1,t_{2,j})$	heat power absorbed by absorption refrigerator
$Q_{AC.heat.max}$	maximum thermal power absorbed by absorption refrigerator, 80 KW
$Q_{AC.heat.min}$	minimum thermal power absorbed by absorption refrigerator, 20 KW
$Q_{AP.cool}(t_1,t_{2,i})$	smoke absorption heat pump outputs cold power
$Q_{AP.cool.max}$	cooling power output gradient constraint of flue gas absorption heat pump, 500 kw
$Q_{AP.heat}(t_1,t_{2,i})$	smoke absorption heat pump outputs heat power
$Q_{AP.heat.max}$	heating power output gradient constraint of flue gas absorption heat pump, 400 kw
$Q_{cool}(t_1)$	cold load
$Q_{EB}(t_1,t_{2,j})$	electric boiler output thermal power
$Q_{EB.max}$	output slope constraints of electric boilers
$Q_{EC}(t_1)$	output cooling power of electric refrigerator
$Q_{EC.max}$	output gradient constraint of electric refrigerator, 60 KW
$Q_{heat}(t_1,t_{2,j})$	thermal load
$Q_{JW}(t_1,t_{2,i})$	output thermal power of cylinder liner water heat exchanger
$Q_{stor.cha}(t_1,t_{2,j})$	heat absorption power of heat storage equipment
$Q_{stor.cha.max}$	maximum heat absorption power of heat storage equipment, 200 KW
$Q_{stor.cha.min}$	minimum heat absorption power of heat storage equipment, 0 KW
$Q_{stor.dis}(t_1,t_{2,j})$	heat release power of heat storage equipment
$Q_{stor.dis.max}$	maximum heat release power of heat storage equipment, 250 KW
$Q_{stor.dis.min}$	minimum heat release power of heat storage equipment, 0 KW
$Q_{stor.dis/cha}(t_1,t_{2,j})$	interactive power of heat storage equipment
t_1	hours of operation in a day
t_2	minutes in an hour
Δt_1	scheduling period of cold energy
$\Delta t_{2,i}$	scheduling period of electric energy
$\Delta t_{2,j}$	scheduling period of heat energy
$T(t_1,t_{2,i})$	inlet temperature of flue gas absorption heat pump
T_{cool}	cold water outlet temperature, 40 °C
T_{heat}	hot water outlet temperature, 100 °C
$T_{out}(t_1,t_{2,i})$	ambient temperature
T_s	reference temperature of generator set, 25 °C
$V_{gas}(t_1,t_{2,i})$	the system consumes natural gas volume
α_{sour}	pollution gas emission coefficient at power supply side of power grid, 0.0009
α_{tran}	emission coefficient of polluted gas transported by power grid lines, 0.000825
α_{PGE}	pollution gas emission coefficient of gas internal combustion engine, 0.0015
$\lambda_{heat}(t_1,t_{2,i})$	smoke heating ratio of smoke absorption heat pump
$\lambda_{cool}(t_1,t_{2,i})$	smoke refrigeration ratio of smoke absorption heat pump
$\eta_{GE.elec}(t_1,t_{2,i})$	power generation efficiency of gas internal combustion engine
η_L	natural loss rate of gas internal combustion engine, 0.08
η_{gas}	natural gas utilization rate of gas internal combustion engine, 0.98
$\eta_{AP.heat}$	thermal efficiency of flue gas absorption heat pump, 0.62

$\eta_{AP.cool}$	refrigeration efficiency of flue gas absorption heat pump, 0.58
η_{JW}	heat transfer efficiency of cylinder liner water heat exchanger, 0.2
$\eta_{batt.cha}$	charging efficiency of power storage equipment, 0.95
$\eta_{batt.dis}$	discharge efficiency of power storage equipment, 0.95
$\eta_{stor.cha}$	heat absorption efficiency of heat storage equipment, 0.98
$\eta_{stor.dis}$	heat release efficiency of heat storage equipment, 0.98

References

1. Xu, X.; Ming, Y.; Chen, L.; Chen, Q.; Hu, W.; Zhang, W.; Wang, X.; Hou, Y. Combined Electricity-Heat Operation System Containing Large Capacity Thermal Energy Storage. *Proc. CSEE* **2014**, *34*, 5063–5072.
2. Qian, A.; Ran, H. Key Technologies and Challenges for Multi-energy Complementarity and Optimization of Integrated Energy System. *Autom. Electr. Power Syst.* **2018**, *42*, 1–8.
3. Sun, H.; Guo, Q.; Zhang, B.; Wu, W.; Wang, B.; Shen, X.; Wang, J. Integrated Energy Management System: Concept, Design, and Demonstration in China. *IEEE Electrif. Mag.* **2018**, *6*, 42–50. [CrossRef]
4. Wang, D.; Liu, L.; Jia, H.; Wang, W.; Zhi, Y.; Meng, Z.; Zhou, B. Tianjin University Review of key problems related to integrated energy distribution systems. *CSEE J. Power Energy Syst.* **2018**, *4*, 130–145. [CrossRef]
5. Sun, H.; Zhaoguang, P.; Guo, Q. Energy Management for Multi-energy Flow: Challenges and Prospects. *Autom. Electr. Power Syst.* **2016**, *40*, 1–8.
6. Cheng, H.; Hu, X.; Wang, L.; Liu, Y.; Yu, Q. Review on Research of Regional Integrated Energy System Planning. *Autom. Electr. Power Syst.* **2019**, *43*, 2–13.
7. You, S.; Song, P.X. A Review of Development of Integrated District Energy System in Denmark. *Distrib. Util.* **2017**, *34*, 2–7.
8. Wu, J. Drivers and State-of-the -art of Integrated Energy Systems in Europe. *Autom. Electr. Power Syst.* **2016**, *40*, 1–7.
9. Zhang, Y.W.; Ding, C.J.; Ming, Y.; Jiang, X.P.; Fan, M.T.; Zhang, Z. Development and Experiences of Smart Grid Projects in Europe. *Power Syst. Technol.* **2014**, *38*, 1717–1723.
10. Feng, H. Research on the development status and business model of comprehensive energy services at home and abroad. *Electr. Appl. Ind.* **2017**, *06*, 34–42.
11. Zhang, Q.; Tezuka, T.; Esteban, M.; Ishihara, K.N. A Study of Renewable Power for a Zero-Carbon Electricity System in Japan Using a Proposed Integrated Analysis Model. In Proceedings of the 2nd International Conference on Computer and Automation Engineering (ICCAE), Singapore, 26–28 February 2010; pp. 166–170.
12. Tan, X. Research and Application of Layout Optimization of Guangzhou Industrial Park. *Buiding Technol. Dev.* **2017**, *44*, 22–23.
13. Liu, J.F. Strong smart grid leads the future. *Power Syst. Commun.* **2012**, *33*, 17.
14. Ming, Z.; Peng, L.L.; Sun, J.H.; Liu, W.; Wu, G. Economic Optimization and Corresponding Algorithm for Distributed Energy System Compatible With Demand-Side Resources. *Power Syst. Technol.* **2016**, *40*, 1650–1656.
15. Amid, P.; Saffaraval, F.; Saffar-avval, M. Feasibility Study of Different Scenarios of CCHP for a Residential Complex. In Proceedings of the 2010 IEEE Conference on Innovative Technologies for an Efficient and Reliable Electricity Supply, Waltham, MA, USA, 27–29 September 2010; pp. 177–183.
16. Cho, H.J.; Mago, P.J.; Luck, R.; Chamra, L.M. Evaluation of CCHP systems performance based on operational cost, primary energy consumption, and carbon dioxide emission by utilizing an optimal operation scheme. *Appl. Energy* **2009**, *86*, 2540–2549. [CrossRef]
17. Wang, Y.; Yu, H.; Yong, M.; Huang, Y.; Zhang, F.; Wang, X. Optimal Scheduling of Integrated Energy Systems with Combined Heat and Power Generation, Photovoltaic and Energy Storage Considering Battery Lifetime Loss. *Energies* **2018**, *11*, 1676. [CrossRef]
18. Wang, Y.; Xing, P.X.; Jiang, H.H. Improved PSO-Based Energy Management of Stand-Alone Micro-Grid Under Two-time Scale. In Proceedings of the 2016 IEEE International Conference on Mechatronics and Automation, Harbin, China, 7–10 August 2016; pp. 2128–2133.
19. Kang, K.; Deng, S.; Wu, S.; Zhong, T.; Wu, X.; Wang, Z. A Multi-time Scale Coordinated Real Time Dispatch Model of CCHP-based Microgrid. In Proceedings of the 2018 International Conference on Power System Technology (POWERCON), Guangzhou, China, 6–8 November 2018; pp. 1613–1621.

20. Xia, T.; Lin, Z.H.; Pan, Z.G.; Sun, H.B. Modeling and Simulation for Multi Energy Flow Coupled Network Computing. In Proceedings of the 2018 International Conference on Power System Technology (POWERCON), Guangzhou, China, 6–8 November 2018; pp. 992–998.

21. Liu, G.; Feng, Y.; Qian, D.; Shen, G.; Shen, X. Research on Multi-objective Optimal Joint Dispatching of Wind-thermal-hydro Power in Multi Time Scales. In Proceedings of the 2016 IEEE PES Asia-Pacific Power and Energy Conference, Xi'an, China, 25–28 October 2016; pp. 1832–1839.

22. Zhang, X.P.; Shahidehpour, M.; Alabdulwahab, A.; Abusorrah, A. Optimal expansion planning of energy hub with multiple energy infrastructures. *IEEE Trans. Smart Grid.* **2017**, *6*, 2302–2311. [CrossRef]

23. Liu, X.Z.; Jenkins, N.; Wu, J.H. Combined Analysis of Electricity and Heat Networks. *Energy Procedia* **2014**, *61*, 155–159. [CrossRef]

24. Ming, L.; Qing, G.; Yan, J.; Qin, G. Thermal Analysis of Underground Thermal Energy Storage UNDER Different Load Modes. In Proceedings of the 2009 International Conference on Energy and Environment Technology, Guilin, China, 16–18 October 2009.

25. Khani, H.; Farag, H.E.Z. Optimal Day-Ahead Scheduling of Power-to-Gas Energy Storage and Gas Load Management in Wholesale Electricity and Gas Markets. *IEEE Trans. Sustain. Energy* **2018**, *9*, 940–951. [CrossRef]

26. Lee, J.; Jeong, S.; Han, Y.H.; Park, B.J. Concept of Cold Energy Storage for Superconducting Flywheel Energy Storage System. *IEEE Trans. Appl. Supercond.* **2011**, *21*, 2221–2224. [CrossRef]

27. Jamrozik, A.; Tutak, W. In Modelling of Combustion Process in the gas Test Engine 2010. In Proceedings of the VIth International Conference on Perspective Technologies and Methods in MEMS Design, Lviv, Ukraine, 20–23 April 2010.

28. Tutak, W. In Modelling of Flow Processes in the Combustion Chamber of IC Engine. In Proceedings of the 2009 5th International Conference on Perspective Technologies and Methods in MEMS Design, Zakarpattya, Ukraine, 22–24 April 2009.

29. Sato, Y.; Taira, T. Discussions on the Pump Absorption Efficiency Under Hot-Band Pumping of Nd:YAG. In Proceedings of the Conference on Lasers and Electro-Optics 2012, San Jose, CA, USA, 9–14 June 2013.

30. Skubienko, S.V.; Yanchenko, I.V.; Babuohkin, A.Y. Using an Absorption Heat Pump in the Regeneration System of Turbine Model K-300-240-2 Manufactured by Kharkov Turbo Generator Plant (KhTGP). In Proceedings of the 2nd International Conference on Industrial Engineering, Applications and Manufacturing (ICIEAM), Chelyabinsk, Russia, 19–20 May 2016.

31. Yongbing, H.; Shijing, W.; Jixuan, W. Research on Improving Unit Thermal Efficiency Based on Absorption Heat Pump By Reducing The Circulating Water Temperature. In Proceedings of the 2011 International Conference on Transportation, Mechanical, and Electrical Engineering (TMEE), Changchun, China, 16–18 December 2011; pp. 2119–2122.

32. Hamed, M.; Fellah, A.; Ben Brahim, A. Performance Evaluation of a Solar Absorption Refrigerator with Dynamic Simulation. In Proceedings of the 2012 IEEE First International Conference on Renewable Energies and Vehicular Technology, Hammamet, Tunisia, 26–28 March 2012; pp. 156–160.

33. Rezgui, W.; Mouss, N.K.; Mouss, L.; Mouss, M.D.; Amirat, Y.; Benbouzid, M. Modeling the PV Generator Behavior Submit to the Open-Circuit and the Short-Circuit Faults. In Proceedings of the IEEE 3rd International Symposium on Environmental Friendly Energies and Applications (EFEA), St. Ouen, France, 19–21 November 2014; pp. 1–6.

34. Hargreaves, N.; Taylor, G.; Carter, A. Smart Grid Interoperability Use Cases for Extending Electricity Storage Modeling within the IEC Common Information Model. In Proceedings of the 2012 47th International Universities Power Engineering Conference (UPEC), London, UK, 4–7 September 2012.

35. Mohamed, H.; Ben Brahim, A. Modeling and Dynamic Simulation of a Thermal Energy Storage System by Sensitive Heat and Latent Heat. In Proceedings of the 2017 International Conference on Green Energy Conversion Systems (GECS), Hammamet, Tunisia, 23–25 March 2017; pp. 1–7.

36. Zhang, W.; Zheng, M.; Wang, X. Modeling and Simulation of Solar Seasonal Underground Thermal Storage in a Solar-Ground Coupled Heat Pump System. In Proceedings of the 2010 IEEE Second International Conference on Computer Modeling and Simulation, Sanya, China, 22–24 January 2010; pp. 27–31.

Article

On Minimisation of Earthing System Touch Voltages

Vaclav Vycital *, Michal Ptacek, David Topolanek and Petr Toman

Department of Electrical Power Engineering, Faculty of Electrical Engineering and Communication,
Brno University of Technology, 601 90 Brno, Czech Republic; ptacekm@feec.vutbr.cz (M.P.);
topolanek@feec.vutbr.cz (D.T.); toman@feec.vutbr.cz (P.T.)
* Correspondence: vycital@feec.vutbr.cz; Tel.: +420-541-146-247

Received: 15 September 2019; Accepted: 3 October 2019; Published: 11 October 2019

Abstract: Finding cost efficient earthing system design with acceptable level of safety might be quite tedious work. Thus, many earthing system engineers try to find the most suitable design either by employing only their best experience or taking advantage of some more complex optimisation programs. Although both approaches might work well under certain circumstances, they might fail either due to counter-intuitiveness of the specific situation or by misunderstanding of the applied optimisation method, its limitations etc. Thus, in this paper, the earthing system design optimisation problem was addressed by analysing optimisation simulation results together with conducted sensitivity analysis. In the paper, a simple double ring earthing system was optimised while using five different optimisation methods. The earthing system was placed in different horizontally stratified soil models and the earthing system was optimised by minimising touch voltages instead of commonly minimised earth potential rise. The earthing system was modelled by Ansys Maxwell software. Apart from using Ansys Maxwell built-in optimisers, the possible solution space has also been mapped by performing sensitivity analysis with changing the earthing system design dimensions and the results of optimisation were compared and validated. It was found out that the Sequential Non-Linear Programming Optimisation technique was quite superior to the other techniques. Additionally, in most cases, the Ansys Maxwell optimiser was able to found optimal solution; however, in some cases, based on the initial conditions, it might get stuck in local minima or the results might be influenced by the solution noise. Additionally, some quite non intuitive dependencies of earthing system electrodes positions had been found when different spatial dimensions constraints are used.

Keywords: earthing; earthing system; optimisation; Ansys Maxwell; sensitivity analysis; touch voltage; non-linear programming

---

1. Introduction

The earthing system (ES) is one of the necessary parts of distribution and transmission power networks. The basic requirements on ES are stated in standards EN 50522:2010 [1] or e.g., in its American counterpart IEEE 80 [2]. Apart from other requirements, the designing engineer must ensure that step voltages (SV) and touch voltages (TV) in the vicinity of the ES shall not exceed the safety limits defined by [1], or [2] respectively. Thus, the proposed design by the engineer is always compliant with safety requirements, e.g., permissible touch voltage curve of EN 50522, however it might not be the best possible design for this desired specific site and its constraints. It might be quite common that the initial approach by the engineer is to apply some universal ES design recommendations (e.g., for distribution transformation station use double ring earthing system). The spatial layout of this initial design is usually the same for every such proposed design. Once this initial design is not compliant with the safety requirements, the designing engineer makes some adjustments to this initial design to find a solution that is compliant with the safety limits. However, the final proposed design by the engineer is not optimised within the specific site constraints and there might be still room for finding better design layout by employing a more complex optimisation approach, i.e., decrease the TV, decrease used material, etc. This designing approach might be especially true in case of the great number of built distribution transformer stations supplying the low voltage distribution networks. In the case of these ESs, the designing engineer very often uses the simplified formulas of an Annex J [1] to get the ES resistance to earth (e.g., ES ring Equation (1))

$$R_r = \frac{\rho}{\pi^2 D} \ln \frac{2\pi D}{d},$$

(1)

where ρ is soil resistivity, D is diameter of earthing ring, and d is the diameter of ring electrode cross section.

Once the resistance to earth is obtained, the safety assessment is quite straight forward, as the failure current might be obtained from simplified equivalent electrical circuit with the equivalent voltage source method [3]. The ES is then assessed based on the prospective permissible touch voltage that is taken as one half of entire earth potential rise (EPR) [1].

Basically, the abovementioned approach is kind of optimization technique with an objective function

$$minimise \; EPR(x)$$

(2)

Subject to design spatial dimension constraints

$$g_1 \cdots g_n \le d_1 \cdots d_n \le h_1 \cdots h_n$$

(3)

where $d_1 \ldots d_n$ are appropriate ES dimensions and $g_1 \ldots g_n$, $h_1 \ldots h_n$ are the spatial dimension constraints. This is because the designing engineer should mainly focus on decreasing the overall EPR to lower the corresponding touch voltages as much as possible. Decreasing the EPR is most effectively done through decreasing the resistance to earth of the ES design. The described procedure of (2) and (3) in all situations would mean lengthening the ES horizontal dimensions up to the outer boundary of the horizontal constraints and burying the ES near the burial depth lower constraint. This procedure might be also expected, regardless of most often used horizontally stratified soil model if the simplified formula of (1) is used.

Usually, the ES horizontal dimensions are constrained by utility owner limited land. Additionally, burying the ES to greater depth might get costly due to more excavation work needed. Thus, many researchers have been dealing with the ES design optimization problem. Some of the papers were dealing with finding/proposing a method to optimize the design by appropriate ES conductor allocation in the soil with reducing earthing system total EPR only [4]. Other works are aimed on the optimization by minimising the total length of used ES horizontal and vertical conductors [5,6]. Some other researchers also included into the optimization the costs that are connected with the installation process, i.e., the excavation costs as well as the ES conductor material costs [5,7,8]. Another, quite a different approach to ES design optimisation can be found in [9,10]. It is proposed by these papers to optimize the ES by using low resistivity materials. This optimization method is based on the fact that by applying different chemicals near or next to ES rods and strips, the fault current is more effectively driven into the ground due to increased effective surface area of the ES rods. By this method, the ES resistance might be influenced either temporarily or permanently. Lastly, but not used very often might be the optimisation by finding optimal distance to some other external grounding grid [11].

Generally, two things are necessary for the optimization procedure. The first thing is some objective function that is used to quantify if the so-called optimum has been reached. The second necessary thing is some kind of step direction function that determines the direction towards the so-called optimum. The usage of some different objective functions has been described in the previous paragraph. However, in the case of the step direction function, the aforementioned papers [4–11] mainly use two different approaches or their combination. The first approach is by performing parametric analysis (i.e., sensitivity analysis). This approach is based on the fact that, if there are only expected smooth changes in the solution, by choosing appropriate parametric step, the whole solution space can be mapped and this solution space is later searched for the so-called optimal solution according to the objective function. The results of this parametric analysis can be also used as a recommendation for the designing engineer if properly depicted [12]. The second approach that was identified in the reference is the use of some more complex optimization approach e.g., using non-linear programming optimization methods, like genetic algorithm optimization, particle swarm optimization, and hybrid particle swarm genetic algorithm optimization. In general, ES design optimization is a spatial problem where the use of the mentioned methods can decrease the overall time that is needed for the parametric analysis. In some cases, performing parametric analysis with fine steps might get quite tedious and time consuming.

Quite a novel approach to earthing system optimization problem might be the optimization through Quantified Risk Analysis [13–15]. The objective function of this approach is the quantified real risk (i.e., the resulting risk of death consisting of the probability of hazardous TVs level and the coincidence probability [13]). This approach is mostly assumed as an alternative to the conventional 'worst case' earthing system approach [1], which might be more strict or more loose, depending on the situation. However, the authors are yet unaware of publication using this approach together with some kind of step direction function and, also, the so-called 'optimum' for this method might be questionable due to diverse risk perception and acceptable risk level laws.

Not many papers have been dealing with the earthing system optimisation through the minimisation of TVs in case of horizontally stratified soil model. For example, in paper [4] the total EPR together with the TV was analysed. The 'compression ratio' function was proposed as a step direction function. The proposed approach is best suited for horizontal earthing mats with parallel and perpendicular conductors and it can be easily used for conducting parametric analysis. In the case of searching for optimal compression ratio, a more sophisticated step direction function different from simple first or second order gradient based approach might be necessary. Additionally, only the horizontal separation of parallel earthing conductors is optimised without any mean of incorporating earthing mat or separate electrodes burying depth.

Thus, in this paper, an optimisation analysis is introduced for the allocation of separate earthing system electrodes in earth through using Ansys Maxwell Optimetrics Component [16]. The Optimetrics Component allows for the user to search for ES design optimal dimensions. The prospective TV was assumed as the source of real associated risk with the earthing system design and it was chosen as the objective function of the optimisation. A comprehensive parametric/sensitivity analysis was also carried out with changing some of the model parameters to assess the correctness of obtained results by the built-in optimisers in Ansys Maxwell [16]. By this approach, the solution space was searched first and the results that were obtained from sensitivity analysis were compared with results from the built-in Maxwell optimisers.

2. Ansys Maxwell Optimisers

Ansys Maxwell [16] is a software module incorporated in the whole simulation software ANSYS. Ansys Maxwell is software for the simulation of low frequency quasi stationary electromagnetic fields (EMF). The EMF can be determined for general spatial electromagnetic problems as well as for only two-dimensional problems. The solution of Maxwell equations is managed through the finite element method. Basically, the user has to define/import the required design model, define sources, boundary and meshing condition, and the solution is determined by approximating the EMF distribution through the mesh of finite elements [17]. Additionally, with the use of the built-in field calculator a user defined quantities might be obtained from postprocessing of solved EMF data.

Ansys Maxwell presents an Optimetrics component that enables the user to perform parametric/sensitivity/sweep analysis. Furthermore, a component for optimisation is available. For the Optimetrics analysis it is necessary to define *user defined variables*, i.e., model length and any other dimensions, and *user defined EMF quantities*, like electric potentials at certain points, etc. The user defined variables are later used in the optimisation analysis as the parameters that are changed in iterative manner to find optimum design. The user defined EMF quantities are used for definition of objective function of the optimisation. The software features six different options for model optimisation [16]:

2.1. Quasi Newton (QN)

The Quasi Newton optimizer is a gradient based optimizer numerically determining the derivatives of quadratically approximated higher order functions of actual local objective function dependency. This optimizer suffers from the possibility of stucking in local minima, and due to finite element method nature of the results also stucking due to high level of solution noise. As the Hessian matrix increase rapidly with the number of optimised design variables the computational time might increase rapidly and thus this optimizer is recommended for optimisation of 1–2 design variables only.

2.2. Pattern Search (PS)

The Pattern Search optimizer uses the grid based simplex search with simplices in the form of tetrahedrons. It is non-gradient based optimisation search, thus it is less dependent on the solution noise. However, it might take longer to find the optimum. The use of only three optimised design variables is recommended.

2.3. Sequential Non-Linear Programming (SNLP)

The SNLP optimizer is more complex one. Basically, the optimizer creates a response surface by using Taylor series approximation of finite element analysis. By this approximation, the optimizer can find locations of improving points and determine next step direction. The SNLP optimizer is supposed to be more accurate and reliable because it is not constrained by the problem of stucking in local minima like QN. That is also caused by the possibility to overcome the local minima by taking larger steps within the optimisation variable constraints limits.

2.4. Sequential Mixed Integer Non-Linear Programming (SMINLP)

SMINLP optimizer is basically same as SNLP. However, this optimizer allows for mixing discrete and continuous optimised design variables.

2.5. Genetic Algorithm (GA)

The GA optimizer is stochastic optimizer. It runs many iterations with random selection of next searched points. It uses an iterative process with initial generation/parents/children and mutated population. The better performing individuals are chosen for the generation of the next generation in each iteration. Through this random search process, the solution space is searched in a structured manner and the optimum is found. However, the disadvantage is also that the non-improving designs are searched in the random selection process and thus this approach takes many more iterations and is thus slow.

2.6. Matlab Optimiser (MO)

Ansys Maxwell also facilitates the possibility to pass parameters to Matlab software (Matlab 2016b, MathWorks, Natick, USA) and invokes Matlab built-in optimisation functions and uses them to optimize the Ansys Maxwell design.

3. Earthing System Optimised Model

The feasibility of using Ansys Maxwell software for earthing system design optimisation was assessed on simple double ring earthing system, as in Figure 1. The first inner ring with a diameter

D_1 was buried at depth h_1 and the outer ring with diameter D_2 was buried at depth h_2. The earthing system was made of FeZn strip with cross section 30×4 mm^2. The earthing system was modelled in Ansoft Maxwell [16,17] with horizontally stratified soil model with two layers—the surface one with resistivity ρ_1 and thickness H and the bottom layer with resistivity ρ_2. The soil was modelled by semi-sphere with radius of 200 m and the analysis percent error was set to 1%. For the illustrative purpose, the earthing system model in Figure 1a,c is depicted with an improper soil scale. The earthing system was excited by a DC current [17] of 30 A (red arrow in bottom right picture of Figure 1c) injected at the centre of the ES, where the ES conductor is brought up above the ground and it is considered as accessible for touch.

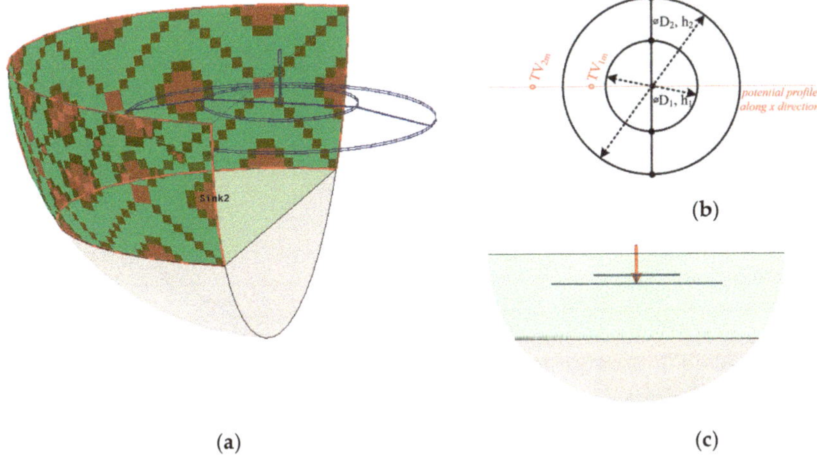

Figure 1. Earthing system model, spatial view (**a**), top view (**b**), and side view (**c**).

4. Earthing System Sensitivity Analysis Results

A comprehensive sensitivity analysis was carried out before the optimisation of ES design by Ansys Maxwell. In this analysis, the ES was placed in seven different soil models, including 'High on Low' (HoL), Uniform, and 'Low on High' (LoH) soil models, as in Table 1.

Table 1. Sensitivity analysis soil models.

Soil Model No.	1 HoL	2 HoL	3 HoL	4 Uniform	5 LoH	6 LoH	7 LoH
ρ_1 (Ωm)	500	1000	500	100	100	100	100
ρ_2 (Ωm)	100	100	100	100	500	1000	500
H (m)	2	2	1	∞	2	2	1

For each of used soil model in the sensitivity analysis, the diameters of the ES were changed as inner ring $D_1 = 1, 2$, and 3 m, and outer ring always greater than inner ring $D_2 = 4$ and 5 m. In the case of burial depths of both rings h_1 and h_2, seven burial depths were used for each ring as 0.3–1.5 m with step of 0.2 m. I.e., for one set combination of ES rings diameters D_1 and D_2 all 49 combinations of ES design burial depths were modelled by Ansys Maxwell and the results were used in the sensitivity analysis. By this way, about 2000 ES design variations were analysed in this analysis. For each of these ES design variations, the potential profile on the earth surface along x direction (Figure 1b) was read and the TV distribution have been determined in the vicinity of the ES. Although the potential profile was read with step of 0.25 m, for the purpose of optimisation evaluation throughout this paper four main performance values have been determined—TV_{1m}, TV_{2m} and TV_{3m} as TV 1, 2, and 3 m apart from the centre of the earthing system, respectively (i.e., potential difference between EPR and point 1, 2 and 3 m apart). Additionally, the total EPR was read. For set diameter D_2 (used here to represent the dimensional constraint), the optimum burial depth was manually searched whilst changing other three dimensions D_1, h_1, and h_2. Incorporating the optimisation of inner ring diameter D_1 whilst manually searching for optimum design is quite challenging as the number of analysed variations increase rapidly. The results of the sensitivity analysis are introduced in Figures 2–5 and Tables 2–5.

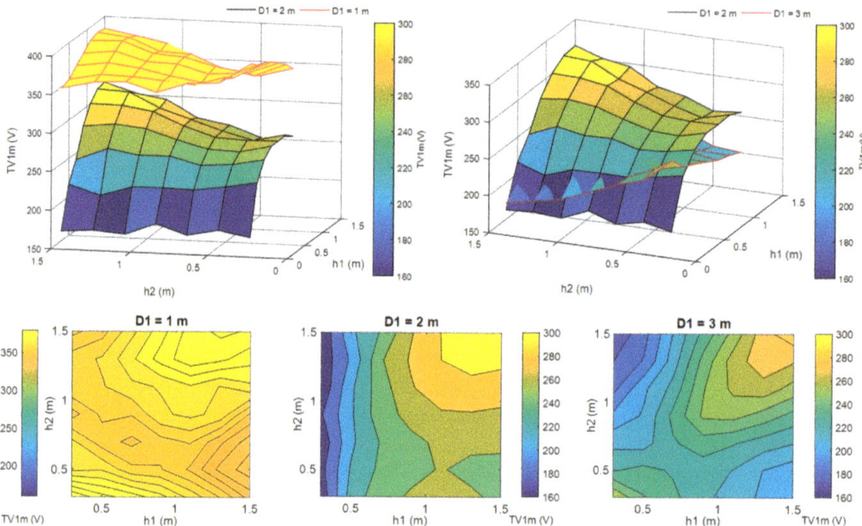

Figure 2. TV1m solution surfaces for ES with diameters $D_1 = 1$–3 m, $D_2 = 4$ m, $\rho_1 = 500\ \Omega$m, $\rho_2 = 100\ \Omega$m, $H = 2$ m.

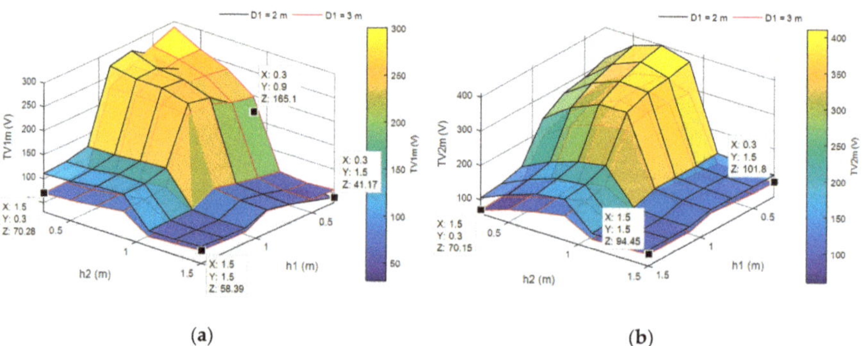

Figure 3. Touch voltage (TV) surfaces, 500 Ωm/100 Ωm/1 m, $D_2 = 4$ m, (**a**) TV_{1m} surface; (**b**) TV_{2m} surface.

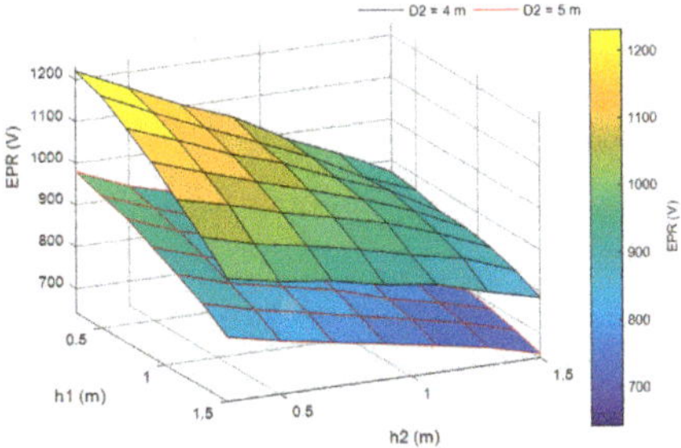

Figure 4. EPR, 500 Ωm/ 100 Ωm/ 2 m, $D_1 = 2$ m, $D_2 = 4$ and 5 m.

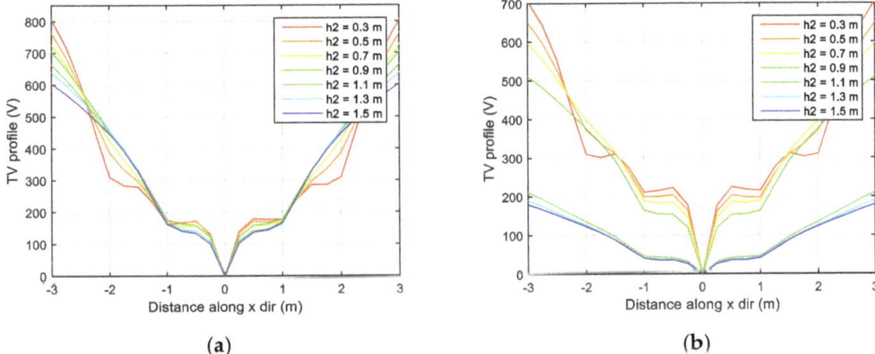

(a) (b)

Figure 5. TV profiles along x direction for HoL model $\rho_1 = 500$ Ωm, $\rho_2 = 100$ Ωm, $D_1 = 2$ m, $D_2 = 4$ m, $h_1 = 0.3$ m; (**a**) surface layer thickness $H = 2$ m; and, (**b**) Surface layer thickness $H = 1$ m.

Table 2. Sensitivity analysis results for $\rho_1 = 500$ Ωm, $\rho_2 = 100$ Ωm, $H = 2$ m, $D_2 = 4$ m, $D_1 = 2$ m.

Objective	Min TV Option				Max TV Option			
	h_1 (m)	h_2 (m)	*TV* (V)	*EPR* (V)	h_1 (m)	h_2 (m)	*TV* (V)	*EPR* (V)
TV_{1m}	0.3	Any	160	870 *	1.5	1.5	315	785
TV_{2m}	1.5	0.3	235	935	0.3	1.3	460	935
TV_{3m}	1.5	1.5	545	785	0.3	0.3	805	1220

* Design with $h = 1.5$ m EPR selected.

Table 3. Sensitivity analysis results for $\rho_1 = 500$ Ωm, $\rho_2 = 100$ Ωm, $H = 2$ m, $D_2 = 4$ m, $D_1 = 3$ m.

Objective	Min TV Option				Max TV Option			
	h_1 (m)	h_2 (m)	*TV* (V)	*EPR* (V)	h_1 (m)	h_2 (m)	*TV* (V)	*EPR* (V)
TV_{1m}	0.3	1.5	170	820	1.5	1.5	290	750
TV_{2m}	1.5	0.3	185	815	1.5	1.5	390	750
TV_{3m}	1.5	0.5	490	805	0.3	0.3	780	1190

Table 4. Sensitivity analysis results for $\rho_1 = 100$ Ωm, $\rho_2 = 500$ Ωm, $H = 2$ m, $D_2 = 4$ m, $D_1 = 2$ m.

Objective	Min TV option				Max TV option			
	h_1 (m)	h_2 (m)	TV (V)	EPR (V)	h_1 (m)	h_2 (m)	TV (V)	EPR (V)
TV_{1m}	0.3	Any	25	510 *	1.5	1.5	75	515
TV_{2m}	1.5	0.3	45	515	1.5	1.5	120	515
TV_{3m}	0.9	0.5	155	515	0.3	0.3	185	550

* Design with $h = 1.5$ m EPR selected.

Table 5. Sensitivity analysis results for $\rho_1 = 500$ Ωm, $\rho_2 = 100$ Ωm, $H = 2$ m, $D_2 = 5$ m, $D_1 = 2$ m.

Objective	Min TV Option				Max TV Option			
	h_1 (m)	h_2 (m)	TV (V)	EPR (V)	h_1 (m)	h_2 (m)	TV (V)	EPR (V)
TV_{1m}	0.3	1.5	135	720	1.5	1.5	255	655
TV_{2m}	1.5	0.3	245	795	0.3	1.5	345	720
TV_{3m}	1.5	0.3	375	795	0.3	0.3	480	980

In Figure 2, the general overview of the results of sensitivity analysis is depicted. For set dimensional constraint $D_2 = 4$ m, three TV_{1m} surfaces are depicted for different inner ring diameter D_1 and each surface is consisting of results of all 49 burial depth variations. In this compact form, the optimum burial depth of both rings can be immediately read. Within the dimensional constraints of h_1, h_2 (being in range of 0.3–1.5 m) and D_2 (being fixed as 4 m), the 'global' minimum of the solution surface can be expected for $D_1 = 2$ m, $h_1 = 0.3$ m, and $h_2 = 1.5$ m. In case of $D_1 = 3$ m, a remarkable saddle point ($h_1 = 0.7$ m, $h_2 = 0.6$ m) is formed by a significant change in slope direction, where two areas with local ($h_1 = 1.5$, $h_2 = 0.5$ m) and global ($h_1 = 0.3$, $h_2 = 1.5$ m) minima are formed. Although this saddle point might not be as much remarkable for all simulated inner ring diameters, it is still present among all the surfaces.

Similar TV surfaces were also obtained for other objective functions in form of TV_{2m} and TV_{3m}. The results of these analysis are summarized in tabular form in Tables 2–5. In this compact tabular form, only the remarkable points of the sensitivity analysis surfaces are listed for different soil models and for different ES dimensions whilst preserving the clarity of the results. The burial depths of ES designs were selected for two extreme cases and for all three analysed TV_{xm} objective functions:

- Min TV where the burial depth of the design is selected for lowest TV_{xm}.
- Max TV where the burial depth of the design is selected for the highest TV_{xm}.

In this way, six different ES designs for set dimensional constraints of D_1, D_2, and selected soil model are listed and they can be compared for different objective functions TV_{xm}, EPR objectives etc.

In Tables 2 and 3, a slight difference in TVs for HoL soil model might be observed when changing the inner ring diameter D_1 from 2 m to 3 m and with set constraint of outer ring $D_2 = 4$ m. Even though the best possible design assessed based on EPR would be with both rings in the bottom layer $h_1 = h_2 = 1.5$ m and $D_1 = 2$ m, based on TV_{1m} the best design would be a different one with only the outer ring in the bottom layer and the inner ring in the surface layer $h_1 = 0.3$ m, $h_2 = 1.5$ m and $D_1 = 2$ m. Additionally, if TV_{2m} and TV_{3m} objectives would be assessed a completely different designs are optimal. These findings might not be as much intuitive. However, with the increasing number of possible assessed option, the situation might get even more tricky.

For example, if additional constraints (i.e., different objective function or less favourable soil model) would be solved, a kind of unexpectable design solution might have been found. This can be apparent from Figure 3a,b, where the TV surface of HoL soil model with surface layer thickness of only 1 m is depicted for objectives TV_{1m} and TV_{2m} and for constraint of outer ring diameter $D_2 = 4$ m. As it might be expected by some experienced engineers, the better designs are with an outer ring burial depth greater than the thickness of the surface layer. The overall best design, as already stated in the previous paragraph, is with $D_1 = 2$ m, $h_1 = 0.3$ m, and $h_2 = 1.5$ m. However, this might not be true if other constraints are applied, e.g., dimensional constraint of both rings maximum burial depth of only 0.9 m and both TV_{1m} and TV_{2m} should be assessed. In the region of constraint burial depth of only 0.9 m, there is an evident change of pattern in optimal burial depths (between (a) and (b) of Figure 3) where the optimum design depending on the objective function is either for inner ring close to the surface with $h_1 = 0.3$ m and h_2 close to bottom layer, or the inner ring close to bottom layer and outer ring close to the surface, respectively. Additionally, design with increased inner ring diameter of 3 m gives better results than 2 m as in the case of TV_{1m} objective. From both figures it is evident that if both TV_{1m} and TV_{2m} objectives need to be evaluated, the contradiction of the solution will need some sort of more complicated addressment, e.g., weighting of risk, etc.

Additionally, based on the results of depicted sensitivity surfaces, as in Figure 3a, the positive effect of earthing electrode grading can be assessed. For example, for the analysed arrangement and objective of TV_{1m}, burying the inner ring only to a shallow depth of 0.3 m led to decreasing the risk. The rest of the designs are either comparable, or even worse. Burying both rings to the bottom layer is worse by slightly more than 40% than the optimal shallow inner ring placement ($h_1 = 0.3$ and $h_2 = 1.5$ m). The TV surface for inner ring diameter of 1 m for HoL has yielded in almost all cases to higher TVs and is thus excluded here.

The Uniform and LoH soil models are almost excluded from presented results, as the optimisation of ES design through changing design dimensions had greater observable dependency in TVs and EPRs for HoL soil model. Table 4 presents the results for LoH soil model. In case of LoH and Uniform soil models the TV differences between the min and max TV options are of small values and thus no big difference was found while searching for optimal solution. Additionally, the EPRs of different designs are almost identical. The abovementioned findings are true if the outer ring diameter is kept constant as a dimensional constraint. If that diameter would be even larger, the differences in TVs would smoothen even more. It is worth pointing out that, in the case of Uniform soil model, the results are even more convenient than in the case of the LoH model.

Increasing the outer ring diameter would of course lead to more suitable ES design, as might be apparent from Table 5, in comparison to Tables 2 and 3, where design with increased outer ring diameter of 5 m is presented (compared to 4 m in Tables 2 and 3). To complete the ES behaviour overview, Figure 4 depicts the EPR surface for different burial depths and outer ring diameters.

Figure 5a,b depict a kind of complementary and expanding overview of ES behaviour. In these figures the potential profiles along the x direction (Figure 1b) on the surface is depicted for ES designs with changing only the outer ring burial depth while keeping the rest of the dimensions constant. From the figures the TV_{1m}, TV_{2m}, and TV_{3m} can be obtained. From Figure 5a, it is obvious that the optimum design in case of objective TV_{1m} is with deep outer ring $h_2 = 1.5$ m and with shallow inner ring $h_1 = 0.3$ m. However, this chosen design is by no means also optimal when also assessed to TV_{2m}.

From Figure 5b the positive effect of burying the outer ring into the bottom layer in HoL soil model with $H = 1$ m is observable, where the TV profile had dramatically flattened.

5. Ansys Maxwell Earthing System Optimisation Results

The optimisation of the ES design was also carried out by Ansys Maxwell Optimetrics component for soil models number 1, 3, 4, and 5 (Table 1). The ES was always modelled with some 'initial' dimensions, denoted by the subscript '*i*' (i.e., the D_1, h_1, and h_2 as D_{1i}, h_{1i} and h_{2i}) and through the optimisation the 'optimised' dimensions have been obtained and are denoted by the subscript '*o*' (i.e., D_{1o}, h_{1o}, h_{2o}). The outer ring diameter D_2 was set as a dimensional constraint equal to 4 m in most of the simulations. The burial depth constraint was set as in region 0.3–1.5 m for both rings and the maximum inner ring diameter constraint was set as $D_{1max} = 0.9 \times D_2$, thus so the inner ring is always smaller than the outer ring. The minimization of TV_{1m} was set as the optimisation objective. The optimisation was carried out by five different optimisers—QN, SNLP, SMINLP, PS, GA. After the optimisation process some other parameters have been also recorded for possible comparison and benchmarking of different optimisers. I.e., TV_{2m}, TV_{3m}, the total EPR of optimised ES design, total time needed for finding the optimal design t, the number of performed simulations in total ItN (optimisation iterations), and a number of iterations (out from total of ItN iterations) where the optimal design was found ItO. As an optimisation analysis stopping criterion was also set the maximum number of optimisation iteration that was in most cases 300, or in some cases only 100. The optimisation results, together with the performed stopping option, are listed in Table 6 (Parts 1 and 2). The stopping option is in the last column 'Status' as either one of the following three options:

- S—for solved, stopped by finding minimum by Ansys Maxwell optimizer, the obtained result is expected to be the global minimum.
- M—for stopped by maximum number of iterations, not necessary global minimum found, the best suiting result was selected.
- F—for simulation failed. Again, the best suiting result was selected, however the optimisation ended prematurely. The failure might had happened due to more different problems. Either Ansys Maxwell adaptive mesher failed to build the mesh of finite elements, or the optimizer failed in finding the dimensions of the next design or the optimizer generally failed without description.

Table 6. Part 1. Ansys Maxwell optimisation analysis results. **Part 2.** Ansys Maxwell optimisation analysis results.

Part 1.

No.	Solver	ρ_1	ρ_2	H	D_2	D_{1i}	h_{1i}	h_{2i}	D_{1o}	h_{1o}	h_{2o}	TV_{1m}	TV_{2m}	TV_{3m}	EPR	t	ItO	ItN	Status
-	-	Ωm	Ωm	m	m	m	m	m	m	m	m	V	V	V	V	m	-	-	-
1	QN	500	100	2	4	2	0.5	0.7	2.45	0.33	0.65	167	376	703	1092	69	24	31	S
2	QN	500	100	2	4	3.4	0.9	0.3	3.57	1.5	0.3	179	157	435	753	120	42	47	S
3	QN	500	100	2	4	3.6	1.5	1.5	3.40	0.63	1.5	199	296	489	785	89	30	31	F
4	PS	500	100	2	4	2	0.5	0.7	2.38	0.37	0.7	170	386	694	1076	61	22	27	S
5	PS	500	100	2	4	3	1.5	0.3	3.53	1.5	0.36	180	178	436	753	60	14	21	S
6	PS	500	100	2	4	3.6	1	1.5	2.43	0.31	1.5	153	394	559	849	112	31	34	S
7	GA	500	100	2	4	2	0.5	0.7	2.34	0.52	0.96	207	418	658	1006	38	13	16	F
8	GA	500	100	2	4	2	0.3	0.6	2.34	0.52	0.96	207	418	658	1006	40	13	16	F
9	GA	500	100	2	4	2.2	0.3	0.6	2.34	0.52	0.96	207	418	658	1006	39	13	16	F
10	GA	500	100	2	4	2	0.5	0.7	2.36	0.35	1.46	164	422	613	748	3714	663	1134	F
11	SMINLP	500	100	2	4	2	0.5	0.7	2.4	0.3	0.8	164	408	693	1064	57	22	23	S
12	SMINLP	500	100	2	4	1	0.3	0.3	2.4	0.3	0.8	165	407	691	1063	64	27	27	S
13	SMINLP	500	100	2	4	1	1.5	1.5	2.4	0.3	0.8	165	407	691	1063	64	27	27	S
14	SMINLP	500	100	2	4	3.6	0.3	0.3	2.4	0.3	0.3	165	407	691	1063	64	27	27	S
15	SMINLP	500	100	2	4	3.6	1.5	1.5	2.4	0.3	0.8	165	407	691	1063	64	27	27	S

** Optimisation ended in local minimum— light red . * The only working GA optimisation— violet . * Global optimum reached— dark green .*

Part 2.

No.	Solver	ρ_1	ρ_2	H	D_2	D_{1i}	h_{1i}	h_{2i}	D_{1o}	h_{1o}	h_{2o}	TV_{1m}	TV_{2m}	TV_{3m}	EPR	t	ItO	ItN	Status
-	-	Ωm	Ωm	m	m	m	m	m	m	m	m	V	V	V	V	m	-	-	-
16	SNLP	500	100	2	4	2	0.5	0.7	2.08	0.3	1.5	160	440	599	872	270	70	100	M
17	SNLP	500	100	2	4	2	0.5	0.7	2.29	0.31	1.5	147	412	580	858	845	240	300	M
18	SNLP	500	100	2	4	1	0.3	0.3	2.36	0.3	1.5	152	404	575	853	433	126	152	S
19	SNLP	500	100	2	4	1	0.3	0.3	2.36	0.3	1.5	152	404	575	853	434	126	152	F
20	SNLP	500	100	2	4	1	0.4	0.4	2.32	0.3	1.5	147	412	581	858	836	251	300	M
21	SNLP	500	100	2	4	1	1.5	1.5	2.26	0.31	1.49	147	415	584	861	616	118	207	S
22	SNLP	500	100	2	4	1	1.5	1.5	2.26	0.31	1.49	147	415	584	861	616	118	207	S
23	SNLP	500	100	2	4	3.6	0.3	0.3	2.38	0.3	1.48	148	403	577	857	930	66	299	S
24	SNLP	500	100	2	4	3.6	1.5	1.5	3.6	1.5	0.33	177	162	430	748	543	74	188	F
25	SNLP	500	100	2	4	3.6	0.4	0.4	2.41	0.31	1.49	150	397	571	581	858	87	300	M
26	SNLP	500	100	2	4	2	0.5	0.7	2	0.31	1.5	164	448	604	876	191	16	66	S
27	SNLP	500	100	2	4	2	1.4	1.4	2	0.30	1.40	172	457	625	907	89	12	28	F
28	SNLP	500	100	2	4	3.6	1.4	1.4	3.6	1.49	0.32	177	163	431	749	473	138	180	F
29	SNLP	500	100	2	5	2	0.5	0.7	2.37	0.34	1.46	125	310	435	718	331	19	99	F
30	SNLP	500	100	2	5	2	0.6	0.8	2.39	0.31	1.5	122	311	435	711	391	83	118	F
31	SNLP	500	100	1	4	2	0.5	0.7	2.15	0.3	1.43	37	118	179	349	734	153	300	M
32	SNLP	100	100	NA	4	2	0.5	0.7	3.59	1.29	0.3	29	31	105	265	141	74	89	S
33	SNLP	100	500	2	4	2	0.5	0.7	2.6	0.37	1.45	24	86	152	518	81	62	78	F

* In case of Uniform soil model the surface layer thickness is inappropriate so denoted in the table as NA—not available. ** Significant values defining optimisation, e.g., D_1 not optimised, increased D_2 = 5 m, different soil model—red text. * Optimisation ended in local minimum— light red , * Global optimum reached— dark green . * Only burial depths optimised— light green , * Iteration limit set to 100 only— light orange . * Uniform and LoH soil models— light blue .

For HoL soil model 1, it was expected that the optimum burial depth for both rings with fixed outer ring diameter $D_2 = 4$ m (Figure 2) should be 0.3 and 1.5 m for h_1 and h_2, respectively, and inner ring $D_1 = 2$ m (as was found in sensitivity analysis Table 2, Table 3 minimum TV_{1m}). For simplicity and comparability, first two optimisation analysis had been performed just with only burial depths h_1 and h_2 being optimised (lines 26, 27 of Table 6). This is also due to the fact that the simulated designs in sensitivity analysis were simulated with quite great step of inner ring diameter of 1 m, as the possible number of designs rapidly increase when a lower step would had been used. These two optimisation analysis were performed with different initial conditions. In both cases, Ansys Maxwell found ES design close to the expected optimal solution. In the second case, the outer ring depth was only 1.4 m where the optimiser evaluated visiting the depth of 1.5 m as unnecessary. This might be caused by the level of solution noise, i.e., with the set analysis percent error of 1% (Section 3) the changes in solution data of 5–15 V might be expectable due to randomness of creation of finite element mesh and also due to not sufficiently fine mesh elements. Better results might be expected for percent error of about 0.3% [17], however this will cause about three times or even greater increase in solution time.

In the rest of the simulation, the inner ring diameter was also optimised. As the sensitivity analysis was only done with 1 m step of the inner ring diameter, the results of optimisation cannot be really matched to sensitivity analysis results. However, from the comparison of TV_{1m} surfaces from Figure 2 it can be expected that the optimal design should has inner ring diameter in 2–3 m region, and the burial depths as $h_1 = 0.3$ m and $h_2 = 1.5$ m. This expectation had been met in most of the results for SNLP optimiser. However, in case of the other optimisers, kind of invalid and different results had been found. This might have happened in some cases due to stucking in actual local minima, stucking in local minima caused by the solution noise, failing due to Ansys Maxwell Adaptive Mesher, etc.

In the case of GA optimiser, more extensive analysis of correct setting of the optimiser might lead to better results. For example, out of four different settings of the optimiser, the optimal design was found in only one (line 10). However, this optimisation had been stopped by user after more than two days of continuous simulation. From the optimisation results it had been found that the optimiser visited the location of global minimum ($D_1 \sim 2$ m, $h_1 \sim 0.3$ m, $h_2 \sim 1.5$ m) only about three times out of total 1134 optimisation iterations. The optimal design was already found after about the first half of all iterations in the 664th iteration, however the optimiser did not evaluate it as the global optimum and was further searching for better solution.

The QN and SMINLP optimisers only found a local minima or did not find the expected optimum design at all in their all 8 performed simulation. Even though the suggested results by the optimisers are quite close to the global optimum, the global optimum still was not reached. The PS optimiser reached the global optimum in only one case out of three and the results seemed to be heavily reliable on the design initial condition. In case of all these three QN, PS, and SMINLP optimisers, there are not many options in Ansys Maxwell software that the user can change to improve the performance of the optimiser. One possible way might be to decrease the analysis percent error, which might help to overcome the optimiser problem of stucking in local minima also caused by solution noise. However, a corresponding increase in solution time is expectable.

The SNLP optimiser tended to give more reliable results, so a more extensive analysis was conducted for this optimiser only focusing mainly on the optimiser immunity to setting different initial conditions of the simulation. An outer ring diameter constraint of 4 m was set in most of the simulations. It was found out that the local minimum becomes significant once the inner ring diameter is close to 3.5 m when a saddle point (Figure 2, comments in Section 4) separates it from the global minimum. This assumption was also clarified by performing additional sensitivity analysis with inner ring diameters of 1.5, 2.5 and 3.5 m (Figure 6 solution surface for $D_1 = 3.5$ m). However, in case of this local minimum the TV_{1m} value is about 15–25 V higher that the expected global minimum in the region of $D_1 = 2$–2.5 m. Thus, from results of lines 24 and 28 in (and also 2 QN and 5 PS) Table 6, it can be seen that, in the case of unfavourable initial conditions, the global minimum is not always found and the optimiser might be unable to overcome this problem on its own.

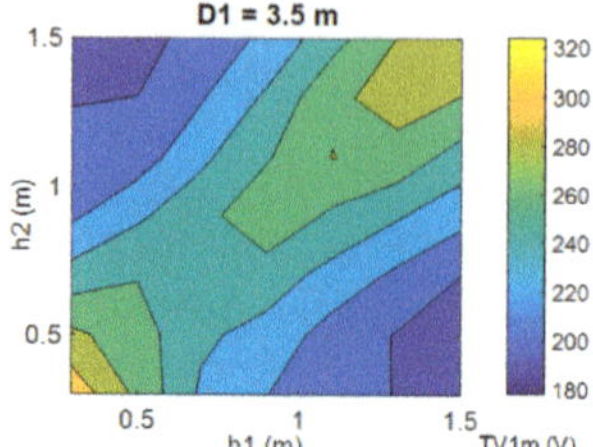

Figure 6. TV1m solution surfaces for ES: 500 Ωm/100 Ωm/2m, D_1 = 3.5 m, D_2 = 4 m.

Lastly, in most cases, the maximum number of 300 optimisation iterations was enough to find the optimal design. In the case of SNLP optimiser, usually slightly more than 100 iterations were necessary to find the global minimum of the optimisation problem. However, if for example the maximum number was set only to 100 iterations, only a solution close to global optimum was found and the 100 iterations limit might be too restricting, i.e., in case of line 16.

6. Conclusions

In this paper, the appropriateness of using the Ansys Maxwell optimizer to optimize ES design was demonstrated on couple of examples. The use of the SNLP optimiser might be assessed as satisfactory and superior to other Ansys Maxwell optimisers. In most cases, the SNLP method found the optimum design, however in the case of unfavourable initial conditions it is still prone to stucking in local minima. This problem can be overcome by conducting the optimisation analysis for different initial conditions to see whether the found result is kind of reliable. In the case of other compared optimisers, different settings might lead to more reliable results, however the increase in solution time as well as more experienced engineer (user) might be necessary.

Another novel approach that is presented in this paper is the optimisation of the design by minimising touch voltages instead of most often used EPR. Unlike simple EPR minimisation would lead to burying of ES in greater depth it was pointed out that this approach might not necessarily lead to optimal layout also assessed based on actual touch voltages. As the real risk that is associated with the ES is driven by the touch voltages instead of EPR, using of touch voltages is more precise than simple optimisation of EPR. However, when the design is optimised to minimise TVs, it was found out that the optimum ES design might get quite counter intuitive considering the wide variety of options, like different dimensional constraints, different soil models, or even different objective functions ~ transferred voltages, step voltages. Thus, as was presented in this paper, the use of software with more complex optimisation methods (non-linear programming, like SNLP method) might be beneficial, if not even necessary.

Author Contributions: Conceptualization, V.V.; Funding acquisition, P.T.; Investigation, V.V.; Supervision, D.T. and P.T.; Validation, M.P. and D.T.; Writing—original draft, V.V.; Writing—review & editing, M.P., D.T. and P.T.

Funding: Authors gratefully acknowledge financial support from the Technology Agency of the Czech Republic (project No. TN01000007).

Acknowledgments: Authors gratefully acknowledge the Centre for Research and Utilization of Renewable Energy (CVVOZE) where this research work was carried out.

Conflicts of Interest: The authors declare no conflict of interest.

References

1. CENELEC. *Earthing of Power Installation Exceeding 1 kV a.c.*; CENELEC Standard EN 50522:2010; European Committee for Electrotechnical Standardization (CENELEC): Brussels, Belgium, 2010.
2. *IEEE Guide for Safety in AC Substation Grounding*; IEEE Std 80-2013; IEEE Standard: Piscataway, NJ, USA, 2015.

3.   *Short-Circuit Currents in Three Phase a.c. Systems: Part 0: Calculation of Currents*; IEC 60909-0; International Electrotechnical Commision (IEC): Geneva, Switzerland, 2016.

4.   Lee, H.-S.; Kim, J.-H.; Dawalibi, F.P.; Ma, J. Efficient Ground Grid Designs in Layered Soils. *IEEE Trans. Power Deliv.* **1998**, *13*, 745–751.

5.   Taher, A.; Said, A.; Eliyan, T.; Hafez, A. Optimum Design of Substation Grounding Grid Based on Balancing Parameters using Genetic Algorithm. In Proceedings of the (MEPCON)—Twentieth International Middle East Power Systems Conference, Cairo, Egypt, 18–20 December 2018; pp. 352–360.

6.   Silva, C.L.B.; Oliveira, D.n.; Pires, T.G.; Nerys, J.W.L.; Calixto, W.P.; Alves, A.J. Optimization of Grounding Grid's Multidesign Geometry. In Proceedings of the IEEE (EEEIC) 16th International Conference on Environment and Electrical Engineering, Florence, Italy, 7–10 June 2016; pp. 1–6.

7.   Alik, B.; Teguar, M.; Mekhaldi, A. Minimization of Grounding System Cost Using PSO, GAO, and HPSGAO Techniques. *IEEE Trans. Power Deliv.* **2015**, *30*, 2561–2569. [CrossRef]

8.   Perng, J.-W.; Kuo, Y.C.; Lu, S.-P. Grounding System Cost Analysis Using Optimization Algorithms. *Energies* **2018**, *11*, 1–19. [CrossRef]

9.   Al-Arainy, A.A.; Khan, Y.; Quereshi, M.I.; Malik, N.H.; Pazheri, F.R. Optimized Pit Configuration for Efficient Grounding of Power System in High Resistivity Soils using Low Resistivity Materials. In Proceedings of the Fourth International Conference on Modeling, Simulation and Applied Optimization, Kuala Lumpur, Malaysia, 19–21 April 2011; pp. 1–5.

10.   Boualegue, A.; Ghodbane, F. A new Parametric Approach for Grounding Analysis in Multilayered Soils with Finite Volumes. In Proceedings of the First International Conference on Renewable Energies and Vehicular Technology, Hammamet, Tunisia, 26–28 March 2012; pp. 274–278.

11.   Pan, W.; Sun, H.; Chai, S.; Zhou, J. Research of Optimal Distance of External Grounding Grid. In Proceedings of the IEEE (POWERCON) International Conference on Power System Technology, Wollongong, Australia, 28 September–1 October 2016; pp. 1–4.

12.   Datta, A.J.; Taylor, R.; Ledwich, G. Earth Grid Safety Criteria Determination with Standards IEEE-80 and IEC-60479 and Optimization of Installation Depth. In Proceedings of the (AUPEC) Australian Universities Power Engineering Conference, Wollongong, Australia, 28 September–1 October 2015; pp. 1–5.

13.   Carman, B.; Palmer, S.; Fickert, L.; Griffiths, H.; Moller, C.; Toman, P. *Substation Earthing System Design Optimisation Through the Application of Quantified Risk Analysis*, 1st ed.; Cigre—International Council on Large Electric Systems: Paris, France, 2018; ISBN 978-2-85873-451-1.

14.   Topolanek, D.; Vycital, V.; Toman, P.; Carman, B. Application of the probabilistic approach for earthing system evaluation in distribution network. *Int. J. Electr. Power Energy Syst.* **2018**, *20*, 268–279. [CrossRef]

15.   *National Annexes NA and NB of British Adopted European Standard Earthing of Power Installations Exceeding 1 kV a.c.*; BS EN 50522:2010; British Standards Instituiton (BSI): London, UK, 2012.

16.   *EMF Simulating Software*, Version 18, Ansys Maxwell Package; Ansys: Canonsburg, PA, USA, 2017.

17.   Vycital, V.; Toman, P. Modelling of Electrical Installations Earthing Systems in Ansoft Maxwell. In Proceedings of the 24th Conference STUDENT EEICT 2018, Brno, Czech Republic, 26 April 2018; pp. 487–491.

Article

Using the Thermal Inertia of Transmission Lines for Coping with Post-Contingency Overflows

Xiansi Lou, Wei Chen and Chuangxin Guo *

College of Electrical Engineering, Zhejiang University, Hangzhou 310027, China; louxiansi@zju.edu.cn (X.L.); chenwei_ee@zju.edu.cn (W.C.)
* Correspondence: guochuangxin@zju.edu.cn

Received: 31 October 2019; Accepted: 17 December 2019; Published: 20 December 2019

Abstract: For the corrective security-constrained optimal power flow (OPF) model, there exists a post-contingency stage due to the time delay of corrective measures. Line overflows in this stage may cause cascading failures. This paper proposes that the thermal inertia of transmission lines can be used to cope with post-contingency overflows. An enhanced security-constrained OPF model is established and line dynamic thermal behaviors are quantified. The post-contingency stage is divided into a response substage and a ramping substage and the highest temperatures are limited by thermal rating constraints. A solving strategy based on Benders decomposition is proposed to solve the established model. The original problem is decomposed into a master problem for preventive control and two subproblems for corrective control feasibility check and line thermal rating check. In each iteration, Benders cuts are generated for infeasible contingencies and returned into the master problem for adjusting the generation plan. Because the highest temperature function is implicit, an equivalent time method is presented to calculate its partial derivative in Benders cuts. The proposed model and approaches are validated on three test systems. Results show that the operation security is improved with a slight increase in total generation cost.

Keywords: security-constrained OPF; preventive control; corrective control; post-contingency overflow; transmission line; thermal inertia

---

1. Introduction

The security-constrained optimal power flow (SCOPF) is an essential tool for making a day-ahead and real-time generation plan [1]. From the perspective of control measures, the SCOPF model can be divided into the preventive SCOPF (PSCOPF) model and the corrective SCOPF (CSCOPF) model. The former model is widely used in current operations of large-scale [2] and island power systems [3]. The operating point obtained by this model can guarantee the safe operation of systems under all credible contingencies. How to filter the contingency set determines to a great extent the performance of PSCOPF model [4]. For given contingencies, results of the PSCOPF model are conservative with the high generation cost. The risk conception has been introduced in [5] to enhance its economics. In [6], an identification method of superfluous constraints has been proposed and the model of PSCOPF can be simplified and efficiently solved. With the large-scale access of sustainable energy and frequent occurrences of natural disasters, the operation environment of power systems becomes more complex and the feasible region of PSCOPF model may not exist in some difficult operating conditions.

For seeking the lower operating cost and more flexible dispatch plans, the CSCOPF model has been put forward. In this model, corrective measures such as the unit rescheduling and load shedding are utilized after the contingency occurrence to maintain the transmission security of systems [7]. There have been some trails about using the CSCOPF model in energy management systems to help operators make timely and reasonable dispatch decisions [8]. However, two main problems hinder its wide application. The first problem is the heavy computational burden caused by the huge number of contingencies especially for large-scale systems and coupling constraints between the preventive control and corrective control. Many efforts have been devoted to shortening the solving time. In [9], Benders decomposition has been applied to solve the CSCOPF model, and the computational complexity has been analyzed. Results showed that the computation speed was significantly improved without sacrificing the accuracy, whether in the serial or parallel computing environment. In [10], an iterative approach that comprises four modules has been proposed, and its performance was better than the direct approach and Benders decomposition. A hybrid method has been used to solve the CSCOPF model in [11] where the maximum feasible region was randomly searched through evolutionary algorithms and then deterministic solutions were provided by the interior-point method. A similar solving strategy has also been adopted in [12], and the optimal coordination of the preventive control and the corrective control was obtained.

A more significant problem of the CSCOPF model is the insecurity of power systems during the post-contingency stage. After the contingency such as line failures or generator outages, conventional corrective measures like the unit rescheduling cannot be immediately implemented due to the limit of ramping rates. The system still operates under the preventive control while violations of line power flow and bus voltages may happen. Once cascading failures are triggered, the effectiveness of pre-made corrective plan is influenced, and the scope of the original accident could be extended. Some previous work has been done to improve the security of the CSCOPF model. An optimal locating method for support generator units has been proposed in [13] to improve ramping abilities and enhance the robustness of systems. In [14], quick-start units have been utilized in corrective actions for reducing the duration of post-contingency stage. References [15,16] have discussed that the fast-response distributed battery energy storage can be used to alleviate post-contingency overloads and reduce the power flow below the short-term emergency rating. In [17], the state of charge has been considered in the distributed energy storage model and results showed that the generation cost increased as the initial charging state declined. The post-contingency demand response has been used to relieve overflows incurred by the renewable generation fluctuation and "N−1"contingencies in [18]. References [19,20] have analyzed the risk of cascading failures caused by post-contingency overloads in the alternating current (AC) and direct current (DC) power transmission network, while the thyristor controlled series capacitor and the multi-terminal direct current have been applied to minimize the load shedding and generation rescheduling, respectively. In addition, the model predictive control (MPC) provides a higher perspective beyond static and open-loop approaches. It is a class of strategies that utilize a process model to establish a control sequence for controlling the future behavior of systems over a horizon [21]. Under this framework, the generation redispatch, load shedding, and regulating transformers have been applied to alleviate emergency thermal loads in [22,23]. Moreover, the energy storage and curtailment of renewables have also been identified as corrective actions in [24].

In this paper, the DC power flow is utilized while post-contingency line overloads are the insecurity problem discussed and solved. The essence of this problem can be described as the lack of short-term transfer capacities of power grids. Compared with approaches such as using quick-start generators, distributed battery energy storage, demand response, and other power flow controllers, the method of improving line capacities could be more direct and effective. The line thermal rating provides a novel perspective instead of conventional power flow limits. The static power flow rating obtained under the worst environment condition is usually conservative. Reference [25] has provided a report of two pioneering schemes in the U.S. and the U.K. where the real-time thermal rating was applied. A new power flow formulation considered the line electro-thermal coupling effect has been introduced in [26], and the error in power transfer evaluation decreased by about 20% by using this method. In [27–29], the thermal rating of transmission lines has been utilized to increase the threshold value of wind power integration and avoid the unnecessary tripping of pre-selected generation assets. Besides the static rating difference, transmission lines have the thermal inertia and corresponding time constants range between several mins and ten mins [30]. This means that dynamic variations of conductor temperatures cannot be ignored especially for the short-term post-contingency stage discussed in this paper. Reference [31] has demonstrated that the neglect of transient thermal behaviors of lines caused the underestimation of power transfer capabilities and led to misoperations. In [32,33], the concept of electrothermal coordination has been proposed and the numerical analysis through a number of case studies validated its benefits in augmenting power transfer capability, emergency control, congestion management, and improving the system loadability. A temperature dependent power flow model has been established in [34] where the dependence of line resistances on conductor temperature was taken into account. In [35], the dynamic electrothermal effect has been brought into safety constraints of the post-contingency power flow control to excavate the overload endurance capability of lines.

When the dynamic thermal process of lines is considered in optimal power flow models, the heat balance equation (HBE) which describes variations of conductor temperatures must be integrated. Because the radiated heat loss rate is proportional to the fourth power of the conductor temperature, the HBE is a nonlinear differential equation. How to solve those electrothermal coupling optimization problems becomes a challenge. In [32,33], a modified Euler method has been used to discretize the HBE and then the problem was broken down into several linearized subproblems whose solutions were iteratively refined and linked. An approximate quadratic relationship between the temperature and the square of current has been derived by some reasonable approximations in [34], and the analytical solution of HBE was obtained for a step change in currents. Reference [35] has proposed that, through transforming the HBE into algebraic difference equations by a numerical integration method, the optimal solution can be realized by combining those transformed equations with other algebraic equations.

This paper proposes that the thermal inertia of transmission lines can be used to cope with post-contingency overflows in the conventional CSCOPF model. According to the system behavior, the post-contingency stage is divided into a response substage and a ramping substage in time sequence. In the previous substage, corrective measures have not been implemented while the power flow shows a step change. In the latter substage, the power flow variation is approximately described by a linear function with the adjustment of unit active power outputs. Based on the assumption of constant resistances and the linearization of the radiated heat loss function, analytical solutions about real-time temperatures in both substages are obtained from the HBE. An enhanced security-constrained OPF (ESCOPF) model is established on the basis of CSCOPF model where thermal rating constraints are integrated in the post-contingency stage. The objective of the established model is to minimize the generation cost. In order to effectively deal with this multi-stage optimization problem containing the quadratic objective function and nonlinear constraints, a solving strategy based on Benders decomposition is proposed. The original problem is decomposed into a master problem of the preventive control and two subproblems of the corrective feasibility check and the thermal rating check. Due to the independence between each contingency, the parallel algorithm can be implemented in solving those two subproblems. An equivalent time method is presented to build an explicit relationship between the highest conductor temperature and unit power outputs. In each iteration, corresponding Benders cuts are generated for infeasible contingencies and returned into the master problem. The final generation plan is obtained until those two checks are satisfied for all contingency scenarios. Numerical simulation results validate that the system security is markedly improved by considering the line thermal inertia in the post-contingency stage and the generation cost is just slightly higher than the result of CSCOPF model.

The main contributions of this paper are summarized as follows:

(a) Two typical temperature variations of post-contingency overload lines are analyzed and their analytical solutions are obtained. If the post-contingency power flow is significantly higher than the value in the corrective stage, the stationary point of the conductor temperature exists. Otherwise, its variation is monotonous.

(b) An ESCOPF model is established to minimize the system generation cost. Thermal rating constraints for line conductors are integrated in the post-contingency stage to avoid cascading failures.

In addition, then, the unit rescheduling is implemented as the corrective measure to remove long-term overloads on transmission lines.

(c) A Benders decomposition based solving strategy is proposed to solve the ESCOPF model, which is divided into a master problem and two subproblems. In order to derive the partial derivative of the highest conductor temperature with respect to the unit active power output, an equivalent time method is presented.

The remainder of this paper is organized as follows. Section 2 analyzes temperature variations of post-contingency overload lines and solutions of HBE in different stages are deduced. In Section 3, the detailed formulation of ESCOPF model, which consists of the preventive stage, post-contingency stage and corrective stage, is presented. In Section 4, a solving strategy based on Benders decomposition is proposed and the ESCOPF model is effectively solved by an iterative and parallel algorithm. Validities of the proposed model the solving strategy are verified on a 6-bus test system, a modified IEEE RTS-96 system, and a case 2383wp system in Section 5. The main conclusions of this paper are drawn in Section 6.

2. Temperature Variations of Post-Contingency Overload Lines

The preventive control and corrective control are utilized to guarantee the operation security of power systems in the base case and eliminate long-term power flow violations caused by random failures, respectively. Because corrective measures cannot be implemented without time delay after the contingency, there is a post-contingency stage between the preventive stage and the corrective stage. This time delay comes from two aspects. First, automatic dispatch systems or dispatchers need time to conduct fault locations and determine the fault type. If the automatic reclosing fails, corresponding corrective measures like the unit rescheduling and emergency load shedding should be matched or remade. The time spent in this process is denoted as the response time in this paper. Then, it takes time for conducting those corrective measures such as the unit rescheduling. The specific duration of ramping time depends on adjusted power outputs and ramping rates of units. Therefore, the system post-contingency stage is divided into the response substage and ramping substage while the power flow shows different variations that are presented in Figure 1, and detailed formulations are derived in the following subsections.

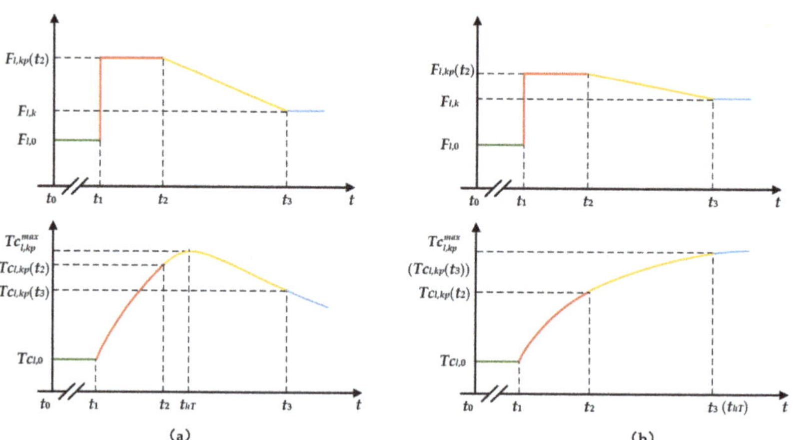

Figure 1. Two typical power flow and temperature variation curves of the post-contingency overload line *l*: (**a**) the concave curve of the conductor temperature variation; (**b**) the monotonous curve of the conductor temperature variation.

2.1. Preventive Stage

The preventive stage is the time interval before t_1. For the discussed line l, the power flow $F_{l,0}$ in this stage is limited by power flow constraints. Since the contingency happens randomly, the conductor temperature of the line l in the preventive stage $Tc_{l,0}$ is conservatively assumed to equal its steady-state value. This temperature can be calculated by the steady-state heat balance equation (SHBE) [36].

$$R\left(Tc_{l,0}\right) I_{l,0}^2 + q_s - q_c(Tc_{l,0}) - q_r(Tc_{l,0}) = 0, \tag{1}$$

where $R(Tc_{l,0})$ indicates the AC resistance of conductor at the temperature $Tc_{l,0}$ and $I_{l,0}$ denotes the preventive load current which is proportional to the power flow $F_{l,0}$:

$$I_{l,0} = I_{l,Rate}\frac{F_{l,0}}{F_{l,Rate}}, \tag{2}$$

where $I_{l,Rate}$ and $F_{l,Rate}$ denote the rated load current and the rated power flow of line l, respectively.

In Equation (1), q_s presents the heat gain rate per unit length from the sun which is regarded as a constant in the short term. q_c and q_r are the convected heat loss rate and radiated heat loss rate per unit length of the line l which are formulated as functions of $Tc_{l,0}$ in Equations (3) and (4), respectively:

$$q_c(Tc_{l,0}) = \left[1.01 + 0.0372\left(\frac{D\rho_f V_W}{\mu_f}\right)^{0.52}\right] k_f K_{angle}\left(Tc_{l,0} - Ta\right), \tag{3}$$

$$q_r(Tc_{l,0}) = 0.0178D\varepsilon\left[\left(\frac{Tc_{l,0} + 273}{100}\right)^4 - \left(\frac{Ta + 273}{100}\right)^4\right], \tag{4}$$

where D indicates the conductor diameter and V_W denotes the speed of air steam around the line l. ρ_f and μ_f are the air density and air dynamic viscosity, respectively. k_f is the thermal conductivity of air. K_{angle} and Ta present the wind direction factor and the ambient air temperature, respectively, while ε indicates the emissivity. It can be seen from the above two equations that q_c is the linear function of $Tc_{l,0}$ while the relationship between q_r and $Tc_{l,0}$ is more complex. Numerical simulations such as the Runge–Kutta method can be utilized to solve the SHBE and obtain $Tc_{l,0}$.

2.2. Response Substage

After the occurrence of contingency at t_1, the preventive control still works and the power flow redistributes immediately with the extremely little electric time constant of the system. For the post-contingency overload line, the power transferred on the line l increases straight to $F_{l,kp}(t_1)$, which is higher than its rated value. Based on the definition of response time t_{resp}, it is expressed as:

$$t_{resp} = t_2 - t_1. \tag{5}$$

The dynamic thermal behavior of the line l during this system response substage can be quantified by the transient heat balance equation (THBE) [36].

$$\frac{dTc_{l,kp}}{dt} = \frac{1}{mC_p}\left[R\left(Tc_{l,kp}\right) I_{l,kp}^2 + q_s - q_c(Tc_{l,kp}) - q_r(Tc_{l,kp})\right], \tag{6}$$

where $Tc_{l,kp}$ indicates the real-time conductor temperature of the line l in the post-contingency stage, and mC_p denotes the total heat capacity of the conductor. In order to obtain the analytical solution of Equation (6), some approximations are needed to be made. First, the resistance-temperature effect is neglected, and R is regarded as a constant. Then, the radiated heat loss rate q_r is locally linearized at the median of the ambient temperature Ta and the rated conductor temperature $Tc_{l,Rate}$:

$$q_r(Tc_{l,kp}) \approx R_1 Tc_{l,kp} + R_2(Ta), \tag{7}$$

where R_1 and R_2 are equivalent parameters. THBE can be simplified as a linear differential equation:

$$\frac{dTc_{l,kp}}{dt} = K_1 Tc_{l,kp} + K_2 I_{l,kp}^2 + K_3 \quad t \in [t_1, t_2], \tag{8}$$

where K_1, K_2 and K_3 denote parameters depending on the line type and surrounding environment conditions. The variation of load current $I_{l,kp}$ during this stage is described by a step function:

$$I_{l,kp}(t) = I_{l,Rate} \frac{\left(F_{l,kp}(t_2) - F_{l,0}\right) H(t - t_1) + F_{l,0}}{F_{l,Rate}} \quad t \in [t_1, t_2], \tag{9}$$

where $H(t)$ presents the Heaviside step function. The analytical expression of $Tc_{l,kp}$ are derived from Equations (8) and (9):

$$Tc_{l,kp}(t) = -\frac{K_2 I_{l,kp}^2(t_1) + K_3}{K_1} + \exp\left[K_1 \left(t - t_1\right)\right] \left(\frac{K_2 I_{l,kp}^2(t_1) + K_3}{K_1} + Tc_{l,0}\right) \quad t \in [t_1, t_2]. \tag{10}$$

It can be observed that the temperature variation of the line l in the system response process is an exponential function of the time t.

2.3. Ramping Substage

Corrective measures start to be taken at t_2 and the post-contingency power flow $F_{l,kp}$ gradually drops below its rated value. At the beginning of ramping substage, all rescheduled units adjust their active power outputs while the power flow of the line l decreases fast. After a few mins, some units finish adjustments and the decline becomes slow. Therefore, the curve of actual power flow variation is multiplied piecewise and convex. In order to simplify the problem, it is assumed that $F_{l,kp}$ linearly decreases from $F_{l,kp}(t2)$ to the corrective value $F_{l,k}$ during the ramping substage. In Figure 1, t_3 denotes the time point when the last unit accomplishes its rescheduling plan. The load current $I_{l,kp}$ in this stage can be approximately formulated as:

$$I_{l,kp}(t) \approx \frac{I_{l,Rate}}{F_{l,Rate}} \left[\frac{F_{l,k} - F_{l,kp}(t_2)}{t_{ramp}}(t - t_2) + F_{l,kp}(t_2) \right] \quad t \in [t_2, t_3], \tag{11}$$

where the ramping time t_{ramp} is defined as:

$$t_{ramp} = t_3 - t_2. \tag{12}$$

Equation (11) can be further compacted and represented as:

$$I_{l,kp}(t) \approx I_{l,kp}(t_2) - \Delta I_{l,kp} (t - t_2) \quad t \in [t_2, t_3], \tag{13}$$

where $I_{l,kp}(t_2)$ and $\Delta I_{l,kp}$ are expressed by the power flow $F_{l,kp}(t_2)$ and $F_{l,k}$:

$$I_{l,kp}(t_2) = \frac{I_{l,Rate}}{F_{l,Rate}} F_{l,kp}(t_2), \tag{14}$$

$$\Delta I_{l,kp} = \frac{I_{l,Rate}}{F_{l,Rate}} \frac{F_{l,k} - F_{l,kp}(t_2)}{t_{ramp}}. \tag{15}$$

Because the actual load current of the line l at any time from t_2 to t_3 does not exceed the approximate value provided by Equation (13), the above simplification makes the thermal rating constraint stricter.

The temperature variation can be obtained through Equations (8) and (13):

$$
\begin{aligned}
Tc_{l,kp}(t) = {}&\exp\left[K_1(t - t_2)\right]\left(Tc_{l,kp}(t_2) + K_4\right) - K_4 - \frac{\Delta I^2 K_2(t - t_2)^2}{K_1} - \\
&\frac{2\Delta I K_2(t - t_2)\left(\Delta I_{l,kp} - I_{l,kp}(t_2)K_1\right)}{K_1^2} \quad t \in [t_2, t_3],
\end{aligned}
\tag{16}
$$

where

$$
K_4 = \frac{K_2 I_{l,kp}^2(t_2)K_1^2 - 2K_2 I_{l,kp}(t_2)\Delta I_{l,kp}K_1 + 2K_2\Delta I_{l,kp}^2 + K_3 K_1^2}{K_1^3}.
\tag{17}
$$

In the initial period of the ramping stage, the real-time conductor temperature $Tc_{l,kp}(t)$ is lower than the steady-state temperature of the real-time power flow $F_{l,kp}(t)$, hence the temperature continues increasing. There are two different curves of the subsequential temperature variation.

(a) If $F_{l,kp}(t_2)$ is significantly higher than the corrective value $F_{l,k}$, $F_{l,kp}(t)$ decreases rapidly; there is a stationary point at t_{hT} when $Tc_{l,kp}(t_{hT})$ equals the steady-state temperature of $F_{l,kp}(t_{hT})$. The temperature reaches its maximum $Tc_{l,kp}^{max}$ and then declines with the continuing decrease of the power flow $F_{l,kp}$. This temperature variation is shown in Figure 1a.

(b) If $F_{l,k}$ is slightly lower than $F_{l,kp}(t_2)$, while $F_{l,kp}(t)$ declines slowly between t_2 and t_3. The ramping stage may end before the temperature $Tc_{l,kp}(t)$ climbs to the steady-state value of the real-time power flow $F_{l,kp}(t)$. Therefore, the temperature variation is monotonous in this case, which is presented in Figure 1b.

2.4. Corrective Stage

After t_3, corrective measures are finished while the power flow falls below the limit specified by security constraints. The load current of the line l is expressed as:

$$
I_{l,k} = I_{l,Rate}\frac{F_{l,k}}{F_{l,Rate}}.
\tag{18}
$$

The corrective conductor temperature $Tc_{l,k}$ can be derived from Equations (8) and (18):

$$
Tc_{l,k}(t) = -\frac{K_2 I_{l,k}^2 + K_3}{K_1} + \exp\left[K_1(t - t_3)\right]\left(\frac{K_2 I_{l,k}^2 + K_3}{K_1} + Tc_{l,kp}(t_3)\right) \quad t \in [t_3, +\infty).
\tag{19}
$$

It can be seen from the above equation that the conduction temperature variation is monotonous. Due to power flow constraints existing in the corrective stage, the security operation of line l can be guaranteed as long as $Tc_{l,kp}(t_3)$ is lower than the rated temperature.

Based on the electrothermal analysis in the four stages above, the key point of improving the system security is controlling the post-contingency conductor temperature within the permissible scale. The highest temperature could be in the initial period of the ramping substage or at the end of this stage. However, the specific variation type cannot be judged by a succinct algebraic criterion before plotting the curve according to temperature functions. Meanwhile, due to the combination of an exponential function and a quadratic function on the right side of Equation (16), explicit formulations of t_{hT} and $Tc_{l,kp}^{mac}$ cannot be obtained.

3. Enhanced Security-Constrained OPF

The enhanced security-constrained OPF model is established on the basis of the PSCOPF model. Thermal rating constraints are integrated in the post-contingency stage to improve the system security and the objective is to minimize the generation cost. Detailed formulations are presented as follows:

$$\min \sum_{i=1}^{N_G} \left(a_i PG_{i,0}^2 + b_i PG_{i,0} + c_i \right), \tag{20}$$

$$s.t. \sum_{i=1}^{N_G} PG_{i,0} = \sum_{j=1}^{N_D} PL_j, \tag{21}$$

$$PG_{i,min} \leq PG_{i,0} \leq PG_{i,max} \quad i = 1, 2, ..., N_G, \tag{22}$$

$$- F_{Rate} \leq F_0 = T_0 \left[PG_0 - PL \right] \leq F_{Rate}, \tag{23}$$

$$F_{kp} = T_k \left[PG_0 - PL \right] \quad k = 1, 2, ..., N_K, \tag{24}$$

$$Tc_{l,kp}^{max} (F_{l,0}, F_{l,kp}, F_{l,k}, t_{resp}, t_{ramp}) \leq Tc_{l,Rate} \quad l = 1, 2, ..., N_L \quad k = 1, 2, .., N_K, \tag{25}$$

$$PG_{i,min} \leq PG_{i,k} \leq PG_{i,max} \quad i = 1, 2, ..., N_G \quad k = 1, 2, ..., N_K, \tag{26}$$

$$PG_{i,0} - \Delta PG_i \leq PG_{i,k} \leq PG_{i,0} + \Delta PG_i \quad i = 1, 2, ..., N_G \quad k = 1, 2, ..., N_K, \tag{27}$$

$$- F_{Rate} \leq F_k = T_k \left[PG_k - PL \right] \leq F_{Rate} \quad k = 1, 2, ..., N_K. \tag{28}$$

In the objective function (20), a_i and b_i denote the second order and the first order cost coefficients of unit i, respectively. c_i indicates the fixed cost. $PG_{i,0}$ is the active power output of the unit i determined by the preventive control. N_G is the number of units. Because probabilities of contingencies are quite low and the rescheduling amount of power output is limited by unit ramping capacities, the cost of corrective control can be neglected.

Constraints (21)–(23) are utilized to guarantee the system operation security in the preventive stage. Equation (21) implies the active power balance where PL_j is the load demand at the bus j and N_L represents the number of load buses. Constraints (22) are power output limits where $PG_{i,min}$ and $PG_{i,max}$ denote the minimum and maximum power output of the unit i, respectively. In power flow constraints (23), PG_0 and PL are the preventive power output vector and the load demand vector, respectively. F_{Rate} and F_0 indicate the rated power flow vector and the preventive power flow vector, respectively. T_0 is the shift matrix in the base case without any failures. In the post-contingency stage, due to the change of grid topology caused by device outages, the power flow distribution is needed to be recalculated for each contingency. Constraint (24) indicates the power flow equation where F_{kp} and T_k are post-contingency power flow vector and the shift matrix in the contingency k, respectively. Conductor temperatures are limited by the constraint (25). According to the analysis of temperature variation in Section 2, the highest temperature $Tc_{l,kp}^{max}$ in the post-contingency stage k depends on $F_{l,0}$, $F_{l,kp}$, $F_{l,k}$, t_{resp} and t_{ramp}, and their functional relationships are determined by Equations (5)–(17).

Corrective control constraints consist of inequations (26)–(28). Constraints (26) and (27) indicate unit output limits in the corrective stage, where $PG_{i,k}$ is the rescheduled power output of the unit i in the contingency k and ΔPG_i is the adjustable power output determined by its reserve capacity. In the power flow constraint (28), F_k denotes the corrective power flow vector and PG_k indicates the rescheduled power output vector in the contingency k. Line overloads must be completely eliminated by corrective measures and the power system enters a new safe operation status.

The established ESCOPF model is a multi-stage dispatch problem with the quadratic objective function (20) and nonlinear constraints (25). Due to the complex form of the temperature variation function in Equation (16), $Tc_{l,kp}^{max}$ is presented as an implicit function about $F_{l,0}$, $F_{l,kp}$, $F_{l,k}$, t_{resp} and t_{ramp}. This implies that conventional optimization methods could not be directly utilized to solve the established model, and some pretreatment measures need to be made. Moreover, different unit rescheduling plans are made for each contingency scenario. Therefore, the computational burden increases as the number of contingencies grows. It is necessary to control the solving time when applying the proposed model in large-scale power systems

4. Solving Strategy Based on Benders Decomposition

Benders decomposition is one of the commonly used decomposition techniques in power systems. J. F. Benders first introduced this decomposition algorithm for solving large-scale mixed-integer programming (MIP) problems [37]. At present, it has been successfully applied to handle various optimization problems including the unit commitment [38], economic dispatch [39], and transmission planning [40]. In applying Benders decomposition, the optimal solution may require iterations between the master problem and several subproblems. The lower bound and upper bound solution are provided by the master problem and feasible subproblems, respectively. If any of the subproblems is infeasible, a Benders cut will be introduced to the master problem. When the upper bound and the lower bound are sufficiently close, the iteration stops. In this section, a solving strategy based on Benders decomposition is presented to solve the established model with nonlinear thermal rating constraints.

The ESCOPF model is first decomposed into a master problem optimizing the preventive control and two subproblems for the corrective control feasibility check and the line thermal rating check, respectively. In the second subproblem, conductor temperatures are calculated based on the determined power flow distribution in the preventive, post-contingency and corrective stage. Meanwhile, the independence between each contingency is taken into account and both subproblems can be processed in parallel. For those contingencies which cannot pass the corrective feasibility check or the line thermal rating check, corresponding Benders cuts are generated and returned into the master problem. The master problem and subproblems are sequentially solved until both checks are satisfied for all contingency scenarios. The specific solving flow is shown in Figure 2 and detailed formulations for the master problem and two subproblems are presented as follows.

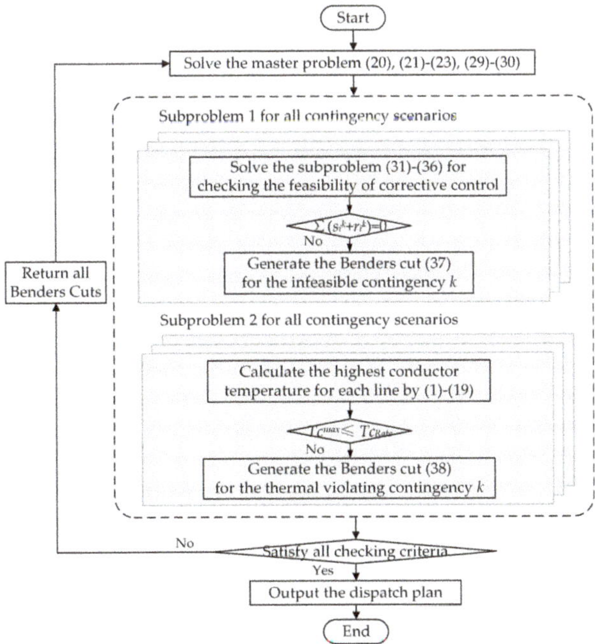

Figure 2. Solving flow based on Benders decomposition.

4.1. Master Problem (Preventive Control Optimization)

The preventive control problem is optimized separately in the master problem which consists of the objective function (20), constraints (21)–(23). The preventive power flow is obtained and the power flow redistribution in each contingency during the response stage can be calculated by Equation (24). Corresponding temperature variations from t_0 to t_2 are quantified by Equations (1) and (10). In each iteration, Benders cuts (29) and (30) are generated for those check violating contingencies and added into the master problem as constraints:

$$\text{Corrective control feasibility check Benders cuts : Constraints (37)}, \tag{29}$$

$$\text{Thermal rating check Benders cuts : Constraints (38)}. \tag{30}$$

The feasible region of the master problem is continuously shrunk by returned Benders cuts and the most economical operating point is searched again in the new feasible region. According to following generating approaches for Benders cuts, above constraints (29) and (30) are linear. Therefore, the master problem in each iteration is a linearly constrained quadratic optimization problem which can be directly solved by current optimizers such as Cplex, Gurobi and so on.

4.2. Subproblem 1 (Corrective Control Feasibility Check)

Due to the existence of correlated constraints (27) between the preventive control and corrective control, the feasibility of corrective control needs to be checked for the operating point determined by the master problem in each iteration. In the credible contingency k, slack variables $s_{i,k}$ and $r_{i,k}$ are introduced for power output rescheduling constraints. Detailed formulations of the corrective control feasibility check subproblem are presented as follows:

$$\min \sum_{i=1}^{N_G} (s_{i,k} + r_{i,k}), \tag{31}$$

$$s.t. \quad \text{Constraints (26) and (28)}, \tag{32}$$

$$PG_{i,k} - PG_{i,0}^* - s_{i,k} \leq \Delta PG_i \quad i = 1, 2, ..., N_G, \tag{33}$$

$$PG_{i,k} - PG_{i,0}^* + r_{i,k} \geq -\Delta PG_i \quad i = 1, 2, ..., N_G, \tag{34}$$

$$s_{i,k} \geq 0 \quad i = 1, 2, ..., N_G, \tag{35}$$

$$r_{i,k} \geq 0 \quad i = 1, 2, ..., N_G. \tag{36}$$

In the constraints (33) and (34), the determined preventive power output $PG_{i,0}^*$ is a constant provided by results of the master problem in the current iteration. After solving the subproblems (31)–(36), the corrective power flow distribution is obtained and temperature variations after t_2 can be quantified by Equations (16) and (19).

Once the optimized result of the objective function (31) equals 0, overflows in the contingency k can be eliminated by a feasible unit rescheduling. Otherwise, corresponding Benders cuts are needed to be generated and returned to the master problem. The Benders cut of the contingency k which cannot be corrected is deduced as:

$$\sum_{i=1}^{N_G} (s_{i,k} + r_{i,k}) + \sum_{i=1}^{N_G} \left[(-\lambda_{i,k} + \gamma_{i,k}) \left(PG_{i,0} - PG_{i,0}^* \right) \right] \leq 0, \tag{37}$$

where $\lambda_{i,k}$ and $\gamma_{i,k}$ are the Lagrange multipliers of the constraint (33) and (34), respectively. The Benders cut (37) is a linear constraint about $PG_{i,0}$ where $s_{i,k}$, $r_{i,k}$, $\lambda_{i,k}$, $\gamma_{i,k}$ and $PG_{i,0}^*$ are constants. Only if the optimized result of this subproblem equals 0 for all contingencies, the operating point set by the master problem can pass the feasibility check for the corrective control.

4.3. Subproblem 2 (Line Thermal Rating Check)

Temperature variations for four sequential stages are calculated based on the power flow variation determined by the master problem and the corrective control feasibility check subproblem. In each contingency, only if the highest temperature of each line conductor is below its rated value, the thermal rating check can be satisfied. Otherwise, the Benders cut is needed to be derived and added into the master problem:

$$\sum_{l \in L_k} \left(Tc_{l,kp}^{max} - Tc_{l,Rate} \right) + \sum_{i=1}^{N_G} \left(\sum_{l \in L_k} \frac{\partial Tc_{l,kp}^{max}}{\partial PG_{i,0}} \bigg|_{PG_{i,0}=PG_{i,0}^*} \right) \left(PG_{i,0} - PG_{i,0}^* \right) \leq 0, \tag{38}$$

where L_k denotes the set of transmission lines which violate thermal rating constraints. The key point of obtaining the thermal rating Benders cut is calculating the partial derivative of $Tc_{l,kp}^{max}$ with respect to $PG_{i,0}$. In this paper, an equivalent time method is proposed to build an explicit relationship between the highest conductor temperature and the preventive power output.

The basic idea is using an extended response process to simulate the original two-stage process of the post-contingency temperature variation. The equivalent principle is that those two processes have the same highest temperature and the equivalent time $t_{l,kp}^{equal}$ of the line l in the contingency k can be calculated from the following equation:

$$Tc_{l,kp}^{max} = -\frac{K_2 I_{l,kp}^2(t_1) + K_3}{K_1} + \exp\left(K_1 t_{l,kp}^{equal} \right) \left(\frac{K_2 I_{l,kp}^2(t_1) + K_3}{K_1} + Tc_{l,0} \right). \tag{39}$$

Due to the monotonous variation of the conductor temperature in the extended response process, the partial derivative of $Tc_{l,kp}^{max}$ with respect to $PG_{i,0}$ can be derived by the chain rule:

$$\frac{\partial Tc_{l,kp}^{max}}{\partial PG_{i,0}} = \left\{ -1 + \exp\left(K_1 t_{l,kp}^{equal} \right) \right\} \frac{K_2}{K_1} 2I_{l,kp}^* \frac{\partial I_{l,kp}(t_1)}{\partial PG_{i,0}} + \exp\left(K_1 t_{l,kp}^{equal} \right) \frac{\partial Tc_{l,0}}{\partial PG_{i,0}}, \tag{40}$$

where $I_{l,kp}^*$ indicates the load current of the line l at t_1 in the contingency k which can be obtained from results of the master problem. The partial derivative of $I_{l,kp}$ with respect to $PG_{i,0}$ can be calculated from the power flow Equation (24):

$$\frac{\partial I_{l,kp}}{\partial PG_{i,0}} = \frac{dI_{l,kp}}{dF_{l,kp}} \frac{\partial F_{l,kp}}{\partial PG_{i,0}} = \frac{I_{l,Rate}}{F_{l,Rate}} T_k(l,i). \tag{41}$$

Similarly, based on the power flow equation in the preventive stage (23), the partial derivative of $Tc_{l,0}$ with respect to $PG_{i,0}$ can be deduced as:

$$\frac{\partial Tc_{l,0}}{\partial PG_{i,0}} = \frac{dTc_{l,0}}{dI_{l,0}} \frac{dI_{l,0}}{dF_{l,0}} \frac{\partial F_{l,0}}{\partial PG_{i,0}} = \frac{dTc_{l,0}}{dI_{l,0}} \frac{I_{l,Rate}}{F_{l,Rate}} T_0(l,i). \tag{42}$$

Using a quadratic function to fit the SHBE, the above partial derivation can be further derived as:

$$Tc_{l,0} = M_1 I_{l,0}^2 + M_2 I_{l,0} + M_3, \tag{43}$$

$$\frac{dTc_{l,0}}{dI_{l,0}} = 2M_1 I_{l,0}^* + M_2. \tag{44}$$

In Equation (44), $I_{l,0}^*$ denotes the determined preventive load current of the line l. It can be seen from the Equations (41)–(44) that the partial derivative of $Tc_{l,kp}^{max}$ with respect to $PG_{i,0}$ is a constant for the determined operating point and the unit rescheduling plan. This means the return Benders cut (38) is also a linear constraint.

The pseudocode of the solving strategy based on Benders decomposition is shown in Algorithm 1. It may help to reproduce and implement the approach proposed in this paper. The NFC and NTC indicate the number of contingenies which cannot pass the corrective control feasibility check and the line thermal rating check.

Algorithm 1 The solving strategy based on Benders decomposition

---

Input: Power system parameters and environment parameters. Durations of the response time and ramping time. The rated conductor temperature. The contingency set.

1: **while** $NFC \neq 0$ and $NTC \neq 0$ **do**
2:      solve the master problem consisting of Equations (20), (21)–(23) and (29), (30);
3:      obtain unit active output PG_0^* and power flow distribution F_0;
4:      $NFC = 0$;
5:      **for** each $k \in [1, N_K]$ **do**
6:          solve the subproblem 1 consisting of Equations (31)–(36);
7:          **if** $\sum s_{i,k} + r_{i,k} \neq 0$ **then**
8:              generate the Benders cut by Equation (37) and return it into th master problem;
9:              obtain the power flow distribution F_k; $NFC = NFC + 1$;
10:         **end if**
11:     **end for**
12:     $NTC = 0$;
13:     **for** each $k \in [1, N_K]$ **do**
14:         **for** each line l operating in the contingency k **do**
15:             calculate the $Tc_{l,kp}^{max}$ and $t_{l,kp}^{equal}$ by Equations (1)–(19) and (39), repsectively;
16:             **if** $Tc_{l,kp}^{max} > Tc_{l,Rate}$ **then**
17:                 add line l into the set L_k
18:             **end if**
19:         **end for**
20:         **if** $L_k \neq \phi$ **then**
21:             $NTC = NTC + 1$; $sumdTc = 0$;
22:             **for** each $l \in L_k$ **do**
23:                 $sumdTc = sumdTc + (Tc_{l,kp}^{max} - Tc_{l,Rate})$;
24:             **end for**
25:             **for** each $i \in [1, N_G]$ **do**
26:                 $sumpd_i = 0$;
27:                 **for** each $l \in L_k$ **do**
28:                     calculate the partial derivative $\frac{\partial Tc_{l,kp}^{max}}{\partial PG_{i,0}}$ by Equations (40)–(44);
29:                     $sumpd_i = sumpd_i + \frac{\partial Tc_{l,kp}^{max}}{\partial PG_{i,0}}$;
30:                 **end for**
31:             **end for**
32:             return $sumdTc + \sum_{i=1}^{N_G} [sumpd_i(PG_{i,0} - PG_{i,0}^*))] \leq 0$ into the master problem;
33:         **end if**
34:     **end for**
35: **end while**

Output: The active power output of each unit in the preventive stage and its adjusted amount in the corrective stage. The power flow distribution and variations of line conductors.

---

5. Case Studies

The model and solving strategy proposed in this paper are validated on a 6-bus test system, a modified IEEE RTS-96 system, and a case 2383wp test system. It assumed that all transmission lines in those three test systems use the 400 mm^2 Drake 26/7 ACSR conductor (Nexans S.A., Paris, France) whose parameters (resistance R, diameter D, heat capacity mCp and emissivity ε) are given by Reference [30] and shown in Table 1.

Table 1. The parameters of transmission lines.

R (Ω/m)	D (mm)	mCp (J/(m*°C))	ε ($-$)
8.688×10^{-5}	28.1	399.0	0.5

Another assumption is that lines operate in the same environment condition, although it is difficult to be satisfied for large-scale power systems. The related environment parameters can be divided into two kinds. The air density ρ_f, dynamic viscosity μ_f and the thermal conductivity of air k_f may not have obvious variations and could be regarded as constants. The other parameters including the ambient air temperature Ta, speed of air steam V_W, wind direction factor K_{angle} and the power gain rate from the sun q_s can be provided by online micro-meteorology monitoring systems. With the development of smart grids, those systems have a wide range of applications [41]. The more accurate measured parameters are obtained, the more effective control for the line thermal rating can be achieved. The data collected by distributed monitoring systems are transmitted through the high-speed optical networks or wireless networks like ZigBee, GPRS and 4G to energy management systems (EMS). Through data preprocessing, those data can be used to solve the electro-thermal coupling OPF model. Values of two kinds of parameters are presented in Table 2.

Table 2. The environment parameters.

ρ_f (kg/m^3)	μ_f (Pa $*$ s)	k_f (W/(m*°C))	Ta (°C)	V_W (m/s)	K_{angle} (°)	q_s (W/m)
1.029	2.04×10^{-5}	0.0295	40.0	0.61	90.0	14.1

In addition, for protection systems, the highest threshold value of line load current should be appropriately raised and the time delay of triggering for overload lines needs to be increased. It will let protections not take action immediately and allow transmission lines to operate under the overload condition for a short period of time. In addition, then the ESCOPF model proposed in this paper can work in practical applications.

Under the condition of given parameters, the time constant of line thermal process is about 14 min which has been verified in Reference [30]. Meanwhile, the rated load current of lines is set at 992 A while the corresponding steady-state temperature is 100.0 °C. The widely used "N$-$1" criterion is adopted for the former two test systems to filter line contingencies. This means, in each contingency, there is just a single transmission line outages. All simulations are performed on a personal computer with 4 Intel (R) Core (TM) i5-6200U CPU (2.3 GHz) (Lenovo, Beijing, China) and 8 GB memory. The programs are implemented using a Matlab (R2016a, The MathWorks, Natick, MA, USA) environment and underlying optimization problems are solved by Cplex.

5.1. 6-Bus Test System

The 6-bus test system consists of three units, three load bus, and 11 transmission lines. Its wiring diagram is presented in Figure 3. Detailed parameters of units, load buses and lines are given by Tables A1–A3 (Appendix A), respectively. There are 11 "N$-$1" contingencies where the line k failures in the contingency k. The response time and the ramping time of this system are 5 min and 7 min, respectively, and the rated conductor temperature is conservatively set at 100.00 °C. The mean program executing time of 10 tests with the consistent convergence solution is 10.07 s. Simulation results in each iteration are shown in Table 3.

In the initial iteration, the preventive control is optimized separately with the lowest generation cost \$861.92. Four contingencies have post-contingency overflows which cannot be eliminated by the feasible corrective control. The thermal rating check is violated in three contingencies, and the highest temperature reaches 136.64 °C. Unit active power outputs are adjusted by returned Benders cuts as the iteration number grows. Combining with generation cost coefficients in Table A1, it can be found that the output of the most expensive unit G2 increases, which results in the growth of *TGC*. It is worth noting that, after iteration 3, *NFC* increases from 0 to 1, while *NTC* decreases from 2 to 1. This implies that the corrective control feasibility check and the thermal rating check may be conflicting in some situations, and this conflict is solved by further shrinking the feasibility region of the operating point. After five iterations, both *NFC* and *NTC* drop to 0 and the highest conductor temperature is strictly controlled within the permissible scale 100.00 °C. This means that both short-term and long-term safe operation of this 6-bus system can be guaranteed. The final generation cost rises to $917.51, which is 6.5% higher than its initial results.

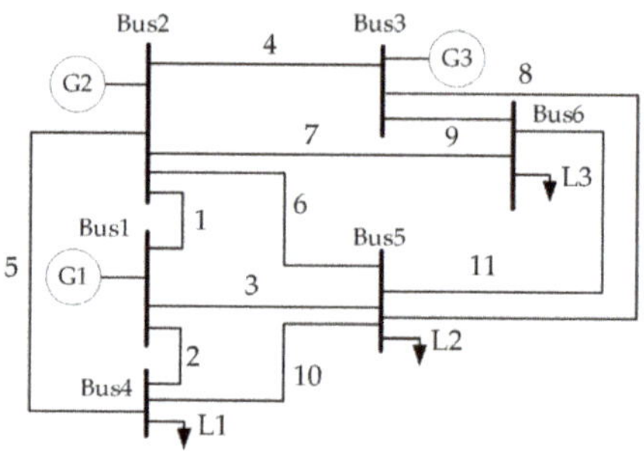

Figure 3. The wiring diagram of 6-bus test system.

Table 3. The simulation results in each iteration for a 6-bus system.

Iteration	$PG_{1,0}$ (MW)	$PG_{2,0}$ (MW)	$PG_{3,0}$ (MW)	NFC [a]	NTC [b]	Tc^{max} (°C)	TGC [c] ($)
0	160.84	0.00	109.16	4	3	136.64	861.92
1	140.67	11.26	118.07	2	2	113.66	889.38
2	130.10	23.97	115.93	0	2	102.69	911.14
3	127.77	26.31	115.93	1	1	100.56	916.07
4	127.29	26.92	115.79	0	1	100.12	917.21
5	127.18	27.04	115.77	0	1	100.03	917.46
6	127.16	27.07	115.77	0	0	100.00	917.51

[a] The number of contingencies which cannot pass the corrective control feasibility check, [b] The number of contingencies which cannot pass the thermal rating check, [c] The total generation cost.

In addition, temperature variations of lines that have the highest conductor temperature in each contingency are shown in Figure 4. Displayed durations of the preventive stage and the corrective stage are both 5 min. The post-contingency stage starts at the 5th min and ends at the 17th min. It is further confirmed that conductor temperatures in all contingencies are below the allowable value 100.00 °C. The highest temperature occurs on line 1, which connects two units G1 and G2 in the "N−1" contingency with the failure of line 2. The post-contingency load rate of line 1 reaches 1.35 at the 5th min and drops to 1.00 from the 10th min to the 17th min. The temperature rapidly rises to the peak 100.00 °C at around the 15th min and then slowly declines. This result verifies that the power flow constraint and the thermal rating constraint are not equivalent, and the former constraint is over conservative from the short-term perspective.

The conductor temperature varies monotonously under the condition that the corrective power flow is higher or slightly lower than it in the post-contingency stage. This corresponds to cases of the highest temperature line in the contingencies 1, 3, 5, 6, and 7. For those lines with highest temperature in the contingencies 2, 4, 8, 9, 10, and 11, the post-contingency power flow is significantly higher than the corrective value while concave temperature curves are presented. In those cases, the maximum absolute difference between load rates in the post-contingency and the corrective stage reaches 0.35, which is obviously higher than the value 0.14 in former cases. This result validates the analysis about two typical temperature variations in Section 2.

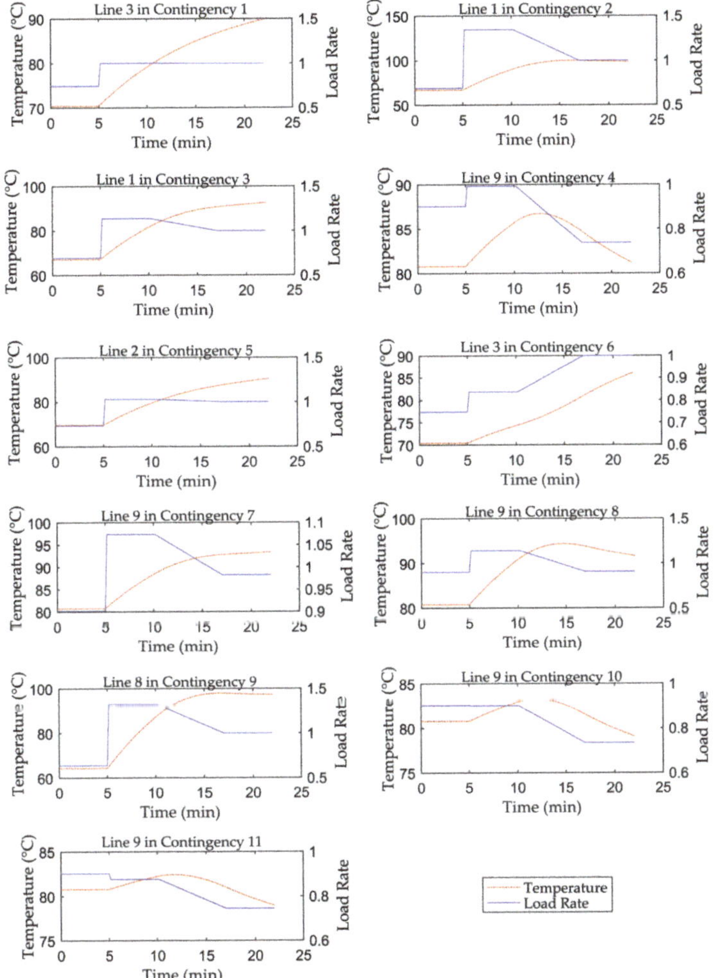

Figure 4. The temperature variations of the line with the highest temperature in each contingency.

It can be found from Figure 4 that the highest temperature occurs five times on line 9 in 11 credible contingencies. Line 9 directly connects the unit G3 and the load bus L3 while the load rate in the preventive stage is 0.90. It is markedly higher than the second highest load rate 0.75 of line 3, which becomes the highest temperature line for the remaining two contingencies. From this perspective, line 9 can be regarded as the key transmission line in the 6-bus system while the security and economy of the system could be improved by appropriately raising its transmission capacity.

5.2. IEEE RTS-96 System

The modified RTS-96 system consists of three interconnected RTS-79 systems and has 96 units, 73 buses, and 120 branches in total. Its topology is presented in Figure 5. The unit data and load data can be found in Reference [42] while the modified line data are provided by Reference [15]. Because line 52 in area 2 and line 90 in the area 3 are single-circuit lines connecting to the bus 207 and bus 307, respectively, their outages will cause a power imbalance in the post-contingency stage. Other measures such as more frequent inspections and maintenance could be used to improve their secure operations. Therefore, the final number of "N−1" line contingencies is 118.

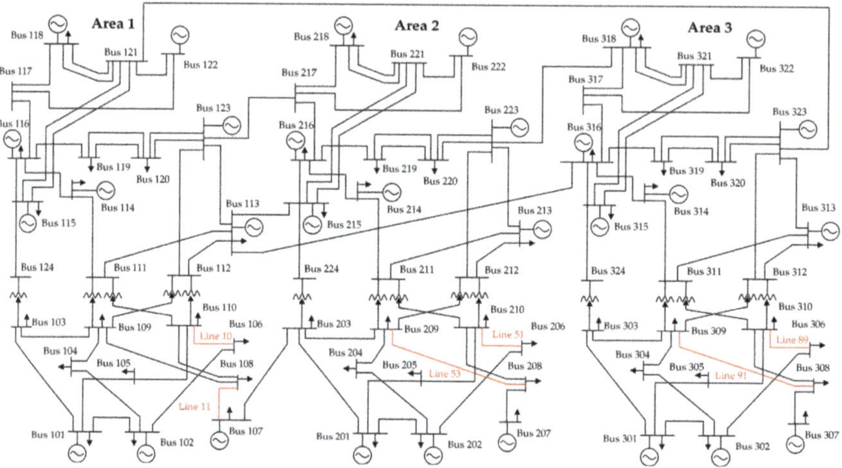

Figure 5. The wiring diagram of the IEEE RTS-96 test system.

In the base case, the response time and ramping time of this system are also set at 5 min and 7 min, respectively, while the rated conductor temperature is 100.00 °C. The mean value of the program executing time for 10 simulations with the same convergence solution equals 53.87 s and detailed simulation results in each iteration are shown in Table 4.

Table 4. The simulation results in each iteration for the IEEE RTS-96 system.

Iteration	$NFC^{\,a}$	$NPC^{\,b}$	LR^{max}	$NTC^{\,c}$	Tc^{max} (°C)	TGC^{d} (\$)
0	10	118	1.78	19	136.04	5058.8
1	2	53	1.46	8	109.35	5069.6
2	2	51	1.37	6	102.56	5071.9
3	2	51	1.35	3	100.73	5072.5
4	0	51	1.35	3	100.22	5072.5
5	0	51	1.34	2	100.07	5072.5
6	0	51	1.34	1	100.02	5072.5
7	0	51	1.34	0	100.00	5072.5

[a] The number of contingencies which cannot pass the corrective control feasibility check, [b] The number of contingencies which violate power flow limits in the post-contingency stage, [c] The number of contingencies which cannot pass the thermal rating check, [d] The total generation cost.

There are 10 contingencies which cannot satisfy the corrective control feasibility check in the initial solution. Post-contingency overflows occur in all contingencies, and the highest load rate reaches 1.78. From the perspective of thermal rating, 19 contingencies are not secure. The highest temperature is 136.04 °C, and the lowest total generation cost is \$5058.8. After the first iteration, NFC, NPC, and NTC decrease significantly to 2, 53, and 8, respectively. In results after iteration 3, TGC approaches its final optimized value. This is because the same type of units have the same generation cost coefficients. Actually, the generation plan is still adjusted, which can be observed from variations of NFC, LR^{max}, NTC, and Tc^{max}. After seven iterations, long-term overflows can be totally removed by the feasible corrective control while the highest temperature equals the rated value 100 °C, which confirm that the algorithm has converged. It is worth noting that there are still 51 contingencies having post-contingency overload lines. However, according to the analysis in Section 2, those short-term power flow violations will not affect the operation security of transmission lines.

In order to further verify the effective control for conductor temperatures through the ESCOPF model, the highest conductor temperature of each transmission line in all credible contingencies is presented in Figure 6.

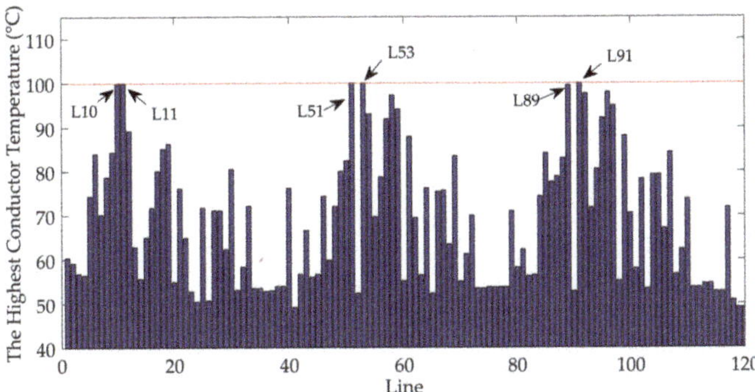

Figure 6. The highest conductor temperature of each line in all credible contingencies.

It can be seen that most of temperatures are in the interval between 50 °C and 90 °C. The highest post-contingency temperature 100 °C occurs on line 10 and line 11, while temperatures of line 51, line 53, line 89, and line 91 are also very close to this limit. Based on the system diagram, it can be found that those lines are at similar topological locations in corresponding areas. For instance, line 10, line 51, and line 89 connect the bus 106 and bus 110, bus 206 and bus 210, bus 306 and bus 310, respectively. When focusing on the area 1, the 136 MW load demand is located at the bus 106. Once line 5 connecting bus 102 and bus 106 fails, 136 MW of power is needed to be transferred by line 10 alone. Meanwhile, the load rate of line 10 in the preventive stage 0.82 is relatively higher. The system operation security can be improved by controlling conductor temperatures of those key transmission lines in the post-contingency stage.

The performance of the proposed ESCOPF model is further compared with the conventional PSCOPF model and CSCOPF model from the perspective of security and economy, while detailed results are listed in Table 5.

Table 5. The comparison of results of PSCOPF, ESCOPF, and CSCOPF models.

Model	NPC^a	LR^{max}	NTC^b	Tc^{max} (°C)	TGC^c ($)
PSCOPF	0	1.00	–	–	5145.4
ESCOPF	51	1.34	0	100.00	5072.5
CSCOPF	88	1.49	17	115.21	5063.6

[a] The number of contingencies which violate power flow limits in the post-contingency stage, [b] The number of contingencies which cannot pass the thermal rating check, [c] The total generation cost.

It can be seen that the PSCOPF is the most conservative model with the highest generation cost $5145.4. For the operating point obtained through this model, power flow constraints are satisfied for all contingencies. Since there is no post-contingency stage, the values of NTC and Tc^{max} do not exist. The CSCOPF model tries to eliminate overflows by the corrective control while effects of the time delay are ignored. Both power flow constraints and thermal rating constraints are violated in the post-contingency stage. The highest load rate and conductor temperature are 1.49 and 115.21 °C, respectively. Total generation cost $5063.6 is the lowest. In results of the proposed ESCOPF model, NPC declines from 88 to 51 and LR^{max} is limited to 1.34. All credible contingencies can pass the thermal rating check. From the viewpoint of line dynamic thermal behavior, the ESCOPF model and the PSCOPF model have the equivalent security. However, the former total generation cost $5072.5 is just slightly higher than the result of CSCOPF model.

In addition, influences of the rated conductor temperature on the highest line load rate in all contingency scenarios and the total generation cost of the system are studied and results are shown in Figure 7.

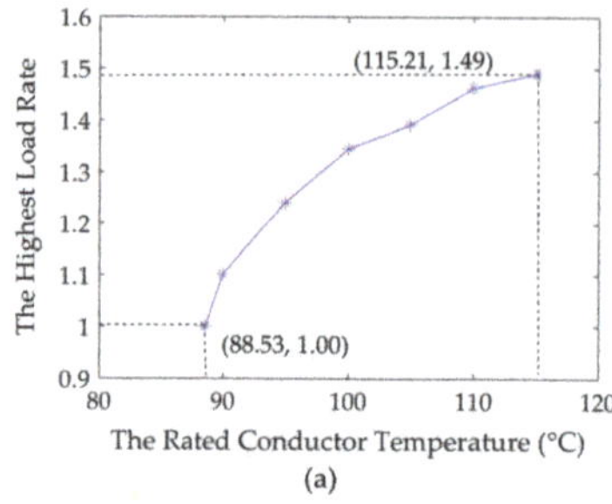
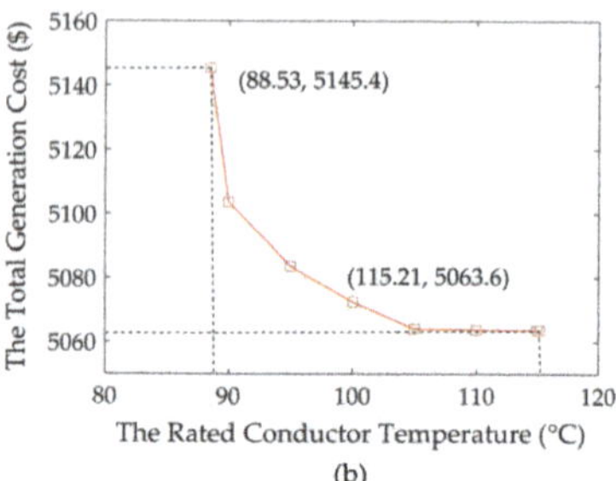

(a)

(b)

Figure 7. The highest load rate and total generation cost under various rated temperatures: (**a**) the highest load rate variation; (**b**) the total generation cost variation.

The highest load rate in the post-contingency stage increases while the total generation cost decreases as the rated conductor temperature grows. It can be observed that both rates of change decline. When the rated temperature is set at 88.53 °C, there is no contingency having overload lines, and the total generation cost equals the result of the PSCOPF model. Actually, the corrective control is no more needed while the ESCOPF model degenerates into a PSCOPF model. When the preset rated temperature is higher than 115.21 °C which is over the highest temperature occurring in results of the CSCOPF model, thermal rating constraints in the post-contingency stage become redundant. This means that the established ESCOPF model is converted into a CSCOPF model and their final optimized generation costs $5063.6 are consistent.

Moreover, Figure 8 presents the surface of total generation costs under various combinations of the response time and the ramping time. The generation cost remains unchanged while the response time is below 3 min and the ramping time is within 5 min. Due to the thermal inertia, lines can operate under extremely high load rates and do not violate thermal rating limits for a short duration. The operating point of the system is not affected by post-contingency thermal rating constraints. As both growths of the response time and the ramping time, the generation cost continues increasing. It can be observed that the variation of the response time has a greater influence. This is because the post-contingency power flow keeps up a high level in this duration. While the response time is in the interval of 5–8 min and the ramping time is within the interval of 6–9 min, the increase of the total generation cost is the most significant. Beyond this time interval, the growth becomes slow. The above analysis implies that shortening the system response time and implementing units with the quick ramping capacities can improve the security of power systems and decline its operating cost.

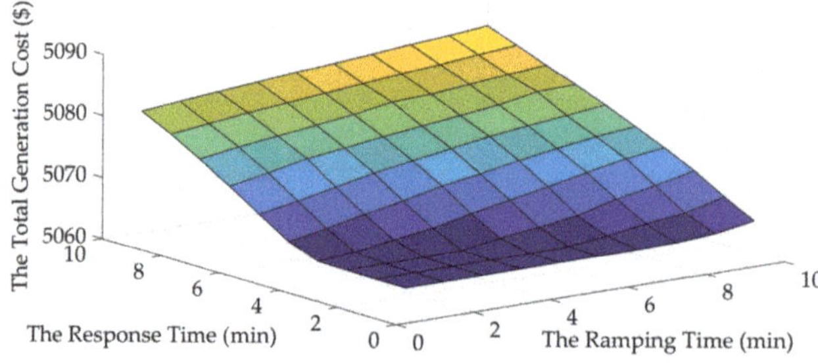

Figure 8. The total generation cost under various combinations of the response time and ramping time.

5.3. Case 2383wp Test System

In order to verify the validities of the proposed ESCOPF model and solving strategy on large-scale power systems, the case 2383wp test system that consists of 2383 buses, 2896 branches, and 327 units is adopted. Its detailed system parameters can be found in the file provided by Matpower (6.0, Cornell University, Ithaca, NY, USA). In addition, the "N−5" contingencies are randomly generated for simulations. It means that there are five failure transmission lines in each contingency. The rated conductor temperature is set at 100.00 °C for all lines. Simulations are conducted for 10, 20, 50 and 100 "N−5" contingencies and the executing time is 367.83 s, 547.76 s, 1817.73 s, and 4540.95 s, respectively. The detailed results are shown in Table 6–9.

Table 6. The simulation results for case 2383wp test system under 10 "N−5" contingencies.

Iteration	NFC^a	NPC^b	LR^{max}	NTC^c	Tc^{max} (°C)	TGC^d ($)
0	6	9	8.17	6	1342.3	21,033
1	0	9	3.23	6	259.5	25,527
2	0	9	2.03	6	128.0	26,624
3	0	9	1.67	6	104.8	26,949
4	0	9	1.60	1	100.8	27,014
5	0	9	1.59	1	100.1	27,025
6	0	9	1.58	0	100.0	27,027

[a] The number of contingencies which cannot pass the corrective control feasibility check, [b] The number of contingencies which violate power flow limits in the post-contingency stage, [c] The number of contingencies which cannot pass the thermal rating check, [d] The total generation cost.

Table 7. The simulation results for case 2383wp test system under 20 "N−5" contingencies.

Iteration	NFC^a	NPC^b	LR^{max}	NTC^c	Tc^{max} (°C)	TGC^d ($)
0	14	19	8.58	14	1473.7	21,033
1	1	19	3.00	15	231.4	26,613
2	0	16	1.93	1	120.6	27,709
3	0	16	1.66	1	102.9	27,988
4	0	16	1.61	1	100.4	28,033
5	0	16	1.60	0	100.0	28,039

[a] The number of contingencies which cannot pass the corrective control feasibility check, [b] The number of contingencies which violate power flow limits in the post-contingency stage, [c] The number of contingencies which cannot pass the thermal rating check, [d] The total generation cost.

Table 8. The simulation results for case 2383wp test system under 50 "N−5" contingencies.

Iteration	NFC^a	NPC^b	LR^{max}	NTC^c	Tc^{max} (°C)	TGC^d ($)
0	35	49	7.97	35	1279.9	21,033
1	2	50	3.14	34	248.6	28,222
2	0	49	1.99	5	124.9	30,732
3	0	49	1.60	3	106.8	31,611
4	0	49	1.64	5	102.4	31,893
5	0	49	1.59	2	100.8	32,000
6	0	49	1.60	2	100.2	32,027
7	0	49	1.59	0	100.0	32,037

[a] The number of contingencies which cannot pass the corrective control feasibility check, [b] The number of contingencies which violate power flow limits in the post-contingency stage, [c] The number of contingencies which cannot pass the thermal rating check, [d] The total generation cost.

It can be found that the algorithm can converge quickly and steadily for the four simulations above within 8 iterations. The executing time increases with the growth of the number of contingencies. Those results are obtained on a person computer with series computation. According to the analysis in Section 4, both subproblems can be processed in parallel due to the independence between each contingency. Therefore, when the model and algorithm are applied on more large-scale contingency sets, the parallel computing architecture can be used to effectively decline the executing time.

Table 9. The simulation results for case 2383wp test system under 100 "N−5" contingencies.

Iteration	NFC^a	NPC^b	LR^{max}	NTC^c	Tc^{max} (°C)	TGC^d ($)
0	70	96	8.58	70	1473.7	21,033
1	3	92	3.52	3	297.7	29,058
2	5	96	3.00	70	231.6	29,978
3	3	92	2.12	6	144.2	31,304
4	0	92	1.63	4	108.8	33,012
5	0	92	1.61	2	102.9	33,160
6	0	92	1.61	3	102.7	33,547
7	0	92	1.61	2	100.3	33,555
8	0	92	1.61	0	100.0	33,593

[a] The number of contingencies which cannot pass the corrective control feasibility check, [b] The number of contingencies which violate power flow limits in the post-contingency stage, [c] The number of contingencies which cannot pass the thermal rating check, [d] The total generation cost.

In initial solutions, maximum load rates reach around 8.00 and corresponding highest conductor temperatures are over 1200 °C. Both indicators are gradually controlled and decrease significantly as the iterations progress. In final convergence solutions, maximum load rates decline to around 1.6 while the highest temperatures are below the rated value 100.0 °C. For optimized operating points, all contingencies can pass the corrective control feasibility check and line thermal rating check. It also needs to be noted that the total generation cost increases from $27,027 to $33,593 with the growth of the number of "N−5" contingencies. This is because the feasible region of the power system is shrunk by adding more credible contingencies.

6. Conclusions

In this paper, an ESCOPF model where thermal rating constraints are integrated to limit post-contingency conductor temperatures is established. After the contingency occurrence, the system first experiences a response stage and a ramping stage in time sequence. Due to the thermal inertia, lines can transfer the power flow that is much higher than the rated value while conductor temperatures are still within the safe scale. The corrective control is implemented with a time delay and long-term overflows are eliminated.

A solving strategy based on Benders decomposition is proposed to deal with the ESCOPF model. The original dispatch problem is divided into a master problem and two subproblems. The system operating point is determined in the master problem while the corrective control feasibility check and the line thermal rating check are separately conducted in two subproblems. The partial derivative of the highest temperature with respect to unit power output in the thermal rating Benders cut is calculated by a presented equivalent time method.

Simulation results on a 6-bus test system, a modified IEEE RTS-96 system, and a case 2383wp test system demonstrate that proposed approaches can offer the following advantages:

(a) The solving strategy based on Benders decomposition steadily converges within eight iterations for three test systems. The finally obtained operating point can pass the feasibility check and thermal rating check and has the lowest generation cost.

(b) The explicit relationship between the highest temperature and unit power output can be built by the proposed equivalent time method. Post-contingency temperatures of line conductors are strictly controlled below the prescribed limit.

(c) The proposed ESCOPF model can find a better balance between the security and economy compared with the conventional SCOPF model. As rated temperatures decline, the model gradually degenerates into the PSCOPF model. Conversely, the obtained operating point approaches the result of the CSCOPF model.

(d) The generation cost increases rapidly as the duration of the response stage rises. Shortening the system duration in the response stage can significantly extend the safe operation region of systems and decrease the total generation cost.

The proposed ESCOPF model is based on the DC power flow which will cause the underestimation of actual conductor temperatures of transmission lines and a more accurate model is needed to be established. In addition, just like the conventional PSCOPF model and CSCOPF model, the ESCOPF model proposed in this paper is a tool that can be used to make the day-ahead, real-time generation plan, and so on. Therefore, a lot of work such as applying this model into the electricity market operation and power systems with the larger-scale renewable energy integration need to be done in future research.

Author Contributions: Conceptualization, C.G.; Methodology, X.L.; Validation, W.C. All authors have read and agreed to the published version of the manuscript.

Funding: This work conducted by Xiansi Lou, Wei Chen, and Chuangxin Guo was supported in part by the National Key R&D Program of China under Grant 2017YFB0902600 and in part of the State Grid Corporation of China Project under Grant SGJS0000DKJS1700840.

Conflicts of Interest: The authors declare no conflicts of interest.

Appendix A

Table A1. The parameters of units.

Number	a_i ($/MW^2)	b_i ($/MW)	c_i ($)	$PG_{i,min}$ (MW)	$PG_{i,max}$ (MW)	ΔPG_i (MW)
1	0.005	10	0	0	200	35
2	0.008	15	0	0	150	30
3	0.007	12	0	0	180	35

Table A2. The parameters of load buses.

Number	Bus Number	Load Demand (MW)
1	4	80
2	5	100
3	6	90

Table A3. The parameters of lines.

Number	"From" Bus Number	"To" Bus Number	Resistence (p.u.)	Rated Power Flow (MVA)
1	1	2	0.20	50
2	1	4	0.20	70
3	1	5	0.30	55
4	2	3	0.25	55
5	2	4	0.10	80
6	2	5	0.30	40
7	2	6	0.20	70
8	3	5	0.26	50
9	3	6	0.10	80
10	4	5	0.40	40
11	5	6	0.30	40

References

1. Ott, A.L. Experience with PJM Market Operation, System Design, and Implementation. *IEEE Trans. Power Syst.* **2003**, *18*, 528–534. [CrossRef]
2. Martinez-Crespo, J.; Usaola, J.; Fernandez, J.L. Security-Constrained Optimal Generation Scheduling in Large-Scale Power Systems. *IEEE Trans. Power Syst.* **2006**, *21*, 321–332. [CrossRef]

3. Wu, Y.K.; Ye, G.T.; Tang, K.T. Preventive Control Strategy for an Island Power System That Considers System Security and Economics. *IEEE Trans. Ind. Appl.* **2017**, *53*, 5239–5251. [CrossRef]

4. Capitanescu, F.; Glavic, M.; Ernst, D.; Wehenkel, L. Contingency Filtering Techniques for Preventive Security-Constrained Optimal Power Flow. *IEEE Trans. Power Syst.* **2007**, *22*, 1690–1697. [CrossRef]

5. Wang, Q.; McCalley, J.D.; Zheng, T.X.; Litvinov, E. A Computational Strategy to Solve Preventive Risk-Based Security-Constrained OPF. *IEEE Trans. Power Syst.* **2013**, *28*, 1666–1675. [CrossRef]

6. Ardakani, A.J.; Bouffard, F. Identification of Umbrella Constraints in DC-Based Security-Constrained Optimal Power Flow. *IEEE Trans. Power Syst.* **2013**, *28*, 3924–3934. [CrossRef]

7. Shahidehpour, M.; Tinney, W.F.; Fu, Y. Impact of Security on Power System Operation. *Proc. IEEE* **2005**, *93*, 2013–2025. [CrossRef]

8. Lenoir, L.; Kamwa, I.; Dessaint, L.A. Overload Alleviation with Preventive-Corrective Static Security Using Fuzzy Logic. *IEEE Trans. Power Syst.* **2009**, *24*, 134–145. [CrossRef]

9. Li, Y.; McCalley, J.D. Decomposed SCOPF for Improving Efficiency. *IEEE Trans. Power Syst.* **2009**, *24*, 494–495.

10. Capitanescu, F.; Wehenkel, L. A New Iterative Approach to the Corrective Security-Constrained Optimal Power Flow Problem. *IEEE Trans. Power Syst.* **2008**, *23*, 1533–1541. [CrossRef]

11. Zhang, R.; Dong, Z.Y.; Xu, Y.; Wong, K.P.; Lai, M. Hybrid Computation of Corrective Security-Constrained Optimal Power Flow Problems. *IET Gener. Transm. Distrib.* **2014**, *8*, 995–1006. [CrossRef]

12. Xu, Y.; Dong, Z.Y.; Zhang, R.; Wong, K.P. Solving Preventive-Corrective SCOPF by a Hybrid Computational Strategy. *IEEE Trans. Power Syst.* **2014**, *29*, 1345–1355. [CrossRef]

13. Qiu, Y.W.; Luo, Z.H.; Chen, B.; Zhao, Q.; Xia, B.Q.; Wu, H.; Song, Y.H. Location Optimal Support Generation Units Based on Polynomial Approximation of Post-Contingency Static Stability and Security Regions Boundaries. *J. Eng.* **2017**, *13*, 1857–1861.

14. Wu, L.; Shahidehpour, M.; Liu, C. MIP-Based Post-Contingency Corrective Action with Quick-Start Units. *IEEE Trans. Power Syst.* **2009**, *24*, 1898–1899.

15. Wen, Y.F.; Guo, C.X.; Kirsche, D.S.; Dong, S.F. Enhanced Security-Constrained OPF with Distributed Battery Energy Storage. *IEEE Trans. Power Syst.* **2015**, *30*, 98–108. [CrossRef]

16. Wen, Y.F.; Guo, C.X.; Dong, S.F. Coordination Control of Distributed and Bulk Energy Storage for Alleviation of Post Contingency Overloads. *Energies* **2014**, *71*, 1599–1620. [CrossRef]

17. Cao, J.; Du, W.; Wang, H.F. An Improved Corrective Security Constrained OPF with Distributed Energy Storage. *IEEE Trans. Power Syst.* **2016**, *31*, 1537–1545. [CrossRef]

18. Zheng, Q.Z.; Ai, X.M.; Fang, J.K.; Wen, J.Y. Data-Adaptive Robust Transmission Network Planning Incorporating Post-Contingency Demand Response. *IEEE Access* **2019**, *7*, 100296–100304. [CrossRef]

19. Bi, R.Y.; Chen, R.S.; Ye, J.; Zhou, X.M.; Xu, X.L. Alleviation of Post-Contingency Overloads by SOCP Based Corrective Control Considering TCSC and MTDC. *IET Gener. Transm. Distrib.* **2018**, *12*, 2155–2164. [CrossRef]

20. Cao, J.; Du, W.; Wang, H.F. An Improved Corrective Security Constrained OPF for Meshed AC/DC with Multi-Terminal VSC-HVDC. *IEEE Trans. Power Syst.* **2016**, *31*, 485–495. [CrossRef]

21. Otomega, B.; Marinakis, A.; Glavic, M.; Cutsem, T.V. Model Predictive Control to Alleviate Thermal Overloads. *IEEE Trans. Power Syst.* **2007**, *22*, 1384–1385. [CrossRef]

22. Otomega, B.; Marinakis, A.; Glavic, M.; Cutsem, T.V. Emergency Alleviation of Thermal Overloads Using Model Predictive Control. In Proceedings of the 2007 IEEE Lausanne Power Tech, Lausanne, Switzerland, 1–5 July 2007.

23. Carneiro, J.S.A.; Ferrarini, L. Preventing Thermal Overloads in Transmission Circuits via Model Predictive Control. *IEEE Trans. Control Syst. Technol.* **2010**, *18*, 1406–1412. [CrossRef]

24. Martin, J.; Hiskens, I.A. Corrective Model-Predictive Control in Large Electric Power Systems. *IEEE Trans. Power Syst.* **2017**, *32*, 1651–1662.

25. Greenwood, D.M.; Gentle, J.P.; Myers, K.S.; Davison, P.J.; West, I.J.; Bush, J.W.; Ingram, G.L.; Troffaes, M.C.M. A Comparison of Real-Time Thermal Rating Systems in the U.S. and the U.K. *IEEE Trans. Power Deliv.* **2014**, *29*, 1849–1858. [CrossRef]

26. Dong, X.M.; Wang, C.F.; Liang, J.; Han, X.S.; Zhang, F.; Sun, H.; Wang, M.X.; Ren, J.G. Calculation of Power Transfer Limit Considering Electro-Thermal Coupling of Overhead Transmission Line. *IEEE Trans. Power Syst.* **2014**, *29*, 1503–1511. [CrossRef]

27. Dong, M.X.; Kang, C.Q.; Ding, Y.Y.; Wang, C.F. Estimating the Wind Power Integration Threshold Considering Electro-Thermal Coupling of Overload Transmission Lines. *IEEE Trans. Power Syst.* **2019**, *34*, 3349–3358. [CrossRef]

28. Wang, M.X.; Wang, M.Q.; Huang, J.X.; Jiang, Z.; Huang, J.Y. A Thermal Rating Calculation Approach for Wind Power Grid-Integrated Overhead Lines. *Energies* **2018**, *11*, 1523. [CrossRef]

29. Cong, Y.H.; Regulski, P.; Wall, P.; Osborne, M.; Terzija, V. On the Use of Dynamic Thermal Line Ratings for Improving Operational Tripping Schemes. *IEEE Trans. Power Syst.* **2016**, *31*, 1891–1900. [CrossRef]

30. Douglass, D.; Reding, J. *IEEE Standard for Calculating the Current-Temperature of Bare Overhead Conductors*; IEEE Standard 738-2006; IEEE: Piscataway, NJ, USA, 2007.

31. Wang, M.X.; Yang, M.; Wang, J.H.; Wang, M.Q.; Han, X.S. Contingency Analysis Considering the Transient Thermal Behavior of Overhead Transmission Lines. *IEEE Trans. Power Syst.* **2018**, *33*, 4982–4993. [CrossRef]

32. Banakar, H.M.; Alguacil, N.; Galiana, F.D. Electrothermal Coordination Part I: Theory and Implement Schemes. *IEEE Trans. Power Syst.* **2005**, *20*, 798–805. [CrossRef]

33. Alguacil, N.; Banakar, H.M.; Galiana, F.D. Electrothermal Coordination Part II: Case Studies. *IEEE Trans. Power Syst.* **2005**, *20*, 1738–1745. [CrossRef]

34. Ngoko, B.; Sugihara, H.; Funaki, T.; Suita, Y. A Temperature Dependent Power Flow Model Considering Overhead Transmission Line Conductor Thermal Inertia Characteristics. In Proceedings of the 2019 IEEE International Conference on Environment and Electrical Engineering and 2019 IEEE Industrial and Commercial Power Systems Europe, Genova, Italy, 11–14 June 2019.

35. Hu, J.; Wang, J.; Xiong, X.F.; Chen, J. A Post-Contingency Power Flow Control Strategy for AC/DC Hybrid Power Grid Considering the Dynamic Electrothermal Effects of Transmission Lines. *IEEE Access* **2019**, *7*, 65288–65302. [CrossRef]

36. Wang, Y.L.; Mo, Y.; Wang, M.Q.; Zhou, X.F.; Liang, L.K.; Zhang, P. Impact of Conductor Temperature Time-Space Variation on the Power System Operational State. *Energies* **2018**, *11*, 760. [CrossRef]

37. Benders, J.F. Partitioning Procedures for Solving Mixed-Variables Programming Problems. *Numer. Math.* **1962**, *4*, 238–252. [CrossRef]

38. Wu, L. An Improved Decomposition Framework for Accelerating LSF and BD Based Methods for Network-Constrained UC Problems. *IEEE Trans. Power Syst.* **2013**, *28*, 3977–3986. [CrossRef]

39. Li, Z.G.; Wu, W.C.; Zhang, B.M.; Wang, B. Decentralized Multi-Area Dynamic Economic Dispatch Using Modified Generalized Benders Decomposition. *IEEE Trans. Power Syst.* **2016**, *31*, 526–538. [CrossRef]

40. Huang, S.J.; Dinavahi, V. A Branch-and-Cut Benders Decomposition Algorithm for Transmission Expansion Planning. *IEEE Syst. J.* **2019**, *13*, 659–669. [CrossRef]

41. Arroyo, A.; Castro, P.; Martinez, R.; Manana, M.; Madrazo, A.; Lecuna, R.; Gonzalez, A. Comparison between IEEE and CIGRE Thermal Behaviour Standards and Measured Temperature on a 132-kV Overhead Power Line. *Energies* **2015**, *8*, 13660–13671. [CrossRef]

42. Grigg, C.; Wong, P.; Albrecht, P.; Allan, R.; Bhavaraju, M.; Billiton, R.; Chen, Q.; Fong, C.; Haddad, S.; Kuruganty, S.; et al. The IEEE Reliability Test System-1996. A report prepared by the Reliability Test System Task Force of the Application of Probability Methods Subcommittee. *IEEE Trans. Power Syst.* **1999**, *14*, 1010–1020. [CrossRef]

Article

Voltage Regulation Planning for Distribution Networks Using Multi-Scenario Three-Phase Optimal Power Flow

Antonio Rubens Baran Junior *, Thelma S. Piazza Fernandes and Ricardo Augusto Borba

Electrical Engineering Department, Federal University of Paraná (UFPR), Curitiba 81531-980, Brazil; thelma@eletrica.ufpr.br (T.S.P.F.); borba6@gmail.com (R.A.B.)
* Correspondence: juniorbaran@gmail.com

Received: 29 October 2019; Accepted: 25 December 2019; Published: 29 December 2019

Abstract: Active distribution networks must operate properly for different scenarios of load levels and distributed generation. An important operational requirement is to maintain the voltage profile within standard operating limits. To do this, this paper proposed a Multi-Scenario Three-Phase Optimal Power Flow (MTOPF) that plans the voltage regulation of unbalance and active distribution networks considering typical scenarios of operation. This MTOPF finds viable operation points by the optimal adjustments of voltage regulator taps and distribution transformer taps. The differentiating characteristic of this formulation is that in addition to the traditional tuning of voltage regulator taps of an active network applied for just one scenario of load and generation, it also performs the optimal adjustment of distribution transformer taps, which, once fixed, is able to meet the voltage limits of diverse operating situations. The optimization problem was solved by the primal-dual interior-point method and the formulation was tested using the IEEE 123-bus system.

Keywords: Three-phase optimal power flow; voltage regulation; distributed generation; distribution transformer taps

1. Introduction

The evolution of distribution networks affected their planning and operational philosophy by requiring load and distributed generation (DG) unbalances to be appropriately represented in the computational analysis tools. Thus, given the complexity of distribution networks, there is a tendency to avoid the simplified single-phase representation of the system in favor of the three-phase representation, which is more in keeping with the reality of 13.8-kV and 34.5-kV lines.

Many previous studies have explored traditional voltage regulation equipment as described by the authors of [1]: Step voltage regulators, switched capacitors, and on-load tap changer (OLTC).

As summarized by the authors of [1], there are advanced methods to realize the voltage regulation, such as generation curtailment during low demand, reactive power control by reactive compensator (VAR compensation), continuously changing the tap changer setting at substation, inverters of smart DG, consumption shifting and curtailing, energy storage, and microgrids providing ancillary service.

In Section 2, the voltage regulation equipment, mathematical formulations, and methods used to confront the challenges of planning and operation of active distribution networks are presented. In Section 2, it is shown that none of the actual studies have used these voltage regulation equipment and methods to make an optimal allocation of distribution transformers taps (DT), which is traditional equipment that can be a better adjustment to better integrate the advanced technologies.

So, before the allocation of these more sophisticated technologies, we proposed a proper selection of DT taps in a way to enjoy more benefits from these technologies, which are important to face the new challenges of voltage profile variations of active networks.

Thus, this research adjusted a single-period Three-Phase Optimal Power Flow (TOPF), proposed by the authors of [1], into a multi-scenario formulation that assists the voltage regulation planning of a distribution network, making not only the traditional adjustments of voltage regulators, but also of the distribution transformers taps (DT) that simultaneously satisfy different configurations of load and DG power injections (that depends on the solar incidence, for example) during a typical day.

The adjustments proposed in this article considered a large insertion of photovoltaic (PV) generation, which required careful monitoring to not exceed operational limits of the network and equipment. Therefore, to address these questions, it was necessary to act on the step voltage regulators and on the distribution transformers taps that existed along the feeders to control the voltage profile.

To do this, the purpose of this article was to propose an optimization problem that applies to a three-phase unbalanced network (besides the conventional control actions, such as voltage regulator taps, as [2]) and DT tap adjustments to monitor the voltage profile, considering not only one point of operation, but a combination of multiple scenarios simultaneously (MTOPF). The consideration of multiple periods (or scenarios) must be made because after the DTs taps are fixed at planned positions, they do not change during the operation time. So, this tap allocation satisfies different conditions of load and DG penetrations (with pre-establishment of a different combination of scenarios) while minimizing the total electrical losses.

Besides the description of the MTOPF proposed, some variations of this main idea were developed in a way to validate its results. We also proposed a parametrization of loads and GD insertions that allowed the execution of each scenario individually. This formulation was named PTOPF, and it can confront the results of a single-period formulation with a multi-period formulation, showing the advantages of the multi-period proposed in this article. Besides these implementations, we also proposed a method that exhaustively tested all the combinations of DT taps to validate the results of the MTOPF and PTOPF.

The results showed that the formulations obtained a configuration of taps for all DTs while optimizing the steps of voltage regulators, finding viable points of operations along a typical day of an active distribution network. Additionally, the formulation found operational points that minimized the total electrical losses.

This work is organized in eight different sections. First, a brief bibliographic review on analysis models for three-phase networks is presented, followed by a description of the three-phase models used in the formulation proposed in this article. Next, the proposed MTOPF formulation is presented with the new considerations introduced to it, followed by a load parametrization modeled to make the validation of the distribution transformers taps optimized. Finally, the results and conclusions are presented for the IEEE 123-bus system.

2. Literature Review

Analytical tools available for the solution of three-phase load flows (LF) have already been developed, such as [3–9]. In addition, the authors of [10] described an open-source system simulator (OpenDSS) that has interface integration with other programs.

All these reported works described conventional LF solutions for three-phase distribution systems. However, more complex strategies to plan voltage control require more attention, especially when there are many single-phase loads connected from the main three-phase trunk, which causes the unbalanced operating voltage across the feeder, increasing losses and hampering voltage regulation, along with the three phases of the circuit.

The voltage control is also important when there are DGs installed at radial feeders, which can induce reverse power flows and change the voltage profile, requiring proper interventions because the original voltage regulation schemes may not meet the requirements after the DG access.

These challenges corroborate the importance of proposing mathematical formulations that consider the networks unbalanced and can make adjustments that are necessary to maintain the operational quality requirements, considering many different scenarios of load and power injection of DGs.

These adjustments can be obtained through optimization problems, such as the Three-Phase Optimal Power Flow (TOPF). Some of them are resolved via the interior point method (IPM), such as:

- Reference [11], which presented a TOPF equated with four-wire current injection, which focused on voltage unbalance analysis and simulates the IEEE 34- and 123-bus systems;
- Reference [12], which analyzed the effects of single-phase and three-phase PV generation on losses and voltage profile; and
- Reference [3], which represented the balance equations through power injections to adjust voltage regulator taps and bank of capacitors.

There are other works that solve TOPF through other methods, such as quasi-Newton, semi-defined programming (SDP), or second-order cone programming, such as:

- Reference [13], which used the quasi-Newton method in conjunction with OpenDSS at Smart Grid applications to redraw the reactive power to reduce electrical losses; and
- References [14–17], which used SDP to solve TOPF when there was a large penetration of DG, focusing on the adjustment of voltage regulator taps.

Despite the advantages of SDP, such as finding an ideal global solution, it was not selected for this study due to the large computational effort required in the simulation of large systems, as it increases the search space of the solution by increasing the number of variables.

The previously described references are related to the mathematical formulation of three-phase systems and methods to solve them, emphasizing the relevance of the IPM, which was also used in this work to solve the optimization problem proposed.

Next, the equipment and methods used to control voltage regulation are cited.

As already mentioned, there are many works that explored traditional voltage regulation equipment, such as step voltage regulators [2], switched capacitors [1], and on-load tap changers (OLTC) [1]. The applications of more sophisticated technologies have also been described by previous works, including low-voltage static var compensators (LV-SVCs) [18], distributed energy-storage systems [19], static synchronous compensator (STATCOM) [20], and microgrids providing ancillary service [21].

But, none of these works optimized the taps using traditional equipment, such as the DTs, which are not yet sufficiently explored to better integrate with the advanced technologies. So, we proposed a proper optimization of the DT taps in a way to enjoy more benefits from them and from the advanced technologies, which are important to face the new challenges of voltage profile variations of active networks.

Some kinds of DT tap adjustments were already performed in [22], which used genetic algorithms to solve the optimizations problem, together with the OpenDSS, which analyzed each solution generated. The authors of [23] proposed a quadratic three-phase transformer model resolved via a mixed-integer quadratically constrained quadratic program model, binary scheme and big-M method.

However, these formulations only considered one operating point, and for different scenarios, they must be executed again, obtaining different DT taps allocations for each individual execution. However, these tap positions remain fixed throughout their operations, and they must simultaneously satisfy several scenarios that cover different load levels and PV generation (several solar radiation profiles), among other parameters that change throughout the day.

Thus, the tuning of DTs must be able to satisfy various operating scenarios of the network to achieve an allocation that meets all of them.

Recent works ([24,25]) presented formulations that used a stochastic method to design high DG penetration in distribution networks, considering various operating scenarios. In [25], the scenarios considered wind generation, and in [24], PV generation was modeled. There is another strand of work that used neural networks [26], which require a long length of time to make the training of data.

Using numerical approaches to simultaneously address several questions about adjustments and three-phase representations, we proposed an improvement to the single-period TOPF modeled by [2] (which used the already well-established IPM), by the addition of DT tap adjustments (to bypass the high-voltage drop at the end of the feeder and avoid exceeding operating limits) and expanded the single-period performance to a multi-period conception that considers simultaneously the most representative scenarios of load levels and DG penetrations. So, the DT taps were allocated in a way to satisfy all the most significant load and DG conditions of a typical day.

This formulation was named Multi-Scenario Three-Phase Optimal Power Flow (MTOPF). It was solved via IPM technique, which was selected over other techniques because it was used to solve nonlinear and non-convex problems, such as an OPF, with good results.

To do it, we proposed a quadratic power injection model that represents the taps of distribution transformers together with a multi-period version of the TOPF [2] to make possible the consideration of many scenarios of load and solar incidence simultaneously. The objective was to find a tap position that was ideal throughout a typical day operation and that minimized the total electrical losses.

Thus, in this article, the MTOPF was proposed. Besides the traditional tuning of voltage regulator taps of an active three-phase network applied for just one period, the MTOPF also has the following functionalities and contributions:

-   Resolve the nonlinear active and reactive power balance equations;
-   Adjust the DT taps (considering the transformer winding connections) to adapt the voltage profile to many periods of operation; and
-   Adjust the voltage regulator taps.

Besides the proposition of the MTOPF, the parametrization of loads and GD insertions was also proposed, which allowed a sequential resolution of each scenario individually. This formulation, named PTOPF, was used to confront the results between a single-period formulation and a multi-period formulation, and showed the advantages of the last one, which was the main differential of this work, as it encompasses the intertemporally of the problem. After these implementations a method was also proposed that exhaustively tested all the combinations of DT taps to validate the results of the MTOPF and PTOPF.

3. Distribution Systems

Distribution networks have predominantly unbalanced connections, radial topology, and different types of loads and lines [27].

In this work, the modeling line used was the circuit π equivalent to a three-phase line, as proposed by the authors of [23], considering the mutual inductances between the phases. The three-phase admittance matrix, $\dot{Y}_{bus}^{A,B,C}$, was organized by blocks by phase because the power balance equations of the problem are also organized by phase (A,B,C) [1]. The three-phase modeling loads used were connected in a star configuration, and their representations can be [28] active and reactive power constant, constant current, or constant impedance models, or any combination of the three models. Capacitors were modeled as a star ground configuration. The capacitive susceptance of a bank of capacitors connected at bus i, phase ph, can be obtained as:

$$c_i^{ph} = \frac{Q_i^{ph}}{\left|\dot{V}_i^{ph}\right|^2} \tag{1}$$

where Q_i^{ph} is the nominal reactive power of the capacitors at bus i, phase ph, and $\left|\dot{V}_i^{ph}\right|$ is the voltage magnitude at bus i, phase ph.

The voltage regulators have an automatic control of taps to adjust the voltage network according to preset parameters. In this work, the voltage regulators were modeled as three single-phase units with a transformation ratio equal to 1:a, where a is the ratio of the magnitudes of the voltages. The transformation ratio *a* affected the elements of the three-phase admittance bus matrix, as described by the authors of [2].

There are many different types of transformer windings. In this work, the wye-grounded/wye-grounded connection was modeled. The equivalent circuit for this kind of DT is presented by the authors of [29] (Figure 1). This DT has an off-nominal tap ratio, $\alpha:\beta$, between the primary and the secondary windings, where α and β are the taps on the primary and secondary sides, respectively. In this work, β was considered at the nominal position ($\beta = 1$) and α was conveniently adjusted. The admittance value of the transformer was $\dot{y}_t$.

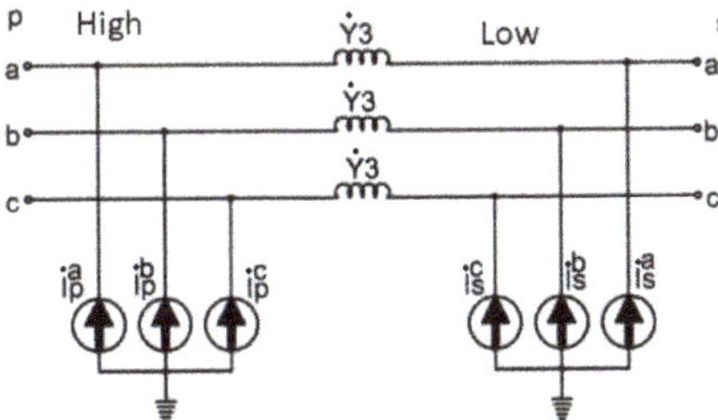

Figure 1. Representation of a distribution transformer with a wye-grounded/wye-grounded connection [28].

The value of $\dot{Y}_3$ and the current injections of Figure 1 are described as follows, in phasor form [28]:

$$\dot{Y}_3 = \frac{\dot{y}_t}{\alpha\beta},\tag{2}$$

$$\overline{I}_p^{ph} = \frac{\dot{y}_t}{\alpha\beta}\left(\frac{\beta-\alpha}{\alpha}\right)\overline{V}_p^{ph},\tag{3}$$

$$\overline{I}_s^{ph} = \frac{\dot{y}_t}{\alpha\beta}\left(\frac{\alpha-\beta}{\beta}\right)\overline{V}_s^{ph}\tag{4}$$

where $\overline{I}_p^{ph}$ is the current injected at the primary of phase *ph*, $\overline{I}_s^{ph}$ is the current injected at the secondary of phase *ph*, $\overline{V}_p^{ph}$ is the voltage at the primary of phase *ph*, and $\overline{V}_s^{ph}$ is the voltage at the secondary phase *ph*.

Hence, the structure of the three-phase admittance bus matrix of the distribution transformer, $\dot{Y}_D^{abc}$, is like the three-phase admittance matrix of a three-phase line whose admittance is equal to $\dot{Y}_3$ without mutual coupling between phases.

It should be pointed out that the authors of [28] proposed this power injection representation to overcome singularity problems. This representation was also used in this work to adjust the taps (α) of the DTs using the MTOPF, which is described in the next section.

4. Formulation of the Multi-Scenario Three-Phase Optimal Power Flow

This section presents the mathematical formulation of the proposed multi-scenario three-phase optimal power flow. The objective function of the proposed method is the minimization of losses. As a differential, the method also has the adjustment of DT taps, considering *np* scenarios at the same time.

As described by the authors of [9,30], due to the known advantageous properties of the rectangular form, instead using the polar form to represent the voltage phasor, this MTOPF uses the rectangular form. So, the nodal phasor voltage, $\dot{V}^{ph,np}$, is represented by the rectangular form as:

$$\dot{V}^{ph,t} = e + j\cdot f,\ t = 1,\ldots,np,\ ph = 1,2,3\tag{5}$$

where e is the real component of the phasor voltage with dimension ($3.nb.np \times 1$), f is the imaginary component of the phasor voltage, nb is the number of buses of the network, and np is the number of scenarios (or periods).

The real and imaginary components of the nodal phasor voltage are expressed as the vector $\mathbf{x}$:

$$\mathbf{x} = \left[\mathbf{e}^T \, \mathbf{f}^T \right]^T \tag{6}$$

where $\mathbf{x}$ has dimension ($6.nb.np \times 1$).

The general formulation of the optimization problem is:

$$\text{Min } Losses = \sum_{ph=1}^{3} \sum_{t=1}^{np} \sum_{i=1}^{nb} \left(Pgh_i^{ph,t} + Pgd_i^{ph,t} - Pd_i^{ph,t} \right) \tag{7}$$

subject to

$$\mathbf{Pg}^{ph,\,t} + \mathbf{Pgd}^{ph,\,t} - \mathbf{Pd}^{ph,t} = \mathbf{P}^{ph,t}\!\left(x, a^{ph,t}, \alpha \right).x, \tag{8}$$

$$\mathbf{Qg}^{ph,\,t} + \mathbf{Qgd}^{ph,\,t} + diag\!\left(\left| \dot{V}^{ph,t} \right|^2 \right).c^{ph,t} - \mathbf{Qd}^{ph,t} = \mathbf{Q}^{ph,t}\!\left(x, a^{ph,t}, \alpha \right).x, \tag{9}$$

$$\mathbf{Pg}_{min} \leq \mathbf{Pg}^{ph,\,t} \leq \mathbf{Pg}_{max}, \tag{10}$$

$$\mathbf{Qg}_{min} \leq \mathbf{Qg}^{ph,\,t} \leq \mathbf{Qg}_{max}, \tag{11}$$

$$V_{min}^2 \leq \left| \dot{V}^{ph,t} \right|^2 \leq V_{max}^2, \tag{12}$$

$$a_{min} \leq a^{ph,t} \leq a_{max}, \tag{13}$$

$$\alpha_{min} \leq \alpha \leq \alpha_{max}, \tag{14}$$

where *Losses* is the objective function that minimizes losses; $Pg^{ph,t}$ is the vector of active power generation of each phase ph and scenario t with dimension ($3.nb.np \times 1$); $Pd^{ph,t}$ is the vector of active power load of each phase ph and scenario t with dimension ($3.nb.np \times 1$); $Qg^{ph,t}$ is the vector of reactive power generation of each phase ph and scenario t with dimension ($3.nb.np \times 1$); $Qd^{ph,t}$ is the vector of reactive power load of each phase ph and scenario t with dimension ($3.nb.np \times 1$); $Pgd^{ph,t}$ is the vector of active power generation of distributed generation of each phase ph and scenario t with dimension ($3.nb.np \times 1$); $Qgd^{ph,t}$ is the vector of reactive power generation of distributed generation of each phase ph and scenario t with dimension ($3.nb.np \times 1$); $a^{ph,t}$ is the ratio voltage magnitudes of voltage regulators of each phase ph and scenario t with dimension ($3.nreg.np \times 1$); $nreg$ is the number of voltage regulators; a_{min} and a_{max} are the minimum and maximum voltage magnitude ratio of the voltage regulators with dimension ($3.nreg.np \times 1$); $c^{ph,t}$ is the capacitive susceptance of capacitor banks installed at nc buses with dimension ($3.nb.np \times 1$); Pg_{min} and Pg_{max} are the minimum and maximum limits of the active generation with dimension ($3.nb.np \times 1$); Qg_{min} and Qg_{max} are the minimum and maximum limits of the reactive generation with dimension ($3.nb.np \times 1$); V_{min} and V_{max} are the minimum and maximum limits of voltage magnitude phasor with dimension ($3.nb.np \times 1$); α is the vector of taps adjusted on the primary of DT with dimension ($ndt \times 1$), where ndt is the number of distributer transformers; and α_{min} and α_{max} are the minimum and maximum tap position of the DTs with dimension ($ndt \times 1$).

The functions $\mathbf{P}^{ph,t}\!\left(x, a^{ph,t}, \alpha \right).x$ and $\mathbf{Q}^{ph,t}\!\left(x, a^{ph,t}, \alpha \right).x$, which represent the active and reactive power injections of each bus, respectively, are quadratic equations due to the rectangular representation of the voltage phasor [31]. To illustrate this, the vector $\mathbf{Pd}^{ph,t}$ (active power load) and the vector $\mathbf{Qd}^{ph,t}$ (reactive power load) have the following layout:

$$\begin{aligned} \mathbf{Pd}^{ph,t} = \big[& Pd_1^{a,1} \ \ldots \ Pd_{nb}^{a,1} \ Pd_1^{b,1} \ \ldots \ Pd_{nb}^{b,1} \ Pd_1^{c,1} \ \ldots \ Pd_{nb}^{c,1} \ \ldots Pd_1^{a,np} \\ & \ldots \ Pd_{nb}^{a,np} \ Pd_1^{b,np} \ \ldots \ Pd_{nb}^{b,np} \ Pd_1^{c,np} \ \ldots \ Pd_{nb}^{c,np} \big]^T, \end{aligned} \tag{15}$$

$$\mathbf{Qd}^{ph,t} = \left[Qd_1^{a,1} \ \cdots \ Qd_{nb}^{a,1} \ Qd_1^{b,1} \ \cdots \ Qd_{nb}^{b,1} \ Qd_1^{c,1} \ \cdots \ Qd_{nb}^{c,1} \ \cdots \right.$$
$$\left. Qd_1^{a,np} \ \cdots \ Qd_{nb}^{a,np} \ Qd_1^{b,np} \ \cdots \ Qd_{nb}^{b,np} \ Qd_1^{c,np} \ \cdots \ Qd_{nb}^{c,np} \right]^T, \tag{16}$$
$$t = 1,\ldots,np$$

where $Pd_i^{k,t}$ represents the active power load at bus i, phase k, and scenario t, and $Qd_i^{k,t}$ represents the reactive power load at bus i, phase k, and scenario t.

The vector of nodal complex voltage has the following layout:

$$\dot{\mathbf{V}}^{ph,t}(\mathbf{x}) = \left[\dot{V}_1^{a,1} \ \cdots \ \dot{V}_{nb}^{a,1} \ \dot{V}_1^{b,1} \ \cdots \ \dot{V}_{nb}^{b,1} \ \dot{V}_1^{c,1} \ \cdots \ \dot{V}_{nb}^{c,1} \ \cdots \right.$$
$$\left. \dot{V}_1^{a,np} \ \cdots \ \dot{V}_{nb}^{a,np} \ \dot{V}_1^{b,np} \ \cdots \ \dot{V}_{nb}^{b,np} \ \dot{V}_1^{c,np} \ \cdots \ \dot{V}_{nb}^{c,np} \right]^T, \tag{17}$$
$$t = 1,\ldots,np.$$

The optimization variables of the MTOPF are as follows: The vector $\mathbf{x}$ represents the nodal voltage phasor modeled by the rectangular form ($\dot{\mathbf{V}}^{ph,t}(\mathbf{x})$), $\mathbf{Pg}^{ph,t}$ represents the active power generation, $\mathbf{Qg}^{ph,t}$ represents the reactive power generation, $a^{ph,t}$ represents the taps of the voltage regulator, and α represents the taps of DTs.

The limits of voltage magnitude values are squared because the voltage phasor is represented in the rectangular form [30].

The multi-scenario three-phase optimal power flow formulated from Equations (7)–(14) is solved by the primal-dual interior-point method. This method obtains the best solution, keeping the search inside the area delimited for restrictions. So, it changes the inequalities into equations of equality, through the introduction of slack variables. The main characteristic of this numerical method is the addition of a logarithmic barrier function to the objective function to guarantee the non-negativity of the slack variables. In sequence, the Karush–Kuhn–Tucker (KKT) conditions that express the first optimality conditions of the optimization problem are resolved by the application of Newton's method to obtain the solution of the non-linear equations (KKT). This method was selected due to its good performance obtained to solve traditional OPF [31,32] of real systems.

4.1. Three-Phase Distribution Transformer Model

The three-phase distribution transformer model can be done in two ways [28]: Using the admittance bus matrix or injecting current into the primary and secondary buses of the transformer as shown in Figure 1.

The distribution transformer model that inserts the taps directly into current injection equations (Equations (2)–(4)) circumvents problems of numerical conditioning in relation to the proposal that inserts the taps directly into the bus admittance matrix [28].

In addition, this strategy facilitates the derivation of the first and second derivatives that must be calculated to optimize the taps (α) of the DTs via the IPM.

Since the active and reactive power balance equations of the optimization problem were modeled using power injections (as described by the authors of [2]), the current injections of the DT were transformed into power injections:

$$\dot{S}_p^{ph,t} = \dot{V}_p^{ph,t} \left[\frac{\dot{y}_3}{\alpha}\left(\frac{1-\alpha}{\alpha}\right)\dot{V}_p^{ph,t} \right]^* \qquad t = 1,\ldots,np, \ ph = 1,\ldots,3 \tag{18}$$

$$\dot{S}_s^{ph,t} = \dot{V}_s^{ph,t} \left[\frac{\dot{y}_3}{\alpha}\left(\frac{\alpha-1}{1}\right)\dot{V}_s^{ph,t} \right]^* \qquad t = 1,\ldots,np, \ ph = 1,\ldots,3 \tag{19}$$

The equations of power injections (18) can be represented as

$$\dot{S}_p = diag(\dot{V}^{ph,t})\left[\frac{\dot{y}_3}{\alpha}\left(\frac{1-\alpha}{\alpha}\right).IncYYp.\dot{V}^{ph,t}\right]^* \qquad t = 1,\ldots,np, \ ph = 1,\ldots,3 \tag{20}$$

where the vector $\dot{S}_p$ represents the power injections at the primary buses of the DT, and *IncYYp* is a matrix of zeros with dimension $(3.nb.np \times 3.nb.np)$. The diagonal positions of *IncYYp* that correspond to the primary transformer buses were assumed to have values equal to 1.

The equations of power injections (19) can be represented as:

$$\dot{S}_s = diag(\dot{V}^{ph,t})[\frac{\dot{y}_3}{\alpha}\left(\frac{\alpha-1}{1}\right).IncYYs.\dot{V}^{ph,t}]^* \qquad t = 1,\ldots,np,\ ph = 1,\ldots,3 \qquad (21)$$

where $\dot{S}_s$ represents the power injections at the secondary buses of the DT, and *IncYYs* is a matrix of zeros with dimension $(3.nb.np \times 3.nb.np)$. The diagonal positions of *IncYYs* that correspond to the secondary transformer buses were assumed to have values equal to 1.

The power injections $\dot{S}_p$ and $\dot{S}_s$ make up the power balance equations related to the DTs, S_{trafo}:

$$\dot{S}_{trafo} = \dot{S}_p + \dot{S}_s \qquad (22)$$

which are quadratic equations because the nodal voltages are represented in a rectangular form via the vector x.

So, the complete power balance equations of the network are:

$$\mathbf{Pg}^{ph,\ t} + \mathbf{Pgd}^{ph,t} - \mathbf{Pd}^{ph,t} = \mathbf{P}^{ph,t}\left(x, a^{ph,t}, \alpha\right).x + real\left[\dot{S}_{trafo}\left(x, a^{ph,t}, \alpha\right).x\right]. \qquad (23)$$

$$\mathbf{Qg}^{ph,\ t} + \mathbf{Qgd}^{ph,t} + diag\left(\left|\dot{V}^{ph,t}\right|^2\right).c^{ph,t} - \mathbf{Qd}^{ph,t} = \mathbf{Q}^{ph,t}\left(x, a^{ph,t}, \alpha\right) + \dot{S}_{trafo}\left(x, a^{ph,t}, \alpha\right).x \qquad (24)$$

The great advantage of this multi-scenario formulation is its ability to obtain a configuration of DT taps which satisfies all operational constraints not only for a given period, but for a range of possible scenarios that can occur.

Normally, a DT has five taps with 2.5% steps. Therefore, the steps that can be selected are 1, 0.975, 0.95, 0.925, and 0.9. So, as the MTOPF formulated provides continuous tap values, they must be discretized. The discretization technique used in this work was based on the proposal formulated in [28] to operate capacitor banks.

So, the continuous values (α_i) need to be discretized (α_{disc_i}), which can assume two values: A tap position immediately above the continuous tap (Tap_{min_i}) or a tap position immediately below the continuous tap (Tap_{max_i}), where α_{disc_i} is selected as the value closest to the continuous tap (α_i):

$$\alpha_{disc\_i} = min([(\alpha_i - Tap_{min\_i}), (Tap_{max\_i} - \alpha_i)]). \qquad (25)$$

5. Load and Solar Incidence Parametrization

The adjustments of the DT taps were also calculated using a parameterized TOPF, which followed the evolution of load and solar incidence of a typical day.

The optimization problem formed by Equations (7)–(14) was adapted in a way that $np = 1$ (named as TOPF), which was repeatedly and separately simulated for only one scenario that corresponds to each moment of a day, obtained by a homotopy function.

The homotopy function allowed the variation of load and solar incidence over $nh = 24$ h (nh is the number of hours) by the variation of the index ε ($\varepsilon = 0, \ldots, nh$).

The proposed homotopy function was:

$$Pd^{ph} = Pd^{ph}_{\varepsilon = 0} + \Delta Pd^{ph}_{\varepsilon} \qquad (26)$$

$$Qd^{ph} = Qd^{ph}_{\varepsilon = 0} + \Delta Qd^{ph}_{\varepsilon} \qquad (27)$$

$$Pgd^{ph} = Pgd^{ph}_{\varepsilon = 0} + \Delta Pgd^{ph}_{\varepsilon} \qquad \varepsilon = 0, \ldots, nh \qquad (28)$$

where each $\Delta Pd_\varepsilon^{ph}$, $\Delta Qd_\varepsilon^{ph}$, and $\Delta Pgd_\varepsilon^{ph}$ were values that provide load and active distribution generation adjustments over a given period of *nh* hours. If the period of operation is a day, the value of *nh* is 24 h (or any other interval of interest).

The TOPF runs sequentially *nh* times, following the parameterization of load and PV generation. For each period ε analyzed, the tap values of each DT_i, a_ε^i, were optimized together with the other optimization variables.

After the parameterization process, different DT taps were obtained for each instant differently from what happens from the MOPT results, which provide a unique tap for each DT and meet all the scenarios simultaneously.

So, a unique tap position must be found from all the taps calculated from the parametrization process. The discretization is made like [33].

So, to discretize the continuous taps, a_ε^i, of each DT_i and each ε, the following equation was used, assuming a study horizon of *nh* hours:

$$\alpha_{min}^i = \min\left(\alpha_\varepsilon^i\right) \tag{29}$$

$$\propto_{disc\_i} = \min\left(\left[\left(a_{min}^i - Tap_{\min\_i}\right)\left(Tap_{\max\_i} - a_{min}^i\right)\right]\right) \tag{30}$$

After the achievement of each $\propto_{disc\_i}$, they were fixed, and the TOPF was calculated again *nh* times to simulate and check if there were no violations of the voltage limits with the taps adjusted.

6. Results

The objective of this section was to present results related to the proposed methodology. The distribution systems used were the IEEE 123-bus (Figure 2), adapted by the inclusion of four voltage regulators (VR) with all loads connected in wye-grounded configuration with constant power representation and two further DTs included: The first between the buses 44 and 47 (DT 1) and the second between the buses 89 and 91 (DT 2). Both distribution transformers had the same impedance parameters as the already connected three-phase transformer between the buses 61 and 124 (DT 3).

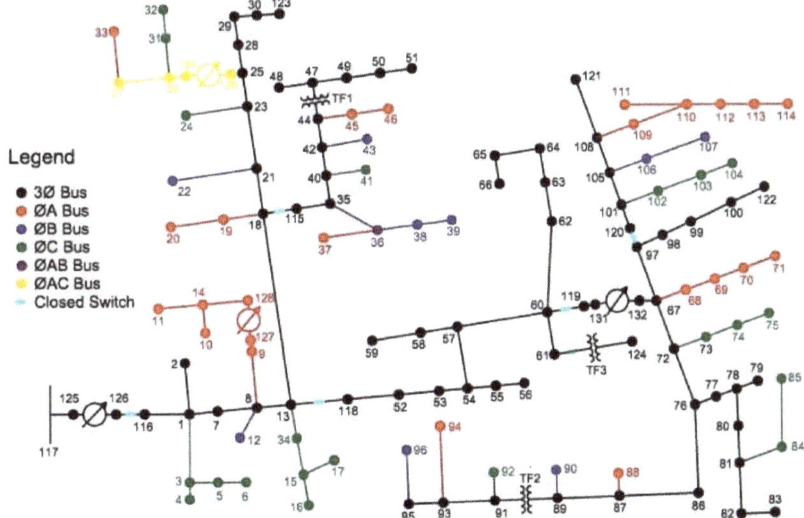

Figure 2. IEEE 123-Bus System (adapted).

Nine distributed equivalent solar generators were also added, which were allocated at buses 47, 48, 49, 50, 51, 91, 93, 95, and 124.

Another change from the original system was the addition of loads to increase the voltage drop in the system, and consequently, force different nominal tap adjustments.

The adopted base was 5 MVA, 4.16 kV with a tolerance of 1×10^{-4}. Nine scenarios were considered, which represent different load configurations and Photovoltaic (PV) generation, as shown in Table 1.

Table 1. Composition of the scenarios, load and PV generation percentage for each period.

%Load \ %PV		0	52.84	96.15
			Scenario	
81.10	Scenario	1	2	3
95.90		4	5	6
110.70		7	8	9

The value %PV is the percentage of power generated by nominal PV power and depends on the solar irradiation levels. The value %Load is the percentage of original power load in a way to represent different load levels. In addition, different PV penetration percentages were also considered: 0%, 20%, and 70% in relation to the nominal heavy load. These penetration values are in relation to the %PV percentage.

According to Table 1, there were three load levels (81.10%, 95.90%, and 110.70%) and three insolation factors (0%, 52.84%, and 96.15%), which were repeated three times for each load level. For example, for scenarios 1, 2, and 3, the load level was 81.10% with insolation factors of 0%, 52.84%, and 96.16%, respectively.

In addition, as the values of DT taps' (α) are continuous variables and the tap positions are discrete values, they must be discretized (as described in Section 4.1). After the discretization, the Multi-Scenario Three-Phase Optimal Flow is simulated again, but with fixed discrete taps to verify the feasibility of the results.

In the sequence, the adjusted taps are presented to 0%, 20%, and 70% of PV penetration using the MTOPF (Sections 6.1–6.3).

To obtain these PV values for different penetration levels, loads from scenarios 7, 8, and 9 were used, which were the largest system loads (heavy load), represented in Figure 3. The corresponding PV generations for a level of 20% and 70% penetration are shown in Figures 4 and 5, respectively. No penetration of 0% is shown because there is no PV generation in this condition.

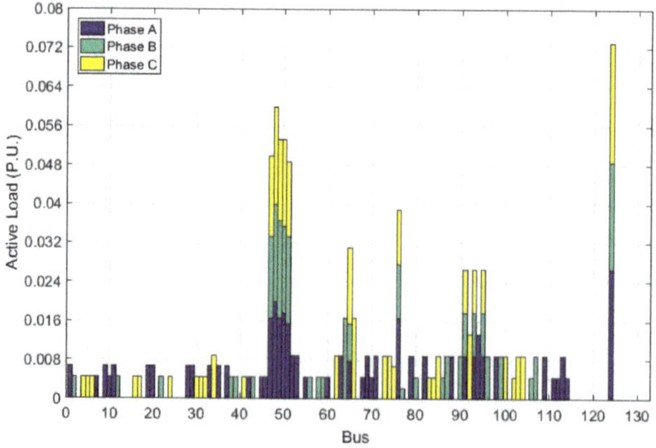

Figure 3. IEEE 123-Bus System Load (adapted).

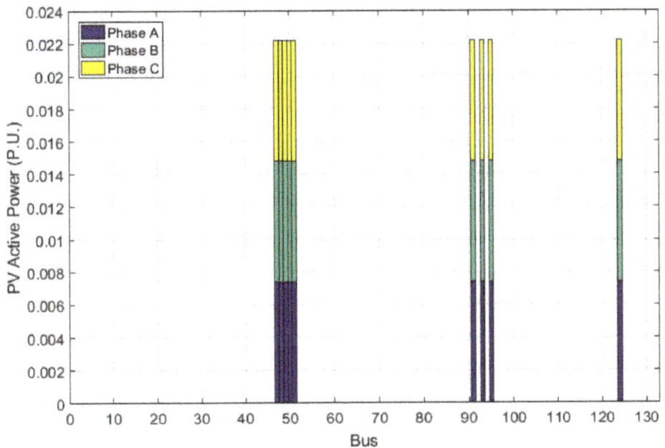

Figure 4. IEEE 123-Bus Maximum PV Active Power, 20% PV Penetration.

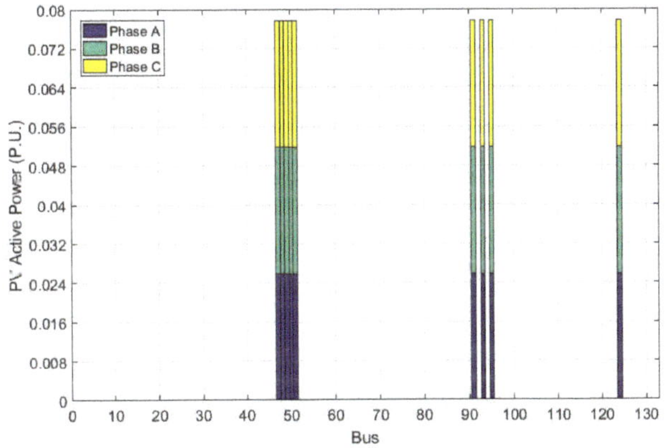

Figure 5. IEEE 123-Bus Maximum PV Active Power, 70% PV Penetration.

The total system of heavy load condition was 0.9985 p.u., with distributions at 0.3941 P.U. in phase A, 0.2756 p.u. in phase B, and 0.3288 p.u. in phase C. The PV generation of the system for 0% penetration was 0 p.u.. For 20% penetration the total PV generation was 0.1997 p.u., being 0.0666 p.u. for each phase. For a 70% penetration the total PV generation was 0.6990 p.u., being 0.2330 p.u. for each of the phases.

In Section 6.4, the results obtained with the tap adjustments by the MTOPF are compared with one simulation fixing all taps at nominal value. This simulation was done to analyze the impacts of the MTOPF results.

In Section 6.5, the results of one exhaustive method, tests all the possible combinations of tap positions of the three distribution transformers of IEEE-123-bus, ae presented. This was done with the purpose of validating the MTOPF results.

In Section 6.6, the results of the Parameterized Three-Phase Optimal Power Flow (PTOPF) are presented and compared with the MTOPF.

All tests were performed on a computer operating with Windows 10 system, with the following configuration: Intel I5-8400 processor, Gigabyte B360M AORUS G3 motherboard, memory g.skill 2x8gb 2400 Mhz (working in dual channel mode), Samsung SSD 500GB 850EVO sata3, VGA Sapphire Radeon R9 270X toxic 2GB.

The convergence of the optimization problem formulated via the interior point method depends on the adjustment of some intrinsic parameters of the method, such as the initial barrier parameter, duality gap acceleration factor, and variable initialization. Once these parameters are well-adjusted, the computational performance depends on the characteristics of the computer, such as floating-point operations per second (flops) of each CPU core and enough memory to encompass all the code and data.

6.1. Results Using 123-Bus System 0% of PV Penetration

For a configuration without generation penetration, according to Table 1, only three scenarios, referring to three load levels, were analyzed.

The continuous and discrete DT taps adjusted are shown in Figure 6.

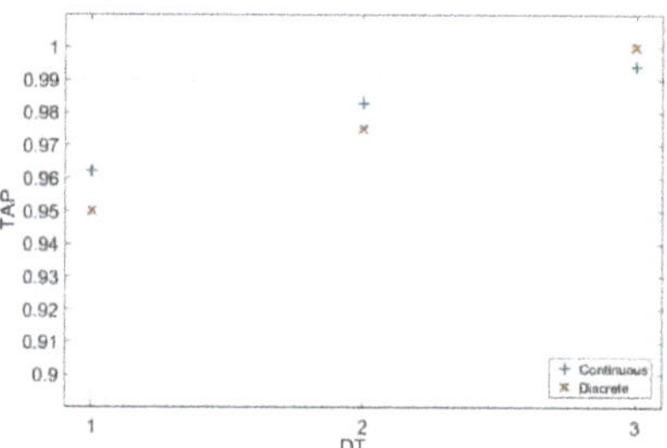

Figure 6. Continuous and discrete adjustment of taps, 0% PV.

Since the results of the voltage magnitudes are similar for each phase, the results are only presented for phase A. Figures 7 and 8 show the voltage magnitudes of phase A using continuous and discrete taps, respectively. All voltage magnitude values were within the established limits.

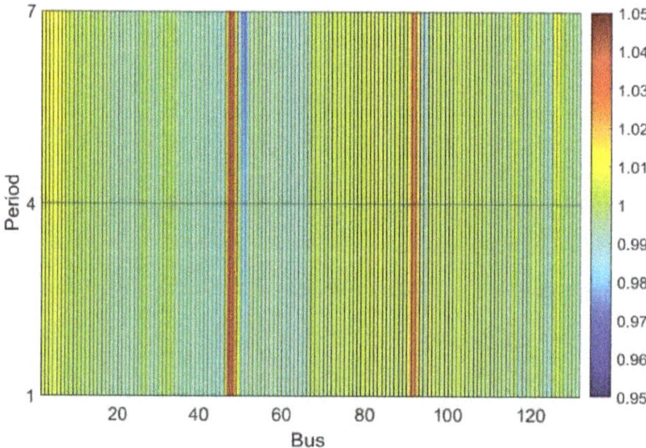

Figure 7. IEEE 123-bus system, voltage magnitudes of phase A, discrete taps, 0% PV.

All graphs of voltage magnitude (Figure 7, Figure 8, Figure 10, Figure 11, Figures 13–15) use a pseudo-color device to enable the representation of a wide range of data generated. X-axis values represent the buses of the system, Y-axis bars represent the periods (or scenarios), and the colors of each coordinate (x, y) are defined by the matrix of voltage magnitude values of each bus and each period. The color of each segment depends on the values assumed at each of its four vertices. The color corresponds to the value of the first vertex, and there is an interpolation between the vertexes of each segment to smooth the color representation.

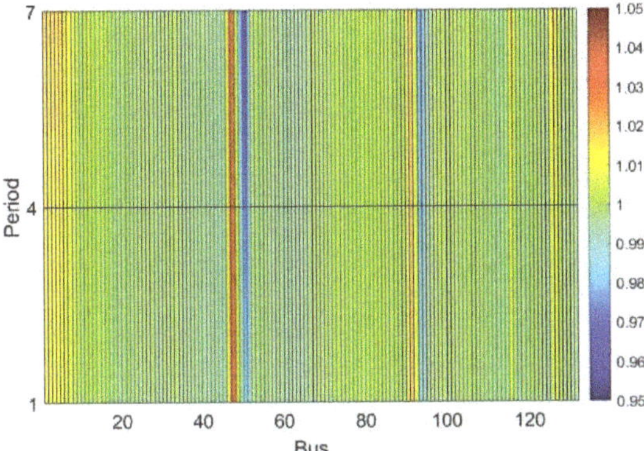

Figure 8. IEEE 123-bus system, voltage magnitudes of phase A, continuous taps, 0% PV.

As the values found in the continuous and discrete modes were not the same, but similar, it is observed, as expected, that the values of voltage magnitude were also slightly different between the two adjustment modes. This was the case except for buses 47 and 91, which showed larger variations because they were precisely at secondary buses of DT 1 and 2 that were adjusted outside the nominal position.

As a result of tap adjustments, the magnitude voltage after the secondary buses of the DTs was higher than the voltage at the primary, thus contributing to an increase in the voltage profile of the buses installed downstream of the DTs. Note that some buses had their voltages reduced, as some buses had high voltages. This was partly due to the voltage regulators, which had their taps adjusted differently in each case.

6.2. Results Using 123-Bus System 20% of PV Penetration

For the configuration with 20% of PV penetration and according to Table 1, nine scenarios were considered.

The taps adjusted in continuous and discrete modes are shown in Figure 9, while Figures 10 and 11 show the voltage magnitudes of phase A using continuous and discrete modes, respectively. All voltage magnitude values were within the established limits.

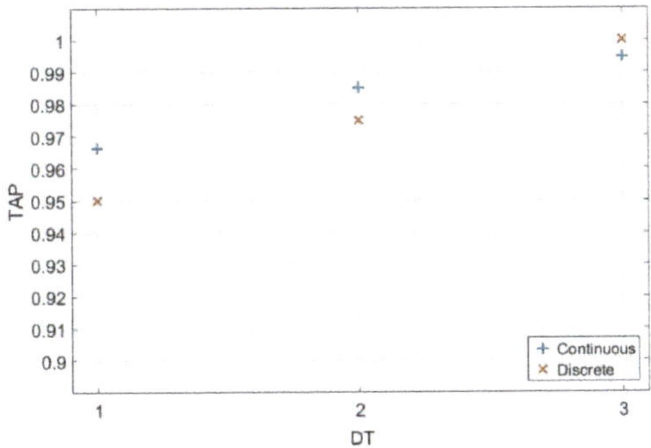

Figure 9. Continuous and discrete adjustment of taps, 20% PV.

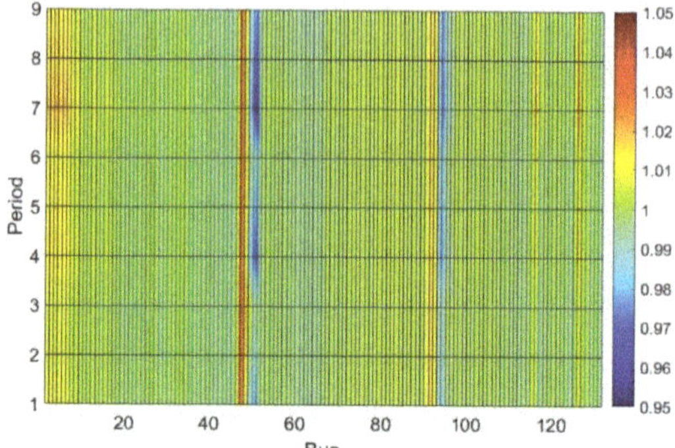

Figure 10. IEEE 123-bus system, voltage magnitudes of phase A, continuous taps, 20% PV.

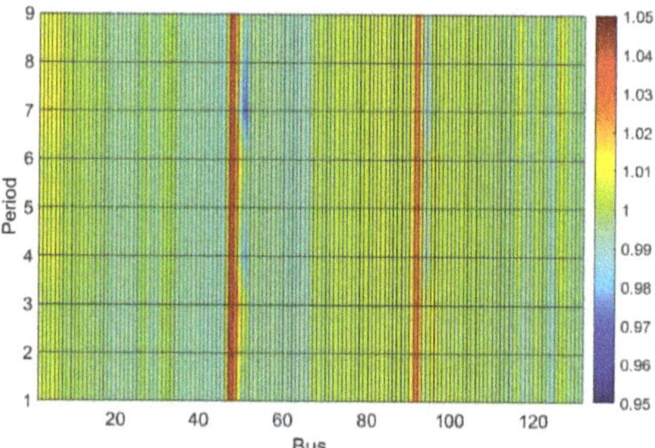

Figure 11. IEEE 123-bus system, voltage magnitudes of phase A, discrete taps, 20% PV.

For example, observing the adjustment of DT 1 in Figure 9, it can be observed that its optimal continuous tap value was approximately equal to 0.966 P.U. Since the discrete taps were set with 0.025 P.U. steps, the discrete tap closest to the optimal continuous value would be 0.975 P.U. However, the discrete value was adjusted to the value equal to 0.95 P.U., which was adequate to keep the voltage magnitude within the operational limits as presented in Figure 10, which shows the voltage magnitudes of phase A. Continuous tap values of period 7 at buses 50 and 51 (secondary downstream buses of DT1) were represented by blue tones very close to the value of 0.95 P.U. If the tap of DT 1 was set to 0.975 P.U., this voltage magnitude would be further reduced, thus violating the minimum voltage limit. Another important fact is that at bus 47, which corresponded to the secondary bus of DT 1, the voltage magnitude was represented by a red tone close to 1.03 P.U., which made it possible to reduce the value of the discrete tap, which caused an increase in voltage.

The increase in voltage profile generated by the discretization of taps can be seen in Figure 11. At bus 41, the color became a stronger red color, indicating that the voltage magnitudes were closer to 1.05 P.U.. The voltage magnitudes at bus 51 for each period had a significant increase, and during period 7, it was a lighter shade of blue than shown in Figure 10.

Another important difference between Figures 10 and 11 is that there was a reduction in the magnitude voltage at several buses. In Figure 10, there are many points where the voltage magnitudes are represented by the colors green and yellow (1 to 1.02 P.U.). In Figure 10, the same buses are represented by the colors cyan and green (0.98 to 1 P.U.).

Even with the inclusion of PV generation, the taps were adjusted with the same discrete adjustments calculated in the case without PV generation. However, the voltage variations at the buses where the DTs were connected were more significant than in the previous case because the discretization imposed a greater deviation from the optimum continuous value.

Another important point to note is the inclusion of PVs. At scenarios when they were activated (2, 5, and 8 with intermediate generation values and at scenarios 3, 6, and 9 with maximum generation values), the voltage at the buses where the PVs were connected tended to increase the voltage profile, especially with light and medium load.

6.3. Results Using 123-Bus System 70% of PV Penetration

For the configuration with 70% of PV penetration and according to Table 1, nine scenarios were considered.

The taps adjusted in continuous and discrete modes are shown in Figure 12, while Figures 13 and 14 show the voltage magnitudes of phase A using continuous and discrete modes, respectively. All voltage magnitude values were within the established limits.

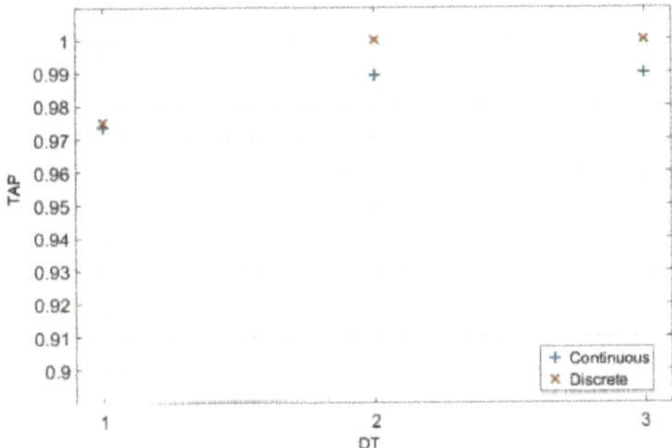

Figure 12. Continuous and discrete adjustment of taps, 70% PV.

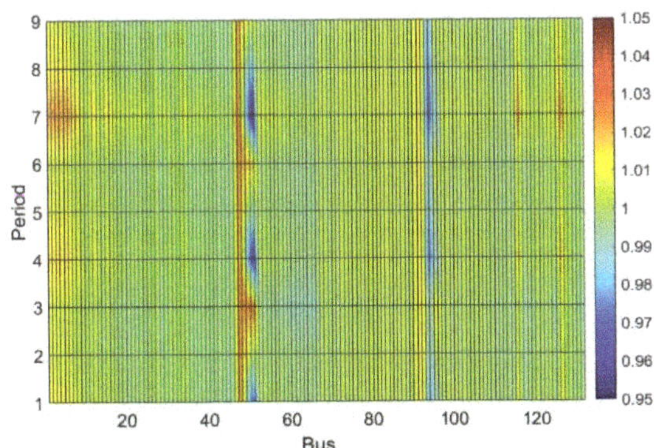

Figure 13. IEEE 123-bus system, voltage magnitudes of phase A, continuous taps, 70% PV.

Analyzing Figure 12, it can be observed that the discrete tap adjustments were close to the continuous values. However, all discrete taps were set to values above the continuous. This implies a reduction of the voltage profile at secondary buses of the DTs, as can be seen when comparing Figures 13 and 14. In Figure 14, buses 50 and 51 (DT 1 secondary) and buses 93 and 96 (DT 2 secondary) have colors tending to a darker blue, indicating lower values, especially in period 7 (with heavy load and without PV generation).

Similar to the case with 20% PV, during the scenarios when PV was activated (2, 5, and 8 with intermediate generation values and scenarios 3, 6, and 9 with maximum generation values), the magnitude of the voltage at buses where PVs were connected increased, especially with light and medium load.

Through the results, the methodology allowed the adjustment of the DT taps, which were equipment already installed in the grid or that will be part of any grid expansion project, to control the voltage profile throughout active network systems, postponing installation of other equipment that, in many cases, may be costly to the utility (such as voltage regulators and static VAR compensators, among others).

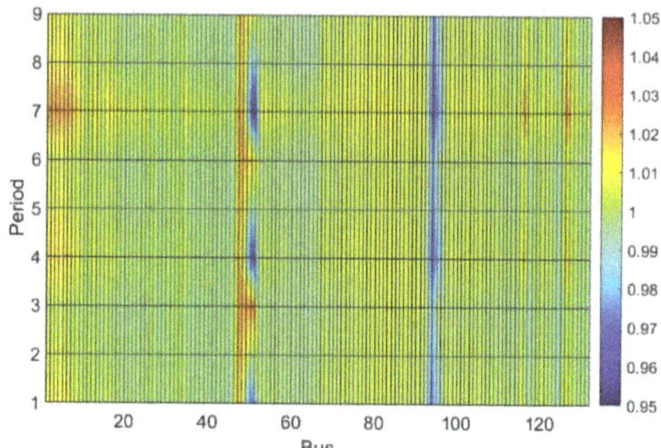

Figure 14. IEEE 123-bus system, voltage magnitudes of phase A, discrete taps, 70% PV.

Analyzing the calculated tap adjustments for different DG penetrations, it is observed for this system that when DG penetration increased, the tap positions tended toward the nominal position. This result was dependent on load level and penetration because as DG was distributed along the feeder, this fact decreased the current supplied by the substation and generally caused smaller voltage drops along the feeder. Thus, the tap adjustments tended to the nominal position to maintain voltages within the operating limits of the network.

6.4. Adjusted TAP vs. Nominal TAP

To show the impact at the voltage profile caused by the tap adjustments obtained by the proposed methodology, a simulation with 0% PV and all taps fixed at the nominal value was performed.

The voltage profile in phase A obtained with the mentioned test is shown in Figure 15. Comparing Figure 15 with Figure 7 (where the taps were adjusted), the increase of the voltage profile was clear, with variations around 0.04 P.U.

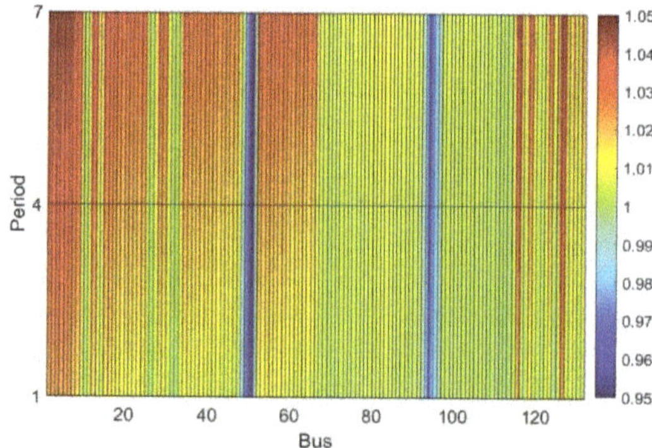

Figure 15. IEEE 123-bus system, voltage magnitudes of phase A, nominal taps, 0% PV.

The voltage magnitude was not the only parameter that was impacted by the optimal adjustment of the taps. The electrical losses were also modified significantly. For example, for scenarios 1, 2, and 3, the losses were reduced by about 40% when the taps were optimized, as shown in Figure 16.

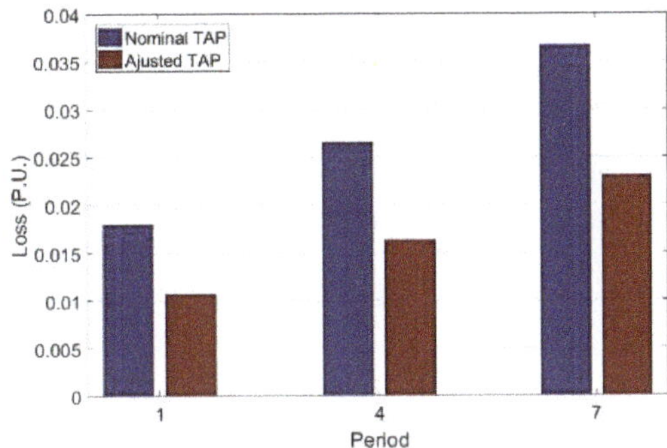

Figure 16. IEEE 123-bus system, electrical losses, 0% PV, with adjusted taps and nominal taps.

6.5. Validation of Multi-Scenario Three-Phase Optimal Flow

There is great difficulty in validating the results obtained through the Multi-Scenario Three-Phase Optimal Flow because there are no commercially available programs that perform the same functions as the proposed method.

One of the possibilities for the validation was to perform all possible combinations of tap positions of the three distribution transformers. So, considering that each DT had five positions available, the complete number of possible combinations of tap positions was 125.

To evaluate which combination of tap positions was the best, four parameters were analyzed: Total electrical loss, tendency to maintain voltage magnitudes around 1 P.U. (sum of voltage deviations from 1 P.U.), and the maximum and the minimum voltage deviations from 1 P.U.

The validation was executed using 0% of PV penetration and the scenarios of Table 1.

After 125 simulations, 85 did not converge, 27 converged with loss values above the losses obtained by the optimal solution provided by the proposed methodology (optimal solution), and 13 converged with losses slightly below the losses obtained by the optimal solution.

Table 2 shows the 13 best candidate solutions and the solution with taps at nominal values. The optimal solution obtained by the optimization problem (MTOPF) was combination 8.

Table 2. The best combination of tap positions.

Combination	Taps of DTs (1–2–3)	Loss [P.U.]	Sum: |ΔV| [P.U.]	ΔV Max [P.U.]	ΔV Min [P.U.]
1	1–1–1	0.24299	45.0706	0.0491	−0.0498
4	0.925–1–1	0.135569	51.2721	0.0447	−0.0448
8	0.95–0.975–1	0.149675	20.816	0.0466	−0.0369
9	0.925–0.975–1	0.072057	54.5962	0.0497	−0.0337
28	0.95–1–0.975	0.098848	24.6425	0.0481	−0.0481
29	0.925–1–0.975	0.055553	58.1011	0.0498	−0.0416
33	0.95–0.975–0.975	0.142254	19.8933	0.0463	−0.037
34	0.925–0.975–0.975	0.08372	22.0082	0.0497	−0.0339
53	0.95–1–0.95	0.115922	26.2772	0.0481	−0.048
54	0.925–1–0.95	0.134602	51.223	0.0493	−0.0449
58	0.95–0.975–0.95	0.135811	24.1333	0.0479	−0.0402
59	0.925–0.975–0.95	0.150734	47.1414	0.0493	−0.0341
79	0.925–1–0.925	0.07205	57.774	0.0498	−0.0466
84	0.925–0.975–0.925	0.10125	53.1383	0.0497	−0.0347

First, analyzing the solution with the taps at the nominal position, it is observed that the losses were much higher than the other candidate solutions and the sum of the voltage deviations modulus in relation to 1 P.U. was among the largest, as well as the maximum and minimum voltage deviations.

Analyzing the best combinations obtained, it is possible to observe that for each analysis parameter, there were distinct optimal solutions. For example, according to the losses, the best solution was combination 29, but the sum of the modulus of voltage deviations was the largest value among the simulations presented. Therefore, the analysis of the optimal solution also considered the incidence of higher maximum and minimum deviations.

The simulations that stand out when the four parameters were analyzed were combinations ϕ = {8, 28, 33, 53, and 58}. For example, the simulation with the lowest value of the electrical loss of the set ϕ was combination 28, the one that presented the smallest sum of the voltage deviation modulus was combination 33, the smallest variation of maximum ΔV was combination 33, and the smallest variation of minimum ΔV was combination 8.

Observing this set of discrete-tap solutions, the combination with the smallest ΔV and the smallest sum of the voltage deviation was combination 33, but the proposed methodology indicated that combination 8 was the best solution. The difference between simulations 8 and 33, considering the maximum and minimum ΔV, was 0.03% and 0.01%, respectively. The sum of the modulus of voltage deviations was 4.43%, and the loss difference was 4.96%. So, the two solutions were very close. These results can be more easily visualized when all the parameters are summed, which is presented in Table 3. So, the combination 33 was 0.33% better than solution 8.

Table 3. Best adjustment of tap positions.

Combination	Taps of DTs (1–2–3)	Sum
8 (continuous)	0.9622–0.9828–0.9941	14.1116
8 (discrete)	0.95–0.975–1	21.049175
33	0.95–0.975–0.975	21.118854

Comparing the values obtained by the optimal continuous solution, also presented in Table 3, the fitness of the solution was 33% better than the discrete solution. Then, the discretization caused slight distortion in the final result. However, it is an insignificant price to pay compared to executing all possible combinations.

For the nine scenarios considered, the MTOPF expended 3167 s to obtain the convergence. For each combination using the DT taps fixed, the MTOPF spent around 135 s (only for the cases that had feasible solutions). As previously described, 85 combinations of taps were not viable (in these cases, the simulations end after exceeding the number of iterations). So, to finish the 125 combinations of taps, the exhaustive method spent about 10 h in the total.

This combinatorial solution technique is completely unfeasible for larger systems. For example, if the network has 20 DTs, then 95 trillion combinations will be required to obtain the best configuration of taps. So, the advantage of the proposed method is that in a unique simulation, it obtains a viable solution that satisfactory assists a range of probable load and generation throughout the entire operation time.

To execute larger systems, the feeder can be divided into equivalent zones as proposed by the authors of [34]. Then, the MTOPF can be applied in equivalent areas.

Besides the application of exhaustive combination, which consumes a lot of time, the uses of other solution techniques, such as parametric processes, artificial intelligence technique, or mixed-integer nonlinear programming technique, are also costly.

Even so, the parameterized three-phase optimal power flow, presented in Section 5, was also executed in order to reinforce the results obtained by the proposed method, MTOPF, as presented in Section 6.5.

6.6. Parameterized Three-Phase Optimal Power Flow

Unlike the MTOPF, which solved multiple scenarios at the same time, the parameterized TOPF solved multiple periods individually and separately, yielding a result for each period.

It can be emphasized that, as the discretization processes of each method are different (as described in Sections 4 and 5), they also led to slightly different adjustment results of DT taps, as shown in Table 4. Table 4 presents the results obtained by parameterization (PTOPF), discrete MOPOF, and exhaustive simulations.

Table 4. Best continuous adjustment of tap positions.

Method	TAPs of DTs (1–2–3)	Sum [P.U.]
PTOPF (discrete)	0.975–1–1	33.8697
MTOPF (discrete)	0.95–0.975–1	21.9409
Exhaustive Combinations	0.95–0.975–0.975	21.118854

According to Table 4, each method presented different, but adherent, tap settings. Consequently, each method also presented different, but very close, total losses. These results validated the three strategies described to plan the allocation of distribution transformers taps that must support different levels of loads and levels of GD insertion.

The lowest loss values were obtained by the MTOPF (21.9409 P.U.) and by the exhaustive combinations (21.118854 P.U.). Although the result obtained by the exhaustive method was slightly better, the computational time spent by it was impeditive (around 10 h).

The difference in adjustments between PTOPF and MTOPF occurred because PTOPF simulated each scenario individually and then adjusted the appropriate taps. That is, each scenario had its taps individually optimized only for that period, so this approach could not see the temporal connections between the periods, and its combinations results were not the best.

On the other hand, the MTOPF simulated all the scenarios at the same time, adjusting a unique combination of taps that satisfy all of them. So, each DT tap was adjusted to obtain the best positions for all periods considered simultaneously. Because of that, MTOPF is the method that provides the best results with good computational performance because it comprises a systematic process that encompasses several operating conditions simultaneously and executes the program only one time and not repeated for each individual scenario.

7. Conclusions

This article proposed an optimization problem, which was then applied to a three-phase unbalanced network. Besides the conventional control actions, such as voltage regulator taps, the optimization problem also optimizes DT taps to monitor the voltage profile considering not only one point of operation, but multiple combinations of load and photovoltaic generation, simultaneously (MTOPF).

The consideration of multiple periods (or scenarios) must be made, as after the DTs taps are fixed at planned positions, they do not change during the operation time. So, this tap allocation must satisfy a different combination of scenarios.

The MTOPF, which minimizes the total electrical loss, is solved by the primal-dual interior-point method. This method can be applied to planning studies to obtain the fixed adjustment of the taps of distribution transformers before of the application of more advanced methods and technologies useful to face the new challenges of voltage profile variations of active networks.

Some variations of the MTOPF were also developed in a way to validate its results as the parametrization of loads and GD insertions, which allowed the execution of each scenario individually (PTOPF). This formulation confronts the results of a single-period formulation (TOPF) with a multi-period formulation, showing the advantages of the multi-period.

One of these advantages is the calculation of an optimal DT tuning plan that meets all load conditions and levels of generation during the operation time.

Another advantage is the computational time that MTOPF requires to obtain a solution, which is much faster and more efficient than the parameterized method (PTOPF).

Besides these implementations, a method that exhaustively tests all the combinations of DT taps was also proposed. The method successfully validated the results of the MTOPF and PTOPF.

Author Contributions: A.R.B.J. is responsible for the conception and implementation of the methodology, the generation of results was made by R.A.B. and T.S.P.F. coordinated the activities and reviewed the work. All authors have read and agreed to the published version of the manuscript.

Funding: Araucária Foundation and CAPES (Higher Education Personnel Improvement Coordination – Brazil): 88882.168617/2018-01

Acknowledgments: The authors are grateful for the support for this research provided by CNPq (National Council for Scientific and Technological Development – Brazil), CAPES (Higher Education Personnel Improvement Coordination – Brazil) and Araucária Foundation.

Conflicts of Interest: The authors declare no conflict of interest.

References

1. Mahmud, N.; Zahedi, A. Review of control strategies for voltage regulation of the smart distribution network with high penetration of renewable distributed generation. *Renew. Sustain. Energy Rev.* **2016**, *64*, 582–595. [CrossRef]
2. Baran, A.R.; Fernandes, T.S.P. A three-phase optimal power flow applied to the planning of unbalanced distribution networks. *Int. J. Electr. Power Energy Syst.* **2016**, *74*, 301–309. [CrossRef]
3. Cheng, C.S.; Shirmohammadi, D. A three-phase power flow method for real-time distribution system analysis. *IEEE Trans. Power Syst.* **1995**, *10*, 671–679. [CrossRef]
4. Shirmohammadi, D.; Hong, H.W.; Semlyen, A.; Luo, G.X. A compensation-based power flow method for weakly meshed distribution and transmission networks. *IEEE Trans. Power Syst.* **1988**, *3*, 753–762. [CrossRef]
5. Broadwater, R.P.; Chandrasekaran, A.; Huddleston, C.T.; Khan, A.H. Power flow analysis of unbalanced multiphase radial distribution systems. *Electr. Power Syst. Res.* **1988**, *14*, 23–33. [CrossRef]
6. Garcia, P.A.N.; Pereira, J.L.R.; Carneiro, S.; da Costa, V.M.; Martins, N. Three-phase power flow calculations using the current injection method. *IEEE Trans. Power Syst.* **2000**, *15*, 508–514. [CrossRef]
7. Teng, J.-H. A direct approach for distribution system load flow solutions. *IEEE Trans. Power Deliv.* **2003**, *18*, 882–887. [CrossRef]
8. Ramos, E.R.; Exposito, A.G.; Cordero, G.A. Quasi-Coupled Three-Phase Radial Load Flow. *IEEE Trans. Power Syst.* **2004**, *19*, 776–781. [CrossRef]
9. Penido, D.R.R.; de Araujo, L.R.; Carneiro, S.; Pereira, J.L.R.; Garcia, P.A.N. Three-Phase Power Flow Based on Four-Conductor Current Injection Method for Unbalanced Distribution Networks. *IEEE Trans. Power Syst.* **2008**, *23*, 494–503. [CrossRef]
10. Dugan, R.C. Reference Guide—The Open Distribution System Simulator (OpenDSS). Electric Power Researsch Institute, 2013. Available online: https://spinengenharia.com.br/wp-content/uploads/2019/01/OpenDSSManual.pdf (accessed on 28 December 2019).
11. Araujo, L.R.; Penido, D.R.R.; Carneiro, S.; Pereira, J.L.R. A Three-Phase Optimal Power-Flow Algorithm to Mitigate Voltage Unbalance. *IEEE Trans. Power Deliv.* **2013**, *28*, 2394–2402. [CrossRef]
12. Ayikpa, M.E. Unbalanced distribution optimal power flow to minimize losses with distributed photovoltaic plants. *Int. J. Electr. Comput. Energ. Electron. Commun. Eng.* **2017**, *11*, 181–186.
13. Bruno, S.; Lamonaca, S.; Rotondo, G.; Stecchi, U.; La Scala, M. Unbalanced Three-Phase Optimal Power Flow for Smart Grids. *IEEE Trans. Ind. Electron.* **2011**, *58*, 4504–4513. [CrossRef]
14. Liu, Y.; Li, J.; Wu, L.; Ortmeyer, T. Chordal Relaxation Based ACOPF for Unbalanced Distribution Systems With DERs and Voltage Regulation Devices. *IEEE Trans. Power Syst.* **2018**, *33*, 970–984. [CrossRef]
15. Dall'Anese, E.; Zhu, H.; Giannakis, G.B. Distributed Optimal Power Flow for Smart Microgrids. *IEEE Trans. Smart Grid* **2013**, *4*, 1464–1475. [CrossRef]

16. Bazrafshan, M.; Gatsis, N.; Zhu, H. Optimal Tap Selection of Step-Voltage Regulators in Multi-Phase Distribution Networks. In Proceedings of the 2018 Power Systems Computation Conference (PSCC), Dublin, Ireland, 11–15 June 2017; IEEE: Dublin, Ireland, 2018; pp. 1–7.

17. Robbins, B.A.; Zhu, H.; Dominguez-Garcia, A.D. Optimal Tap Setting of Voltage Regulation Transformers in Unbalanced Distribution Systems. *IEEE Trans. Power Syst.* **2016**, *31*, 256–267.

18. Padullaparti, H.V.; Nguyen, Q.; Santoso, S. Optimal Placement and Dispatch of LV-SVCs to Improve Distribution Circuit Performance. *IEEE Trans. Power Syst.* **2019**, *34*, 2892–2900. [CrossRef]

19. Wang, Y.; Tan, K.T.; Peng, X.Y.; So, P.L. Coordinated Control of Distributed EnergyStorage Systems for Voltage Regulation in Distribution Networks. *IEEE Trans. Power Deliv.* **2016**, *31*, 1132–1141.

20. Rohouna, W.; Balog, R.; Peerzada, A.A.; Begovi, M.M. D-STATCOM for harmonic mitigation in low voltage distribution network with high penetration on nonlinear loads. *Renew. Energy* **2020**, *145*, 1449–1464. [CrossRef]

21. Wang, X.; Wang, C.; Xu, T.; Guo, L.; Li, P.; Yu, L.; Meng, H. Optimal voltage regulation for distribution networks with multi-microgrids. *Appl. Energy* **2018**, *210*, 1027–1036. [CrossRef]

22. Vitor, T.S.; Vieira, J.C.M. Optimal voltage regulation in distribution systems with unbalanced loads and distributed generation. In Proceedings of the 2016 IEEE Innovative Smart Grid Technologies—Asia (ISGT-Asia), Melbourne, Australia, 28 November 2015–1 December 2016; IEEE: Melbourne, Australia, 2016; pp. 942–947.

23. Ju, Y.; Wu, W.; Lin, Y.; Ge, F.; Ye, L. Three-phase optimal load flow model and algorithm for active distribution networks. In Proceedings of the 2017 IEEE Power & Energy Society General Meeting, Chicago, IL, USA, 16–20 July 2017; IEEE: Chicago, IL, USA, 2017; pp. 1–5.

24. Xu, J.; Wang, J.; Liao, S.; Sun, Y.; Ke, D.; Li, X.; Liu, J.; Jiang, Y.; Wei, C.; Tang, B. Stochastic multi-objective optimization of photovoltaics integrated three-phase distribution network based on dynamic scenarios. *Appl. Energy* **2018**, *231*, 985–996. [CrossRef]

25. Galdi, V.; Vaccaro, A.; Vallacci, D. Voltage regulation in MV networks with dispersed generations by a neural-based multiobjective methodology. *Electr. Power Syst. Res.* **2008**, *78*, 785–793. [CrossRef]

26. Mokryani, G.; Hu, Y.F.; Pillai, P.; Rajamani, H.-S. Active distribution networks planning with high penetration of wind power. *Renew. Energy* **2017**, *104*, 40–49. [CrossRef]

27. Chen, T.-H.; Chen, M.-S.; Hwang, K.-J.; Kotas, P.; Chebli, E.A. Distribution system power flow analysis-a rigid approach. *IEEE Trans. Power Deliv.* **1991**, *6*, 1146–1152. [CrossRef]

28. Stagg, G.W.; El-ablad, A. *Computer Methods in Power System Analysis*; McGraw-Hill Book Co.: New York, NY, USA, 1968; ISBN 978-0-07-060658 6.

29. Chen, M.-S.; Dillon, W.E. Power system modeling. *Proc. IEEE* **1974**, *62*, 901–915. [CrossRef]

30. Torres, G.L.; Quintana, V.H. An interior-point method for nonlinear optimal power flow using voltage rectangular coordinates. *IEEE Trans. Power Syst.* **1998**, *13*, 1211–1218. [CrossRef]

31. Fernandes, T.S.P.; Almeida, K.C. A methodology for optimal power dispatch under a pool-bilateral market. *IEEE Trans. Power Syst.* **2003**, *18*, 182–190. [CrossRef]

32. Wright, S.J. *Primal-Dual Interior-Point Methods*; Society for Industrial and Applied Mathematics: Philadelphia, PA, USA, 1997; ISBN 978-0-89871-382-4.

33. Dahlke, D.B.; Steilein, G.; Fernandes, T.S.P.; Aoki, A.R. A heuristic to adjust automatic capacitors using parameterization of load. *Int. J. Electr. Power Energy Syst.* **2012**, *42*, 16–23. [CrossRef]

34. Zhou, X.; Liu, Z.; Zhao, C.; Chen, L. Accelerated Voltage Regulation in Multi-Phase Distribution Networks Based on Hierarchical Distributed Algorithm. *Math. Optim. Control* **2019**, *1903*, 00072. [CrossRef]

Article

Comparing Corrective and Preventive Security-Constrained DCOPF Problems Using Linear Shift-Factors

Victor H. Hinojosa

Department of Electrical Engineering, Universidad Técnica Federico Santa María, Valparaíso 2390123, Chile; victor.hinojosa@usm.cl; Tel.: +56-32-265-4354

Received: 13 December 2019; Accepted: 17 January 2020; Published: 21 January 2020

Abstract: This study compares two efficient formulations to solve corrective as well as preventive security-constrained (SC) DC-based optimal power flow (OPF) problems using linear sensitivity factors without sacrificing optimality. Both SCOPF problems are modelled using two frameworks based on these distribution factors. The main advantage of the accomplished formulation is the significant reduction of decision variables and—equality and inequality—constraints in comparison with the traditional DC-based SCOPF formulation. Several test power systems and extensive computational experiments are conducted using a commercial solver to clearly demonstrate the feasibility to carry out the corrective and the preventive SCOPF problems with a reduced solution space. Another point worth noting is the lower simulation time achieved by the introduced methodology. Additionally, this study presents advantages and disadvantages for the proposed shift-factor formulation solving both corrective and preventive formulations.

Keywords: linear OPF problem; shift-factors; line outage distribution factors; security-constrained; corrective formulation; preventive formulation

---

1. Introduction

Carpentier introduced the optimal power flow (OPF) problem in 1960 [1]. The optimal result obtains the economic dispatch and transmission power flows. The power flow solution must meet technical power generation and transmission network limits. Several optimization algorithms have been presented in the technical literature to solve operation and planning problems based on this conventional mathematical formulation [2–4].

In real-time operation, power system operators (Independent System Operators (ISOs) and Regional Transmission Organizations (RTOs)) must execute a lot of N−1 power flows very quickly taking into account generation and transmission (lines or tranformers) failures to obtain a safe condition after a contingency event [5]. Security studies should guarantee that not only the power flow result will be maintained below the thermal limit but also overloading conditions in the transmission elements will be mitigated to avoid undesirable operational effects. This study is only focused on transmission contingencies.

1.1. Technical Literature Analysis

Power system planners and operators have typically used the direct-current OPF (DCOPF) problem. A DCOPF problem is a simplified linear approach modelling nonlinear transmission constraints based on approximations regarding voltage (magnitudes and angles), admittances and reactive power [6]. Nowadays, the most common transmission network formulation is the so-called DC model [7].

Another point worth noting is the transmission network modelling. In the technical literature, there are two methodologies using linear factors—(1) the classical DC formulation [8] and (2) the linear shift-factor (SF) formulation [6]. In the first case, power unit and voltage phase angles are decision variables to model the transmission network. According to References [6,7,9], another problem has been introduced using power unit generation as decision variable. Although decision variables, as well as equality and inequality constraints, are lower in comparison with the classical DC formulation, the SF-based formulation does not affect OPF optimality.

In electrical power systems, security studies must guarantee that transmission modifications and failures do not impact the transmission network with overloading conditions. Furthermore, ISOs must reduce the amount of power flow simulations that should be executed to verify the power system security. For very large-scale power systems, linear DC models are typically employed to solve contingency events [10].

Researchers have found that not all transmission failures originate an alert condition. Consequently, electrical power engineers must obtain a safety and economic post-contingency steady-state system with respect to the worst or a list (ranking) of contingencies to adequately solve the SCOPF problem without affecting the power system security. This problem is known as security-constrained OPF (SCOPF) problem [8,11]. For more information about SCOPF, read the following References [6,12,13].

In the state-of-the-art, two mathematical frameworks have been developed to solve the SCOPF problem—(i) the corrective formulation and (ii) the preventive formulation.

i   The first approach is larger than the classical DC-based OPF problem because new variables (power unit generation) and constraints (power flow post-contingency conditions) are added in the optimization problem to model the pre-contingency and the post-contingency (N−1) power system states. Nonetheless, the optimal solution usually requires that the ISO redispatches power generation of several units in a very short time to avoid operational overcost as well as overloading conditions. For this reason, the ISO must be able to handle many corrective generation actions without affecting power system security [14–17].

ii   For the preventive approach, one set of decision variables (power unit generation) is only needed to model the SCOPF formulation. With this assumption, the ISO does not need to redispatch the power generation because the preventive model avoids power generation changes between pre- and post-contingency power system states. However, there is an overcost in the pre-contingency state in comparison with the previous formulation. For more information, References [5,6,11,18,19] could be reviewed in detailed.

1.2. Contributions

References [6,7] introduce the N−1 preventive security analysis by using shift-factors and line outage distribution factors. While the security problem only includes the worst contingency, a ranking of contingencies should be modelled in the security-constrained analysis to avoid risky operational conditions and technical transmission problems to supply adequately the load of the customers for an unlikely line or transformer failure. Based on our knowledge, the corrective formulation is not applied in the technical literature using shift-factors. Hence, it would be very attractive to accomplish several analyses and comparisons with short- and large-scale power systems to validate scalability and simulation performance using both CL- and SF-based formulations and a commercial solver (Gurobi). Therefore, the following issues have been solved in this study—(1) the corrective SCOPF problem is solved using shift-factors and a comparative analysis for both corrective and preventive formulations has been carried out using different-scale test power systems; and (2) a realistic case is successfully solved (National Electric Power System of Chile). Simulation results have demonstrated superior performance when the SF-based formulation is applied to the SCOPF problem in comparison with the CL-based formulation. Notice that the introduced formulation could bring better performance and practical advantages solving large-scale problems as well as complicated problems such as stochastic OPF, stochastic unit commitment and generation planning methodologies.

This study is organized as follows. In Section 2, a detailed study shows different OPF problems applying the classical formulation and the introduced formulation. Section 3 presents the corrective and preventive SCOPF simulations using shift-factors. Besides, results and comparisons are achieved using several test power systems. Section 4 concludes the study.

2. Security-Constrained Optimal Power Flow Problem

In this section, corrective and preventive SCOPF problems are mathematically presented in detailed. In addition, the optimization problem does not include the DC power losses.

2.1. Corrective SCOPF Problem Using the Classical DC-Based Formulation

This optimization problem is modelled using the following objective function:

$$\min(C_{total}) = \min(C_{pre} + C_{post}) \tag{1a}$$

$$C_{pre} = \sum_{g}(A_g + B_g \times p_g^{pre} \times S_{base} + C_g \times p_g^{pre^2} \times S_{base}^2) + VoLL \sum_{v} v_b^{pre} \times S_{base} \qquad \forall\, g \in G, \forall\, v \in V \tag{1b}$$

$$C_{post} = \sum_{g}(A_g + B_g \times p_g^{post} \times S_{base} + C_g \times p_g^{post^2} \times S_{base}^2) + VoLL \sum_{v} v_b^{post} \times S_{base} \qquad \forall\, g \in B, \forall\, v \in V \tag{1c}$$

Subject to:

$$(p_b^{pre} + v_b^{pre}) - D_b - \sum_{b-l} f_{b-l}^{pre} = 0 \qquad \forall\, b \in B, \forall\, b-l \in L \tag{2}$$

$$(p_b^{post} + v_b^{post}) - D_b - \sum_{b,l} f_{b-l}^{post} = 0 \qquad \forall\, b \in B, \forall\, b-l \in L-1 \tag{3}$$

$$f_{b-l}^{pre} = B_{b-l} \times (\delta_b^{pre} - \delta_l^{pre}) \qquad \forall\, b-l \in L, \forall\, b,l \in B \tag{4}$$

$$f_{b-l}^{post} = B_{b-l} \times (\delta_b^{post} - \delta_l^{pre}) \qquad \forall\, b-l \in L-1, \forall\, b,l \in B \tag{5}$$

$$|f_{b-l}^{pre}| \le F_{b-l}^{max} \qquad \forall\, b-l \in L \tag{6}$$

$$|f_{b-l}^{post}| < F_{b-l}^{max} \qquad \forall\, b-l \in L-1 \tag{7}$$

$$-R_g^{down} \le p_g^{post} - p_g^{pre} \le R_g^{up} \qquad \forall\, g \in G \tag{8}$$

$$P_g^{min} \le p_g^{pre} \le P_g^{max} \qquad \forall\, g \in G \tag{9}$$

$$P_g^{min} \le p_g^{post} \le P_g^{max} \qquad \forall\, g \in G \tag{10}$$

$$|\delta_b^{pre}| \le \pi/2 \qquad \forall\, b \in B \tag{11}$$

$$|\delta_b^{post}| \le \pi/2 \qquad \forall\, b \in B \tag{12}$$

$$\delta_b^{pre} = 0 \qquad b = SL \tag{13}$$

$$\delta_b^{post} = 0 \qquad b = SL. \tag{14}$$

Equations (2) and (3) represent nodal power balance constraints for the pre- and the post-contingency conditions, respectively. Equations (4) and (5) define power flows in the transmission elements and (6) and (7) limit these power flows for both conditions. Equation (8) models the ramp-up and ramp-down power unit generation limits and (9) and (10) limit the minimum and the maximum power unit generation. Constraints (11) and (12) limit voltage bus angles for both pre- and post-contingency conditions, respectively. Last, slack reference is defined for both conditions using (13) and (14).

In this mathematical formulation, the decision variables (n) are active power generation, voltage angles and transmission power flows for pre- and post-contingency analyses.

$$n = 2n_B + 2n_G + n_L + n_{L-1}$$

The equality constraints (n_e) are the following:

$$n_e = 2n_B + n_L + n_{L-1} + 2$$

The inequality constraints (n_i) are the following:

$$n_i = 2(2n_B + 3n_G + n_L + n_{L-1}).$$

2.2. Preventive SCOPF Problem Using the Classical DC-Based Formulation

In this mathematical formulation, the post-contingency condition is the same that the pre-contingency condition. With this assumption, the ramp-up and ramp-down constraints (8) are not necessary to add in the optimization problem. Nevertheless, transmission power flows are different because these values represent the pre-contingency power system state and the post-contingency state. As a result, there is only one set of decision variables as follows: $p^{pre} = p^{post} = p$, $v^{pre} = v^{post} = v$ and $\delta^{pre} = \delta^{post} = \delta$. Hence, the optimization problem considers a different objective function and it is subject to the following constraints:

$$\min(C_{total}) = \sum_{g}(A_g + B_g \times p_g \times S_{base} + C_g \times p_g^2 \times S_{base}^2) + VoLL \sum_{v} v_b \times S_{base} \qquad \forall\, g \in B, \forall\, v \in V \tag{15}$$

$$(p_b + v_b) - D_b - \sum_{b-l} f_{b-l}^{pre} = 0 \qquad \forall\, b \in B, \forall\, b-l \in L \tag{16}$$

$$(p_b + v_b) - D_b - \sum_{b-l} f_{b-l}^{post} = 0 \qquad \forall\, b \in B, \forall\, b-l \in L-1 \tag{17}$$

$$f_{b-l}^{pre} = B_{b-l} \times (\delta_b - \delta_l) \qquad \forall\, b-l \in L, \forall\, b,l \in B \tag{18}$$

$$f_{b-l}^{post} = B_{b-l} \times (\delta_b - \delta_l) \qquad \forall\, b-l \in L-1, \forall\, b,l \in B \tag{19}$$

$$|f_{b-l}^{pre}| \le F_{b-l}^{max} \qquad \forall\, b-l \in L \tag{20}$$

$$|f_{b-l}^{post}| \le F_{b-l}^{max} \qquad \forall\, b-l \in L-1 \tag{21}$$

$$P_g^{min} \le p_g \le P_g^{max} \qquad \forall\, g \in G \tag{22}$$

$$|\delta_b| \le \pi/2 \qquad \forall\, b \in B \tag{23}$$

$$\delta_b = 0 \qquad b = SL. \tag{24}$$

The number of decision variables n and equality n_e and inequality n_i constraints are calculated using $n = n_B + n_G + n_L + n_{L-1}$, $n_e = 2n_B + n_L + n_{L-1} + 1$ and $n_i = 2(n_B + n_G + n_L + n_{L-1})$.

2.3. Corrective SCOPF Problem Using the SF-Based Formulation

In this section, the corrective SCOPF problem is introduced using shift-factors. For this optimization problem, the classical DC-based set of transmission network constraints is reformulated using the inverse of admittance matrix avoiding to compute voltage bus angles and nodal transmission constraints. Therefore, nodal balance constraints are turned into only one equality constraint modelling both pre-contingency (25) and post-contingency states (26). Additionally, transmission power flows for both (27) and (28) conditions are obtained using shift-factors and net power injected.

The objective function is presented in (1a) and the optimization problem is subject to the following constraints:

$$\sum_{g} (p_g^{pre} + v_g^{pre}) - D^{total} = 0 \qquad \forall\, g \in G \tag{25}$$

$$\sum_{g} (p_g^{post} + v_g^{post}) - D^{total} = 0 \qquad \forall\, g \in G \tag{26}$$

$$|\sum_{k} SF_{l-b,k}^{pre} \times (p_k^{pre} - D_k)| \leq F_{b-l}^{max} \qquad \forall\, k \in B,\, k \neq SL,\, \forall\, b-l \in L \tag{27}$$

$$|\sum_{k} SF_{l-b,k}^{post} \times (p_k^{post} - D_k)| \leq F_{b-l}^{max} \qquad \forall\, k \in B,\, k \neq SL,\, \forall\, b-l \in L-1 \tag{28}$$

$$- R_g^{down} \leq p_g^{post} - p_g^{pre} \leq R_g^{up} \qquad \forall\, g \in G \tag{29}$$

$$P_g^{min} \leq p_g^{pre} \leq P_g^{max} \qquad \forall\, g \in G \tag{30}$$

$$P_g^{min} \leq p_g^{post} \leq P_g^{max} \qquad \forall\, g \in G, \tag{31}$$

where $SF^{pre} = Bf^{pre} \times Xbus_r^{pre}$ and $SF^{post} = Bf^{post} \times Xbus_r^{post}$ are the SF for pre- and post-contingency conditions and $Xbus_r^{pre} = (A_r^T \times Bbus_r^{pre})^{-1}$ and $Xbus_r^{post} = (A_r^T \times Bbus_r^{post})^{-1}$ are the reactance bus matrix for both conditions.

For this mathematical formulation, decision variables are exclusively the power unit generation. As a result, there are $n_e = 2$ and $n_i = 2(3\,n_G + n_L + n_{L-1})$. According to the lower number of variables and constraints, this formulation achieves a very compact SCOPF optimization problem.

2.4. Preventive SCOPF Problem Using the SF-Based Formulation

With previous assumptions given in Section 2.2, the OPF problem is subject to the following constraints:

$$\sum_{g} (p_g + v_g) - D^{total} = 0 \qquad \forall\, g \in G \tag{32}$$

$$|\sum_{k} SF_{l-b,k}^{pre} \times (p_k - D_k)| \leq F_{b-l}^{max} \qquad \forall\, k \in B,\, k \neq SL,\, \forall\, b-l \in L \tag{33}$$

$$|\sum_{k} SF_{l-b,k}^{post} \times (p_k - D_k)| \leq F_{b-l}^{max} \qquad \forall\, k \in B,\, k \neq SL,\, \forall\, b-l \in L-1 \tag{34}$$

$$P_g^{min} \leq p_g \leq P_g^{max} \qquad \forall\, g \in G. \tag{35}$$

For this formulation, there are $n_e = 1$ and $n_i = 2(n_G + n_L + n_{L-1})$.

2.5. Computing the Post-Contingency Shift-Factors

To compute more efficiently the SF^{post} matrix, this study implements previous definition given in Reference [6] to determine post-contingency transmission constraints using Equation (36).

$$|SF_b^{pre} + LODF_{b,k} \times SF_b^{pre} \times (P - P^d)| \leq F^{postmax} \qquad \forall\, b \in L-1. \tag{36}$$

Lastly, LODF factors are computed as follows:

$$LODF_{b,k} = SF_b^{pre} \times A_k^{pre^T} \times [1/(1 - SF_k^{pre} \times A_k^{pre^T})]. \tag{37}$$

Even though line outage distribution factors are only computed with pre-contingency shift-factors; that is, network data before the transmission outage, islanding conditions could be detected finding out states where the denominator of (37) is zero.

3. Simulation Results

Several simulations are conducted using short- and large-scale electrical power systems to find out the performance of both corrective and preventive SCOPF formulations. To formulate the optimization problem, Python [20] has been used. Moreover, Gurobi [21] is used as a commercial solver on a computer with the following characteristics: Intel Core i7 3930 (3.20 GHz) six core with RAM 32 GB.

3.1. SCOPF Formulation Applied to an Example Power System (6-Bus)

The first security-constrained simulation takes into account the 6-bus power system presented by Wood and Wollenberg [8]. Transmission network data can be seen in [8] or in MatPower [22].

In comparison with the original generation system, three new power units are located at each load bus (G_4, G_5 and G_6). Figure 1 shows the electrical power system.

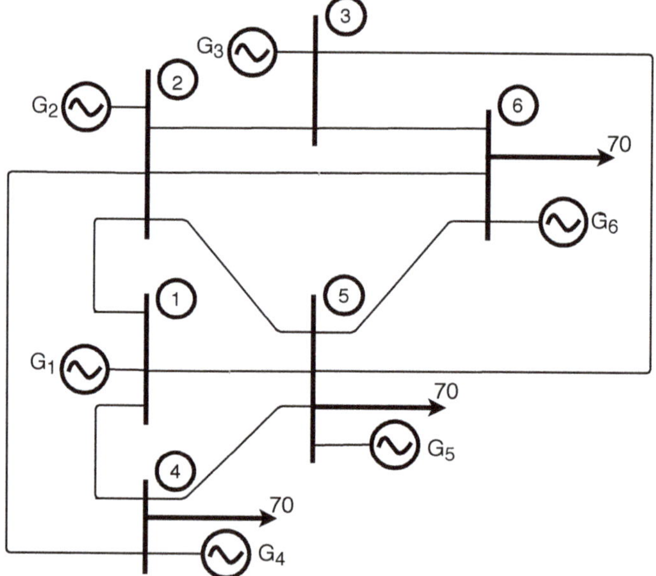

Figure 1. Test system I: A new 6-bus electrical power system is proposed in this study.

Technical data for the power generation system are given in Table 1. Besides, ramp-up and ramp-down characteristics are incorporated in the last two columns for each power unit.

Table 1. Power generation technical parameters.

Power Unit	A ($/MW^{2}h)	B ($/MWh)	C ($/h)	P^{min} (MW)	P^{max} (MW)	R^{up} (MW/min)	R^{down} (MW/min)
G_1	0.00533	11.669	213.1	200	50	9.0	8.5
G_2	0.00889	10.333	200.0	150	37.5	12.0	12.0
G_3	0.00741	10.833	240.0	180	45	11.0	10.1
G_4	0.00301	14.198	40.0	70	5	2.5	5.0
G_5	0.00111	4.955	300.0	70	5	4.0	2.0
G_6	0.00876	18.003	10.0	70	5	3.5	5.0

Solving the classical OPF problem, the optimal cost is C_{total} = 3003.17 \$/h. The power generation solution is P_1 = 50.0 MW, P_2 = 37.5 MW, P_3 = 45.0 MW, P_4 = 5.0 MW, P_5 = 67.5 MW and P_6 = 5.0 MW. Figure 2 shows the power flow solution. According to the power flow solution, there is no congestion in transmission elements. PyPower [23] is used to validate the solution and both solutions are the same.

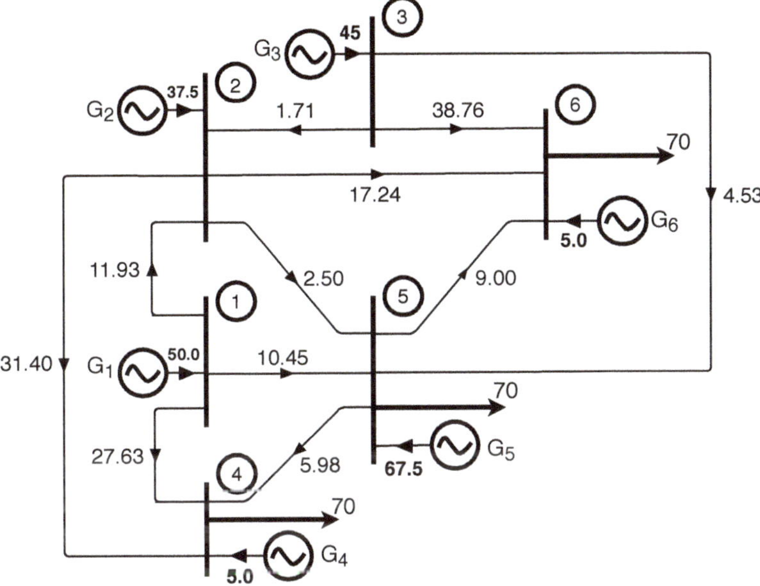

Figure 2. Classical optimal power flow (OPF) solution.

To realize possible transmission effects, the maximum transmission limits are changed to 40 MW in lines 1–4, 2–4 and 3–6. With this new capacity, the OPF problem also obtains the same solution (operational cost, power unit generation and transmission power flows). This new transmission capacity is used to develop the N−1 power system security.

DC-Based and SF-Based SCOPF Formulations

In the first analysis, both classical (CL) and introduced (SF) problems are implemented to model the SCOPF. These formulations are subject to one pre- and eleven post-contingency constraints.

- Corrective SCOPF problem—for this analysis, power generation data are incorporated in the formulation assuming that the ISO has 10 min to eliminate overloading transmission effects and to recover the steady-state power system security. The main advantage of the corrective formulation is related to the operator clearly knows the post-contingency economic dispatch. This solution considers not only technical generation constraints but the variable fuel cost of each power unit. Indeed, the new power generation setting will achieve a safety and robust security-constrained N−1 solution no matter which transmission element failure.

Two cases are conducted to determine effects when ramps constraints are added in the SCOPF problem. These optimization problems have been solved using the SF-based formulation. Results are the following:

Case 1   Without ramp constraints—the operational cost is 3003.17 \$/h for the pre-contingency condition and 3487.87 \$/h for the post-contingency condition and the optimal total cost is C_{total} = 6491.04 \$/h. Actually, the pre-contingency cost is the same than the traditional OPF problem (3003.17 \$/h).

Case 2   Including ramp-up and -down constraints—the cost is 3146.35 $/h for the pre-contingency condition and 3487.87 $/h for the post-contingency condition and the total cost is C_{total} = 6634.22 $/h. For this case, the number of decision variables is 12 and the number of constraints is 280.

Table 2 shows the results, operational cost and power dispatch for each SCOPF solution. Positive values imply that the power unit generation must decrease its dispatch. With this information, the ISO rapidly decides (10 or 20 min) which generator would economically change the power unit setting to optimally eliminate overloading conditions.

Table 2. Power generation solution for pre- and post-contingency states.

Power	Without-Ramps			With-Ramps		
Unit	Pre-Contingency	Post-Contingency	Difference	Pre-Contingency	PosT-Contingency	Difference
#	Power, MW	Power, MW	MW	Power, MW	Power, MW	MW
P_1	50.00	50.00	0	50.00	50.00	0
P_2	37.50	37.50	0	60.86	37.50	+23.36
P_3	45.00	45.00	0	45.00	45.00	0
P_4	5.00	27.24	−22.24	5.00	27.24	−22.24
P_5	67.50	24.14	+36.36	44.14	24.14	+20.00
P_6	5.00	26.12	−21.12	5.00	26.12	−21.12

Because PyPower does not formulated the security-constrained analysis, it is not possible to contrast this power generation solution.

Because of ramp constraints, the ISO will be not able to achieve a safe post-contingency steady-state with the solution obtained in Case 1). Therefore, the system operator will need to decrease the load of the customers or/and turn-on fast reserve power units.

Notice that the ramp-down constraint of power unit 5 (44.14 − 24.14 = 20 MW) is active with a shadow price of 6.36 $/MWh. Therefore, the pre-contingency cost is higher (143.18 $/h) than the traditional OPF problem. Additionally, the power generation solution for the post-contingency condition is the same for both cases.

In comparison with the SF-based formulation, simulation results obtained by the CL-based formulation are the same. Therefore, both formulations achieve the same optimal solution; that is, these optimization problems are equivalent.

Incorporating the ramp constraints in the CL SCOPF problem, the number of decision variables is 205 and the number of constraints is 483. Comparing with the SF formulation, there is a very significant reduction of 94.1% and 42.0%, respectively.

Figure 3 displays the pre- and the post-contingency solutions when line 1–2 is outage. Both solutions could be compared to realize redispatch effects in the transmission network for generation two, four, five and six (grey color).

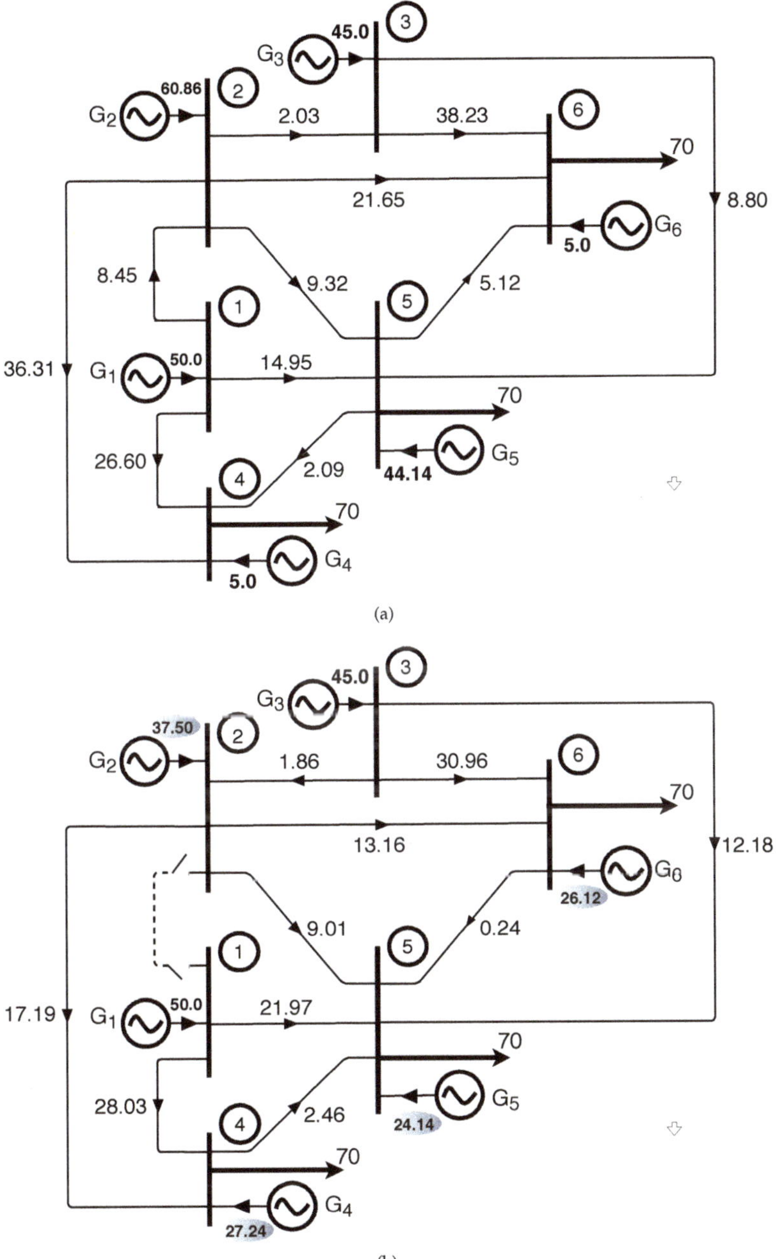

Figure 3. Corrective OPF solution considering the outage of line 1–2. (**a**) Pre-contingency power flow solution. (**b**) Post-contingency power flow solution.

Using the post-contingency power generation solution (Table 2), eleven N−1 power flows are computed. For detailed information, Table 3 shows the power flow solution for each outage.

Table 3. Power flow solution for each N−1 transmission contingency.

Line	1–4 MW	1–5 MW	2–3 MW	2–4 MW	2–5 MW	2–6 MW	3–5 MW	3–6 MW	4–5 MW	5–6 MW
-	24.21	19.95	10.38	−1.41	7.99	8.53	8.46	10.70	10.31	10.11
f_{1-4}	-	30.05	21.94	36.14	21.24	21.39	21.37	22.03	21.13	21.86
f_{1-5}	25.79	-	17.68	15.26	20.78	20.08	20.17	17.26	18.56	18.03
f_{2-3}	−1.51	2.37	-	2.09	1.67	5.91	−6.15	−16.98	−0.39	−0.98
f_{2-4}	**40.00**	20.20	23.12	-	26.49	25.73	25.83	22.66	21.63	23.49
f_{2-5}	9.65	16.73	10.76	16.20	-	14.39	14.54	10.13	11.68	11.29
f_{2-6}	13.57	18.15	14.00	17.81	17.32	-	11.74	32.40	14.89	13.81
f_{3-5}	12.58	17.02	13.69	16.69	16.22	10.91	-	28.02	13.86	13.96
f_{3-6}	30.91	30.36	31.31	30.40	30.46	**40.00**	38.85	-	30.75	30.06
f_{4-5}	−2.76	7.49	2.29	−6.62	4.96	4.36	4.44	1.93	-	2.59
f_{5-6}	−0.60	−4.63	−1.43	−4.33	−3.90	3.87	−6.71	11.47	−1.76	-

Results showed there are two post-contingency congestions in the following lines: (a) transmission line 2–4 when line 1–4 is out; and (b) transmission line 3–6 when line 2–6 is out. We have highlighted these values with bold. Based on the OPF solution, shadow prices (Lagrangian multipliers) for each congestion are the following: (a) for line 2–4 is 28.60 \$/MWh; and (b) for line 3–6 is 41.98 \$/MWh. The OPF overcost and the bigger Lagrangian multipliers are produced by transmission congestion and ramp-down constraint of unit 5.

- Preventive SCOPF problem—the main advantage of this formulation is related to the operator does not need to modify the post-contingency power dispatch.

Using the SF-based formulation, the number of decision variables is 6 and the number of constraints is 280. The optimal cost obtained by Gurobi is C_{total} = 3487.87 \$/h which was previously obtained. In the solution, there is no congestion in the pre-contingency condition. However, there is congestion in transmission line 2–4 (when line 1–4 is out) and line 3–6 (when line 2–6 is out).

Gurobi achieves the same solution solving the CL-based SCOPF formulation. Nevertheless, the number of decision variables is 199 and the number of constraints is 459.

To compare corrective and preventive SCOPF solutions, the power dispatch obtained in the pre-contingency condition must be compared for both solutions. The preventive overcost is 342.52 \$/h (9.82%). Therefore, there is an important operational saving cost obtained by the corrective formulation.

3.2. Corrective SCOPF Using a Ranking of Contingencies

In real power systems, all contingencies do not result in a post-contingency alert condition [8]. To figure out the SCOPF problem, several authors have used as security criterion the most severe contingency [6].

Regarding the N−1 power flow solution obtained previously by the initial OPF (3003.17 \$/h), we solves eleven power flows for each N−1 post-contingency event. Simulations have shown there are five overloading states: (a) when line 1–4 is out, the power flow in line 2–4 is 131.32%; (b) when line 2–4 is out, the power flow in line 1–4 is 117.11%; (c) when line 2–6 is out, the power flow in line 3–6 is 124.41%; (d) when line 3–5 is out, the power flow in line 3–6 is 103.70%; and (e) when line 5–6 is out, the power flow in line 3–6 is 109.42%. Indeed, the ranking of overloading contingencies is the following: (1) line 1–4, (2) line 2–6, (3) line 2–4, (4) line 5–6 and (5) line 3–6.

The outage of these transmission elements produce a risky operational condition and probably technical problems to supply adequately the load of the customers. In the next analysis, each line of this ranking will be added in the SCOPF problem. Table 4 displays the number of decision variables and constraints and the objective function solving the SCOPF problem.

Table 4. Security-constrained operational cost considering the ranking of contingencies.

Optimal Solution	(i) N−1: line$_{1-4}$ $/h	(ii) N−1: line$_{2-6}$ $/h	(iii) N−1: line$_{2-4}$ $/h	(iv) N−1: line$_{5-6}$ $/h	(v) N−1: line$_{1-4}$ and line$_{2-6}$ $/h
n	12	12	12	12	12
$n_e + n_i$	80	80	80	80	100
$Cost_{pre}$	3003.17	3003.17	3003.17	3003.17	3146.36
$Cost_{post}$	3178.12	3241.03	3122.32	3091.56	3487.87
$Cost_{total}$	6181.29	6244.21	6125.49	6094.74	6634.22

- For the first case, the pre-contingency and the worst contingency (line 1–4) constraints are simultaneously incorporated in the SCOPF problem.
  Solving the SCOPF problem, the post-contingency power unit solution is used to validate electrical system security using eleven N−1 power flow problems. Simulation results show there are three overloading conditions: (a) when transmission line 2–6 is out, the power flow in line 3–6 is 51.18 MW; (b) when transmission line 3–5 is out, the power flow in line 3–6 is 42.74 MW; and (c) when transmission line 5–6 is out, the power flow in line 3–6 is 42.11 MW. Notice that the maximum power flow in line 3–6 is 40 MW. Therefore, a SCOPF problem based on the worst contingency does not guarantee a safe post-contingency operational point.
- In the second case, the SCOPF problem includes the outage of the transmission line 2–6.
  Reviewing the power flow solution, there are two overloading conditions: (a) when transmission line 1–4 is out, the power flow in line 2–4 is 54.34 MW; and (b) when transmission line 2–4 is out, the power flow in line 1–4 is 47.72 MW. Notice that the maximum power flow in line 1–4 and line 2–4 is 40 MW. Even though the operational cost is bigger than the previous case, this solution does not guarantee a safe power system as well. Additionally, this solution obtains the most overloading condition in line 2–4 (136%).
- In the third case, failure of transmission line 2–4 is added in the optimization problem.
  Simulating the N−1 post-contingency power flow method, the overloading conditions are the following: (a) when transmission line 1–4 is out, the power flow in line 2–4 is 43.97 MW; (b) when transmission line 2–6 is out, the power flow in line 3–6 is 50.73 MW; (c) when transmission line 3–6 is out, the power flow in line 2–6 is 41.62 MW; and (d) when transmission line 5–6 is out, the power flow in line 3–6 is 42.63 MW. Furthermore, this is an unacceptable post-contingency overloading condition.
- For the fourth case, transmission constraints for the outage of transmission line 5–6 is included in the N−1 optimization problem.
  Simulating the post-contingency power flow method, the overloading conditions are the following: (a) when transmission line 1–4 is out, the power flow in line 2–4 is 53.20 MW; (b) when transmission line 2–4 is out, the power flow in line 1–4 is 47.17 MW; (c) when transmission line 2–6 is out, the power flow in line 3–6 is 46.11 MW; and (d) when transmission line 3–5 is out, the power flow in line 3–6 is 40.16 MW.
- We do not show the last contingency because power flow results also display an overloading condition.

Finally, the post-contingency analysis including two N−1 candidate lines is developed. The SCOPF problem incorporates two major contingencies: line 1–4 and line 2–6. This framework adds in the optimization problem simultaneously one set of pre-contingency constraints and two sets of post-contingency constraints. For this optimization problem, the achieved solution is the optimal (the same solution considering the outage of all lines).

Table 4 also shows a reduction of 35.7% in constraints with respect to the original problem. In addition, both problems have the same number of decision variables.

As a result, the SCOPF problem should be modelled not only with the worst contingency but also with two or more contingencies to avoid risky operational conditions and technical problems to supply the load of the customers.

3.3. SCOPF Methodology Applied to Different-Scale Power Systems

We employ different electrical power test systems to determine the performance of both corrective and preventive SCOPF formulations using the SF-based as well as the CL-based frameworks: 14-bus system (five generators and twenty transmission lines); 57-bus system (seven generators and eighty transmission lines); 118-bus system (fifty-four generators and one hundred eighty-six transmission lines); and 300-bus system (sixty-nine generators and four hundred eleven transmission lines). PyPower shows technical information.

3.3.1. Corrective and Preventive Formulations

Because PyPower does not include ramp generation constraints, we have selected these limits using random numbers provided by a normal distribution function with a mean of 7 MW/min and a variance of 1.

Table 5 shows simulation results for both corrective and preventive analyses applying the CL and the SF formulations. For each SC problem, we have included the objective function (OF), the number of decision variables (n), constraints ($n_i + n_e$) and the average simulation time considering 200 trials.

For both 14-bus and 57-bus power systems, the security-constrained problem models the outage of all transmission lines.

Considering the higher simulation time obtained by a complete $N-1$ analysis, a reduced number of contingencies should be used to solve the security problem applied to large-scale power systems. In the 118-bus and the 300-bus systems, we have modelled the $N-1$ analysis using 176 transmission and 100 transmission candidate lines, respectively. This assumption is accomplished not only by the bigger number of variables and constraints but also by the higher simulation time.

Table 5. Simulation results using different power systems.

Formulation	Corrective				Preventive			
System	OF, \$/h	n	$n_i + n_e$	ts, s	OF, \$/h	n	$n_i + n_e$	ts, s
14-bus (CL)	15,285.19	704	1545	0.0040	7642.59	699	1525	0.0036
14-bus (SF)	15,285.19	10	832	0.0004	7642.59	5	811	0.0009
57-bus (CL)	82,013.47	10,895	23,645	0.0821	41,006.74	7760	16,832	0.0414
57-bus (SF)	82,013.47	14	6366	0.0048	41,006.74	7	9023	0.0049
118-bus (CL)	260,788.55	55,558	123,669	0.9203	130,394.28	55,807	124,127	0.8459
118-bus (SF)	260,788.55	108	62,312	0.3057	130,394.28	54	62,095	0.3106
300-bus (CL)	1,422,058.21	71,849	155,048	0.8777	711,029.11	71,780	154,772	0.9468
300-bus (SF)	1,422,058.21	138	21,110	0.1350	711,029.11	69	20,833	0.1349

For all power systems, there is an important improvement in the simulation time. Comparing both classical and SF-based formulations applied to the 300-bus system, SF-based formulation achieves the best performance with an efficiency of 6.97 times using the corrective formulation and 7.02 times using the preventive formulation.

Simulation results have shown the corrective formulation obtains the lower OF in comparison with the preventive formulation.

Notice that the reactance bus matrix needs an inverse method. Therefore, the main disadvantage of the proposed methodology is the time computing to obtain the SF matrix. However, the security-constrained analysis could be carried out using an off-line SF matrix; that is, the SF matrix does not need to compute each $N-1$ time. Therefore, an external file could be used to save these matrixes improving the formulation time.

The LODF matrix could be computed using basic mathematical operations—Equation (37). Therefore, the computation time is not significant.

3.3.2. Another Classical DC-Based Preventive Formulation Applied to the 300-Bus Power System

In the following analysis, we use a different preventive SCOPF formulation. For this approach, power flow variables are eliminated from the optimization problem using two admittance bus matrixes

($Bbus^{pre}$ and $Bbus^{post}$). Consequently, there are only two sets of nodal balance constraints for the pre- and post-contingency analyses. Besides, the primitive-admittance incidence matrix for the pre-contingency Bf^{pre} and the post-contingency Bf^{post} conditions is substitute to determine transmission power flows.

Simulating 200 trials, the average simulation time is 0.7648 s for the preventive SCOPF problem which is 19.22% lower than the original CL-based formulation. For this mathematical problem, the number of decision variables is 30,369 and the number of constraints is 113,361. Therefore, the new classical formulation causes a reduction of 57.69% in the number of decision variables and a reduction of 26.76% in the number of equality and inequality constraints. Although there is an improvement in the simulation performance, this time is still higher (5.67 times) than the SF-based formulation.

Based on simulations, the introduced SF-based formulation achieves the best simulation time to carry out the corrective formulation and the preventive formulation applied to the SCOPF problem. This improvement is obtained by the lower number of decision variables as well as by the lower number of equality and inequality constraints. Therefore, this framework could be recommended to solve very large-scale power systems.

3.4. Corrective and Preventive SCOPF Methodology Applied to the Chilean Electrical Power System

The Chilean Interconnected Power System (in Spanish, SEN: Sistema Eléctrico Nacional) is used to validate both SCOPF formulations. Concerning the lower simulation time accomplished previously, the SF-based methodology is only applied to the SCOPF problem.

In January, 2019, the SEN generation capacity was 21,189.95 MW. In 2019, the peak load was 9382.7 MW. Review the Regulatory Company webpage (CNE: www.cne.cl) or the ISO webpage (CEN: www.coordinador.cl) to obtain more technical information.

The electrical power system contains one hundred fifty-nine buses, three hundred twenty-one transmission lines from 110 to 500 kV and two hundred sixty-seven generation power units (thermal, hydro and renewable energy). The webpage link https://infotecnica.coordinador.cl displays detailed technical and economic information about the Chilean study case.

For the power generation system, there are one hundred three renewable power units. For hydro, wind and solar, a capacity factor of 20%, 20% and 30% are selected, respectively. For thermal units, random numbers model ramp generation constraints for each power unit. Additionally, one hundred transmission lines are randomly selected to formulate the N−1 security-constrained analysis.

- Corrective SCOPF problem—this optimization problem has 534 decision variables and 22,082 constraints. The cost is 326,948.31 \$/h for the pre-contingency condition and 327,087.37 \$/h for the post-contingency condition. The simulation time is 1.0218 s using 100 trials.

  There is congestion in Cardones-Maintencillo 220 kV in the pre-contingency solution. On the contrary, there is post-contingency congestion in the following transmission elements—(1) Andes-Nueva Zaldivar 220 kV when Antucoya-Antucoya aux. 220 kV is out; (2) Antofagasta-Desalant 110 kV when Atacama-Esmeralda 220 kV is out; (3) Capricornio-Mantos Blancos 220 kV when CD Arica-Tap Quiani 66 kV is out; (4) Carrera Pinto-Carrera Pinto aux. 220 kV when Charrúa-Lagunillas 220 kV is out; and (5) CD Arica-Arica 66 kV when Charrúa-Mulchen 220 kV is out.

- Preventive SCOPF problem—this optimization problem has 267 decision variables and 21,013 constraints. The optimal cost is 329,981.43 \$/h and the simulation time is 0.6085 s. For the pre-contingency condition, there is congestion in the same line (Cardones-Maintencillo 220 kV). Furthermore, there are seven lines with congestion in the post-contingency condition: (1) Andes-Nueva Zaldivar 220 kV; (2) Antofagasta-Desalant 110 kV; (3) Capricornio-Mantos Blancos 220 kV; (4) Carrera Pinto-Carrera Pinto aux. 220 kV; (5) CD Arica-Arica 66 kV; (6) Charrúa-Hualpen 220 kV (when Colbún-Ancoa 220 kV is out); and (7) Crucero-Nueva Crucero/Encuentro 220 kV (when Don Goyo-Don Goyo aux. 220 kV is out).

Even though the security cost is similar, the OPF formulation only considers one load time-period. Moreover, the unit commitment solution obtained by the ISO should be incorporated in the OPF problem to realize real saving costs.

4. Conclusions

This study analyzed two very effective methodologies to solve both corrective and preventive security-constrained DC optimal power flow problems using shift-factors. We used these factors to formulate pre- and post-contingency steady-state conditions. The optimal solution was found with lower number of decision variables and constraints in comparison with the traditional (CL) SCOPF formulation. To validate simulation time, a lot of analyses were conducted to determine which methodology obtained the best performance solving the SCOPF problem with different-scale test power systems.

The research group believes that the introduced security-constrained methodology could be applied to formulate deterministic and stochastic power system issues where the transmission topology does not change. For instance, unit commitment, generation planning, among other power system problems.

Funding: This research was funded by Universidad Técnica Federico Santa María, Chile, grant number USM PI_M_18_14 and The APC was funded by the same research project.

Acknowledgments: The author would like to thank the associate editor and the anonymous reviewers for their valuable comments.

Nomenclature

Parameters

S_{base}	Power system base value S_{base} = 100 MW
C^{total}	Total security-constrained operational cost ($/h)
C^{pre}	Pre-contingency variable cost ($/h)
C^{post}	Post-contingency variable cost ($/h)
A_g, B_g and C_g	Constants of the quadratic cost function for the g-th thermal unit ($/h), ($/MWh) and ($/MW^{2}h), respectively
$VoLL$	Value of lost load ($/MWh)
D_b	Load demand at bus b (pu)
D^{total}	Total power system demand (pu)
B_{b-l}	Susceptance value for transmission element $b-l$ (pu)
F^{max}_{b-l}	Maximum power flow for transmission element $b-l$ (pu)
R^{up}_g	Ramp-up for the g-th power unit (pu)
R^{down}_g	Ramp-down for the g-th power unit (pu)
P^{max}_g	Maximum power generation for unit g (pu)
P^{min}_g	Minimum power generation for unit g (pu)
$SF^{pre}_{b-l,k}$	Linear power distribution factor of element $b-l$ with respect to bus k for the pre-contingency condition
$SF^{post}_{b-l,k}$	Linear power distribution factor of element $b-l$ with respect to bus k for the pre-contingency condition
Bf^{pre}	Primitive-admittance incidence matrix for the pre-contingency condition
Bf^{post}	Primitive-admittance incidence matrix for the post-contingency condition
$Bbus^{pre}_r$	Reduced admittance bus matrix for the pre-contingency condition
$Bbus^{post}_r$	Reduced admittance bus matrix for the post-contingency condition
$Xbus^{pre}_r$	Reduced impedance bus matrix for the pre-contingency condition
$Xbus^{post}_r$	Reduced impedance bus matrix for the post-contingency condition
A_r	Reduced incidence power system matrix
SL	Reference (slack) bus

SETS

G Set of power generator units
V Set of virtual power units
B Set of voltage bus angles
L Set of pre-contingency transmission elements
$L-1$ Set of post-contingency transmission elements

VARIABLES

p_g^{pre} Active pre-contingency power dispatched by unit g-corrective formulation
p_g^{post} Active post-contingency power dispatched by unit g-corrective formulation
v_g^{pre} Pre-contingency unserved energy at bus g-corrective formulation
v_g^{post} Post-contingency unserved energy at bus g-corrective formulation
f_{b-l}^{pre} Transmission pre-contingency power flow on element $b-l$-corrective formulation
f_{b-l}^{post} Transmission post-contingency power flow on element $b-l$-corrective formulation
δ_k^{pre} Voltage angle at bus k for the pre-contingency condition—corrective formulation
δ_k^{post} Voltage angle at bus k for the post-contingency condition—corrective formulation
p_g Power generation dispatched by unit g for the preventive formulation
v_g Unserved energy at bus g for the preventive formulation
δ_k Voltage angle at bus k for the preventive formulation

References

1. Carpentier, J. Contribution a L'etude du Dispatching Economique. *Bull. Soc. Fr. Electr.* **1962**, *8*, 431–447.
2. Bai, W.; Lee, D.; Lee, K.-Y. Stochastic dynamic AC optimal power flow based on a multivariate short-term wind power scenario forecasting model. *Energies* **2017**, *10*, 2138. [CrossRef]
3. Momoh, J.; Adapa, R.; El-Hawary, M. A review of selected optimal power flow literature to 1993. *IEEE Trans. Power Syst.* **1993**, *14*, 96–104. [CrossRef]
4. Pizano, A.; Fuerte, C.R.; Ruiz, D. A new practical approach to transient stability constrained optimal power flow. *IEEE Trans. Power Syst.* **2011**, *26*, 1686–1696. [CrossRef]
5. Mohagheghi, E.; Alramlawi, M.; Gabash, A.; Li, P. Incorporating charging/discharging strategy of electric vehicles into security-constrained optimal power flow to support high renewable penetration. *Energies* **2017**, *10*, 729.
6. Hinojosa, V.; Gonzalez-Longatt, F. Preventive security-constrained DCOPF formulation using power transmission distribution factors and line outage distribution factors. *Energies* **2018**, *11*, 1497. [CrossRef]
7. Gutierrez-Alcaraz, G.; Hinojosa, V. Using generalized generation distribution factors in a MILP model to solve the transmission-constrained unit commitment problem. *Energies* **2018**, *11*, 833. [CrossRef]
8. Wood, A.J.; Wollenberg, B.F. *Power Generation, Operation and Control*; John Wiley & Sons: Hoboken, NJ, USA, 1996.
9. Gonzalez-Longatt, F.; Rueda, J.L. *Power Factory Applications for Power System Analysis*; Springer: Berlin, Germany, 2014.
10. Hinojosa, V.H.; Velásquez, J. Improving the mathematical formulation of security-constrained generation capacity expansion planning using power transmission distribution factors and line outage distribution factors. *Electr. Power Syst. Res.* **2016**, *140*, 391–400. [CrossRef]
11. Alsac, O.; Stott, B. Optimal Load Flow with Steady-State Security. *IEEE Trans. Power Appar. Syst.* **1974**, *93*, 745–751. [CrossRef]
12. Capitanescu, F.; Martinez-Ramos, J.L.; Panciatici, P.; Kirschen, D.; Marano-Marcolini, A.; Platbrood, P.; Wehenkel, L. State-of-the-art, challenges, and future trends in security constrained optimal power flow. *Electr. Power Syst. Res.* **2012**, *81*, 275–285. [CrossRef]
13. Sass, F.; Sennewald, T.; Marten, A.; Westermann, D. Mixed AC high-voltage direct current benchmark test system for security constrained optimal power flow calculation. *IET Gener. Transm. Distrib.* **2017**, *11*, 447–455. [CrossRef]

14. Martínez-Lacañina, P.; Martínez-Ramos, J.; de la Villa-Jaén, A. DC corrective optimal power flow base on generator and branch outages modelled as fictitious nodal injections. *IET Gener. Transm. Distrib.* **2014**, *8*, 401–409.

15. Monticelli, A.; Pereira, M.V.F.; Granville, S. Security-constrained optimal power flow with post-contingency corrective rescheduling. *IEEE Trans. Power Syst.* **1987**, *2*, 175–180. [CrossRef]

16. Dzung, P.; Kalagnanam, J. Some efficient optimization methods for solving the security-constrained optimal power flow problem. *IEEE Trans. Power Syst.* **2014**, *29*, 863–872.

17. Phan, D.; Sun, X. Minimal impact corrective actions in security-constrained optimal power flow via sparsity regularization. *IEEE Trans. Power Syst.* **2015**, *30*, 1947–1956. [CrossRef]

18. An, K.; Song, K.; Hur, K. A survey of real-time optimal power flow. *Energies* **2018**, *11*, 3142.

19. Wu, X.; Zhou, Z.; Liu, G.; Qi, W.; Xie, Z. Preventive security-constrained optimal power flow considering UPFC control modes. *Energies* **2017**, *10*, 1199. [CrossRef]

20. Python. Available online: http://www.python.org (accessed on 1 December 2019).

21. Gurobi Optimization. Available online: http://www.gurobi.com (accessed on 1 December 2019)

22. Matpower. Available online: http://www.pserc.cornell.edu/matpower (accessed on 1 December 2019).

23. Pypower. Available online: https://pypi.org/project/PYPOWER/ (accessed on 13 December 2019).

Review

New High-Efficiency Resonant O-Type Devices as the Promising Sources of Microwave Power

Andrei Baikov [1,*] and Olga Baikova [2,3]

[1] Basic Technologies and Components of Vacuum Devices, Ltd., Moscow 117342, Russia
[2] Institute of Nanoengineering in Electronics, Spintronics and Photonics, National Research Nuclear University MEPhI, Moscow 115409, Russia; obkv@mail.ru
[3] Federal State Educational Establishment of Professional Medical College, Moscow 117105, Russia
[*] Correspondence: a_yu_baikov@mail.ru; Tel.: +7-916-175-9100

Received: 9 April 2020; Accepted: 11 May 2020; Published: 15 May 2020

Abstract: New O-type high-power vacuum resonant microwave devices are considered in this study: COM klystrons, CSM klystrons and resotrodes. All these devices can output a large amount of power (up to units of MW and higher) with an efficiency of up to 90%. Such devices are promising microwave sources for industrial microwave technologies as well as for microwave energy. The principle of GSP-equivalence for klystrons is described herein, allowing a complete physical analog of this device with other parameters to be created. The existing mathematical and computer models of klystrons are analyzed. The processes of stage-by-stage optimization and the embedding procedure, which leads to COM and to CSM klystrons, are considered. Resotrodes, IOT-type devices with energy regeneration in the input circuit, are also considered. It is shown that these devices can combine high power with an efficiency of up to 90% and a gain of more than 30 dB. Resotrodes with 0-regeneration can be effective sources of radio frequency (RF) power in the range of 20 to 200 MHz. Resotrodes with 2π-regeneration are an effective source of RF/microwave energy in the range of 200 MHz to 1000 MHz.

Keywords: klystron; IOT; resotrode; efficiency; microwave; bunching; computer simulation; global optimization

---

1. Introduction

Powerful vacuum microwave devices have existed and been developed since the 1930s. From the moment of their introduction until recently, such devices have mainly been used for radar and communication. Other applications of one of these devices (the magnetron) are defrosting food and cooking in household microwave stoves.

Some applications of vacuum microwave devices have been discovered relatively recently. In the 1960s, charged particle accelerators appeared that used microwave fields for acceleration. In the 1970s, medical microwave devices began to appear, including medical accelerators, microwave scalpels, among others.

Recently, industrial technologies using microwave energy have begun to be developed. These include, for example, technologies for defrosting large volumes of products in defrosters [1], for the radiation sterilization of products and materials, for sintering ceramics, for drying wood [2], and so on.

The development potential of industrial microwave technologies is very large. For example, several years ago, successful experiments were conducted to produce artificial sandstone from crushed sand. This is a building material that can compete with concrete in mass production. Foam glass is another building material that can be efficiently produced using microwave technology. There is also the possibility of deep oil refining using microwave technologies.

Another area with great potential for development is microwave energetics. This is a set of technologies utilizing microwave radiation for wireless energy transmission. Energy can be transferred to hard-to-reach places, to mountainous areas, to remote islands, to polar stations, and so on. In addition, energy can be transferred from the earth to aircrafts and spacecrafts. In the long term, the reverse process is possible via the transfer of energy to Earth from solar panels located in space.

Classification of existing high-power vacuum electronic devices (Figure 1), amplifiers and generators of microwave power, can be carried out by taking two basic characteristics into account: the type of interaction of the microwave field with the electronic flow (O-type, M-type, gyrotron type) and the nature of the microwave vibrations (traveling wave or standing wave). As examples, Figure 1 shows the most typical representatives of each class.

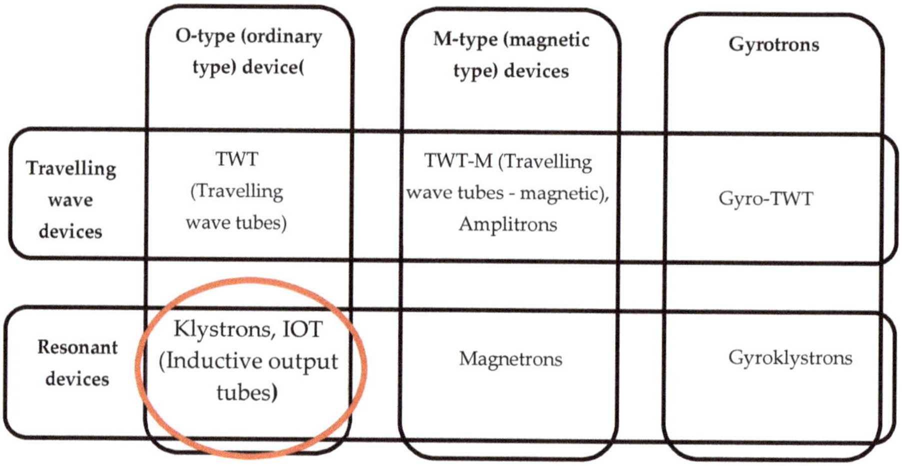

	O-type (ordinary type) device(	M-type (magnetic type) devices	Gyrotrons
Travelling wave devices	TWT (Travelling wave tubes)	TWT-M (Travelling wave tubes - magnetic), Amplitrons	Gyro-TWT
Resonant devices	Klystrons, IOT (Inductive output tubes)	Magnetrons	Gyroklystrons

Figure 1. Classification of high-power vacuum electronic devices.

In contrast to devices with a traveling wave, the interactions of electron beams with microwave fields in O-type resonant devices are local and realized in the gaps of resonators, the values of which are significantly less than the wavelength.

O-type resonant devices can be divided into two classes: devices with velocity modulation (klystron type) and devices with emission modulation (IOT type). The first class includes various modifications of klystrons, while devices of the second class have several names: klystrodes, inductive output tubes (IOTs), tristrodes, among others.

Figure 2 represents the simplified schism of standard klystrons and IOTs.

Klystrons and IOTs are the powerful microwave amplifiers (i.e., transformers of a small input microwave signal to a large output). The output signal receives energy from a DC source or a pulsed electrical power source (modulator). The electron beam is accelerated via the voltage of the electrical source, then bunched and decelerated by the microwave field. Slowing down, the electron beam gives away its kinetic energy to the microwave field. This process can be considered the transformation of DC or pulsed power into microwave power, and therefore klystrons and IOTs can be considered microwave power sources. Klystrons and IOTs are currently widely used for powering particle accelerators and in medical equipment, but their promotion in the field of mass industrial technologies and energy is hindered by their lack of efficiency.

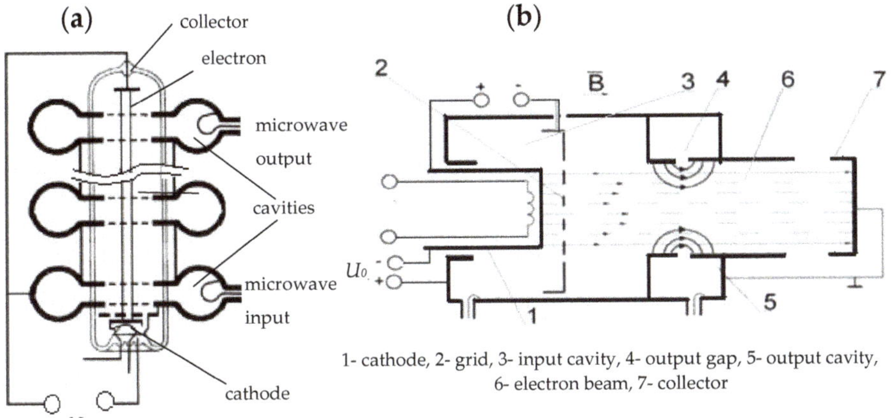

Figure 2. (**a**) Simplified schemes of klystrons and (**b**) IOTs.

Recently, several publications have proposed and investigated new variants of O-type resonant devices based on computer models: COM klystrons [3–9], CSM klystrons [9–12] and resotrodes [13–17]. It has been shown that the efficiency of such devices can reach 90%.

Tables 1 and 2 compare the main characteristics of the standard and new devices.

We use the term "standard length" instead of a specific numeric value, since the length depends on the operating frequency, current and voltage. If the values of these parameters are set, the "standard length" becomes a specific number.

A big length is a problem for frequencies under 1000 MHz (standard length is recognized as about 2 m or more in these cases). Thus, devices can be very large and expensive. COM klystrons are 60% longer than the standard length (see Table 1), meaning that they are even longer and even more expensive. As the operating frequency increases, the standard length decreases proportionally, reaching values of 15–20 cm at 10 GHz. For such high frequencies, increasing the length in COM klystrons is an advantage, since it allows the problem of heat removal to be solved.

The main disadvantage of CSM klystrons is their more complex design, which includes cavities of the second and third harmonics. This disadvantage becomes more significant with increases in work frequency.

Thus, in the lower part of the microwave range (up to about 5 GHz), it is advisable that CSM klystrons be used, while COM klystrons are optimal for higher frequencies.

In general, the main disadvantages of COM and CSM klystrons are their larger dimensions (for COM), complicated designs, and increased costs. The main advantage of these klystrons is their heightened efficiency.

Unlike IOT, resotrodes use the regeneration of microwave power in the input circuit. This inevitably narrows the gain band. Therefore, the main drawback of resotrodes is the inability to use them as broadband amplifiers (Table 2). When using a resotrode as the source of microwave power, this drawback does not manifest itself. The main advantage of resotrodes in comparison with IOTs is the ability to provide high gain at high output power with high efficiency.

The purpose of this paper is to review the latest achievements in the field of modeling and to research new O-type resonant microwave devices that could be used for the development of industrial microwave technologies and microwave energetics.

Table 1. Comparison of standard klystrons with COM and CSM ones.

Parameter	Standard	COM	CSM
Number of cavities	5–8	7–8	10–12
Harmonic cavities	No or one 2nd harmonic cavity	No	3–4 cavities of 2nd harmonic and 1–2 cavities of 3rd harmonic
Length	standard length	~1.6·(standard length)	~standard length
Bandwidth	0.5–5%	0.5–5%	0.5–5%
Work frequency	400–30,000 MHz	400–30,000 MHz	400–30,000 MHz
Gain	40–60 dB	40–60 dB	40–60 dB
Efficiency	<70%	up to 90%	up to 90%

Table 2. Comparison of IOTs with resotrodes.

Parameter	IOT	Resotrode with 0-Regeneration	Resotrode with 2π-Regeneration
Number of cavities	2	2	3–4
Multi gap cavities	No	2-gaps cavity	3-gaps cavity with far-spaced gaps
Harmonic cavities	No	No	2nd or 3rd harmonic cavity
Broadband amplifier	yes	no	no
Work frequency	400–1000 MHz	40–200 MHz	200–1000 MHz
Gain	<20 dB	35 dB and more	35 dB and more
Efficiency	<70%	up to 90%	up to 90%

2. Efficiency of Resonant Microwave Devices of O-Type

Both in industrial microwave technologies and in microwave power engineering, it is essential to achieve high efficiency values.

For example, in power systems, no more than 10–12% is lost during transportation and transmission (i.e., the total efficiency of power systems is about 90%). When implementing microwave power systems, efficiency must be at least 90%, which means that the efficiency of devices and devices included in such systems must be at least 90%.

In industrial microwave technologies, an increase in efficiency from 60% to 90% (1.5 times) will reduce the cost of electricity consumed by the same 1.5 times. This can turn some promising, but currently unprofitable, technologies into cost-effective ones. At the same time, the power of additional cooling equipment will be reduced by four times, which will result in additional savings.

For all these applications, high efficiency must be combined with high power output.

Efficiency is one of the main output parameters of the device and is defined as the ratio of useful to consumed power:

$$\eta = \frac{P_{useful}}{P_{supply}} \tag{1}$$

Depending on the interpretation of the concepts "useful power" and "spent power", different definitions of efficiency are possible. In the Equation (1), P_{useful} refers to the microwave power transmitted to the load, and P_{suply} refers to the power of the power source spent on beam acceleration. This definition corresponds to the term "load efficiency".

In this study we also use the term "electronic efficiency", in which P_{suply} is defined as above, while P_{useful} refers to the microwave energy transmitted by the electron beam to the microwave field in the output cavity. Electronic efficiency is always higher than load efficiency. In resonant microwave devices of the O-type, as a rule, these two values differ by less than 1%, but sometimes the difference can be several percent.

For a long time during the development of O-type resonant microwave devices, efficiency was given less priority compared with other output parameters such as output power and bandwidth. The 2005 SLAC report [18] showed a graph of the increase in the pulsed power of klystrons, revealing that the power grew exponentially over the period from 1940 to 2000, reaching a level of 1 GW by 2000. This process is still ongoing.

As for efficiency, there have been no similar dynamics in improving this parameter. The efficiency of a multi-cavity klystron of 50% was achieved at the turn of the 1950s–60s, and since then then no significant increase in the average efficiency of manufactured klystrons has been achieved.

During the second half of the 20th century, several single klystrons with a fairly high efficiency were created. Among them, we should note the 50 kW CW klystron developed by E. L. Lien in 1970 [19,20], which has an efficiency of 75%, and a klystron with an efficiency of more than 80% (electronic efficiency of about 90%) that was developed by S. V. Lebedinsky in 1979 [21].

Devices with emission modulation have had an efficiency of 50–60% since the beginning of their development in the 1980s. Among these devices, we should note the tristrod [22,23] (an IOT with an additional bunching cavity), created by V. A. Tsarev (RF, NGO "Contact") in 1998. This device had a measured efficiency of about 90%.

These devices were unique, and these results have not been repeated by anyone until now.

Most of the currently manufactured klystrons do not reach an efficiency level of 65%, and only three of the manufactured devices (Figure 3) have an efficiency of about 70%. These are the TH-1801 from Thales, E-3736 from Toshiba, and VKL-8301 from CPI.

Figure 3. Modern industrial klystrons with maximum efficiency.

3. Klystron Parameters and the GSP Equivalence Principle

Before we talk about the possibility of increasing the efficiency of klystrons, we will analyze the main parameters that define the device design and power mode.

The set of these parameters can be divided into two groups, which we will call "group A" and "group B".

Group A includes parameters that are selected a priori based on the technical requirements for the device and general physical considerations. These are the main frequency f_0, the accelerating voltage U_0, the total current I_0, the number of electron beams N_b, the number of cavities n, the working harmonics, the gap values l_g, the R/O factors of the cavities ρ, the proper Q-factors of the cavities Q_{0i}, the beam radii r_b, the canal radii r_c, among others. The index i means the number of the cavity or the stage. If all the parameters of group A are set, we will say that the basic design of the device is set.

Group B consists of all other parameters of the device, which include the natural frequencies f_i and the loaded Q-factors of the cavities Q_i, the length of the drifts l_{ci} and the input power P_{in}. Group B can be narrowed if the values of some of the listed parameters must be fixed in accordance with technical requirements. For example, in a narrow-band klystron, the Q-factors of the intermediate cavities can be set equal to their own Q-factors, and, accordingly, they will not be included in group B.

If complex electrodynamic systems and multi-band cavities are used in klystron stages, the number of parameters will increase.

A complete set of input parameters of groups A and B that are enough for the development of the device will be called the complete set of parameters.

We will now determine the number of optimization parameters for an n-stage klystron consisting only of the standard stages (with a single single-gap cavity that is excited at only one frequency). The parameters of group B are n detunings, n Q-factors (including the loaded Q–factors of the input and output cavities), $n-1$ drifts and the input power. As such, there are $3n$ parameters. Some of these parameters may not be included in the optimization or, accordingly, in group B. For example, if all the intermediate cavities are detuned far beyond the band, then their Q-factors can be set equal to their own

Q-factors and fixed. The number of parameters disabled from optimization will not exceed *n2*, since the loaded *Q*-factors of the input and output cavities should remain in optimization. Thus, the number of parameters in group B for the *n*-cavity klystron are in the range from $2n + 2$ to $3n$. That is, from 12 to 15 parameters are obtained for the five cavity klystron, and for the 16 to 21 parameters are obtained for the seven cavity klystron.

The number of parameters is reduced after reducing them to a dimensionless form.

When writing formulas, we will use two types of normalization, the first of which is the fundamental normalization [24], which is obtained by dividing physical quantities by the quantities of the same dimension made up of fundamental physical constants: the charge of the electron, the mass of the electron, the speed of light and the dielectric constant. Values for this normalization are shown in Table 3.

Table 3. Values for fundamental normalization.

Name	Designation/Formula	Value
Charge of the electron	e	1.60×10^{-19} Kl
Mass of the electron	m_e	9.11×10^{-31} kg
Speed of light	c	3.0×10^{8} m/c
Dielectric constant	ε_0	8.85×10^{-12} F/m
Fundamental voltage	$U_e = \frac{m_e \cdot c^2}{e}$	511 kV
Fundamental current	$I_e = \frac{m_e \cdot c^3 \cdot \varepsilon_0}{e}$	1.36 kA
Fundamental perveance	$K_e = \varepsilon_0 \cdot \sqrt{\frac{e}{m_e}}$	$3.7 \ \frac{\mu A}{V^{\frac{3}{2}}}$
Fundamental impedance	$Z_e = \frac{1}{c \cdot \varepsilon_0}$	377 Ohm

We will denote the fundamental-normalized (f. n.) values by underlining them from below. For example, $\underline{U} = \frac{U}{U_e}$ (f. n. voltage), $\underline{I} = \frac{I}{I_e}$ (f. n. current), $\underline{v} = \frac{v}{c}$ (f. n. speed), and so on.

The second, "functional" normalization is obtained by dividing physical quantities by quantities of the same dimension, which are made up of the main functional parameters of the device: output power, accelerating voltage and main frequency (Table 4).

Table 4. Values for functional normalization.

Name	Designation/Formula	Dimention
Beam power	P_0	W
Accelerating voltage	U_0	V
Main frequency	f_0 or $\omega_0 = 2\pi f_0$	c^{-1}
Beam speed	$v_0 = c \cdot \sqrt{1 - \dfrac{1}{(\underline{U_0}+1)^2}}$	m/c
Duration of one radian of the microwave field	$t_{norm} = \frac{1}{\omega_0}$	c
Distance that the beam passes during one radian of the microwave field	$l_{norm} = \frac{v_{norm}}{\omega_0}$	m
Beam impedance	$R_{norm} = \frac{U_0^2}{P_0}$	Ohm

Functionally normalized values will be denoted by two asterisks at the bottom for example, $f_{**} = \frac{f}{f_0}$, $z_{**} = \frac{z}{l_{norm}}$, $Z_{**} = \frac{Z}{R_{norm}}$, $t_{**} = \omega_0 \cdot t$, etc.

In the process of functional normalization, the parameters of group A turn into dimensionless complexes [25] (similarity criteria), the number of which is three less than the number of initial parameters.

The first of these dimensionless complexes is the relativistic perveance assigned to a single beam of a multi-beam (MB) klystron:

$$\underline{K}_{rel} = \frac{\underline{P}_0}{\underline{U}_0^{5/2}\left(1 + \frac{\underline{U}_0}{2}\right)^{3/2} N_b} \tag{2}$$

In addition, dimensionless complexes are the normalized length of the gap (the angle of passage through the gap),

$$\Theta = \frac{\omega_0 \cdot l_g}{c\sqrt{1 - \frac{1}{(1+\underline{U}_0)^2}}} \tag{3}$$

the gap form factor,

$$\mu = \frac{l_g}{2r_c} \tag{4}$$

the coefficient of filling the channel with a beam,

$$\alpha = \frac{r_b}{r_c} \tag{5}$$

and the excitation parameter [24], which has the meaning of the "natural" bandwidth of the device,

$$v = \frac{\rho \cdot P_0}{U_0^2} \tag{6}$$

The parameters of group "B" are converted to dimensionless lengths of drifts,

$$l_{ci}^{**} = \frac{l_{ci}}{l_{norm}} \tag{7}$$

to the relative detuning of cavities,

$$\delta_{f\,i} = \frac{f_i - f_0}{f_0} \tag{8}$$

and to the normalized input power,

$$P_{in}^{**} = \frac{P_{in}}{P_0} \tag{9}$$

The number of parameters in group B does not change during the normalization process.

The general scaling principle (GSP; transcription as "global scaling principle" was also used in publications [25]) is the statement that two klystrons with the same values of dimensionless Parameters (2)–(9) are physically equivalent. All processes of the grouping and selection of energy occur in these two devices in the same way. These two devices will have the same dimensionless output characteristics, including the same efficiency. At the same time, such klystrons can differ greatly in the level of output power, the number of beams, the accelerating voltage, the operating frequency and other basic dimensional characteristics.

The GSP principle splits the set of all existing klystrons and all klystrons that may be developed in the future into equivalence classes. Each equivalence class is defined by five parameters (Parameters (2)–(6)).

Note that of Parameters (2)–(6), only Parameters (2) and (6) are free. Although optimization for Parameters (3)–(5) is not usually performed, their values are chosen close to the known conditionally optimal values $\mu \simeq 1$, $\alpha \simeq 0.5$ and $\Theta \simeq 1.2 \div 1.4$, and do not deviate significantly from these value, since such a deviation leads to a deliberately suboptimal construction. Thus, Parameters (3)–(5) can be considered fixed.

Consider an equivalence class given by two parameters $\underline{K}_{rel}$ and v, assuming that each of the parameters μ, α and Θ is in some reasonable range near the optimal value. This class of klystron equivalence will be called the $K - v$ class.

Each $K - v$ class is represented by a dot on the $K - v$ diagram.

In Figure 4, klystrons that are currently produced by various manufacturers are divided into three groups according to level of efficiency, and are marked on two $K - v$ diagrams. In the right-hand diagram, the abscissas axis is replaced by the normal perveance $K = \frac{P_0}{U_0^{5/2} \cdot N_b}$ instead of the relativistic perveance (Parameter (2)).

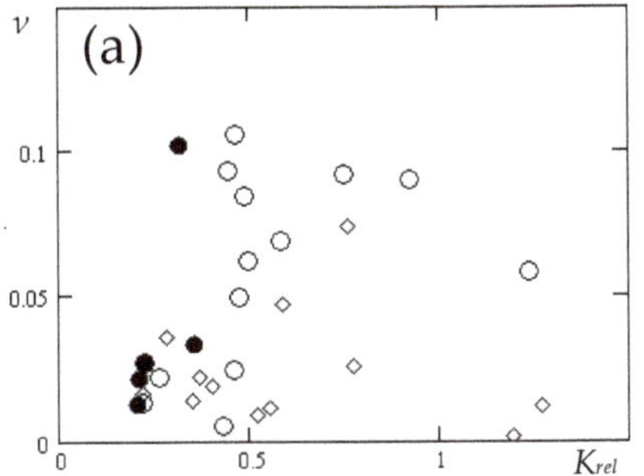
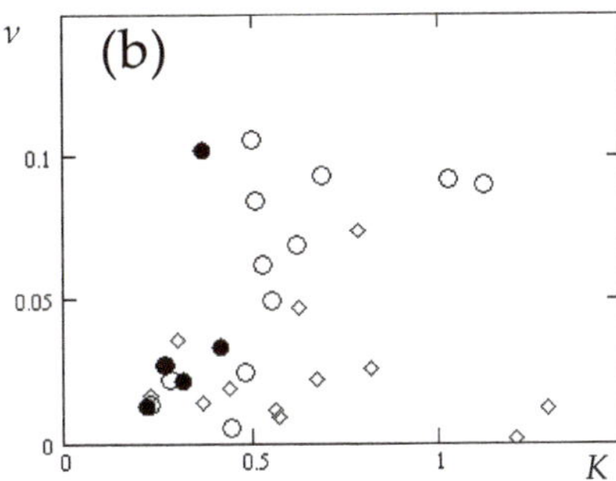

Figure 4. Produced klystrons: (**a**) in the coordinates (K_{rel}-v,), (**b**) in the coordinates (K-v), divided into three groups by efficiency level: ◊ $\eta < 40\%$, ○ $\eta = 40$–60%, • $\eta > 60\%$.

The sample includes devices of different frequency ranges (from 400 MHz to 30 GHz), different levels of output power (from 3 kW to 80 MW) and different numbers of beams (from 1 to 42).

All klystrons are divided into three groups: high efficiency (60–70%), medium efficiency (40–60%) and low efficiency (less than 40%).

Klystrons with low and medium efficiency are distributed evenly across the diagram, while klystrons with high efficiency are concentrated in the region of relativistic perveance 0.2–0.3. This range is optimal in terms of efficiency.

The GSP principle makes it easier to develop new high-performance klystrons if their analogues already belong to the same class. To do this, it is enough to recalculate all the parameters, leaving the dimensionless values (Parameters (2)–(9)) constant. In this case, GSP modeling avoids a long, complex and expensive process of computer optimization of the device. The method of such recalculation is described in [24]. The limitations of using GSP modeling are also discussed there.

4. Methods of Klystron Modeling and Optimization

If there are no suitable GSP analogues (and no analogues for efficiency above 70%), it is necessary to optimize the parameters of group B using a mathematical model implemented in the form of a computer code.

Currently, developers use several computer codes that simulate the operations of klystrons. The most appropriate codes are based on particles-in-cell (PIC) models [26], which were initially developed for modeling processes in plasma, then adapted for electrovacuum devices.

Klystron developers actively use two universal PIC packages: the MAGIC code [27], developed by the American firm Orbital ATK, and the CST studio suit code [28], produced by the German company CST.

However, it should be noted that universal PIC codes do not consider the specifics of microwave devices, so the calculation process for them are extremely expensive. On an office PC, the calculation of one specific device variant with fixed frequency and input power values is a day or more. This process can only be performed on high-performance supercomputers. At the same time, the computing process itself not only requires a lot of resources, but also a lot of data preparation, and any changes to the data are associated with significant difficulties. For example, in order to change a cavity's detuning, it is necessary to repeatedly change its size, select a new frequency value, rebuild the computational grid, and only then perform the calculation process with the new frequency. This complexity of changing data makes it impossible to optimize the parameters of devices using universal PIC codes, even on supercomputers.

Note that, despite the huge computational complexity of universal PIC codes, their use does not guarantee a reliable result. Although the methodological error in such codes can be practically reduced to zero by shredding the computational grid and increasing the number of particles, this cannot be said about the computational error, which, on the contrary, often begins to increase under these conditions. The computational error is shown as "numerical noise", the presence of which is usually estimated visually. At the same time, there are neither sufficiently reliable mechanisms for recognizing computational noise, nor universal ways to suppress it. Usually such noise is suppressed by heuristic manipulations such as changes in boundary conditions (deterioration of wall conductivity), adding virtual absorbers, and so on. It is obvious that such actions change the model of the system, and, consequently, lead to a deterioration in the adequacy of modeling the original system.

However, universal non-stationary 2D/3D PIC codes are currently the main recognized criterion for the accuracy of results obtained by other computational methods.

In addition to universal PIC codes, quite a lot of special computing codes have been developed that are adapted to microwave devices of one type or another.

Let's look at the features of such codes on the example of the TESLA (USA) code, created by a group of specialists from several scientific organizations in the United States. As follows from the description in [29], for example, the TESLA code uses the separation of the microwave fields of cavities and fields (coulomb and microwave) in the region of beam motion. The crosslinking of these fields in the gap area occurs on the continuation of the inner surface of the drift pipe. This separation corresponds to the specifics of the klystron and allows us to separate the problems of electrodynamics (modeling of microwave fields in the cavity) and electronics (beam bunching, interaction of the beam with microwave fields). This, in turn, leads to the possibility of using different scales of computational grids in the cavity region and in the interaction region, while accordingly leading to the possibility of a significant reduction in the number of particles and a consequent reduction in calculation time. The calculation time for one option under the TESLA code is 10–20 min on a personal computer. This is too much to optimize on a PC, but on multiprocessor supercomputers, optimization can be performed using this code. It should be noted, however, that due to the simplifying assumptions used in the construction of the model, the question of the methodological error of the results remains open. For example, in the results given in [29], the discrepancy between the results of TESLA and MAGIC seems significant.

One of the latest developments of adapted codes is the KlyC/1.5 code (CERN, Geneva, Switzerland) [30], developed by I. Syrachev and J. Cai. This code allows you to model the operation of a klystron in approximations 1D and 1.5 D (one-dimensional motion with a bundle). According to the authors, the difference between the results of KlyC-1.5 D and CST is no more than 1%. The calculation time for a single variant using the KlyC-1.5 D code with seven layers depends very much on the device being modeled, but is usually at least 30–40 min on a PC. This is too much for full multiparametric optimization, which requires tens or even hundreds of thousands of calculations [31].

In order to combine high performance with high adequacy, a new class of klystron models was developed called "discrete-analytical models" [32].

Discrete-analytical models appeared as a reasonable compromise between analytical and numerical approaches in the construction of a mathematical model of the device. This compromise is achieved by splitting the structure of the device or processes occurring in it into conditionally autonomous parts, building models for these parts in the form of analytical dependencies and coupling these dependencies.

Thus, the core of the mathematical model is a set of analytical dependencies, while the rest of the model structure, including the method of coupling these dependencies, is formed on the basis of numerical algorithms.

The numerical algorithms can also be used in the process of forming the core to determine the numerical parameters that are included in the core of model. In this case, the gain of time is achieved because the process of the numerical determination of parameters (the process of "learning") is separated from the model operation process.

In addition to the speed and adequacy of the model, the optimization method itself is an important factor for successful optimization of klystron parameters.

The main feature of any klystron goal function constructed on the basis of efficiency is the existence of a multidimensional "quasi-optimal" variety of an unknown a priori structure, characterized by a weak and non-monotonic change in the objective function with a very large (10^{12} or more) number of local extremes [31].

As one of the simplest examples of movement through such a variety, we can cite the process of tuning the klystron input cavity while increasing the input power so that the amplitude of the microwave voltage in the input gap does not change. In this case, the efficiency of the klystron due to changes in the phase of the input voltage will change, but very slightly and non-monotonously. The trajectory of changes in the detuning and the input power in the parameter space forms a curve, a one-dimensional quasi-optimal variety. If you add a loaded Q-factor to these two parameters and change all three parameters at once so that the microwave amplitude in the input gap does not change, you will get a two-dimensional quasi-optimal variety. As the number of parameters increases, there are more and more opportunities to change them without significantly changing the efficiency, but it is very difficult to find all such variations a priori (i.e., the structure and dimension of a quasi-optimal manifold are unknown in advance).

This structure of the target function makes it useless to use any optimization method aimed at finding a single local extremum, which explains the fact that most researchers using optimization have always achieved local results without getting a procedure that would allow them to approach the global extremum with a guarantee (even for a very long time).

To overcome these difficulties, a method of macro-steps was developed [31] that includes probing a relatively large working area in the parameter space, finding the local extremes closest to the best points obtained as a result of probing, selecting one of these local extremes and moving to a new working area. The macro-steps method allows for multi-parameter optimization and provides hope for achieving a global maximum of the goal function. To increase the guarantee of achieving the global extremum, the macro-steps method must be supplemented with an appropriate method of structural optimization, as will be discussed in more detail later.

The KlypWin computer code was developed based on the discrete-analytical klystron model and the macro-steps method [33]. It takes less than 1 s to calculate one version of the device using KlypWin.

A comparison of the calculation results for the KlypWin and Magic codes conducted by David Constable in 2013–2016 for 10 klystrons synthesized as part of the work of the international High Efficiency Klystron Activity (HEIKA) group [7] is shown in Figures 5–7.

Note that each point by the MAGIC code in Figures 6 and 7 requires more than a day of computer time, while a similar point by the KlypWin requires only 0.3 s. As you can see from the figures, the difference in results for all 10 devices does not exceed 5%. The results show that KlypWin can be considered an adequate code for modeling klystrons. At the same time, its speed makes it possible to perform global multi-parameter optimization.

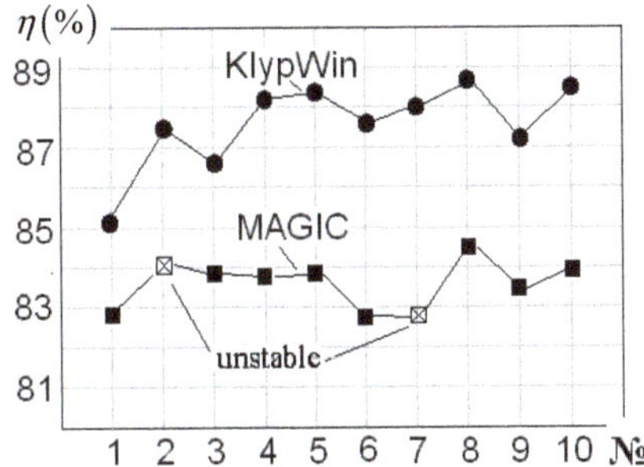

Figure 5. Results of COM-klystron modeling by codes KlypWin and MAGIC.

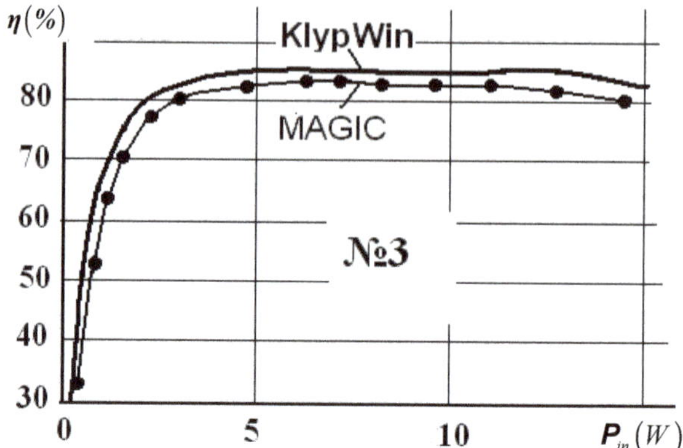

Figure 6. Comparison of the transfer curves of the klystron №3 Figure 5.

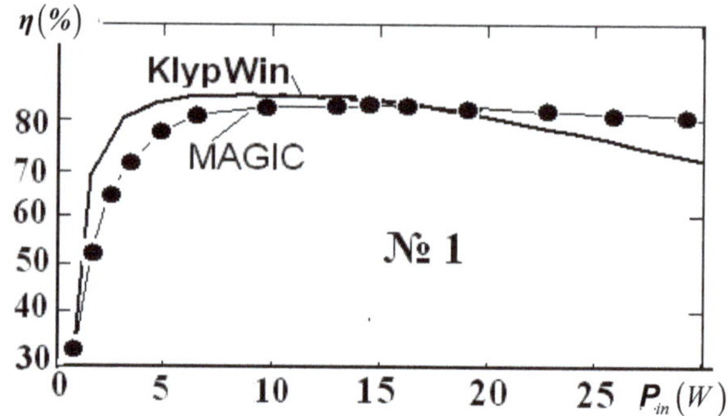

Figure 7. Comparison of the transfer curves of the klystron №1 Figure 5.

5. COM and CSM Klystrons are Promising Power Sources in the Centimeter Range

Based on the KlypWin code, a large set of studies has been conducted over the past decade on the possibility of achieving the maximum efficiency values in klystrons [4–12,24]. In the course of these studies, it was found that maximum efficiency values were achieved with two bunching modes: the core oscillation method (COM) [4–9] and the core stabilization method (CSM) [9–12].

As was found in the course of research, in order to reliably achieve the global extremum of the target function, the macro-steps method must be supplemented with a suitable structural optimization method. Structural optimization refers to the process of gradually complicating the structure of a device, starting with the simplest structure. After each stage of structural complexity, global optimization is performed using the macro-steps method. The process that includes structural optimization of this type and global optimization by the method of macro-steps is called full optimization [34].

Klystrons with extreme efficiency values are obtained as a result of two possible processes of full optimization.

The first full optimization process is called stage-by-stage optimization [24] and was first described in [4]. The process begins with the formation of a two-cavity klystron based on a given prototype (a set of parameters of group A). This two-cavity klystron is optimized, after which another stage is added to the optimal two-cavity klystron, and it turns into a three-cavity klystron. For the resulting three-cavity klystron, the optimal input power is found, then global optimization is performed for all parameters in a wide range of their changes. The resulting optimal three-cavity klystron is converted into a four-cavity klystron by adding another stage, the global optimization process is started again, and so on. Starting with a three-cavity klystron, adding a new stage is performed by duplicating the second stage. The process stops when, after adding a new stage and performing optimization, the efficiency stops increasing (it comes to saturation).

The stage-by-stage optimization process is shown for a C-band klystron with an output power of 8.5 MW, as shown in Figure 8.

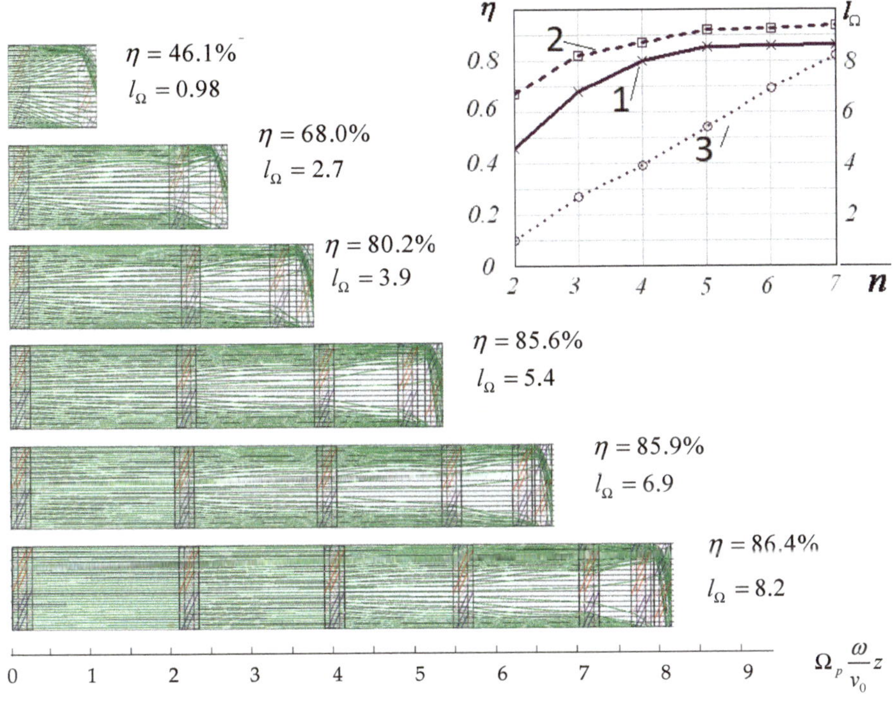

Figure 8. Stage-by-stage optimization of the klystron.

In addition to efficiency, the figure shows two other key parameters. The first one is the saturation of the bunch [5], which reflects the proportion of electrons caught in the bunch at the entrance to the output gap. The second parameter is lumped bunching length (LBL) [4], which is describe in the Table 5.

Table 5. Bunching dimensionless values.

Name	Designation/Formula	Description
Total normalized length of the bunching [4]	$l_{**} = \frac{\omega \cdot l}{v_0}$	Normalized total length of all drifts and gaps
Parameter of the convection wave [24]	Ω_p	Ratio of the electron beam convective wave frequency and the microwave frequency
Lumped bunching length (LBL)	$l_\Omega = \Omega_p \cdot l_{**}$	Number of radians of the convective wave that fit over the bunching length

The left part shows the change in the bunching process (phase trajectories) as the number of stages increases. The upper-right corner shows the change in efficiency (1), bunch saturation (2) and LBL (3) depending on the number of stages.

In the stage-by-stage optimization process, only standard stages are added, which include a simple single-gap cavity and a single drift. Accordingly, all klystrons obtained during this process have a standard configuration (i.e., they do not contain multi-gap cavities and cavities of higher harmonics).

All klystrons resulting from stage-by-stage optimization are characterized by the same non-monotonic character of bunching.

For the core particles of a bunch, this mode is oscillatory: these particles approach the center of the bunch, then move away from it. For particles of anti-bunch [5], the oscillatory nature of the motion is changed to a monotone. It is in this bunching mode that the klystron of the standard configuration reaches the maximum saturation of the bunch at a given frequency.

In turn, the saturation of the bunch is one of the main and most difficult factors to achieve in order to obtain maximum efficiency.

The key parameter that ensures the maximum saturation of the bunch is the LBL, which must be increased 1.6–1.8 times compared to known prototypes to increase the bunch saturation. The length of the device increases accordingly.

The fact that the stage-by-stage optimization of various prototypes that differ significantly from each other leads to the same nature of bunching and to similar values of the LBL and efficiency, allows us to speak about the fundamental nature of both the bunching mode itself and the LBL as the main parameters that characterize the bunching mode. This bunching mode was named core oscillation method (COM) [5].

All COM klystrons synthesized during full stage-by-stage optimization are characterized by efficiency from 85% to 90%, and the result practically does not depend on the prototype used, in particular, on its power. Thus, powerful COM klystrons with high efficiency can be developed for use in industrial microwave technologies and in microwave power engineering.

The second method of full optimization is called the embedding procedure. In this case, the optimal three-cavity COM klystron is selected as the initial option. Then, the total length is fixed, and new cavities are inserted between the existing ones. As shown by the research, only the first harmonic does not affect the efficiency in this case during the insertion of cavities, and efficiency remains at the level of 70%. To increase efficiency, it is necessary to alternate the cavities of the first harmonic with the cavities of the second harmonic, and to achieve an efficiency of about 90% it is necessary to include at least one cavity of the third harmonic. An explanation of this fact is given in [9,24].

Note that second harmonic cavities are used in the development of many klystrons. As a rule, developers have been limited to one such cavity, but in rare cases their number has increased to two or three. In this regard, we can mention the works of I. A. Guzilov, which were devoted to the development of BAK-bunching [35].

We could not find information about the use of third harmonic cavities in commercially produced klystrons. High-efficiency klystron models based on third harmonic cavities were constructed in [36,37] by Chiara Marelli (ESS) and Igor Syratcev (CERN).

The embedding procedure results differ significantly from COM bunching. There are no fluctuations in the core of the bunch; instead, the core stabilizes and attains a monospeed. Peripheral electrons due to voltage harmonics acquire a high relative velocity, which is quickly extinguished when these electrons approach the core of the bunch. After the speed is extinguished, these electrons replenish the stabilized core of the bunch. This process is called the core stabilization method (CSM). It provides a high saturation of the bunch, but at a much shorter length than COM. At the same time, the total number of cavities increases in comparison with COM klystrons. The optimal COM klystron has seven to eight cavities, while in the optimal CSM klystron the number of cavities increases to 10 or 11. The problem of arranging cavities in this case turns out to be solved, because the cavities of the second and third harmonics are relatively small.

The results of the synthesis of CSM klystron at a frequency of 800 MHz and a pulse power of 20 MW are shown in Figure 9. The numbers of working harmonics of the cavities are signed above the diagram of phase trajectories.

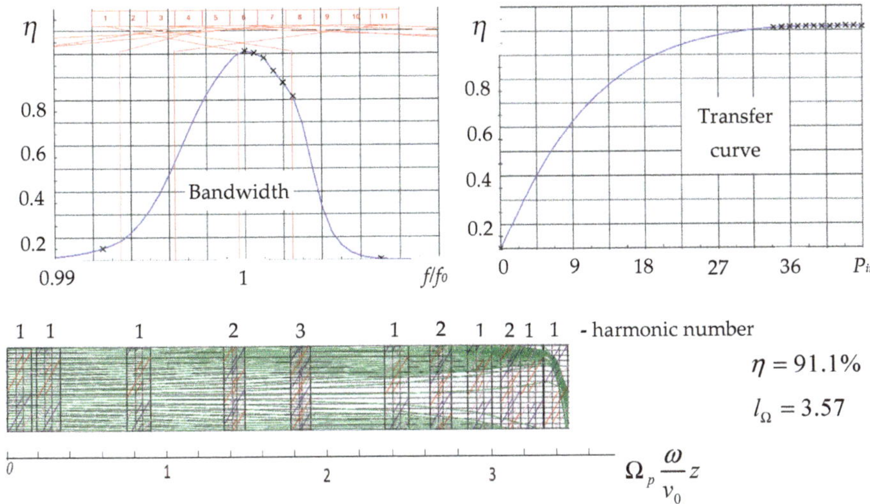

Figure 9. Result of syntheses of CSM klystron by embedding procedure.

Another high-efficiency bunching mode is possible, which combines the characteristics of COM and CSM modes and makes it possible to realize an efficiency of about 90%. This mode is called COM2. It is implemented only when an additional second harmonic signal is applied to the input. The klystron buncher in this case contains either two-frequency cavities operating on both the first and second harmonics, or alternating cavities of the first and second harmonics. The length of the LBL is less than that of COM klystrons, but longer than that of CSM klystrons. COM2 klystrons are described in more detail in [24].

Some synthesized COM and CSM klystrons are shown in Figure 9, where the numbering of the prototypes from [24] are used. As can be seen from the *K-v* diagram, high efficiency values are obtained for klystrons with a beam perveance from 0.2 to 0.4. The synthesized klystrons do not correspond to the empirical dependence [38] of the maximum efficiency on the perveance, shown by the inclined straight line in the right part of Figure 10.

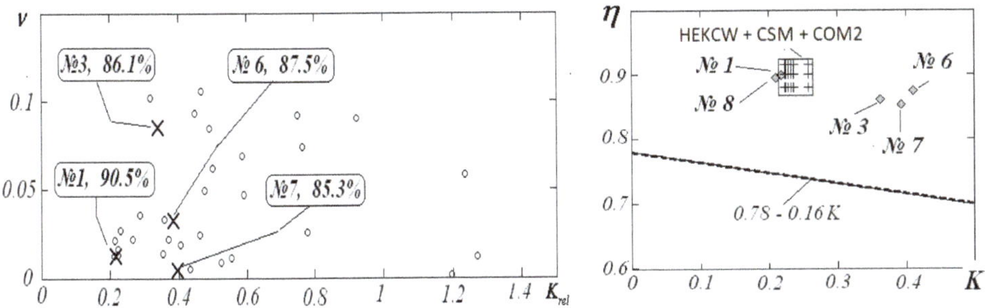

Figure 10. COM and CSM klystrons in *K-v* (right) and *K-η* diagrams.

COM and CSM klystrons with an efficiency of about 90% can be considered as promising sources of microwave power for industrial applications and for microwave energetics.

6. Resotrode with 0-Regeneration

In devices with emission modulation (IOTs), it is possible to realize sufficiently high efficiency values by reducing the length of the clot (cutoff angle). This reduces both the output power and the gain.

A new class of devices called resotrodes can achieve high efficiency with high output power and a high gain value [13]. These devices are sources of RF/microwave power for the upper RF range (20–200 MHz), for the lower microwave range (300–1000 MHz) and for the boundaries of RF/microwave bands (200–300 MHz).

In devices with emission modulation, the electronic gun and input cavity are combined within the same design.

There are three possible variants of this combination, as shown in Figure 11.

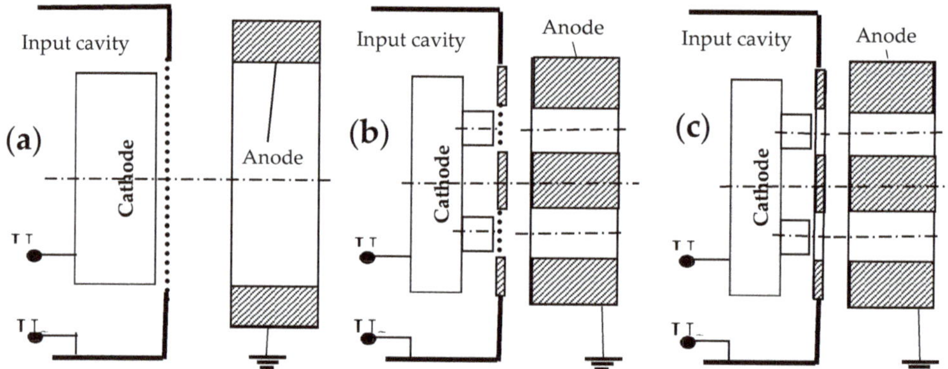

Figure 11. Types of near-cathode parts of O-type resonant microwave devices with emission modulation: (**a**) single-beam grid design, (**b**) multi-beam (MB) grid design, (**c**) multi-beam grid-free design.

Classical IOTs use variant (a), multi-beam IOTs use variant (b).

One of the disadvantages of IOTs is the low gain, which does not usually exceed 20 dB. Another disadvantage is the need to place the control grid very close to the cathode to obtain even this gain, which significantly complicates the technology and reduces the reliability of the device. This problem is particularly significant in multi-beam devices (MB IOTs), since it is necessary to achieve the same cathode-grid distance for all cathodes at a distance of 0.2 mm or less. Moreover, this same small distance must be implemented in the "hot mode", considering the thermal expansion.

Most commercially produced IOT devices have a single-beam grid design (Figure 11a). The first multi-beam grid structures appeared in the late 1990s (Figure 11b).

The constructions in Figure 11c, as far as is known, were not used for the development of industrial designs of IOTs. This is because the ratio of the absolute values of the grid voltage to anode in this design cannot be less than an amount equal to about 1/4 to obtain a sufficiently high efficiency. Thus, the ratio of the amplitude of the input signal to the amplitude of the output signal must also be about 1/4, which leads to a very low gain (5–6 dB).

For relatively low frequencies (20–200 MHz), a new device with emission modulation has been proposed, corresponding to the scheme Figure 11c and using energy regeneration in the input circuit. This device is called a resotrode (names "resotrod" and "rezotrod" [13] have also been used in publications).

An essential feature of the resotrode that allows the disadvantages of IOTs to be overcome is the ability to implement a high gain at a relatively high value of the voltage amplitude in the input gap (in the cathode-grid gap).

This possibility appears because the control electrode–anode gap is the second gap of the two-gap input cavity, which operates on antiphase-type vibrations. However, the degree of compensation (regeneration) in such a design depends significantly on the angle of passage of the cathode–anode: the higher the angle of passage, the smaller the regeneration. At frequencies above 200 MHz, the compensation becomes incomplete and quickly drops as the frequency increases.

The general scheme of a resotrode with 0-regeneration (assuming that the angle of passage through a pair of input gaps, including the regenerating gap, is close to zero) is shown in Figure 12.

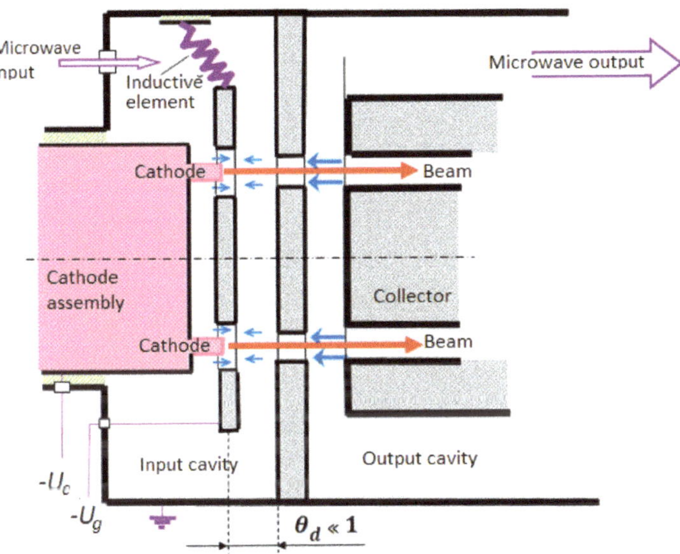

Figure 12. Scheme of a resotrode with 0-regeneration.

The anode-collector gap is the gap of the output cavity. The control electrode is current-less, because the instantaneous dynamic value of the potential on it is always negative (c- or d-lamp operation mode).

Thus, the current flows only in the cathode-collector circuit and, therefore, only this circuit should be designed for high power. It should be noted that when working in amplification mode at time intervals when the signal is absent, the cathode is locked, and the current does not flow to the collector, which ensures the efficiency of the device and reduces the thermal load on the collector.

In a general sense we will use "a resotrode" to refer to any devices with emission modulation that has full or excessive electronic load compensation in the input circuit.

Thus, name "resotrode" can apply not only to the construction in Figure 11c, but also to the type in Figure 11b with the appropriate compensation.

7. Evaluation of the Efficiency and Gain of a Resotrode

Unlike the klystron output parameters, a resotrode can be quantified with relatively simple analytical relations. Further refinement of these estimates is possible within the framework of computer 2D/3D modeling only.

We assume that the angles of passage of the electrons through all the gaps are small $\Theta_d \ll 1$, and that the electronic load operates at full compensation, in which the amplitudes of the microwave voltage in the first and second gaps of the input cavities are the same. In this case, the electron velocities will correspond to the static potential of the anode after entering the anode hole (i.e., the bunch will be strictly monospeed).

As shown by klystron studies, a narrow monospeed bunch can be slowed down to zero speed, and, if the gap is narrow, full braking is provided by zero detuning and a microwave voltage amplitude equal to the accelerating voltage.

The instantaneous voltage at the control electrode must be negative, which implies a condition of $U_1 \leq U_g$, where U_1 is the amplitude of the microwave voltage in the input gap, and U_g is the absolute value of the constant voltage at the control electrode (offset voltage) relative to the cathode. On the other hand, a decrease in the microwave amplitude at a given offset leads to a sharp decrease in the current and, consequently, the power of the device. It follows that the next equality is optimal:

$$U_1 = U_g \tag{10}$$

In the output gap at small angles of flight, the amplitude of the microwave voltage U_2 at zero detuning cannot be made greater than the accelerated voltage U_0, because this will lead to electron reflections. On the other hand, if $U_2 < U_0$, then the bunch will not be completely stopped (i.e., under this condition, the efficiency will not be maximum). This leads to the conclusion that for zero detuning, equality $U_2 = U_0$ is optimal. Next, we assume that the output cavity detuning is zero.

The input cell of resotrode (Figure 12) is a three-electrode electron gun, which is characterized by the cutoff potential U_{sh} (i.e., the value of the control electrode negative potential $-U_{sh}$ at which the anode current ceases to go). The relative value of the cutoff potential is determined by the shape of the electrodes and the relative distances between them. In the grid-free constructions (Figure 11c), the value $U_{sh} = \frac{U_{sh}}{U_0}$ is bound from below by the values 0.12–0.15. Next, we assume that the value U_{sh} is set.

Under Equation (10), the instantaneous voltage at the gap between the cathode and the control electrode has the form $u_{in}\left(t\right) = U_g \cdot \left(1 - \cos\left(t\right)\right)$, and the width of the bunch is determined by the ratio

$$k_{sh} = \frac{U_g}{U_{sh}} \tag{11}$$

which will then be called the cutoff coefficient.

From the bottom of Figure 13, it follows that the normalized width τ_b of the bunch is related to the cutoff coefficient k_{sh} by the ratio

$$\tau_b = 2\arccos\left(1 - \frac{1}{k_{sh}}\right) \tag{12}$$

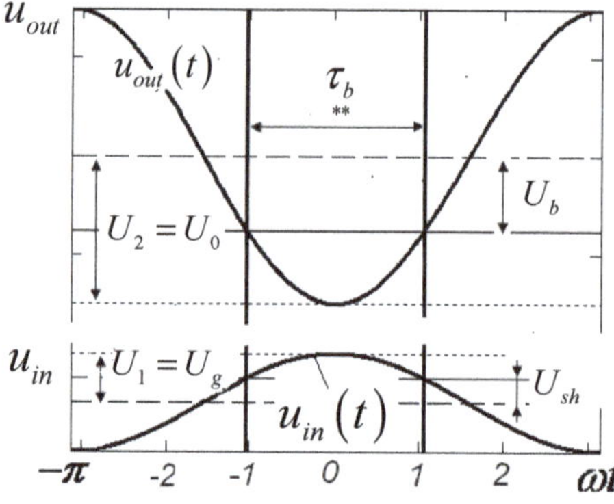

Figure 13. Schemes of formation of the bunch in the entry gap and its inhibition in the output gap of the resotrode.

Let us consider the deceleration of a single-speed bunch of width (Equation (12)) by the field of an infinitely thin gap of the output cavity with zero detuning. When such a bunch passes through the gap, the instantaneous value of the microwave voltage changes from the amplitude value U_0 corresponding to the center of the bunch, to some value U_b corresponding to the edges of the bunch (Figure 13, upper). The electronic efficiency can be calculated as the ratio of the interaction power to the beam power,

$$\eta_e = \frac{1}{U_0 I_0} \cdot \frac{1}{\tau_b} \cdot \int_{-\tau_b/2}^{\tau_b/2} U_0 \cdot \cos(s) \cdot I(s) ds \tag{13}$$

If we bring the current outside the integral, using the averaging theorem and assuming approximately that this average current is equal to I_0 (calculations for various forms of bunches show that the error of this approximation does not exceed 2%), then from Equation (13) we get an expression for the electronic efficiency of the resotrode $\eta_e = \dfrac{\sin\left(\tau_b/2\right)}{\tau_b/2}$.

Given Equation (12), we finally get

$$\eta_e = \frac{\sqrt{2k_{sh}-1}}{k_{sh}\cdot\arccos\left(1-\frac{1}{k_{sh}}\right)} \tag{14}$$

The dependencies defined by Equations (12) and (14) are shown in Figure 14. From these dependencies, it follows that to obtain the maximum efficiency values (90% and higher), the cutoff coefficient values $k_{sh} > 4$ are required, which correspond to the value of the normalized width of the bunch $\tau_b < \frac{\pi}{2}$.

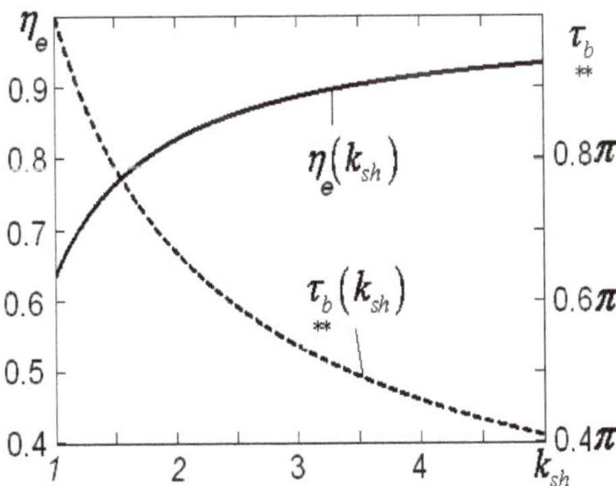

Figure 14. Dependences of the normalized bunch width and resotrode efficiency on the cutoff coefficient.

Now let us estimate the gain of the resotrode.

In the mode of full electronic load compensation, the input power is equal to the power of the input cavity's own losses (i.e., $P_{in} = \dfrac{U_1^2}{2\cdot\rho_1\cdot Q_1}$, where ρ_1 is the R/Q factor, and Q_1 is the input cavity's own O-factor).

Representing the output power as $P_{out} = \eta_e \cdot U_0 \cdot I_0$, we find that

$$\frac{P_{out}}{P_{in}} = 2\eta_e \cdot v \cdot Q_1 \cdot \left(\frac{U_0}{U_g}\right)^2 \tag{15}$$

where v is the output gap excitation parameter defined by Equation (6). Substituting Equations (11) and (14) into Equation (15) and proceeding to the logarithmic recording of the gain, we finally get

$$Ku = 10 \cdot \log\left(\frac{2 \cdot v \cdot Q_1}{U_{sh}^2} \frac{\sqrt{2k_{sh} - 1}}{k_{sh}^3 \cdot \arccos\left(1 - \frac{1}{k_{sh}}\right)} \right) \tag{16}$$

Since the resotrode is essentially a high-perveance device, the characteristic values of the excitation parameter v are 0.1 and higher. The own Q-factor can be obtained from 2000 to 5000.

Let's compare the resulting formula with the gain formula for the classical IOT design, which can be written as

$$Ku = 10 \cdot \log\left(\frac{1}{U_{sh} \cdot k_{sh}} \frac{1}{1 + \frac{k_{sh} U_{sh}}{v \cdot Q_1}} \right) \tag{17}$$

or, for large values of its own Q-factor, in the form

$$Ku = 10 \cdot \log\left(\frac{U_0}{U_g} \right) \tag{18}$$

Graphs of the gain dependence for resotrode with parameters $U_{sh} = 0.15$, $Q_1 = 2000$, $v = 0.1$ and for IOTs with the same values Q_1 and v are shown in Figure 15. As you can see from the graphs, the resotrode has a gain of about 40 dB at $k_{sh} = 1$ (a bunch of duration π), the gain decreases as k_{sh} increases, but even at $k_{sh} = 5$ (a bunch of duration 0.4 π) the gain remains more than 25 dB. In a grid-less IOT (lower graph), the gain is equal to 9 dB if $k_{sh} = 1$, but falls to almost zero if $k_{sh} = 5$. Only when using a very fine-grained grid ($U_{sh} = 0.001$) can the gain exceed 20 dB.

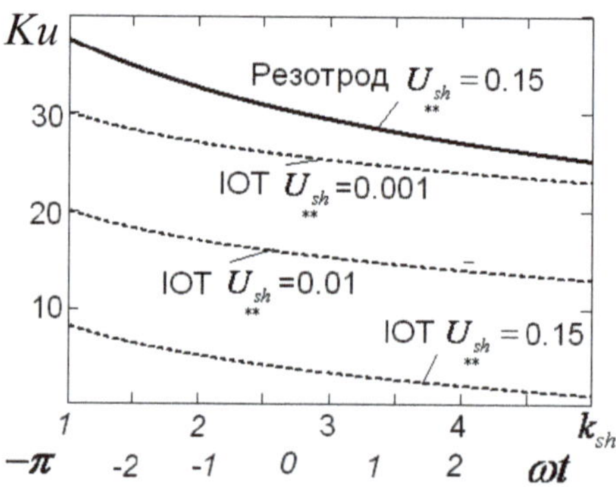

Figure 15. Dependences of the gain on the cutoff factor for resotrodes and IOTs.

The dependence of electronic efficiency of a resotrode on gain is shown in Figure 16, from which it follows that at a gain of 30 dB in a resotrode you can achieve an electronic efficiency of about 90%.

It should be noted that when the cutoff increases due to a decrease in the pulse duration, the output power of the device also decreases. In Figure 16 the output power is normalized to the output power of $k_{sh} = 1$ (i.e., when the offset voltage of the gain is equal to the cutoff voltage).

From the graphs, we can conclude that the optimal choice is a gain of 25–30 dB, which corresponds to a cutoff coefficient from 3 to 5.

In general, the obtained analytical estimates suggest that the combination of high efficiency with high gain is possible in resonant microwave devices with emission modulation.

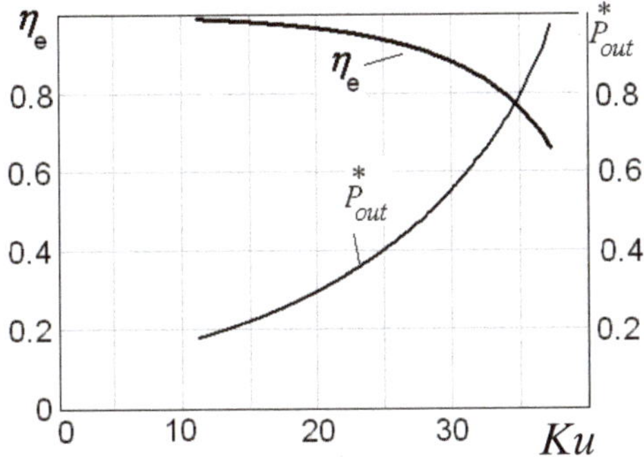

Figure 16. The dependence of the maximum electronic efficiency of resotrodes and normalized output power $P_{out}^* = \frac{P_{out}(k_{sh})}{P_{out}(1)}$ on the gain.

8. Possible Designs and Applications of Resotrodes with 0-Regeneration

Resotrodes are modular devices that can be represented as a set of functional cells.

One of the first studied designs of a resotrode for a range of 27–40 MHz included flat cells with tape cathodes. These cells can form various devices that differ in the number of cells and their layouts within a single design.

Research has been conducted to determine the parameters of the optimal functional electron-optical cell of a device, including the sizes and shapes of the electrodes, and their potentials.

Cell synthesis was performed using an electron-optical code according to the "gun-anti-gun" scheme (Figure 17). First, a gun was synthesized that provides a laminar uniform beam in the center of the anode hole (Figure 17a). Next, a "mirror" system was synthesized containing a virtual supposed "issue" of the beam from the collector surface. This beam was gradually transformed into a uniform laminar beam in the process of moving to the center of the anode hole (Figure 17B), where these two beams were sewed. The functional cell synthesized as a result of calculations (Figure 17c) consisted of a linear cathode 0.8 mm thick and 50 mm long with a 1.3-mm radius of the curvature of the emitting surface, a control electrode protruding 0.3 mm above the cathode surface, an anode with a narrow (0.7 mm) input groove and an external part expanding (up to 3 mm) to the collector and a collector with an anti-dynatron grid.

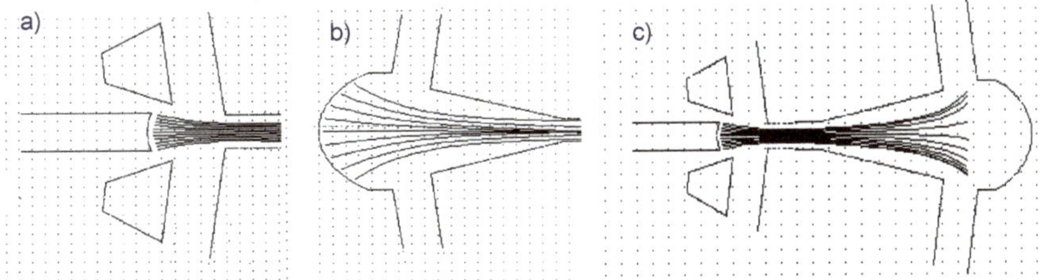

Figure 17. Synthesis of a functional resotrode cell according to the "gun-anti-gun" scheme: (**a**) cathode-anode part (gun), (**b**) collector-output part (anti-gun), (**c**) full synthesized cell.

The synthesized cell was quite versatile and, depending on the basic requirements for the device, provided a possibility for implementing various electrostatic potentials on the electrodes and various values of RF voltage amplitudes. For example, by setting the potential of the control electrode at −300 V, the anode potential at +1.2 kV and the collector potential at +300 V, an HF power source of 1–2 kW can be created at a frequency of 27 or 40 MHz and powered directly from the supply mains (without a power transformer) based on the 12 cells considered. Such a source would be very useful for creating high-volume household RF stoves of 60–80 L or more with uniform volumetric heating. Having realized the potentials of 1 kV on the control electrode and +4 kV on the anode and the collector, it is possible to create a power source of 100 kW for industrial heating installations on the basis of the same 12 cells, as well as a continuous power source of 1 MW on the basis of 120 cells (when the voltage increases, the number of cells decreases by the law of degree 5/2).

Based on the synthesized optimal cell, the design of the vacuum part of the resotrode was developed, containing 12 azimuth-oriented functional cells. The control electrode was a molybdenum cylinder with slotted longitudinal grooves, the anode was made of a copper cylinder and the collector had a special prefabricated anti-dynatron structure that absorbed almost all secondary electrons [24].

As a result of the research, the following patterns were established:

1. The synthesized electron-cell is minimally critical to small changes in the shape of the electrodes, the distances between them and the applied potentials (i.e., with small (up to 5%) changes in the specified parameters, the electron trajectory practically does not change and the current passage remains complete). This stability ensure the stable operation of the device despite errors in the manufacture of parts, errors in the assembly of components and thermal care.

2. The linearity of the amplitude characteristic of the device in the average RF range (tens of MHz) is preserved up to the maximum value of the input signal amplitude.

3. The efficiency of the device in monoharmonic mode with a normalized width of the clot about 2 rad. more than 80%. It is possible to obtain maximum efficiency values (close to 100%) when the cut-off angle is reduced.

4. The conditions of monotron excitation of the common-phase type of vibrations (the main parasitic type), as well as other (higher) types of parasitic vibrations are not fulfilled. Thus, the device does not generate spurious vibrations.

In addition to the flat design, a resotrode design with cylindrical beams has also been studied. This design uses magnetics, and focuses on permanent magnets.

The design of such a device with the possibility of wide-range frequency tuning from 20 to 150 MHz has been considered, for which the general design is described in [24].

As a result of numerical simulation, V. N. Kozlov [24] showed that a functional cell with a convex cathode is optimal for such a device, since it reduces the locking potential (Figure 18).

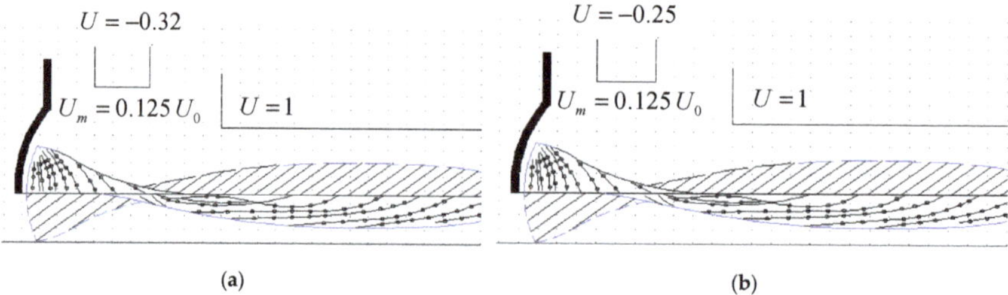

Figure 18. Trajectories of particles in the input part of the resotrode (**a**) with a concave cathode, (**b**) with a convex cathode.

Currently, there are no powerful and efficient microwave power sources in the 50–150 MHz range. Traditional microwave devices—magnetrons, klystrons, and TWTs—would be too large in this range, and standard IOTs would have too small a gain. However, this range is the most attractive for solving problems of uniform volumetric heating in industrial processes (e.g., oil distillation, production of building materials, water desalination, etc.).

For effective uniform volumetric heating, it is necessary that the size of the object being heated be somewhat (but not much) less than half the wavelength. Shorter waves inevitably lead to the appearance of zeros in the spatial distribution of the microwave field and, consequently, to uneven heating. Longer waves lead to the need to increase the amplitude of the RF (microwave) field strength and, consequently, to the occurrence of breakouts. The characteristic size of objects that need to be heated during technological processes is about a meter, requiring the wavelength of the microwave field for effective heating to be 2–3 m in order to achieve the frequency range mentioned above. This uniform volumetric heating can be used, for example, for the manufacture of new environmentally friendly materials (foam glass, artificial Sandstone, etc.), for deep oil refining, and for other applications. A design with a direct connection of the resotrode collector to the capacitive electrode of the heating chamber is possible. In this case, heating will be produced not only by the microwave field, but also by the heat of the heated collector (i.e., double microwave and convection heating will be provided).

9. Resotrode with 2π-Regeneration

Based on the given design, low-frequency resotrode can be moved it to a higher-frequency region using scaling, but this procedure will lead to a significant decrease in output power.

Indeed, if all dimensions are reduced in proportion to the wavelength, the emitting area will decrease in proportion to the square of the wavelength, and the total current and power of the device will decrease accordingly. In addition, correct scaling requires the saving value of the beam's perveance, which leads to an additional decrease in voltage and, consequently, in power.

Attempts to compensate for decreases in input power by increasing the voltage while maintaining the distance will lead to an increase in the current density from the cathode (and therefore to a decrease in the durability of the device) and to a risk of electrical breakdowns in the area of the input cavity.

An increase in the emission area without changing the voltage can lead to a decrease in the R/Q factor of the cavity, and will runs into restrictions on the transverse dimensions of the capacitive part of the cavity of the permissible fraction of the wavelength.

If you increase the voltage while increasing the gaps, then the 0-regeneration condition is violated, which requires that the angle of passage from the cathode to the anode be close to zero.

The relationship between the frequency and the maximum output power of the resotrode with a 0-regeneration is determined by a scaling formula, which implies that using such a device at frequencies less than 200 MHz is impractical.

Thus, to create a powerful microwave source at a frequency greater than 200 MHz, the ideas underlying the resotrode must be corrected.

For operation in the area of hundreds of MHz, a different device scheme was proposed [15], as shown in Figure 19.

This device is called a resotrode with 2π-regeneration, because the optimal angle of passage between the main (accelerating) and regenerating (braking) gaps of the input cavity become 2π instead of 0. The input cavity turns out to be a three-gap cavity, in which the first gap is the main one, and the third one is the regenerating one. The microwave voltage in the second gap is common to the voltage in the first gap, but its amplitude must be small (the gap is "off"). This shutdown is provided by an additional large capacity that provides the microwave closure of this gap to the central bushing.

Between the second and third gaps there is a drift distance enough for the location of at least one additional bunching cavity. When using a two-gap bunching cavity or higher harmonics cavities (i.e., second and third harmonics), two or even three bunching gaps can be placed here. This makes it possible to significantly improve the quality of the bunch, reduce its length and ensure the speed distribution corresponds to the converging bunch [24].

The idea of additional bunching in devices with emission modulation was previously proposed and successfully implemented by V. A. Tsarev [22,23], A. D. Sushkov and V. K. Fedyaev [39] in devices called a tristrod and a tryston, respectively.

Note the main features of the resotrode with 2π-regeneration.

The distance between the first (forming) and third (regenerating) gaps of the input cavity does not depend on the "cathode–anode" distance, so the beam perveance is not related to the pass angle (i.e., the electron-optical parameters (current and voltage) and the pass angle that provide regeneration do not depend on each other). This is the main feature of the design, which makes it possible to work in the high-frequency area.

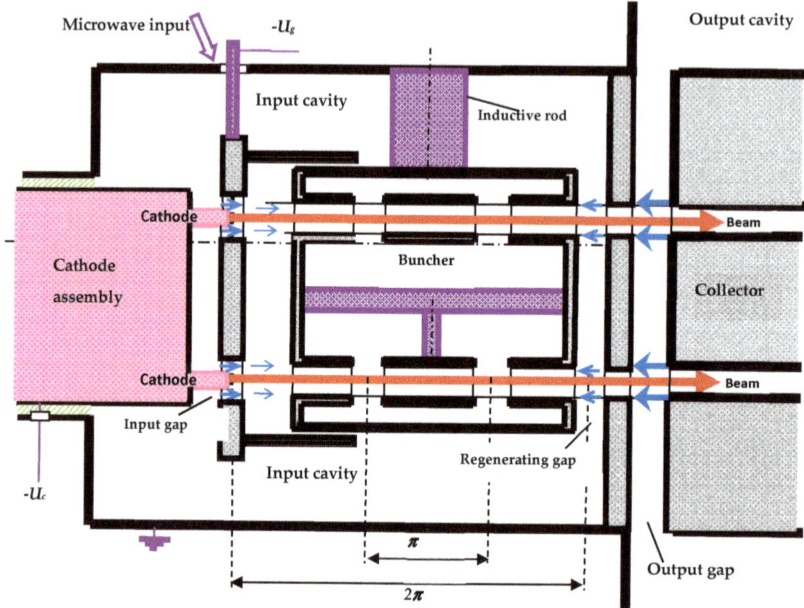

Figure 19. Scheme of a multi-beam resotrode with 2π-regeneration. The blue arrows indicate the electric microwave field, and their thickness illustrates the size of the field.

On the other hand, in contrast to a resotrode with 0-regeneration, the bunch at the entrance to the anode hole is not monospeed, since its speed is modulated by the input signal. This circumstance does not allow such a device to be implemented in a completely grid-free version (Figure 11c), because in this case, even with a minimum offset voltage equal to the cutoff voltage, the speed modulation would be too large and would lead to a scattering of the bunch when it moves to the regenerating gap.

Thus, for a resotrode with 2π-regeneration, it is necessary to use the scheme of Figure 11b. However, in contrast to the classic MB IOTs, the grid can be much more large-scaled and, for flat cathodes, a set of monocrystalline graphite "whiskers" laid in parallel in a row on the control electrode can be used as such a grid. This design of the grid (common to all beams) will significantly simplify the technology and avoid misaligned deformations due to thermal care.

The size of the grid cells is determined by the allowable value of the input signal amplitude, which in resotrodes, unlike IOTs, is not the main factor determining the gain.

The estimates of the efficiency and gain of a resotrode with 0-regeneration made in Section 7 require adjustments to apply them to a resotrode with 2π-regeneration.

In a resotrode with 2π-regeneration, the width of the bunch during its formation and at the entrance to the regenerating gap will differ due to additional bunching. The result of this bunching can be estimated based on the results obtained for klystrons. Under the condition that the width of the bunch, in accordance with Equation (12) and Figure 14, becomes smaller (i.e., the bunch, according to the terminology used for klystrons, becomes fully saturated (FS)). Further bunching of the FS-bunch is primarily related to the transformation of its velocity function. It is necessary to acquire the bunch that comes together and ensures the most effective braking in the exit gap. In this case, the output cavity itself must be detuned to the left. In the klystron, the necessary transformation of the velocity function is performed in the gap of last cavity of buncher, which is located close enough to the output gap.

However, the bunch comes into this gap close to the monospeed. In a resotrode with 2π-regeneration, the original velocity function has a different form: the center of the bunch has a maximum velocity that falls to the edges (Figure 20). Further transformation of such a bunch in the drift without additional impact would lead to additional bunching of the front part of the bunch and to the unbunching and lagging of its rear part (i.e., to the formation of a kind of "comet").

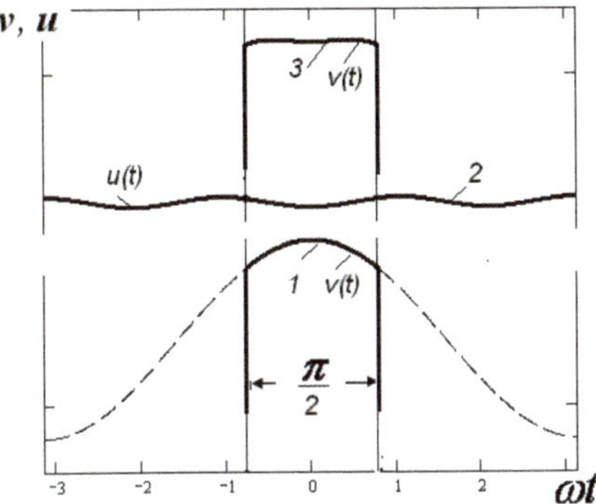

Figure 20. Scheme for correcting the shape of the velocity function by a resonator of the third harmonic with zero detuning: (1) original speed function, (2) microwave voltage in the resonator of the third harmonic, (3) corrected speed function.

To prevent this, it is necessary to correct the shape of the velocity function (e.g., by using a cavity of the second or third harmonic tuned to the proper resonance (Figure 20)). Thus, the buncher must contain at least one higher harmonic cavity and one main harmonic cavity. If the bunch width is small, the main harmonic cavity can be replaced with a second harmonic cavity that is detuned to the right (Figure 21).

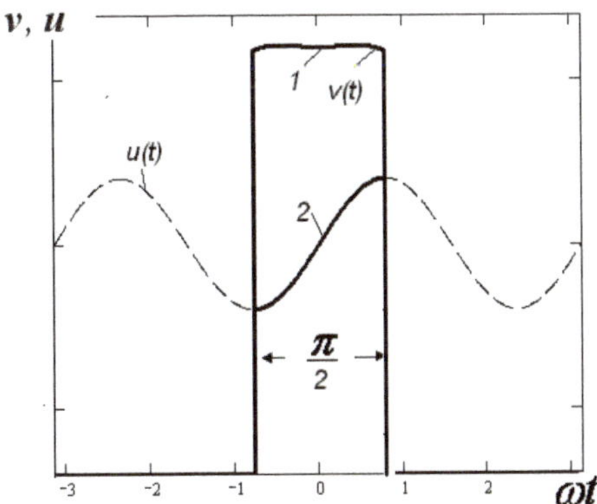

Figure 21. Scheme for bunching of an FS-bunch with a second harmonic cavity detuned to the right: (1) monospeed bunch, (2) microwave voltage in the cavity of the second harmonic.

Thus, in a resotrode with 2π-regeneration, a converging FS-bunch can be formed, which allows to an efficiency of 90% or more to be obtained.

As for the gain factor, the rating (Equation (15)) applies for it in the mode of full electronic load compensation. Since the ratio is at least 10 in the grid design, the gain factor should be at least 30 dB even for small gain factors.

For the currently active large hadron collider (LHC), the frequency of 400 MHz is the main frequency, and it is assumed that this frequency will also be used for the FCC (future circular collider) and CLIC (compact linear collider). Currently, klystrons are used as microwave sources for the LHC. Resotrodes with 2π-regeneration have significant advantages over such a klystron: lower supply voltage, higher efficiency and significantly smaller dimensions. You can lower the voltage of a klystron by switching to a multi-beam design. As shown in Section 5, this can significantly increase efficiency, as seen with the COM and CSM klystrons. But the size (and, consequently, the weight and cost) of the klystron cannot be significantly reduced.

A resotrode with 2π–regeneration at the frequency of 400 MHz seems to be a much more promising microwave source than the klystron.

Currently, the first preliminary stage of designing a resotrode with 2π-regeneration at a frequency of 400 MHz and an output power of 300 kV in a continuous mode has been carried out.

Dr. Igor Syratchev [16] at CERN conducted a complete electrodynamic simulation of such an instrument using HFSS code and built a 3D model of it. The results of this work are shown in Figure 22.

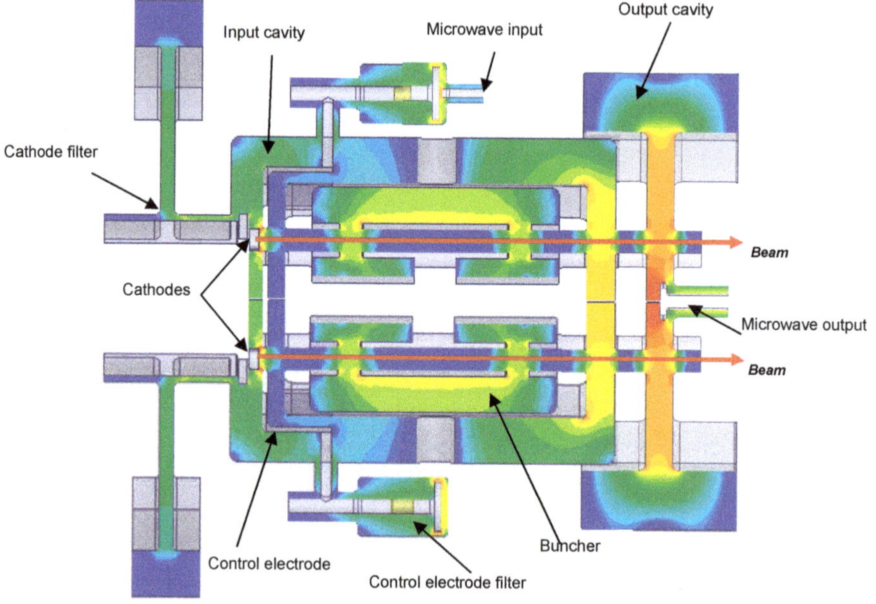

Figure 22. Electrodynamic model of a resotrode (using HFSS code).

The general electrodynamic design of the device (Figure 22) includes an three-phase input cavity that provides bunch formation and energy regeneration, a buncher, an output cavity with a coaxial output of energy through a capacitive connection, as well as blocking filters in the cathode circuit and in the control electrode circuit.

The input cavity has four inductive rods that provide a working antiphase (relative to the third gap to the first two) type of oscillation.

The buncher is modeled with two possible variants, both of which have a two-gap cavity of the main frequency and cavities of harmonics. In the first case, an antiphase type of oscillation is excited in the two-gap cavity, and the gaps are located at a distance of π from each other, which makes it possible to implement bunching phases of the microwave voltage in both gaps. In the second case, the cavity of the third harmonic is tuned to a resonance that corrects the velocity function in accordance with

Figure 20, while the cavity of the second harmonic forms a converging bunch in accordance with Figure 21.

Important electrodynamic elements of the device are the resonant choke filters developed by I. Syratchev [16], which ensure the absence of microwave radiation through the power supply chain on the working-type vibrations, simultaneously suppressing parasitic types of vibrations. Note that the filter in the control electrode circuit is also a communication element for the input signal, the suppression of parasitic vibrations is provided by their low loaded Q-factor and grid-less and gridded designs of the resotrode were both considered.

A grid-less design, as already noted, requires a microwave amplitude that is too large for the input gap.

As shown by preliminary calculations, an amplitude of at least 30% of the accelerating voltage requires to ensure the duration of the bunch $\pi/2$, which is an unacceptably high value. However, even with this amplitude, a gain of at least 25 dB is obtained, although a preferable option is a sparse grid that provides a gain of more than 35 dB in full-compensation mode. Note that if you switch from the full compensation mode to the excess compensation mode, making the amplitude of the microwave voltage in the regenerating gap greater than the total amplitude in the first gap, you can increase the gain indefinitely until you switch to the generation mode.

The question of the possibility of using a resotrode in the generation mode remains open until studies of the stability of such a mode are conducted.

A general 3D model of a resotrode with 2π-regeneration at 400 MHz and 300 kW in a continuous mode was made by I. Syrachev [16], as shown in Figure 23.

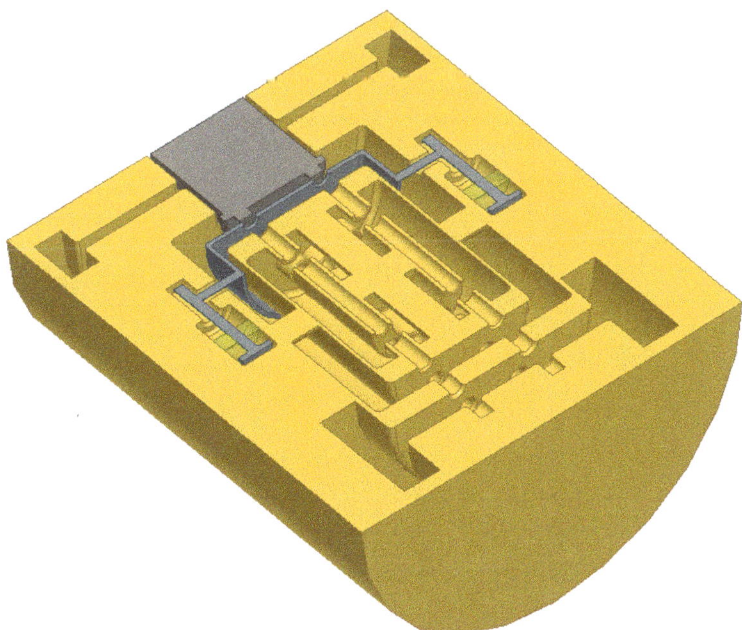

Figure 23. A 3D model of a resotrode with 2π-regeneration.

This device is described in more detail in [16,24]. Although this device was modeled as a microwave source for the LHC, it can also be used for other purposes: for defrosters, for wood drying devices and for other industrial technologies, as well as for advanced microwave energetic devices.

10. Conclusions

In this study, new types of resonant microwave O-type samples are considered including COM klystrons, CSM klystrons and resotrodes, which can combine high output power with an efficiency of up to 90%.

The modeling and research of COM and CSM klystrons has been made possible by the development of mathematical modeling and optimization methods; in particular, by the development of a discrete-analytical klystron model, the macro-steps method for global multiparametric optimization, the stage-by-stage method and the embedding procedure for full structural optimization.

It was shown that the synthesized optimal variants of high-performance multipath klystrons can be transformed into equivalent devices corresponding to other values of output power, voltage and number of beams without performing a new optimization, thanks to the GSP-parameter conversion procedure.

The results of the synthesis of COM and CSM klystrons are presented. It is shown that for various prototypes, klystrons were created with an efficiency of about 90%. COM and CSM klystrons can be considered as promising sources of microwave energy for industrial microwave technologies and microwave energy.

COM klystrons are characterized by an increased ratio of the length of the device to the wavelength, so they are optimal microwave sources for the relatively short-wave part of the microwave range from 3 to 10 GHz.

In CSM klystrons, the relative length is fixed and equal to the length of the optimal three-cavity COM klystron. Therefore, CSM klystrons are optimal sources of microwave power for a range from hundreds of MHz to 3 GHz.

For devices with emission modulation, it is shown that it is possible to achieve a combination of high efficiency with high gain within the design of a new microwave device called a resotrode.

Two types of resotrodes are considered: one with 0-regeneration for frequencies of 20–200 MHz and one with 2π-regeneration for frequencies of 200–1000 MHz.

Possible designs of resotrodes with 0-regeneration based on a universal functional cell are shown. Systems of uniform volumetric RF/microwave heating of large-sized objects based on resotrodes with 0-regeneration are proposed.

The design of a resotrode with 2π-regeneration at 400 MHz with an output power of 300 kW in continuous mode is proposed.

All these devices can be used to power modern colliders, for industrial microwave heating installations for various purposes and for advanced wireless power transmission devices.

Currently, leading klystron manufacturers are trying to implement the ideas of COM and CSM bunching by modifying their existing klystron designs. This leads to an increase in efficiency of several percent. The maximum efficiency values of about 90% can only be realized based on a new design initially calculated for the appropriate bunching mode. The development of such a new design is very expensive, so COM and CSM klystrons that fully correspond to the theory have not yet been created. The same problems apply to resotrodes. Their development requires not only the creation of a new design, but also a thorough study of the regenerative amplification mode with a possible transition to the generator mode. This a fundamental experimental investigation which is still waiting in the wings.

Author Contributions: Conceptualization and methodology, A.B.; software, validation, formal analysis, investigation, computer modelling and simulation, review and editing, A.B. and O.B. All authors have read and agreed to the published version of the manuscript.

Funding: This research received no external funding.

Conflicts of Interest: The authors declare no conflict of interest.

References

1. AMTec Microwaves. Available online: https://www.4amtek.com/products/cooking-systems/ (accessed on 4 February 2020).
2. WaveLane. Available online: http://grandtekco.com/ (accessed on 4 February 2020).
3. Baikov, A.Y.; Fuong, K.H.; Petrov, D.M. Possibility making powerful relativist klystron of 3-cm range with electronic efficiency of about 90%. In Proceedings of the International Conference on Actual Problems of Electrons Devices Engineering, APEDE 2006, Saratov, Russia, 20–21 September 2006; Saratov State Technical University pub.: Saratov, Russia, 2006; pp. 106–115.
4. Baikov, A.Y.; Grushina, O.A.; Strikhanov, M.N. Simulation of conditions for the maximal efficiency of decimeter-wave klystrons. *Tech. Phys.* **2014**, *59*, 421–427. [CrossRef]
5. Baikov, A.Y.; Marrelli, C.; Syratchev, I. Toward high-power klystrons with RF power conversion efficiency on the order of 90%. *IEEE Trans. Electron Devices* **2015**, *62*, 3406–3412. [CrossRef]
6. Constable, D.A.; Lingwood, C.; Burt, G.; Syratchev, I.; Marchesin, R.; Baikov, A.Y.; Kowalczyk, R. MAGIG-2D Particle-in-Cell Simulations of High Efficiency Klystrons. In Proceedings of the 17th IEEE International Vacuum Electronics Conference, IVEC-2016, Monterey, CA, USA, 19–21 April 2016; pp. 195–196.
7. Constable, D.; Marrelli, C.; Jensen, A.; Kowalczyk, R.; Baikov, A.; Burt, G.; Hill, V.; Syratchev, I.; Guzilov, I.; Lingwood, C.; et al. High efficiency klystron development for particle accelerators. In Proceedings of the EE-FACT'16, Daresbury, UK, 24–27 October 2016. Paper WET3AH2.
8. Constable, D.A.; Lingwood, C.; Burt, G.; Baikov, A.Y.; Syratchev, I.; Kowalcyzk, R. MAGIC2-D simulations of high efficiency klystrons using the core oscillation method. In Proceedings of the 18th IEEE International Vacuum Electronics Conference, IVEC-2017, London, UK, 24–26 April 2017.
9. Baikov, A.; Baikova, O. Simulation of high-efficiency klystrons with the COM and CSM bunching. In Proceedings of the 20th IEEE International Vacuum Electronics Conference, IVEC-2019, Busun, Korea, 28 April–1 May 2019.
10. Baikov, A.Y.; Baikova, O.A. The possibility of achieving high values of efficiency with small bunching length in the two-frequency klystrons. In Proceedings of the International Conference on Actual Problems of Electrons Devices Engineering, APEDE 2016, Saratov, Russia, 22–23 September 2016; Saratov State Technical University pub.: Saratov, Russia, 2016; pp. 12–14.
11. Baikov, A.Y.; Baikova, O.A. *Simulation of Super Power Two-Frequency Klystrons of L-band with Efficiency of 90%*; Bulletin of the National Research Nuclear University MEPhI: Moscow, Russia, 2017; Volume 6, pp. 83–89. (In Russian)
12. Baikov, A.Y.; Baikova, O.A. On the synthesis of high-Efficiency CSM klystrons by the "embedding" method. In Proceedings of the International Conference on Actual Problems of Electrons Devices Engineering, APEDE 2018, Saratov, Russia, 27–28 September 2018; Saratov State Technical University pub.: Saratov, Russia, 2018; pp. 17–20, ISBN 978-1-5386-4332-7.
13. Baikov, A.Y.; Petrov, D.M. Rezotrod: Opportunities of the solution of a problem of a uniform volumetric heating of large-sized objects. In Proceedings of the International University Conference "Electronics and Radiophysics of Ultra-high Frequencies", Saint Petersburg, Russia, 24–28 May 1999; pp. 430–431.
14. Baikov, A.Y.; Petrov, D.M. Multi Beams Regenerative Amplifier of Electromagnetic Oscillations. Patent RU N2150766 C1, 10 June 2000.
15. Baikov, A.Y. Resotrode with 2 π-regeneration—A promising new source of microwave power. In Proceedings of the International Conference on Actual Problems of Electrons Devices Engineering, APEDE 2016, Saratov, Russia, 22–23 September 2016; Saratov State Technical University pub.: Saratov, Russia, 2016; pp. 15–17.
16. Baikov, A.Y.I.; Syratchev, I. Resotrode—RF Amplifier with Regeneration. Concept. CLIC Workshop 2016, CERN, Geneva, Switzerland, 18–22 January 2016. Available online: https://indico.cern.ch/event/449801/contributions/1945287/ (accessed on 4 February 2020).
17. Baikov, A.Y.; Baikova, O.A. On Achieving the Limit Values of the Efficiency of the Klystrons and Resotrodes. In Proceedings of the Abstracts of XVII International Winter School-Workshop on Microwave Radio Physics and Electronics, Saratov, Russian, 5–10 February 2018; pp. 38–39. (In Russian).

18. Caryotakis, G. High Power Klystrons: Theory and Practice at the Stanford Linear Accelerator Center. SLAC-PUB 10620, January 2005. Available online: https://www.slac.stanford.edu/cgi-bin/getdoc/slac-pub-10620.pdf (accessed on 4 February 2020).

19. Lien, E.L. High Efficiency Klystron Amplifier. In Proceedings of the 8th International Conference Microwave and Optical Generation and Amplification, Amsterdam, The Netherlands, 7–11 September 1970.

20. Lien, E.L. High efficiency klystron amplifiers. In Proceedings of the the MOGA70, Amsterdam, The Netherlands, 7–11 September 1970; pp. 1121–1127.

21. Dokolin, O.A.; Kuchugurny, S.V.; Lebedinsky, S.V.; Malichin, A.V.; Petrov, D.M. Klystron with an electronic efficiency of 90%. *Izv. Vuzov USSR Radio Electron. Kiev.* **1984**, *27*, 47–55. (In Russian)

22. Tsarev, V.A.; Miroshnichenko, A. Reflection Oscillator. *RU 2084042 C1*, 10 July 1997.

23. Tsarev, V.A. High Efficiency IOT with 3 Cavities. CLIC Workshop 2016. 18–22 January 2016, CERN, Geneva, Switzerland. Available online: https://indico.cern.ch/event/449801/contributions/1945286/ (accessed on 4 February 2020).

24. Baikov, A.Y. Methods of Achievement the Limit Values of Efficiency in High-Power Vacuum Resonant Microwave Devices of O-type. Ph.D. Thesis, Moscow University of Finances and Law MFUA, Moscow, Russia; 404p.

25. Baikov, A.Y.; Baikova, O.A. Global scaling principle for simulation of powerful klystrons with high efficiency. In Proceedings of the 2016 International Conference on Actual Problems of Electron Devices Engineering, APEDE 2016, Saratov, Russia, 22–23 September 2016; Saratov State Technical University pub.: Saratov, Russia, 2016; pp. 9–11.

26. Langdon, A.B.; Lasinski, B.F. Electromagnetic and relativistic plasma simulation models. *Meth. Comput. Phys.* **1976**, *16*, 327–366.

27. Orbital ATK. MAGIC Tool Suite. Available online: http://www.orbitalatk.com/magic (accessed on 4 February 2020).

28. CST—Computer Simulation Technology. CST Studio Suit. Available online: https://www.cst.com/products/csts2/ (accessed on 4 February 2020).

29. Cooke, S.J.; Nguyen, K.T.; Vlasov, A.N.; Antonsen, T.M.; Levush, B.; Hargreaves, T.A.; Kirshner, M.F. Validation of the Large-Signal Klystron Simulation Code TESLA. *IEEE Trans. Plasma Sci.* **2004**, *32*, 1136–1146. [CrossRef]

30. Cai, J.C.; Syratchev, I. KlyC: 1.5D Large Signal Simulation Code for Klystrons. *IEEE Trans. Plasma Sci.* **2019**, *47*, 1734–1741. [CrossRef]

31. Baikov, A.Y. Macro Step Method for Global Multi-Parameter Optimization of Powerful Klystrons [Metod makroshagov dlya globaljnoi mnogoparametricheskoi optimizacsii]. *Appl. Math. Math. Phys.* **2015**, *1*, 47–66. (In Russian) [CrossRef]

32. Baikov, A.Y.; Grushina, O.A.; Strikhanov, M.N.; Tishchenko, A.A. Model of electron beam transformation in a narrow tube. *Tech. Phys.* **2012**, *57*, 824–834. [CrossRef]

33. Baikov, A.Y.; Grushina, O.A.; Strikhanov, M.N. Maximal efficiency versus the amplification factor in double-cavity klystrons. *Tech. Phys.* **2013**, *58*, 594–600. [CrossRef]

34. Baikov, A.Y.; Baikova, O.A. Methods for full global optimization of klystron parameters. In Proceedings of the International Conference «Systems of Signal Synchronization, Generating and Processing in Telecommunications, SYNCHROINFO 2019» (IEEE Conference Record #47541), Russia, Russia, 1–3 July 2019.

35. Guzilov, I.A. BAC method of increasing the efficiency in klystrons. In Proceedings of the 10th International Vacuum Electron Sources Conference (IVESC '14), Saint Petersburg, Russia, 30 June–4 July 2014; pp. 101–102.

36. Hill, V.; Marrelli, C.; Constable, D.; Lingwood, C. Particle-in-Cell Simulation of the Third Harmonic Cavity F-Tube Klystron. In Proceedings of the 17th IEEE International Vacuum Electronics Conference, IVEC-2016, Monterey, CA, USA, 19–21 April 2016; pp. 68–69.

37. Hill, V.R.; Burt, G.C.; Constable, D.A.; Lingwood, C.; Marrelli, C.; Syratchev, I. Particle-in-Cell Simulation of Second and Third Harmonic Cavity Klystron. In Proceedings of the 18th IEEE International Vacuum Electronics Conference, IVEC-2017, London, UK, 24–26 April 2017; pp. 97–98.

38. Beunas, A.; Faillon, G.; Choroba, S. A high power long pulse high efficiency multi beam klystron. In Proceedings of the 5th MDK Workshop, Geneva, Switzerland, 26–27 April 2001.

39. Fedyayev, V.K.; Ustinenko, E.A.; Barabanov, O.A. Triode-klystron method of ultra-high band pulse-signal generation. In Proceedings of the 2008 4th International Conference on Ultra Wide Band and Ultra Shot Impulse Signals, UWBUSIS 2008, Sevastopol, Crimea, Ukraine, 15–19 September 2008; pp. 156–157.

Article

Hybrid Machine Learning Models for Classifying Power Quality Disturbances: A Comparative Study

Juan Carlos Bravo-Rodríguez *, Francisco J. Torres and María D. Borrás

Escuela Politécnica Superior, Universidad de Sevilla, c/ Virgen de África 9, 41011 Sevilla, Spain;
fjaviertorres92@gmail.com (F.J.T.); borras@us.es (M.D.B.)
* Correspondence: carlos_bravo@us.es; Tel.: +34-954-552-847

Received: 10 March 2020; Accepted: 25 May 2020; Published: 1 June 2020

Abstract: The economic impact associated with power quality (PQ) problems in electrical systems is increasing, so PQ improvement research becomes a key task. In this paper, a Stockwell transform (ST)-based hybrid machine learning approach was used for the recognition and classification of power quality disturbances (PQDs). The ST of the PQDs was used to extract significant waveform features which constitute the input vectors for different machine learning approaches, including the K-nearest neighbors' algorithm (K-NN), decision tree (DT), and support vector machine (SVM) used for classifying the PQDs. The procedure was optimized by using the genetic algorithm (GA) and the competitive swarm optimization algorithm (CSO). To test the proposed methodology, synthetic PQD waveforms were generated. Typical single disturbances for the voltage signal, as well as complex disturbances resulting from possible combinations of them, were considered. Furthermore, different levels of white Gaussian noise were added to the PQD waveforms while maintaining the desired accuracy level of the proposed classification methods. Finally, all the hybrid classification proposals were evaluated and the best one was compared with some others present in the literature. The proposed ST-based CSO-SVM method provides good results in terms of classification accuracy and noise immunity.

Keywords: power quality disturbances; classification; feature selection; swarm optimization; support vector machine; genetic algorithm; K-NN algorithm; decision tree; S-transform

---

1. Introduction

Power quality (PQ) is essential for electrical systems to operate properly with the minimum possible deterioration of performance. Emerging PQ challenges, such as the growing integration of large power plants based on renewable sources, improvements in nonlinear loads, and the recent requirements of smart grids, must be considered to obtain an optimal operation of the existing power grid. These factors increasingly require constant revisions of the common power quality problems, enhanced standards, further optimization of control systems, and more powerful capabilities for measuring instruments.

The purpose of this research was to contribute to this task, meeting the particular PQ requirements about detection and classification of power quality disturbances (PQDs) through optimal hybrid machine learning approaches.

Usually, the PQDs' identification procedure is carried out in three steps, i.e., signal analysis, feature selection, and classification. In the stage of PQDs' analysis, some advanced mathematical techniques were used to extract the feature eigenvectors that enable disturbance identification. The time-frequency analysis methods include short-time Fourier transform (STFT), Stockwell transform (ST) [1], wavelet transform (WT) [2–4], Hilbert–Huang transform [5,6], Kalman filter [7,8], strong trace filter (STF) [9], sparse signal decomposition (SSD) [10], Gabor–Wigner transform [11], and empirical mode decomposition (EMD) [12,13]. In this study, ST was selected due mainly to its noise immunity, simplicity of implementation, and flexible and controllable—to some extent—time-frequency resolution. These advantages far outweigh the computational cost that may be required.

The selection of suitable feature remains a key challenge that requires developing tools in areas such as statistical analysis, machine learning, or data mining [14]. Valuable efforts have been made in this sense and some techniques are used for a precise selection of features including the principal component analysis [15], K-means-based apriori algorithm [16], classification and regression tree algorithm [17], multi-label extreme learning machine [18], random forest model [19], sequential forward selection [20], and bionic algorithms. This latter group has also been successfully used in classification rule discovery. Particularly significant among bionic algorithms are genetic algorithms (GA) [20–22] and swarm-based approaches like ant colonies [23,24] and, above all, particle swarm optimizers (PSO) [25–28]. For example, recently in [25], a combination of PSO and support vector machine (PSO-SVM) was used to optimize the error of the classifier by selecting the best feature combination. Similarly, in [26], PSO optimizes the noise cut-off threshold of PQD signals working together with a modified ST in the feature extraction stage. However, canonical PSO has some limitations for feature selection. Improved and implemented PSO variants include competitive swarm optimizer [29–31] (CSO), discrete particle swarm optimizer [32], and exponential inertia weight particle swarm optimizer [33]. It should be noted that, as PSO was firstly designed for continuous optimization problems, this may not always be the most appropriate method to solve a combinatorial optimization problem such as feature selection. The CSO algorithm, however, is specifically adapted to perform this type of problem with each particle learning from a pair of randomly selected competitors to elevate both global and local search abilities. In this paper, GA and CSO were selected and compared to minimize the number of selected features.

Regarding the disturbance pattern recognition capability and according to the optimal selection of features provided by the above-mentioned algorithms, numerous machine learning approaches have been widely utilized for classifying power quality disturbances. Common classification techniques include artificial neural network (ANN), K-nearest neighbor (K-NN) algorithm, support vector machine (SVM), and decision tree (DT) methods.

Support vector machine (SVM) is a good option for classification purposes, especially when dealing with small samples, nonlinearity, or high dimension in pattern recognition [2,16,22,34,35]. Among the advantages of SVM are the lack of local extremum, feature mapping of nonlinear separable data, low space complexity, and the capability to adjust only a reduced number of features as compared to, for example, the ANNs [36]. On the contrary, its disadvantages include limitations resulting from speed and size, in both training and testing, as well as those resulting from an improper choice of the kernel. These handicaps involve, in practical terms, high algorithmic complexity and extensive memory requirements. Improved applications of SVMs algorithms include multiclass SVM [M-SVM] [37], directed acyclic graph SVMs [DAG-SVMs] [38] and radial basis function kernel SVM [RBF-SVM] [39].

For its part, the rule-based DT classifier is a good choice when the features are clearly distinguishable from each other [40,41]. PQDs' classifiers based on DT include fuzzy decision tree [42–44] and the aforementioned classification and regression tree algorithm (CART) [17,21]. On the one hand, DT advantages include removing unnecessary computations, a singular set of parameters which allows differentiating between classes and a smaller number of features at each nonterminal node while maintaining performance at an acceptable level. On the other hand, its principal disadvantages include being strongly dependent on the selected features, accumulation of errors from level to level in a large tree, and overlap—which increases the search time and memory space requirements when the number of classes is large. Compared to other approaches, DT flowchart symbols configure a simple and straightforward model in which the control parameters are easy to understand and apply. Thus, DT is easier to set up and interpret and, despite the mentioned dependence of the classification process on the selected features, its execution of data is better than other methods. For example, in [22], a comparative using DT/SVM, wavelet transform (WT) and ST is shown.

Most of the previous works in the literature are mainly focused on pattern recognition issues, so PQDs are generally treated as single-event signals. However, in electrical systems, it is common to find several disturbances consecutively in the same observation window. These combined disturbances are much more difficult to identify and treat than single ones. In this work, complex PQDs were designed through a consecutive or simultaneous combination of two simple ones in the same interval. From ST, time-frequency features were extracted, while feature selection was optimized by using K-NN, GA, and CSO. In the classification stage, K-NN (again) and distinct types of SVM and DT were considered. There were different proposals of classifiers depending on the optimization-classification sequence chosen. All these proposals operated over the same dataset obtained after optimization. A comparative in terms of classification accuracy and noise immunity of the proposed models was planned. In Figure 1, a general block scheme of the proposed classification plan is presented. The main steps included PQD signal processing via ST, feature extraction, optimal feature selection, and classification. A detailed overview of the proposed comparative study including different hybrid methods can be found in Section 5. The MATLAB (Classification Learner Toolbox) software was used to implement the whole machine learning methods required at both optimization and classification stages.

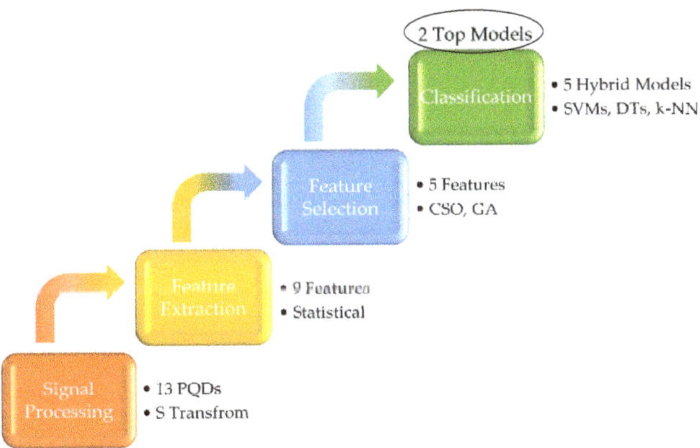

Figure 1. General block scheme of the proposed classification plan.

The rest of this paper is organized as follows: In Section 2, a simplified outline of the extraction of the initial feature set is presented. Section 3 is devoted to the optimal selection of features, describing the optimizers used for this task. Section 4 briefly describes the machine learning methods used to classify. In Section 5, a detailed overview of the proposed classification plan is shown. In Section 6, PQ disturbances' synthesis and the resulting training datasets are detailed. In Section 7, results are discussed. The last section draws conclusions from the results.

2. Initial Feature Set Extraction Based on S-Transform and Statistical Parameters

Each proposed disturbance signal was generated in a discrete form to compute its S-transform, the detailed description of which is given in the Appendix A. ST was chosen for its inherent noise immunity and acceptable time-frequency resolution.

The resulting complex S-matrix provided valuable time-frequency data on which PQD features were extracted by computing several statistics and figures of merit. In this two-dimensional S-matrix, the signal was split into different frequencies (M = 1280 rows) and distinct samples (N = 2560 columns). This extraction of features had a relevant effect on the accuracy of classification because of its great influence on the overall performance of machine learning approaches.

In a first approximation, the chosen initial feature set should have been enough to guarantee a correct identification of every one of the considered disturbed signals. In this work, the extracted set was formed by nine features (k1–k9) and included the introduced disturbance energy ratio (DER) index as well as some of the well-known statistical parameters, such as maximum, minimum, root mean square and mean values, standard deviation, variance, skewness, and kurtosis. These features were calculated following the equations shown in Table 1.

Table 1. Mathematical equations of the initial feature set.

Extracted Features					
K1	Maximum	$M = \max\{A_{jn}\}$ [1]	Standard deviation	$\sigma = \sqrt{\dfrac{\sum_{j=1}^{M} \sum_{n=1}^{N} \left(A_{jn} - \mu_j\right)^2}{(M-1)(N-1)}}$	K6
K2	Minimum	$m = \min\{A_{jn}\}$	Variance	$\sigma^2 = \dfrac{\sum_{j=1}^{M} \sum_{n=1}^{N} \left(A_{jn} - \mu_j\right)^2}{(M-1)(N-1)}$	K7
K3	Mean value	$\mu = \dfrac{\sum_{j=1}^{M} \sum_{n=1}^{N} A_{jn}}{M \cdot N}$	Skewness (phase) [2]	$SK_{(\phi)} = \dfrac{\sum_{j=1}^{M} \sum_{n=1}^{N} \left(\phi_{jn} - \mu_{(\phi)j}\right)^3}{M \cdot N \cdot \sigma_{(\phi)}^3}$	K8
K4	RMS	$RMS = \sqrt{\dfrac{\sum_{j=1}^{M} \sum_{n=1}^{N} A_{jn}^2}{M \cdot N}}$	Kurtosis	$KT = \dfrac{\sum_{j=1}^{M} \sum_{n=1}^{N} \left(A_{jn} - \mu_j\right)^4}{M \cdot N \cdot \sigma^4}$	K9
K5	DER	$DER = \dfrac{RMS_{>50}}{RMS_{50Hz}}$	-	-	

[1] A_{jn}, [2] ϕ_{jn} are the absolute value and phase value of the jn-th element in the S-matrix.

All statistical parameters were calculated from both time samples ($N = 2560$) and frequency ($M = 1280$) intervals. The skewness parameter was computed based on the phase values of the complex elements in the S-matrix. For the rest of the parameters, calculations were done from the absolute values of such elements.

Disturbance Energy Ratio (DER) Index

The introduced DER index represents the ratio between the energy of the signal with frequency components greater than 50 Hz and that one whose components are equal to or less than 50 Hz. Thus, the definition of DER parameter includes the terms

$$RMS_{>50} = \sum_{freq=50.1\ Hz}^{freq=6400\ Hz} RMS_j \tag{1}$$

and

$$RMS_{50Hz} = \sum_{freq=0Hz}^{freq=50Hz} RMS_j. \tag{2}$$

This index is very useful for the characterization of PQ disturbances with high-frequency content as, for example, oscillatory transients.

Sample datasets for training/testing consist of single observations, each of which is computed from features, as shown in Table 1.

3. Optimal Feature Selection: GA and CSO

The main purpose of using an optimizer is to reduce as much as possible the dimension of input feature dataset for the prediction models. Once data have been obtained by S-transform, further analysis is necessary to achieve the optimal feature vector. As seen above, a vector with nine different features was proposed. However, the given feature vector contained attributes whose information was redundant to distinguish the most discriminating features of PQDs. The intraclass compaction could be minimized and the interclass division could be maximized by reducing the number of features. For this purpose, after obtaining the dataset features, it was necessary to select the best optimizer. Wrapper-based techniques are a significant group within feature selection methods that are very accurate and popular and eliminate redundant features by using a learning algorithm with classifier performance feedback. The two main optimization methods used in this work, namely GA and CSO, belong to this group.

3.1. Genetic Algorithm

Darwin's theory of evolution, "Survival of the Fittest", inspired the design of genetic algorithms in the 1960s [45]. GA is an adapted heuristic search algorithm [45] that uses optimization methods based on genetics and rules of natural selection. The flowchart that describes the operation of GA is shown in Figure 2.

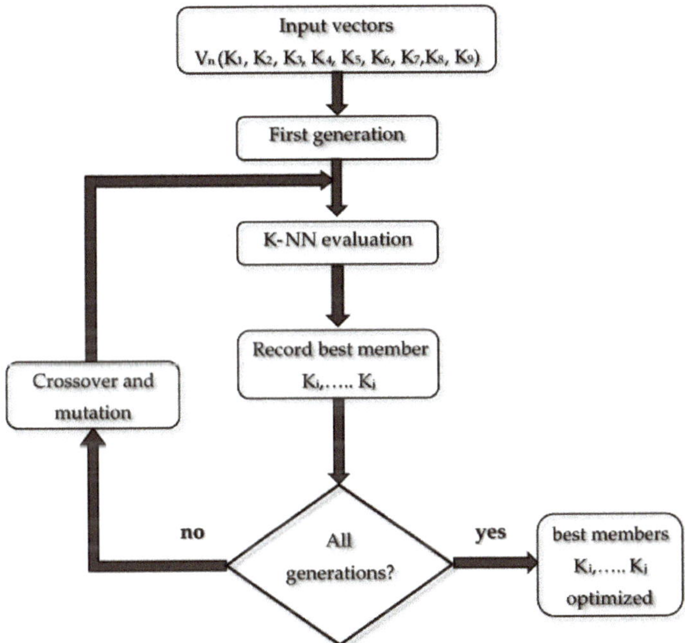

Figure 2. Preprocessing stage using K-nearest neighbor (K-NN) algorithm for the evaluation of genetic algorithms (GA) members.

In GA [46], an optimal feature vector can be represented by a chromosome, which includes the most discriminative features. In turn, chromosomes comprise multiple genes, each one corresponding to a feature. The population is a finite set of chromosomes manipulated by the algorithm in a similar way to the process of natural evolution. In this process, chromosomes are enabled to crossover and to mutate. The crossing of two chromosomes creates two offspring and these two each produce two more, and so on. A genetic mutation in the offspring generates an almost identical copy of the combination of their parents but with some part of the chromosome moved. Generations are the cycles where the optimization process is carried out. Crossover, mutation, and evaluation make it possible to create a set of new chromosomes during each generation. A predefined number of the (best) chromosomes survives to the next cycle of the replica due to the finite size of the population. The population can achieve a fast adaptation despite its limited size, which results in quick optimization of the criterion function (score). The most important step of GA is the crossover, in which exchanges of information among chromosomes are implemented. Once the best individuals are selected, it is necessary to crossover these solutions between themselves. The main purpose of this step is to get a greater differentiation between populations based on new solutions that could be better than the previous ones. A second important step is a mutation, which increases the variableness of the population.

Another key piece is the fitness function. It is necessary to obtain an effective enforcement-oriented version of GA. The fitness function is the procedure or device that is responsible for assessing the quality of each chromosome, specifying which one is the best from the population. Once the fitness function is calculated with each individual of the initial population, the next stage is the so-called selection, in which chromosomes with the best qualities are selected to generate the new evolution of the population using discrimination criteria.

Different GA implementations use specific important parameters to determine the execution and performance of the genetic search. However, some other parameters, including crossover rate, population size, and mutation rate, are usual for all implementations. The probability of taking an eligible pair of chromosomes for crossover is called rate crossover. Conversely, the probability of changing a bit of randomly selected chromosomes is called mutation rate. The crossover rate usually presents high values, close to or equal to 1, while the mutation rate is usually small (1% to 15%).

In the present work, the chromosome consisted of nine genes, each of which represented a feature. As shown in Figure 2, the chromosome is represented as a vector of bits since all the genes could be assigned with either 0 or 1 (0 when the corresponding feature was not selected and 1 when it was). A population of 560 individuals (chromosomes) and 100 iterations (generations) was selected for this problem. The search began initializing the parameters to:

- Initial (parent) population size: 10 (chromosomes).
- Crossover rate: 0.8.
- Mutation rate: 0.01.

The performance of the classifier must be kept above a certain specified level. For this, the least expensive subset of features must be found. For this purpose, the performance is measured using the error of a classifier. The viability of a subset is ensured when the error rate of the classifier is lower than the so-called feasibility threshold. The goal is to find the smallest subset of features among all feasible ones.

In this case, the identification accuracy of the K-NN algorithm was set as the fitness value of the chromosome. In order to assess the quality of the chromosome through the fitness function (accuracy), the k parameter of the K-NN method was adjusted to 10 and the value of cross-validation was set to 8 folds. The K-NN input dataset was exclusively designed for this validation procedure (see Section 6 for details).

3.2. Competitive Swarm Optimization

Competitive swarm optimizer [29] is a particular case of particle swarm optimizer (PSO), thus belonging to evolutionary algorithms inspired by flocking and swarming behavior. Swarm methods try to emulate the adaptive strategy, which considers collective intelligence as behavior without any structure of centralized control over individuals. Usually, the overall structure of swarm optimizers includes different algorithms with each handle a specific task. The critical one is the classification rule discovery algorithm, which is, in essence, a standard GA. Thus, a group of individuals (particles) acts and evolves following the principles of natural selection—survival of the fittest. In PSO, the optimal solution for a problem is obtained from the global interactions among particles. In contrast, the CSO method introduces pairwise interactions randomly selected from the swarm (population). Generations succeed one another after each pairwise competition, in which the fitness value of the loser is updated by learning from the winner that goes directly to the swarm of the next generation. CSO has proven to be better than GA in optimization tasks related to feature selection due to its easy-to-use structure, fewer parameters, and simple concept, even though its computational cost is slightly higher. However, as will be shown below in the conclusions, the superiority of CSO over GA is clear from the solution quality, but in terms of success rate, it is not so.

In this work, particles were defined by the feature set (K1...K9) in the same way as chromosomes (individuals) in GA. They also derived from the same dataset (560 individuals) from which particles were randomly selected. Then, the swarm size was set to 100 and the maximal number of generations (iterations) was set as 200.

Following a parallel process to that carried out in the GA optimizer, a K-NN simple identification model was used to check the efficiency of the CSO-based feature selection, in this case with k = 5. Once again, the accuracy of the K-NN identifier was established as the fitness function of the CSO optimizer.

Both types of optimization methods, GA and CSO, reduced the number of features from nine to five, but they were not the same.

As was mentioned above, K-NN was chosen to act as a fast validation tool in the feature optimal selection stage. At this stage, the aim was to reduce features and high accuracy was not as necessary as simplicity, speed, and efficiency. In these aspects, the K-NN was highly competitive. As shown below, this method is going to be used again in the next stage to compare its classification performance with that of other approaches. In the next section, unlike this one, the aim is to achieve the highest possible accuracy in the classification.

4. Classifiers: K-NN, SVMs, and DTs

Once the optimized set of feature was determined, the next process was the classification of data with these features. In this work, various classification methods were used to find better efficiency and the best behavior with noise signals. These methods included the K-nearest neighbors' algorithm, the support vector machine, and the decision trees.

4.1. K-Nearest Neighbors' Algorithm

One of the proposed classification approaches used the K-NN classifier to identify both single and complex disturbances. K-NN [47], as a supervised learning algorithm, determines the distance to the nearest neighboring training samples in the feature space in order to classify a new object. This Euclidean distance is stated as follows:

$$D_j(x_i, y_j) = \sqrt{\sum_{k=1}^{p}(x_{i,k} - y_{j,k})^2} \tag{3}$$

where $D_j(x_i, y_j)$ is the Euclidean distance-based relationship between the ith p-dimensional input feature vector x_i and the jth p-dimensional feature vector y_j in the training set. A new input vector x_i is classified by K-NN into the class that allows a minimum of k similarities between its members. The parameter k of the K-NN method is a user-specific parameter. Often k is set to a natural number closer to $\sqrt{N_{trsamples}}$ [47], in which $N_{trsamples}$ is the number of samples in the training dataset. In this work, different K-NN classifiers were fit on the training dataset resulting from values of k between 5 and 12. The lowest classification error rate on the validation set permitted selecting the sough-after value of k.

Traditional K-NN approach based on Euclidean distance becomes less discriminating as the number of attributes increases. To improve the accuracy of the K-NN method for PQDs classification, a weighted K-NN classification method can be used [48]. The weight factor is often taken to be the reciprocal of the squared distance, $\omega_i = 1/D_j^2(x_i, y_j)$. Several schemes can be developed to attempt to calculate the weights of each attribute based on some discriminability criteria in the training set.

4.2. Support Vector Machine

SVM is a statistical method of machine learning that uses supervised learning [49]. Although this method was originally intended to solve binary problems, its use was easily extended to multiclass classification problems. The major objective of SVM is the minimization of the so-called structural risk by proposing hypotheses to minimize the risk of making mistakes in future classifications. This method finds optimal hyperplanes separating the distinct classes of training dataset in a high-dimensional feature space and, based on this, test data can be classified. The hyperplane is equidistant from the closest samples of each class to achieve a maximum margin on each side of it. Only the training samples of each class that fall right at the border of these margins are considered to define the hyperplane. These samples are called support vectors [50,51].

Next, a rough sketch of SVM is outlined below in an oversight-specific manner. Consider a dataset containing a data pair defined as $(x_i, y_j)(i = 1, \ldots, M)$, where M is the number of samples, $y_i \in \{-1, 1\}$. Based on an n-dimensional vector w normal to the hyperplane and a scalar b, the issue is to find the minimum value of $\|w\|$ in the objective equation $f(x) = \langle w^T \cdot x + b \rangle$. The position of the separating hyperplane can be determined based on the values of w and b that fulfil the constraint $y_i \cdot (w^T \cdot x_i + b) \geq 1$. The key parameter $b/\|w\|$ gives the distance from the origin (x_0, y_0) to the closest data point along w. Furthermore, to deal with the case of the linear inseparable problem, where empirical risk is not zero, a penalty factor C and slack variables ξ_i are introduced. The optimal separating hyperplane can be determined by solving the following constrained optimization problem [24,52]:

Minimize

$$\frac{1}{2} \cdot \|w\|^2 + C \cdot \sum_{i=1}^{M} \xi_i \tag{4}$$

subject to

$$y_i \cdot (w^T \cdot x_i + b) \geq 1 - \xi_i \ \ for\ i = 1, 2, \ldots, M$$
$$\xi_i \geq 0 \ for\ all\ i \tag{5}$$

where ξ_i is the distance between the margin and wrongly located samples x_i.

Despite SVM being a linear function set, it is possible to solve nonlinear classification problems by using a kernel function. As shown in Figure 3, the mapping translates the classified features onto a high-dimensional space where the linear classification is feasible.

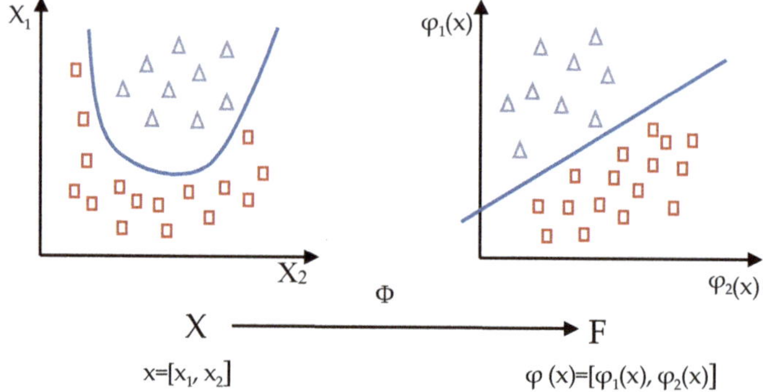

Figure 3. Mapping kernel functions: Effect on the separation hyperplane of two-class datasets.

In SVM method, there are different types of specific kernel functions to improve the classifier, including the linear kernel (the easiest to interpret), Gaussian, or radial basis function kernel (RBF), quadratic, cubic, etc. These kernels differ in the complexity of definition and precision in the classification of different classes. In this work, both quadratic and cubic kernel functions were used.

Two approaches that combine multiple binary SVMs were used to address multiclass classification problems: One versus one (OVO) and one versus all (OVA). The OVO approach needs $m \cdot (m - 1)/2$ SVM classifiers to distinguish between m classes [2]. The classifiers are trained to differentiate the samples of one class from those of another class. Based upon a vote of each SVM, an unknown pattern is classified. Thus, the strategy to accomplish a single class decision follows a majority voting scheme based on $sign\left(y_i \cdot (w^T \cdot x_i + b)\right)$ [52]. The class that wins the most votes is the one predicted for x. This winning class is directly assigned to the test pattern.

4.3. Decision Tree

The decision tree is a classification tool, based on decision rules, which uses a binary tree graph to find an unknown relationship between input and output parameters. A typical tree structure is characterized by internal nodes representing test on attributes, branches symbolizing outcomes of the test, and leaf nodes (or terminal nodes) defining class labels. Decisions at a node are taken with the help of rules obtained from data [43,53].

The DT should have as many levels as necessary to classify the input feature data. Depending on the number of levels of this DT, the classification can be more or less accurate, and more or less calculation complex. A key point, in this sense, is the suitable choice of the maximum number of splits. It is well known that high classification accuracy on the training dataset can be achieved through a fine tree with many leaves. However, such a leafy tree usually overfits the model and often reduces its validation accuracy in respect of the proper training accuracy. On the contrary, coarse trees do not reach such a high training accuracy, but they are easier to interpret and can also be more robust in the sense of approaching the accuracy between both training and representative test dataset.

Based upon the foregoing and in order to achieve the required degree of accuracy, in this work, the maximum number of splits was set to 91 and the so-called Gini's diversity index was chosen as the split criterion. At a node, this index was defined as follows

$$GINI\ index = 1 - \sum_j p_j^2 \tag{6}$$

and it is the probability of class *j* complying with the criteria of the selected node. Gini's diversity index gives an estimation of node impurity since the optimization procedure in tree classifiers tends to nodes with just one class (pure nodes). Thus, a Gini index of 0 is derived from nodes that contain only one class; otherwise, the Gini index is positive. Therefore, the optimal situation for a given dataset is to achieve a Gini index with a value as small as possible.

4.4. Bagged Decision Tree Ensemble

Ensemble classifier compiles the results of many weak learners and combines them into a single high quality ensemble model. The quality of that approach depends on the type of algorithm chosen. In this study, the selected bagged tree classifiers were based on Breiman's random forest algorithm [54]. In the bagged method, the original group of data divides into different datasets by random selection with replacement, and then a classification of each one of them is obtained by a decision tree method. The result of each learner is submitted to a voting process and the winner finally sets the best classification model of the bagged DT Ensemble method.

This method permits obtaining lower data variance than a single DT and also gets a reduced over-adjustment. The model can be improved by properly selecting the number of learners. It should be noted that a large number of them can produce high accuracy but also slow down the classification process. In this work, a compromise solution was found by setting the number of learners to 30.

5. Full Comparative Classification of PQDs: Detailed Overview

A detailed overview of the proposed hybrid classification plan is shown in Figure 4, where the main steps described in previous sections have been highlighted. The first one is the analysis stage, where signal processing of the PQDs was obtained via S-transform. Then, an initial feature set, which was defined by statistical parameters, was extracted. Next, feature vectors were optimized employing both GA and CSO algorithms, including an extra validation provided by K-NN algorithm. The last stage consisted of classification involving the determination of PQ multi-event by using DT (fine tree), bagged decision tree ensemble, weighted K-NN, and both quadratic and cubic SVMs.

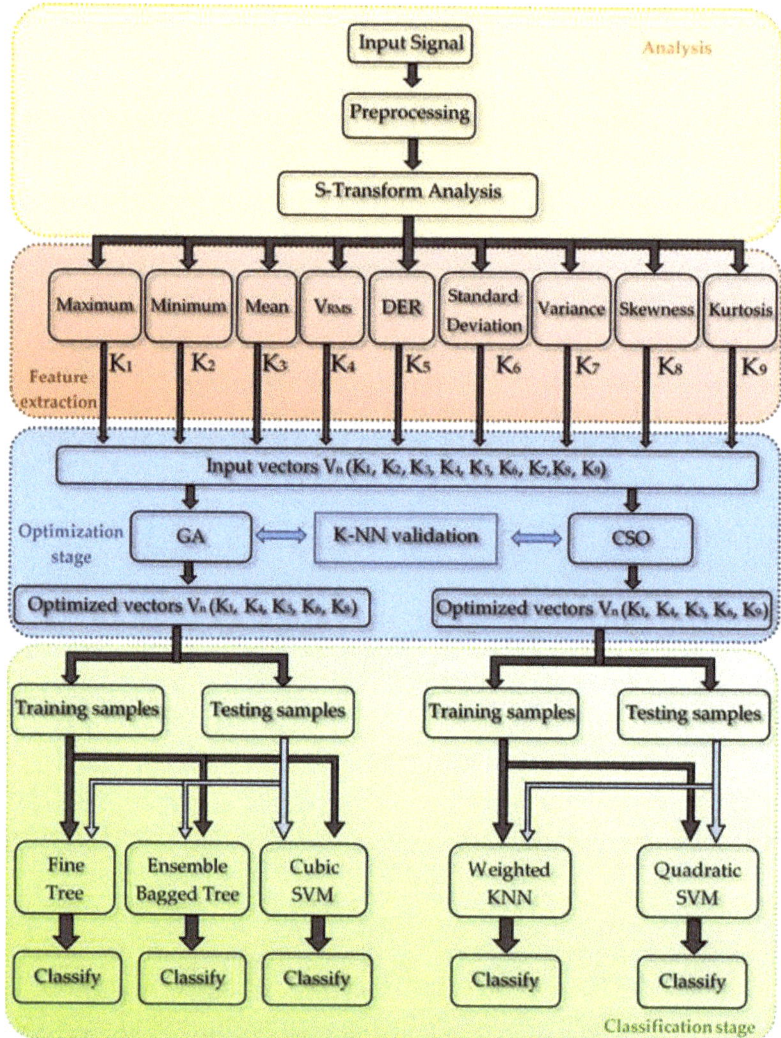

Figure 4. Comparative flowchart of the proposed disturbance classification.

6. Field Data Synthesis of PQDs

In this section, PQDs' synthesis and the resulting training datasets are detailed.

6.1. Generation of PQ Disturbances

A database of 13 PQDs that include both single and multiple events was generated in accordance with both IEEE-1459 and EN-50160 standards [55,56]. The MATLAB R2017a software was used to program a virtual signal generator, called SIGEN, allowing for a customizable setup of the simulated signals required for testing. Thus, SIGEN generated a pattern in each category of events by varying the parameters that were controllable by the user. As a result, both steady-state and transient-state disturbances could be modelled. The graphical user interface of single-phase SIGEN is demonstrated in Figure 5.

The processes of SIGEN were performed to complete the effective generation of electric signals, as follows:

- First, electrical input signals were defined and the user set their parameters.
- Second, signals according to these specifications were built.
- Finally, the synthesized signals could be sent either to a file or a data acquisition board (DAQ).

SIGEN was designed following the guidelines described in the IEEE-1159 standard for monitoring electric power quality [57].

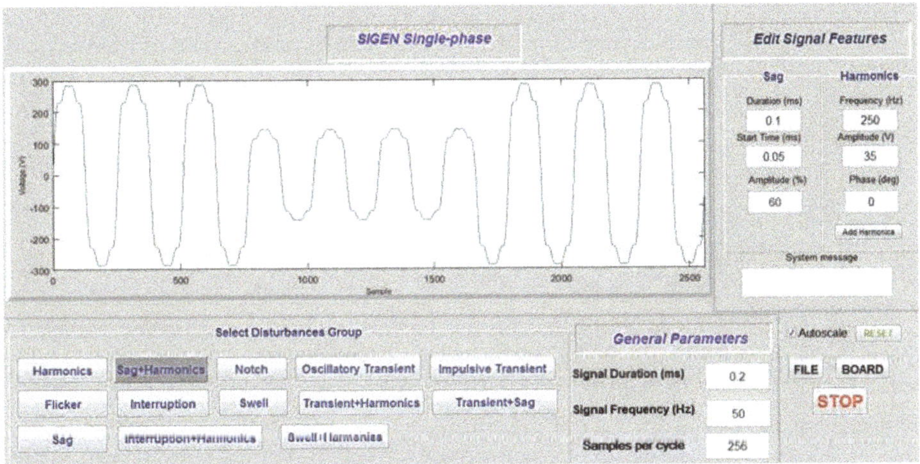

Figure 5. Graphical interface of signal generator SIGEN.

All the signals had in common the following fundamental specifications: Voltage = 230 V RMS (root mean square), frequency = 50 Hz, duration = 0.2 s, sampling frequency = 12.8 kHz, total cycles = 10, total samples = 2560. Stationary and/or transient disturbances, as well as white Gaussian noise, were added to this fundamental signal. The levels of added noise were characterized by the signal-to-noise (SNR) ratio of 20 dB, 30 dB, 40 dB, and 50 dB. For each noise level and each PQD category, 100 signals were generated through SIGEN by varying all the parameters. This set had a total of 5600 simulated signals, 1400 for each noise level involving all categories (13 types of the PQDs plus one pure sinusoidal signal). These signal sets transformed in the dataset intended to verify the classification performance, as will be seen next.

Another set of signals was needed for the feature selection stage. It included 560 simulated signals (40 signals × 14 PQD categories) with random SNR between 20 dB and 50 dB.

6.2. Training/Testing and Validation Datasets

Once statistical features were computed based on the ST matrix, datasets were created. A first dataset contained 560 × 9 data (samples × features) and it was used as input of the K-NN-based fast validation tool for optimal feature selection (see Sections 3.1 and 3.2).

As concluded in Section 3, the choice of either of the selected optimization algorithms (GA and CSO) allowed obtaining a feature set which comprised only five features. Thus, for each of the four specific noise levels considered, the second type of dataset containing 1400 × 5 data (samples × features) was used to fit the different proposed classification models. This training dataset was partitioned into train/test datasets like 80/20%, respectively. Then, while the test dataset was kept aside, the train set split again into the actual train dataset (80% again) and the validation set (the remaining 20%). Data for each subset was randomly selected. This cross-validation procedure iteratively trained and validated the models on these sets, avoiding overfitting. In this study, a 10-folds cross-validation method was used to evaluate the performance of the proposed classifiers.

It must be noted that the mentioned training dataset was different to that one used in the previous feature selection stage. This ensured that no observations that were a part of the feature selection task were a part of the classification task.

6.3. PQDs' Classes

The following 14 classes were tested and labelled as: Harmonic (C1), Flicker (C2), Sags (C3), Sags and harmonic (C4), Interruption (C5), Interruption and harmonic (C6), Notch (C7), Swells (C8), Swells and harmonic (C9), Oscillatory transient (C10), Oscillatory transient and harmonic (C11), Oscillatory transient and sag (C12), Impulsive transient (C13), and pure sinusoidal signal (C14). In Table 2 the simulated PQDs waveforms are depicted.

Table 2. Synthetized PQDs' waveforms.

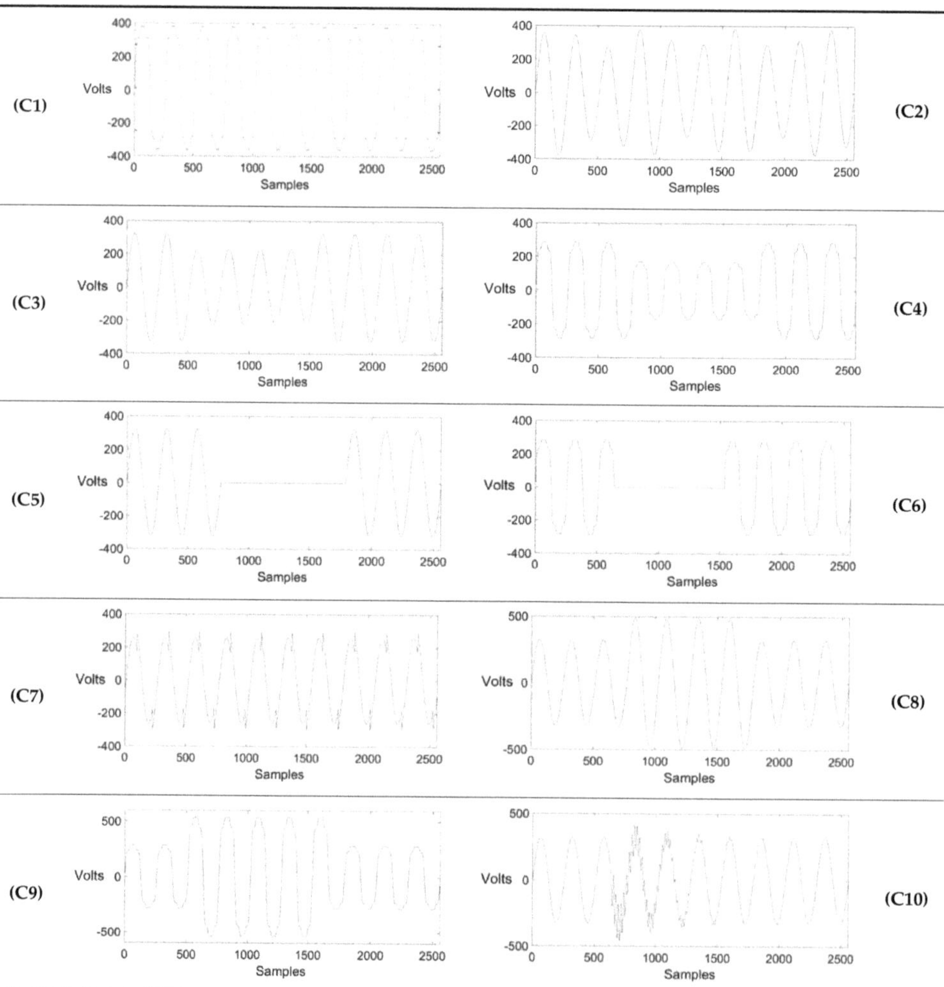

Table 2. *Cont.*

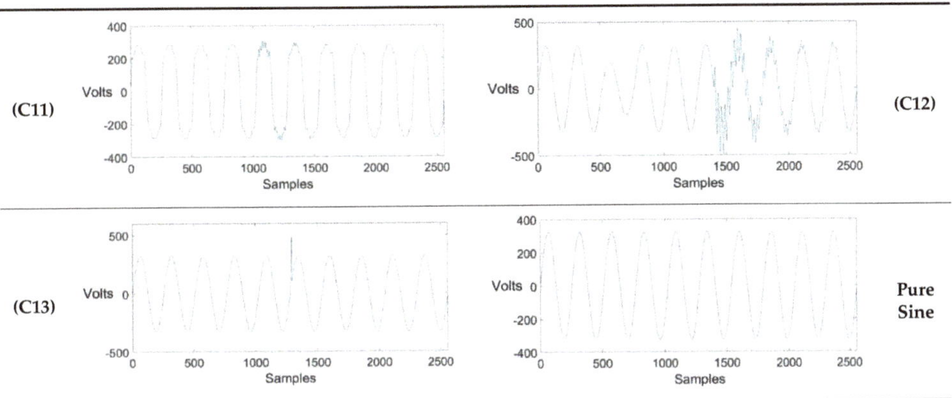

7. Results and Performance Comparison

Depending on the optimization and pattern recognition approach selected, five hybrid classification methods were considered: CSO&QSVM (quadratic SVM), CSO&WK-NN (weighted K-NN), GA&FTree (fine tree), GA&ETree (ensemble tree), and GA&CSVM (cubic SVM). In Table 3, the classification accuracy on training dataset under different noise conditions of the proposed models are listed.

Table 3. Recognition accuracy comparison between proposed methods under noisy conditions.

Accuracy Comparison between Proposed Models (%)				
Optimization &	**SNR**			
Classification	**20 dB**	**30 dB**	**40 dB**	**50 dB**
CSO&QSVM	97.6	99.2	99.85	99.95
CSO&WK-NN	95.7	98.6	99.7	99.75
GA&FTree	93.4	97.7	99.3	99.5
GA&ETree	94.6	97.5	99.4	99.55
GA&CSVM	94.4	97.7	99.0	99.4

The best results of classification accuracy were obtained using CSO&QSVM in any noise conditions. The results for 30 dB SNR and above had high classification accuracy for both CSO&QSVM and CSO&WK-NN, and the rest of the models had acceptable values. In high noisy conditions (20 dB), the accuracy was satisfactory only for CSO&QSVM. The values obtained applying the rest of the methods were lower and required the next analysis with separate PQDs for a better interpretation of the results.

In order to evaluate noise immunity under different accuracy requirements, single-class test datasets (each with 25×5 data of a unique class) were subjected to the trained classifiers separately. The noise threshold (minimum SNR value) for each class was determined through a trial-and-error method, by applying the principle of keeping SNR value as low as possible while maintaining the targeted classification accuracy. Thus, if the classification accuracy on an initial single-class dataset was, for example, lower than 80%, a new dataset with higher SNR value was generated and tested. The iterative procedure was stopped when the pretended accuracy was achieved. In Table 4, a comparison of minimum SNR values using these methods for different accuracy rates and distinct PQDs classes is presented. In each specific case, the maximum level of noise allowing the pretended accuracy rate (80%, 90%, and 100%) is shown. In the last row, the overall SNR average threshold value of each column is determined. This value was just an estimation of the noise immunity for a pretended accuracy rate in the separate classification process.

Table 4. Comparison of noise threshold (SNR minimum) for classification accuracy of the models proposed.

		Minimum SNR (dB) for Individual PQD Classification Using the Models Proposed and According to 3 Levels of Accuracy													
		CSO Optimizer								Genetic Algorithm					
		Quadratic SVM			Weighted K-NN			Fine Tree			Ensemble Tree			Cubic SVM		
PQDs	Accuracy	100%	90%	80%	100%	90%	80%	100%	90%	80%	100%	90%	80%	100%	90%	80%
	Harmonic	26.7	26	25.3	26.65	25.4	25	25.9	25.1	24.8	25.5	25.2	24.9	X	X	X
	Flicker	23	19.3	17.6	24.7	24.3	23.9	14.4	13.2	12.9	13.5	13.1	12.8	19.4	18.5	18.1
	Sag	18.8	18.1	17.2	22.7	21.7	21.35	15.4	14.6	14.3	12.6	12.35	12.1	12.7	12.3	12.1
	Sag + H	19	18	16.3	16.8	15.8	15.3	15.2	14	13.6	19	16.5	15.6	19	17.2	15.8
	Interruption	14	13.2	12.7	17.5	16.3	15.9	11.5	11.15	10.9	11.5	11.1	10.85	18	16.2	14.5
	Interruption + H	10	9.3	8.8	11.3	10.8	10.5	X	X	X	X	X	X	X	X	X
	Notch	13	11.4	10.9	16.5	15.8	14	12.4	11.85	11.6	12.4	10.95	10.65	X	X	X
	Swell	15.9	15.3	14.9	15.4	14.95	14.65	16.25	15.9	15.65	10.95	9.9	9.2	15.4	14.85	14.6
	Swell + H	13.5	12.85	12.6	14.5	14.25	14	X	X	X	X	X	X	12.5	11.6	11
	Osc. Transient	21	19.7	19.35	24.4	22.6	21.7	15	13.5	12.4	13.8	12.7	10.5	12.55	11.7	11.1
	Osc. Transient + H	22.8	20.2	19.6	25.3	23.7	22.45	16.75	16	13.8	13.1	10.6	9.5	17	15.8	13.1
	Osc. Transient + Sag	21.7	19.4	18.9	19.85	19.3	19.1	14.6	13.9	13	11.45	9.9	9	16.8	15.6	13.6
	Impulsive Transient	22	20.4	19.8	19	18.35	18.2	18.4	17.2	15	21.2	18	16.6	15.3	14.5	13.6
	Overall SNR (dB)	18.57	17.16	16.46	19.54	18.71	18.16	15.95	15.13	14.36	15.02	13.66	12.85	15.87	14.82	13.75

From Table 4 it can be seen that the GA-based models offered good SNR values for most PQDs, often even better than those resulting from CSO algorithm, especially when dealing with single events. However, none of the suggested GA-based models was capable to classify properly the multiple interruption plus harmonic disturbances. In the same terms of working with five optimal features, both GA&FTree and GA&ETree also misclassified swell plus harmonic events and GA&CSVM could not resolve notches and harmonics disturbances. This lack of identification of a separated PQDs could justify the modest results obtained by GA-based models in Table 3. On the contrary, the proposed CSO-based models achieved good individual noise rates for a separate classification of classes according to different accuracy targets. As a result, both CSO&QSVM and CSO&WK-NN methods presented an overall SNR average threshold under 20 dB, specifically 18.57 dB and 19.54 dB, respectively. When both QSVM and WK-NN classifiers acted on single-class datasets, this slightly lower SNR value in the CSO&QSVM method suggested a better performance with regard to noise immunity. But this behavior in high noisy conditions was not only valid for single-class datasets, but also for the complete dataset. As shown above, in Table 3, CSO&QSVM obtained the best results also with the complete training dataset.

On the other hand, in Table 5, performances of the best-proposed CSO-QSVM method are compared to those of other classification methods already reported in the literature. It can be seen that some methods reported information of accuracy with noise levels (SNR) up to 50 dB [37,58], 40 dB [27], and 30 dB [42]. Others [17,59,60], although reaching 20 dB, had low accuracy. It is remarkable the high impact of noise in the accuracy of the wavelet-based approaches [37,60]. The remaining works presented acceptable accuracy [15,38,39,41,43] but none was capable to deal with 13 PQDs as the actual proposal did. Only [15] achieved similar performances than those of the proposed CSO-QSVM approach. That studied 12 kinds of single and multiple PQDs through the use of improved principal component analysis (IPCA) and 1-dimensional convolution neural network (1-D-CNN) and achieved an overall accuracy of 99.76% and 99.85% for 20 dB and 50 dB, respectively. That classifier needed six features, distinct from the optimal feature set proposed in this work, which was composed of only five of them.

Table 5. Performance comparison of different methods.

Literature	Processing & Recognition	N° of PQDs	N° of Features	Recognition Accuracy (%)			
				20 dB	30 dB	40 dB	50 dB
[15]	PCA + CNN	12	6	99.76	-	-	99.85
[17]	Fast ST + Embedded DT	12	6	91.50	98.58	98.83	98.92
[27]	Mod. ST + DT + PSO	13	6	-	-	98.38	-
[37]	Wavelet + SVM	16	6	-	-	-	95.56
[38]	ST+DAG-SVM	9	9	97.77	-	-	-
[39]	Modified ST + SVM	8	19	98	-	-	-
[41]	ST + DT	7	6	98.1	-	-	-
[42]	FST + Fuzzy DT	13	5	-	97.95	98.67	-
[43]	ST + DT+ Fuzzy C-M	10	14	99.3	-	-	-
[58]	ST + ANN + DT	13	-	-	-	-	99.9
[59]	ST + PNN	9	4	98.38	98.63	99.13	-
[60]	Wavelet +PNN	14	5	86.86	91.93	93.71	94.57
Proposed	ST + CSO + SVM	13	5	97.6	99.2	99.85	99.95

Table 5 also displays a comparative in terms of feature dimension in each reported approach. The ST-based probabilistic neural network (PNN) approach [59] accepted a set with four features but only dealt with nine PQDs, while PNN based on wavelet [60] was treated with 14 PQDs but with poor noise immunity as indicated by its accuracy rates. This comparative study shows that the proposed CSO-QSVM model, at last, equaled the better results of classification accuracy obtained in the literature, but using only five features per sample and dealing with 13 PQDs classes. These results, together with the comparison between alternative proposals (Table 3) and the detailed analysis of noise immunity (Table 4), constitute the main contributions of this paper.

Although the present work dealt with simulated signals, the results were so good that they could be extrapolated when applied to experimental data. In such a case, a comparative with those studies based on real signals could be applied properly.

As a future extension, an experimental setup would be used to test the effectiveness of the proposed hybrid methods under common real-time working conditions. Emulated PQ incidence on distribution networks could be modelled by low-cost hardware prototyping and software components.

8. Conclusions

The motivation for this work stemmed from challenges facing the electrical systems and equipment in determining optimal, cost-effective, and efficient power quality management. In this way, this paper addressed the optimal hybrid classification methods based on machine learning approaches for meeting detection, identification, and classification of simulated PQDs. Specifically, ST was selected for detection and feature extraction of PQDs, and, following the trend nowadays to further optimize the recognition approach, several optimization algorithms were tested for optimal feature selection. At this step, this work underlined the GA and CSO algorithms since they achieved the best results. The resulting optimal feature sets were fed to several classifiers, highlighting among them the QSVM, CSVM, FTree, ETree, and WK-NN approaches for showing improved performance.

The GA optimization algorithm associated with the FTree, ETree, and CSVM approaches could not classify properly all PQDs under the conditions established in this analysis. However, the results obtained through these approaches were very promising and showed the great potential of these kinds of models when dealing with a certain group of PQDs.

Alternatively, CSO-based methods including CSO-QSVM and CSO-WK-NN achieved high classification accuracy under noisy conditions. A thorough comparative assessment in terms of noise immunity and classification accuracy led us to conclude that the proficiency of CSO-QSVM method is slightly better than CSO-WK-NN method.

It can also be noted that the results found seemed to confirm the current trend by which, despite the optimization based on GA algorithms being highlighted by their efficiency, GA-based methodologies are progressively being replaced by the swarm optimization algorithms.

Finally, performances of CSO-QSVM method were compared to those of other classification methods already reported in the literature, concluding that the proposed method achieved a higher degree of efficiency than most of them, and, based on the results, it may work well under high noise background in practical applications.

Author Contributions: J.C.B.-R. conceived and developed the idea of this research, designed the whole structure of the comparative study, and wrote the paper; F.J.T. generated the data, performed the simulations, and contributed to the methodology; M.D.B. provided the theoretical background of the proposed methodology and contributed to it. All authors have read and agreed to the published version of the manuscript.

Funding: This research was funded by the Universidad de Sevilla (VI Plan Propio de Investigación y Transferencia) under grant 2020/00000596.

Conflicts of Interest: The authors declare no conflict of interest.

Appendix A

The continuous Stockwell transform (ST) [1] of a signal $x(t)$ was originally described as

$$S(\tau, f) = e^{i2\pi\pi f\tau}W(\tau, d) \tag{A1}$$

i.e., a phase factor acting on a continuous wavelet transform (CWT)

$$W(\tau, d) = \int_{-\infty}^{+\infty} x(t)\mu(\tau - t, d)\,dt \tag{A2}$$

where τ is a time displacement factor, d is a frequency-scale dilation factor, and $\mu(t, d)$ is a specific mother wavelet that includes the frequency-modulated Gaussian window

$$\mu(t, d) = \frac{|f|}{\sqrt{2\pi}} e^{-\frac{t^2 f^2}{2}} e^{-i2\pi n ft}. \tag{A3}$$

Instead, in this paper, the *STFT* pathway to address the ST definition was preferred,

$$STFT(\tau, f) = \int_{-\infty}^{+\infty} x(t)w(\tau - t)e^{-i2\pi n ft}\,dt \tag{A4}$$

From this approach, ST can be written as

$$S(\tau, f) = \int_{-\infty}^{+\infty} x(t)w(\tau - t, f)e^{-i2\pi n ft}\,dt \tag{A5}$$

where $w(t, f)$ is the Gaussian window function similar to that proposed by Gabor (1946), but now also introducing the aforementioned added frequency dependence

$$w(t, f) = \frac{|f|}{\sqrt{2\pi}} e^{-\frac{t^2 f^2}{2}} \tag{A6}$$

where the inverse of the frequency $1/|f|$ represents the window width.

From either of the two viewpoints, the complete ST definition can be written as

$$S(\tau, f) = \int_{-\infty}^{+\infty} x(t)\frac{|f|}{\sqrt{2\pi}} e^{-\frac{(\tau-t)^2 f^2}{2}} e^{-i2\pi ft}\,dt \tag{A7}$$

and also can be represented in terms of its relationship with FT and related spectrum $X(f)$ of $x(t)$

$$S(\tau, f) = \int_{-\infty}^{+\infty} X(\alpha + f)e^{\frac{2\pi^2 \alpha^2}{f^2}} e^{i2\pi\alpha\tau}\,d\alpha \quad f \neq 0. \tag{A8}$$

As is well known, the way to recover the original signal from continuous ST is expensive in terms of data storage due to oversampling. This sampled version of the ST permits calculating the widely used ST complex matrix that is obtained as $(\tau \to jT, f \to n/NT)$

$$\begin{cases} S\left[j\tau, \frac{n}{NT}\right] = \sum\limits_{m=0}^{N-1} X\left[\frac{m+n}{NT}\right]G(m, n)e^{\frac{i2\pi mj}{N}} &, n \neq 0 \\ S[j\tau, 0] = \frac{1}{N}\sum\limits_{m=0}^{N-1} X\left[\frac{m}{NT}\right] &, n = 0 \end{cases} \tag{A9}$$

where T denotes the sampling interval, N is the total number of sample points, and both $X\left[\frac{m+n}{NT}\right]$ and $G(m, n)$ result after discrete fast Fourier transform (FFT), respectively, on the PQ disturbance signal $x(t)$ and the Gaussian window function $w(t, f)$:

$$X\left[\frac{n}{NT}\right] = \frac{1}{N}\sum\limits_{k=0}^{N-1} x[kT]e^{-\frac{i2\pi nk}{N}} \tag{A10}$$

$$G(m,n) = e^{\frac{-2\pi^2 m^2}{n^2}} \tag{A11}$$

where j, k, m, and n are integers in the range of 0 to $N - 1$.

The result of discrete ST is a 2D time-frequency matrix that is represented as

$$S(\tau, f) = A(\tau, f)e^{-i\phi(\tau, f)} \tag{A12}$$

where $A(\tau, f)$ is the amplitude and $\phi(\tau, f)$ is the phase. Each column contains the frequency components present in the signal at a particular time. Each row displays the magnitude of a particular frequency with time varying from 0 to $N - 1$ samples.

References

1. Stockwell, R.; Mansinha, L.; Lowe, R. Localization of the complex spectrum: The S transform. *IEEE Trans. Signal Process.* **1996**, *44*, 998–1001. [CrossRef]
2. Borras, M.D.; Bravo, J.C.; Montano, J.C. Disturbance Ratio for Optimal Multi-Event Classification in Power Distribution Networks. *IEEE Trans. Ind. Electron.* **2016**, *63*, 3117–3124. [CrossRef]
3. Bravo, J.C.; Borras, M.D.; Torres, F.J. Development of a smart wavelet-based power quality monitoring system. In Proceedings of the 2018 International Conference on Smart Energy Systems and Technologies (SEST), Seville, Spain, 10–12 September 2018; pp. 1–6.
4. Borras, M.-D.; Montano, J.-C.; Castilla, M.; López, A.; Gutierrez, J.; Bravo, J.-C. Voltage index for stationary and transient states. In Proceedings of the Mediterranean Electrotechnical Conference (MELECON), Valletta, Malta, 26–28 April 2010; pp. 679–684.
5. Sahani, M.; Dash, P. FPGA-Based Online Power Quality Disturbances Monitoring Using Reduced-Sample HHT and Class-Specific Weighted RVFLN. *IEEE Trans. Ind. Inform.* **2019**, *15*, 4614–4623. [CrossRef]
6. Afroni, M.J.; Sutanto, D.; Stirling, D. Analysis of Nonstationary Power-Quality Waveforms Using Iterative Hilbert Huang Transform and SAX Algorithm. *IEEE Trans. Power Deliv.* **2013**, *28*, 2134–2144. [CrossRef]
7. Xi, Y.; Li, Z.; Zeng, X.; Tang, X.; Liu, Q.; Xiao, H. Detection of power quality disturbances using an adaptive process noise covariance Kalman filter. *Digit. Signal Process.* **2018**, *76*, 34–49. [CrossRef]
8. Nie, X. Detection of Grid Voltage Fundamental and Harmonic Components Using Kalman Filter Based on Dynamic Tracking Model. *IEEE Trans. Ind. Electron.* **2020**, *67*, 1191–1200. [CrossRef]
9. He, S.; Zhang, M.; Tian, W.; Zhang, J.; Ding, F. A Parameterization Power Data Compress Using Strong Trace Filter and Dynamics. *IEEE Trans. Instrum. Meas.* **2015**, *64*. [CrossRef]
10. Manikandan, M.S.; Samantaray, S.R.; Kamwa, I. Detection and Classification of Power Quality Disturbances Using Sparse Signal Decomposition on Hybrid Dictionaries. *IEEE Trans. Instrum. Meas.* **2014**, *64*, 27–38. [CrossRef]
11. Cho, S.-H.; Jang, G.; Kwon, S.-H. Time-Frequency Analysis of Power-Quality Disturbances via the Gabor–Wigner Transform. *IEEE Trans. Power Deliv.* **2009**, *25*, 494–499. [CrossRef]
12. Lopez-Ramirez, M.; Ledesma-Carrillo, L.M.; Cabal-Yepez, E.; Rodriguez-Donate, C.; Miranda-Vidales, H.; Garcia-Perez, A. EMD-Based Feature Extraction for Power Quality Disturbance Classification Using Moments. *Energies* **2016**, *9*, 565. [CrossRef]
13. Camarena-Martinez, D.; Valtierra-Rodriguez, M.; Perez-Ramirez, C.A.; Amezquita-Sanchez, J.P.; Romero-Troncoso, R.D.J.; Garcia-Perez, A. Novel Downsampling Empirical Mode Decomposition Approach for Power Quality Analysis. *IEEE Trans. Ind. Electron.* **2015**, *63*, 2369–2378. [CrossRef]
14. Wu, X.; Kumar, V.; Ross Quinlan, J.; Ghosh, J.; Yang, Q.; Motoda, H.; McLachlan, G.J.; Ng, A.; Liu, B.; Yu, P.S.; et al. Top 10 algorithms in data mining. *Knowl. Inf. Syst.* **2008**, *14*, 1–37. [CrossRef]
15. Shen, Y.; Abubakar, M.; Liu, H.; Hussain, F. Power Quality Disturbance Monitoring and Classification Based on Improved PCA and Convolution Neural Network for Wind-Grid Distribution Systems. *Energies* **2019**, *12*, 1280. [CrossRef]
16. Erişti, H.; Yıldırım, Ö.; Erişti, B.; Demir, Y. Optimal feature selection for classification of the power quality events using wavelet transform and least squares support vector machines. *Int. J. Electr. Power Energy Syst.* **2013**, *49*, 95–103. [CrossRef]

17. Huang, N.; Peng, H.; Cai, G.; Chen, J. Power Quality Disturbances Feature Selection and Recognition Using Optimal Multi-Resolution Fast S-Transform and CART Algorithm. *Energies* **2016**, *9*, 927. [CrossRef]

18. Moon, S.-K.; Kim, J.-O.; Kim, C. Multi-Labeled Recognition of Distribution System Conditions by a Waveform Feature Learning Model. *Energies* **2019**, *12*, 1115. [CrossRef]

19. Huang, N.; Lu, G.; Cai, G.; Xu, D.; Xu, J.; Li, F.; Zhang, L. Feature Selection of Power Quality Disturbance Signals with an Entropy-Importance-Based Random Forest. *Entropy* **2016**, *18*, 44. [CrossRef]

20. Jamali, S.; Farsa, A.R.; Ghaffarzadeh, N. Identification of optimal features for fast and accurate classification of power quality disturbances. *Meas. J. Int. Meas. Confed.* **2018**, *116*, 565–574. [CrossRef]

21. Singh, U.; Singh, S.N. Optimal Feature Selection via NSGA-II for Power Quality Disturbances Classification. *IEEE Trans. Ind. Inform.* **2017**, *14*, 2994–3002. [CrossRef]

22. Ray, P.K.; Mohanty, S.R.; Kishor, N.; Catalão, J.P. Optimal Feature and Decision Tree-Based Classification of Power Quality Disturbances in Distributed Generation Systems. *IEEE Trans. Sustain. Energy* **2013**, *5*, 200–208. [CrossRef]

23. Dehini, R.; Sefiane, S. Power quality and cost improvement by passive power filters synthesis using ant colony algorithm. *J. Theor. Appl. Inf. Technol.* **2011**, *23*, 70–79.

24. Singh, U.; Singh, S.N. A new optimal feature selection scheme for classification of power quality disturbances based on ant colony framework. *Appl. Soft Comput.* **2019**, *74*, 216–225. [CrossRef]

25. Wang, J.; Xu, Z.; Che, Y. Power Quality Disturbance Classification Based on DWT and Multilayer Perceptron Extreme Learning Machine. *Appl. Sci.* **2019**, *9*, 2315. [CrossRef]

26. Chen, Z.; Han, X.; Fan, C.; Zheng, T.; Mei, S. A Two-Stage Feature Selection Method for Power System Transient Stability Status Prediction. *Energies* **2019**, *12*, 689. [CrossRef]

27. Huang, N.; Zhang, S.; Cai, G.; Xu, D. Power Quality Disturbances Recognition Based on a Multiresolution Generalized S-Transform and a PSO-Improved Decision Tree. *Energies* **2015**, *8*, 549–572. [CrossRef]

28. Vidhya, S.; Kamaraj, V. Particle swarm optimized extreme learning machine for feature classification in power quality data mining. *Automatika* **2017**, *58*, 487–494. [CrossRef]

29. Cheng, R.; Jin, Y. A competitive swarm optimizer for large scale optimization. *IEEE Trans. Cybern.* **2014**, *45*, 191–204. [CrossRef]

30. Gu, S.; Cheng, R.; Jin, Y. Feature selection for high-dimensional classification using a competitive swarm optimizer. *Soft Comput.* **2018**, *22*, 811–822. [CrossRef]

31. Niu, B.; Yang, X.; Wang, H. Feature Selection Using a Reinforcement-Behaved Brain Storm Optimization. In *Intelligent Computing Methodologies*; Huang, D.-S., Huang, Z.-K., Hussain, A., Eds.; Springer International Publishing: Cham, Switzerland, 2019; Volume 11645, pp. 672–681.

32. Sharaf, A.M.; El-Gammal, A.A.A. A discrete particle swarm optimization technique (DPSO) for power filter design. In Proceedings of the 2009 4th International Design and Test Workshop (IDT), Riyadh, Saudi Arabia, 15–17 November 2009. [CrossRef]

33. Zhao, L.; Long, Y. An Improved PSO Algorithm for the Classification of Multiple Power Quality Disturbances. *J. Inf. Process. Syst.* **2019**, *15*, 116–126. [CrossRef]

34. De Yong, D.; Bhowmik, S.; Magnago, F. Optimized Complex Power Quality Classifier Using One vs. Rest Support Vector Machines. *Energy Power Eng.* **2017**, *9*, 568–587. [CrossRef]

35. Janik, P.; Lobos, T. Automated Classification of Power-Quality Disturbances Using SVM and RBF Networks. *IEEE Trans. Power Deliv.* **2006**, *21*, 1663–1669. [CrossRef]

36. Hajian, M.; Foroud, A.A. A new hybrid pattern recognition scheme for automatic discrimination of power quality disturbances. *Measurement* **2014**, *51*, 265–280. [CrossRef]

37. Thirumala, K.; Pal, S.; Jain, T.; Umarikar, A.C. A classification method for multiple power quality disturbances using EWT based adaptive filtering and multiclass SVM. *Neurocomputing* **2019**, *334*, 265–274. [CrossRef]

38. Li, J.; Teng, Z.; Tang, Q.; Song, J. Detection and Classification of Power Quality Disturbances Using Double Resolution S-Transform and DAG-SVMs. *IEEE Trans. Instrum. Meas.* **2016**, *65*, 2302–2312. [CrossRef]

39. Noh, F.H.M.; Rahman, M.A.; Yaakub, M.F. Performance of Modified S-Transform for Power Quality Disturbance Detection and Classification. *Telkomnika* **2017**, *15*, 1520–1529.

40. Zhong, T.; Zhang, S.; Cai, G.; Huang, N. Power-Quality disturbance recognition based on time-Frequency analysis and decision tree. *IET Gener. Transm. Distrib.* **2018**, *12*, 4153–4162. [CrossRef]

41. Alqam, S.J.; Zaro, F.R. Power Quality Detection and Classification Using S-Transform and Rule-Based Decision Tree. *Int. J. Electr. Electron. Eng. Telecommun.* **2019**, 1–6. [CrossRef]

42. Biswal, M.; Dash, P.K. Measurement and Classification of Simultaneous Power Signal Patterns With an S-Transform Variant and Fuzzy Decision Tree. *IEEE Trans. Ind. Inform.* **2012**, *9*, 1819–1827. [CrossRef]
43. Mahela, O.P.; Shaik, A.G. Recognition of power quality disturbances using S-transform based ruled decision tree and fuzzy C-means clustering classifiers. *Appl. Soft Comput.* **2017**, *59*, 243–257. [CrossRef]
44. Samantaray, S. Decision tree-Initialised fuzzy rule-Based approach for power quality events classification. *IET Gener. Transm. Distrib.* **2010**, *4*, 538. [CrossRef]
45. Goldberg, D.E.; Holland, J.H. Genetic Algorithms and Machine Learning. *Mach. Learn.* **1988**, *3*, 95–99. [CrossRef]
46. Siedlecki, W.; Sklansky, J. A note on genetic algorithms for large-scale feature selection. *Pattern Recognit. Lett.* **1989**, *10*, 335–347. [CrossRef]
47. Kang, M.; Kim, J.; Wills, L.; Kim, J.-M. Time-Varying and Multiresolution Envelope Analysis and Discriminative Feature Analysis for Bearing Fault Diagnosis. *IEEE Trans. Ind. Electron.* **2015**, *62*, 7749–7761. [CrossRef]
48. Fan, G.-F.; Guo, Y.-H.; Zheng, J.-M.; Hong, W.-C. Application of the Weighted K-Nearest Neighbor Algorithm for Short-Term Load Forecasting. *Energies* **2019**, *12*, 916. [CrossRef]
49. Vapnik, V.N. *The Nature of Statistical Learning Theory*; Springer: Berlin/Heidelberg, Germany, 1995.
50. Cortes, C.; Vapnik, V. Support-Vector networks. *Mach. Learn.* **1995**, *20*, 273–297. [CrossRef]
51. Ozgonenel, O.; Yalcin, T.; Guney, I.; Kurt, Ü. A new classification for power quality events in distribution systems. *Electr. Power Syst. Res.* **2013**, *95*, 192–199. [CrossRef]
52. De Yong, D.; Bhowmik, S.; Magnago, F. An effective Power Quality classifier using Wavelet Transform and Support Vector Machines. *Expert Syst. Appl.* **2015**, *42*, 6075–6081. [CrossRef]
53. A Survey on Decision Tree Algorithms of Classification in Data Mining. *Int. J. Sci. Res.* **2016**, *5*, 2094–2097. [CrossRef]
54. Breiman, L. Random Forests. *Mach. Learn.* **2001**, *45*, 5–32. [CrossRef]
55. IEEE Power and Energy Society. *IEEE Standard Definitions for the Measurement of Electric Power Quantities Under Sinusoidal, Nonsinusoidal, Balanced, or Unbalanced Conditions*; IEEE Std 1459-2010 (Revision IEEE Std 1459-2000); IEEE: New York City, NY, USA, 2010; pp. 1–50.
56. European Committee for Electrotechnical Standardization. *Voltage Characteristics of Electricity Supplied by Public Distribution Networks*; EN-50160 2011; Eur. Std: 2010; CENELEC: Brussels, Belgium, 2010.
57. IEEE Power and Energy Society. *IEEE Recommended Practice for Monitoring Electric Power Quality*; IEEE Std 1159-2019 (Revision IEEE Std 1159-2009); IEEE: New York City, NY, USA, 2019; pp. 1–98.
58. Kumar, R.; Singh, B.; Shahani, D.T.; Chandra, A.; Al-Haddad, K.; Garg, R. Recognition of Power-Quality Disturbances Using S-Transform-Based ANN Classifier and Rule-Based Decision Tree. *IEEE Trans. Ind. Appl.* **2014**, *51*, 1249–1258. [CrossRef]
59. Wang, H.; Wang, P.; Liu, T. Power Quality Disturbance Classification Using the S-Transform and Probabilistic Neural Network. *Energies* **2017**, *10*, 107. [CrossRef]
60. Khokhar, S.; Zin, A.A.M.; Memon, A.P.; Mokhtar, A.S. A new optimal feature selection algorithm for classification of power quality disturbances using discrete wavelet transform and probabilistic neural network. *Measurement* **2017**, *95*, 246–259. [CrossRef]

Article

An Evaluation Model of Quantitative and Qualitative Fuzzy Multi-Criteria Decision-Making Approach for Hydroelectric Plant Location Selection

Fengsheng Chien [1], Chia-Nan Wang [2,*], Viet Tinh Nguyen [3], Van Thanh Nguyen [3,*] and Ka Yin Chau [4]

[1] School of Finance and Accounting, Fuzhou University of International Studies and Trade, Fuzhou 350202, China; jianfengsheng@fzfu.edu.cn

[2] Department of Industrial Engineering and Management, National Kaohsiung University of Science and Technology, Kaohsiung 80778, Taiwan

[3] Department of Logistics and Supply Chain Management, Hong Bang International University, HoChiMinh 723000, Vietnam; tinhnv@hiu.vn

[4] Faculty of International Tourism and Management, City University of Macau, Macau 999078, China; gavinchau@cityu.mo

* Correspondence: cn.wang@nkust.edu.tw (C.-N.W.); thanhnv@hiu.vn (V.T.N.)

Received: 27 April 2020; Accepted: 26 May 2020; Published: 1 June 2020

Abstract: Over the past few decades, Vietnam has been one of the fastest growing economies in Asia, experiencing a GDP growth rate of more than 6% per year. The energy industry plays an important role in Vietnam's continuous development, so access to reliable and low-cost energy sources will be an important factor for sustainable economic growth. Achieving the goal of reducing global greenhouse gas emissions, as set out in the Paris agreement on climate change, will depend heavily on the development roadmap of emerging economies, such as Vietnam. Currently, developing hydroelectric plants is Vietnam's optimal choice in order to meet its target of renewable energy development. Hydroelectricity in Vietnam is favorable thanks to the high average annual rainfall, about 1800–2000 mm, and the dense river system, with more than 3450 systems. In addition to providing electricity, hydropower plants are also responsible for cutting and fighting floods for downstream areas in the rainy season, at the same time providing water for production and people's daily needs in the dry season. This work proposes the application of a multicriteria decision-making (MCDM) model to select the best option for the installation of river hydroelectric plants in Vietnam. The use of MCDM techniques in environmental decision-making, including selecting between various alternatives, is important when this involves complex decisions in several disciplinary fields. The most widely used of these techniques are the fuzzy analytical network process (FANP) and the technique for order of preference by similarity to ideal solution (TOPSIS). As a result, Nghe An (LOC05) is found to be the optimal solution for selecting river portions where hydroelectric plants are viable in Vietnam.

Keywords: optimization techniques; hydroelectric; renewable energy; MCDM; fuzzy theory; FANP; TOPSIS

---

1. Introduction

Since the start of the 21st century, the global economy's constant expansion has driven world energy demand to record levels. While thermal energy from fossil fuels provides the world with most of its need—nearly 79.68% [1]—the lack of long-term sustainability and finite nature of these sources encourages the search for more sustainable and renewable energy sources. Among many forms of renewable energy, hydroelectric power is a proven technology, representing 15.90% of all energy generated [2]. However, while hydroelectric power is renewable, it is not without flaws. Damming rivers can cause great environmental and social damage. In addition to the immediate loss of local flora and fauna, building hydroelectric dams causes the displacement of people, a decline in the number of fish (or even the extinction of certain species), and may negatively impact the surrounding food systems and water quality [3]. Sustainable development is defined as development that can satisfy current demand without sacrificing the ability of future generations to satisfy their own needs [4]. The goal of sustainable development is described as the balance between a triple bottom line of social, environmental, and economic elements [5]. As such, it is important to consider social and environmental criteria along with traditional economic criteria when planning for hydroelectricity projects, in order to ensure their sustainability. Hydroelectric power plays an important part in the economic and social development of many countries along the Mekong river. One of the large capacity hydroelectric plants being exploited in Vietnam is Ban Chat hydroelectric plant (Figure 1).

Figure 1. Ban Chat hydroelectric plant in Vietnam [6].

Among these countries, Vietnam considers hydroelectric power to be the focus of its energy security and production strategy, instead of other sources such as nuclear power or thermal power [7]. However, there are many obstacles to developing hydroelectric projects in Vietnam, including environmental, social, economic, and political problems that need to be considered [7].

According to scientists, hydropower accounts for a large proportion of Vietnam's electricity production structure and plays an important role in national energy security. In addition to generating electricity, hydropower plants are also responsible for cutting and fighting floods for downstream areas in the rainy season, and at the same time supplying water to downstream areas to meet people's needs in the dry season. However, according to the Asian Development Bank's (ADB) assessment of electricity planning, the risk to biodiversity is very serious because the hydroelectric dams are located near sensitive and high biodiversity areas. Thus, choosing an optimal location for hydroelectric plants is a multicriteria decision-making (MCDM) problem that requires the decision-maker to incorporate both quantitative and qualitative factors. MCDM refers to making the best choice from a finite set of decision alternatives, in terms of multiple, usually conflicting criteria [8].

The aim of this paper is to propose a hybrid MCDM model for hydroelectric location evaluation and selection in Vietnam. In the first stage of this research, all criteria affecting the hydroelectric plant location evaluation and selection process were defined by experts and a literature review. As many of these criteria were qualitative, it was necessary to solve this problem in a fuzzy environment. Thus, a fuzzy analytical network process (FANP) was used to determine the weight of all the criteria. In the final stage, a technique for order of preference by similarity to ideal solution (TOPSIS) was then applied, to rank all the potential plant locations.

The next part of this paper presents literature reviews to support the building of the MCDM model, methods of determining the parameters of the FANP model, and the TOPSIS model. Discussions and conclusions are presented at the end of the paper.

2. Literature Review

Many literatures have applied an MCDM approach in various fields and disciplines over the years. One of the most popular uses of MCDM is for solving location selection problems and many studies have employed MCDM in energy plant location decisions [9–11]. Farahani and Asgari [12] developed a model for selecting the optimal location of warehouses in a military logistics system. The authors used a TOPSIS model to assess the quality of potential locations, based on a given set of criteria, then used a multiple objective set covering model to obtain the optimal locations. Zak and Weglinski [13] introduced a two-stage procedure for a logistics center location selection problem, based on ELimination Et Choice Translating Reality III/V (ELECTRE III/V). In the first stage of the procedure, a macro-analysis of the regions was performed and multiple criteria affecting the selection process, including sustainability criteria, were obtained. Then, ELECTRE III/V was employed to obtain a complete ranking of the potential locations. Choudhury et al. [14] proposed a hybrid MCDM model, using the analytic hierarchy process (AHP) technique and a multivariant adaptive regression spline (MARS) through a polynomial neural network to find the optimal locations of surface water treatment plants.

Among many MDCM techniques, the TOPSIS and FANP models are frequently employed in solving location selection problems. Suder and Kahraman [15] employed a fuzzy TOPSIS model for selecting the optimal location of a university faculty office. Hanine et al. [16] proposed a hybrid model, using the FAHP (fuzzy analytical hierarchy process) and fuzzy TOPSIS models, to evaluate and select a location for a solid waste landfill site. In this research, the fuzzy AHP model was employed to calculate the weight of the selected criteria and model the linguistic ambiguity, whereas the fuzzy TOPSIS model was used to obtain the ranking of potential locations, with respect to the criteria. Hanine et al. [17] proposed a geophysical information system (GIS)-FAHP-TOPSIS method for location selection problems. In this paper, the authors incorporated a geophysical information system into a hybrid FAHP-TOPSIS multicriteria decision making model to create a method for selecting a landfill site of industrial wastes. Erkeyman et al. [18] developed a fuzzy TOPSIS model for selecting an optimal location for a logistics center in the northeast region of Turkey. In this research, the authors used geographical, physical, socio-economic, and cost factors to form the criteria. Then, a fuzzy TOPSIS model was used to obtain the rankings for three potential alternatives based on these criteria.

Location problems are one of many popular topics in energy production planning and MCDM models are often used to solve these problems. Wang et al. [19] proposed a hybrid MCDM model to assist decision-makers in the evaluation and selection of a solid waste energy plant in Vietnam. The proposed model was built by combining the FANP and TOPSIS techniques. Demirel and Vural [20] suggested FANP as a plausible technique to solve a multicriteria solar energy plant location selection problem. Samanlioglu and Ayag [21] introduced a hybrid FAHP-PROMETHEE II model to evaluate solar power plant potential locations in Turkey. In this paper, FAHP was employed to calculate the weights of criteria, while F-PROMETHEE II provided the rankings of potential solar power plant locations, with respect to the criteria. Jeong and Ramirez-Gomez [22] combined a geographic information system multicriteria decision analysis (GIS-MCDA) and fuzzy decision-making trial and evaluation laboratory (F-DEMATEL) technique to evaluate potential sites for biomass power plants. In this research, the authors used the F-DEMATEL to determine the weights of the criteria, which included environmental, socio-economic, and geophysical constraints. Then, the optimal locations with regards to these criteria are obtained through the use of a weighted linear combination technique.

3. Methodology

In order to build an effective location selection model, the implementation process was carried out in the main steps as shown in Figure 2:

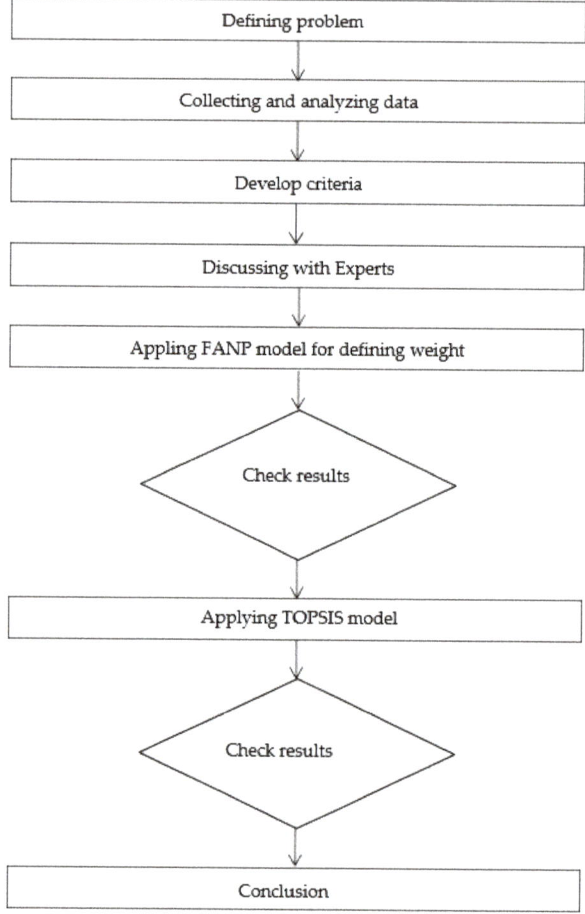

Figure 2. Research graph.

3.1. Fuzzy Theory

The triangular fuzzy numbers (TFN) can be defined as (k, h, g), with k, h and g $(k \leq h \leq g)$ being the parameters that specify the smallest likely value, the promising value, and the largest possible value in TFN. TFN are shown in Figure 3 and can be described as:

$$\mu\left(\frac{x}{\overline{M}}\right) = \begin{cases} 0, & \text{if } x < h, \\ \frac{x-k}{h-k} & \text{if } \quad k \leq x \leq h, \\ \frac{g-x}{g-h} & \text{if } \quad h \leq x \leq g, \\ 0, & \text{if } \quad x > g, \end{cases} \tag{1}$$

A fuzzy number is given as:

$$\widetilde{M} = (M^{o(y)}, M^{i(y)}) = [k + (h-k)y,\ g + (h-g)y],\ y \in [0,1] \tag{2}$$

where $o(y)$ and $i(y)$ represent the left side and the right side of a fuzzy number, respectively. The below shows basic calculations which involve two positive TFN, (k_1, h_1, g_1) and (k_2, h_2, g_2).

$$
\begin{aligned}
(k_1, h_1, g_1) + (k_2, h_2, g_2) &= (k_1 + k_2, h_1 + h_2, g_1 + g_2) \\
(k_1, h_1, g_1) - (k_2, h_2, g_2) &= (k_1 - k_2, h_1 - h_2, g_1 - g_2) \\
(k_1, h_1, g_1) \times (k_2, h_2, g_2) &= (k_1 \times k_2, h_1 \times h_2, g_1 \times g_2) \\
\frac{(k_1, h_1, g_1)}{(k_2, h_2, g_2)} &= (k_1/k_2, h_1/h_2, g_1/g_2)
\end{aligned}
\tag{3}
$$

3.2. Fuzzy Analytic Network Process (FANP) Method

The fuzzy analytic network process (FANP) is commonly used as an alternative to the fuzzy analytical hierarchy process (FAHP) to determine priority weights from fuzzy comparison matrices, due to its simplicity in comparison with FAHP. For instance, Guneri et al. [23] employed FANP for a shipyard location selection process, while incorporating the extent analysis method introduced by Chang [24]. Assume that $X = \{x_1, x_2, x_3, \ldots .x_n\}$ is an object set and $O = \{o_1, o_2, o_3, \ldots .o_n\}$ is a set of goals. According to Chang [24], the process takes each object from the object set and performs an extent analysis of each goal (g_i) of the object. Therefore, m, the extent analysis values of each object, can be written as followed:

$$M_{q_i}^1, M_{q_i}^2, \ldots, M_{q_i}^m, \quad i = 1, 2, \ldots, n \tag{1}$$

where $M_{q_i}^j (j = 1, 2, \ldots, m)$ are the TFN. Chang's extent analysis process is described as follows:

Step 1: The fuzzy synthetic extent value of the ith object is calculated as:

$$S_i = \sum_{j=1}^{m} M_{q_i}^j \otimes \left[\sum_{i=1}^{n} \sum_{j=1}^{m} M_{q_i}^j \right]^{-1} \tag{5}$$

The fuzzy addition operation of m extent analysis values of the object matrix $(\sum_{j=1}^{m} M_{g_i}^j)$ is calculated as:

$$\sum_{j=1}^{m} M_{q_i}^j = \left(\sum_{j=1}^{m} v_j, \sum_{j=1}^{m} u_j, \sum_{j=1}^{m} z_j \right). \tag{6}$$

The fuzzy additional operation of $M_{q_i}^j (j = 1, 2, \ldots, m)$ values $([\sum_{i=1}^{n} \sum_{j=1}^{m} M_{g_i}^j]^{-1})$ are performed as:

$$\sum_{i=1}^{n} \sum_{j=1}^{m} M_{q_i}^j = \left(\sum_{j=1}^{n} v_j, \sum_{j=1}^{n} u_j, \sum_{j=1}^{n} z_j \right). \tag{7}$$

Then, the inversion of the vector in (5) is calculated as:

$$\left[\sum_{i=1}^{n} \sum_{j=1}^{m} M_{g_i}^j \right]^{-1} = \left(\frac{1}{\sum_{i=1}^{n} v_i}, \frac{1}{\sum_{i=1}^{n} u_i}, \frac{1}{\sum_{i=1}^{n} z_i} \right). \tag{8}$$

Step 2: The degree of possibility of $M_2 = (v_2, u_2, z_2) \geq M_1 = (v_1, u_1, z_1)$ is calculated as:

$$V(M_1 \geq M_2) = sup_{y \geq x}[min(\mu_{M_1}(x),),(\mu_{M_2}(y))] \tag{9}$$

Which can be rewritten as follows:

$$V(M_1 \geq M_2) = hgt(M_1 \cap M_2) = \mu_{M_2}(d) = \begin{cases} 1 & \text{if } u_2 \geq u_1 \\ 0 & \text{if } v_1 \geq z_2 \\ \frac{v_1 - z_2}{(u_2 - z_2) - (u_1 - v_1)} & \text{otherwise} \end{cases} \tag{10}$$

where d is the ordinate of the highest intersection point D between μ_{m_1} and μ_{m_2}. To compare M_1 and M_2, we need both the degree of possibility of $(M_1 \geq M_2)$ and $(M_2 \geq M_1)$.

Step 3: The degree of the possibility that a convex fuzzy number is greater than c convex fuzzy number, with $M_i(i = 1, 2, \ldots, c)$, is calculated as:

$$V(M \geq M_1, M_2, \ldots, M_k) = V[(M \geq M_1) \text{ and } (M \geq M_2)]$$
$$\text{and,} \tag{11}$$
$$(M \geq M_c) = \min V(M \geq M_i), \ i = 1, 2, \ldots, c$$

Under the assumption of $d'(A_i) = minV(S_i \geq S_c)$, for $c = 1, 2, \ldots n$ and $c \# I$, the weight vector is calculated as:

$$W' = (d'(A_1), d'(A_2), \ldots d'(A_n))^T, \tag{12}$$

where A_i are n elements.

Step 4: Finally, the normalized weight vectors are calculated as:

$$d(A_i) = \frac{d'(A_i)}{\sum_{i=1}^{n} d'(A_1)} \tag{13}$$

$$W = (d(A_1), d(A_2), \ldots, d(A_n))^T \tag{14}$$

3.3. Technique for Order Preference by Similarity to Ideal Solution (TOPSIS)

The TOPSIS method was introduced by Hwang et al. [25]. A typical TOPSIS procedure can be described as follows [26,27]:

Step 1: Construct a normalized decision matrix:

$$e_{ij} = \frac{X_{ij}}{\sqrt{\sum_{i=1}^{m} X_{ij}^2}} \tag{15}$$

where x_{ij} and e_{ij} are original and normalized scores of a decision matrix, respectively.

Step 2: The normalized weight matrix is determined as follows:

$$s_{ij} = w_j e_{ij} \tag{16}$$

where w_j is the weight for the j criterion.

Step 3: Determine the PIS O^+ matrix and NIS O^- matrix:

$$O^+ = s_1^+, s_2^+, \ldots, s_n^+$$
$$O^- = s_1^-, s_2^-, \ldots, s_n^-; \tag{17}$$

Step 4: Identifying the gap between the performance values of each option with the positive ideal solution (PIS) matrix and negative ideal solution (NIS) matrix:

Distance to PIS.

$$S_i^+ = \sqrt{\sum_{j=1}^{m} \left(s_i^+ - s_{ij}\right)^2} \ ; \ i = 1, 2, \ldots, m \tag{18}$$

Distance to NIS.

$$S_i^- = \sqrt{\sum_{j=1}^{m}\left(s_{ij} - s_i^-\right)^2} \; ; \; i = 1, 2, \ldots, m \tag{19}$$

where D_i^+ is the distance to the PIS and D_i^- is the distance to the NIS for the ith option.

Step 5: Determine the preference value (Pv_i) for each option:

$$Pv_i = \frac{S_i^-}{S_i^- + S_i^+} \qquad i = 1, 2, \ldots, m \tag{20}$$

Pv_i values are used to benchmark and determine the ranking of the potential options.

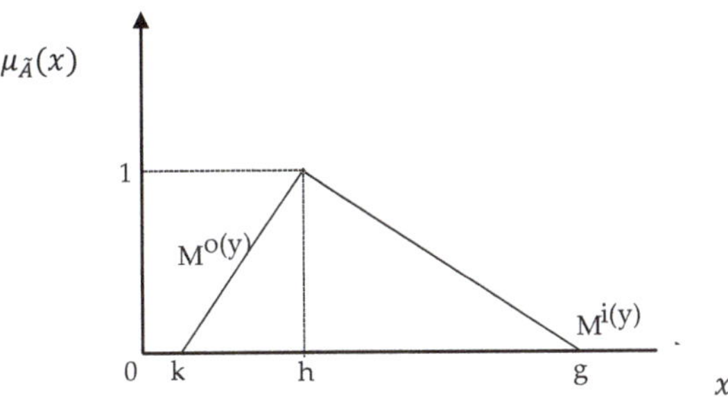

Figure 3. Triangular Fuzzy Number.

4. Case Study

Due to the geographical location of Vietnam, located in a tropical climate, with hot and humid rain, the country has access to relatively rich hydroelectric resources. The topographic distribution stretching from the north to the south, with a coast more than 3400 km long and an elevation change from sea level to more than 3100 m, has created a tremendous source of potential energy, generated by the terrain variation [28].

Many evaluation studies have shown that Vietnam can exploit hydroelectric power sources at about 25,000–26,000 MW, equivalent to about 90–100 billion kWh of electricity. However, the potential of hydroelectricity can be exploited even more: from 30,000 MW to 38,000 MW and an exploitable power of 100–110 billion kWh [28].

In order to serve the needs of industrialization and modernization, this period is very important for exploiting the country's hydroelectric energy. The largest hydropower projects built and completed during this period were Son La Hydroelectricity (2400 MW), Lai Chau Hydroelectricity (1200 MW), and Huoi Quang Hydroelectricity (560 MW) [28].

Recently, the process of inter-reservoir operation for hydropower steps was established and signed by the prime minister to issue decisions for all river basins with hydropower steps. By 2018, a total of 80 large and medium hydropower projects had come into operation, with a total installed capacity of 15.999 MW [28]. It can be said that, to date, many large hydro projects with capacities of over 100 MW have been generated. Vietnam's rivers network is shown in Figure 4.

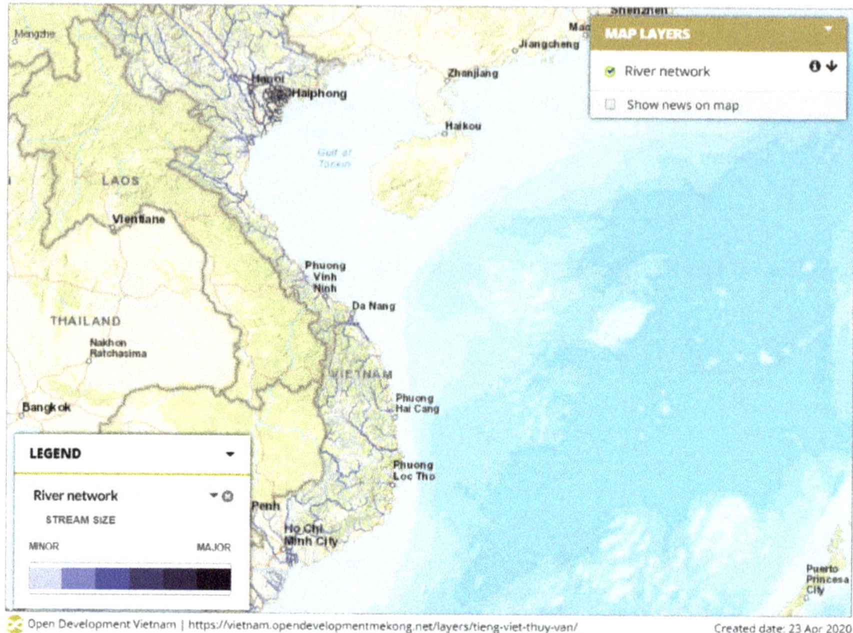

Figure 4. Vietnam's rivers network [29].

Therefore, the study of potential locations to invest in the construction of hydropower plants is essential in order to ensure a stable supply of energy for the socio-economic development of Vietnam. The river system of Vietnam is dense and distributed over many different territories. The potential for small hydroelectricity is concentrated mainly in the northern mountainous areas, the south-central areas, and the central highlands. Hydropower is still the most utilized renewable energy source, contributing about 40% of the total national electricity capacity. The potential of small hydroelectricity is huge, with more than 2200 rivers and streams with a length of over 10 km, 90% of which are small rivers and streams. These offer a favorable basis for developing small hydroelectricity [30].

To prove the research model is feasible, ten potential locations and fourteen criteria were considered. The names of the ten locations and their symbols in the MCDM model are shown in Table 1.

Table 1. Name and symbol of ten location.

No	Location	Symbol
1	Lai Chau	LOC01
2	Son La	LOC02
3	Ha Giang	LOC03
4	Yen Bai	LOC04
5	Nghe An	LOC05
6	Kon Tum	LOC06
7	Dak Lak	LOC07
8	Lam Dong	LOC08
9	Bac Can	LOC09
10	Hoa Binh	LOC10

After conducting a literature review and consulting experts, fourteen criteria were selected to assess each potential location, including: protected fauna, fish population, flow regime, landscape quality, public opposition to project, water quality, vegetation, soil fragility and erosion, accumulation or synergy with other projects, project cost, rated power, distance of power house from grid line, distance of power house from village, and distance of power house from road. All the criteria are shown in Figure 5.

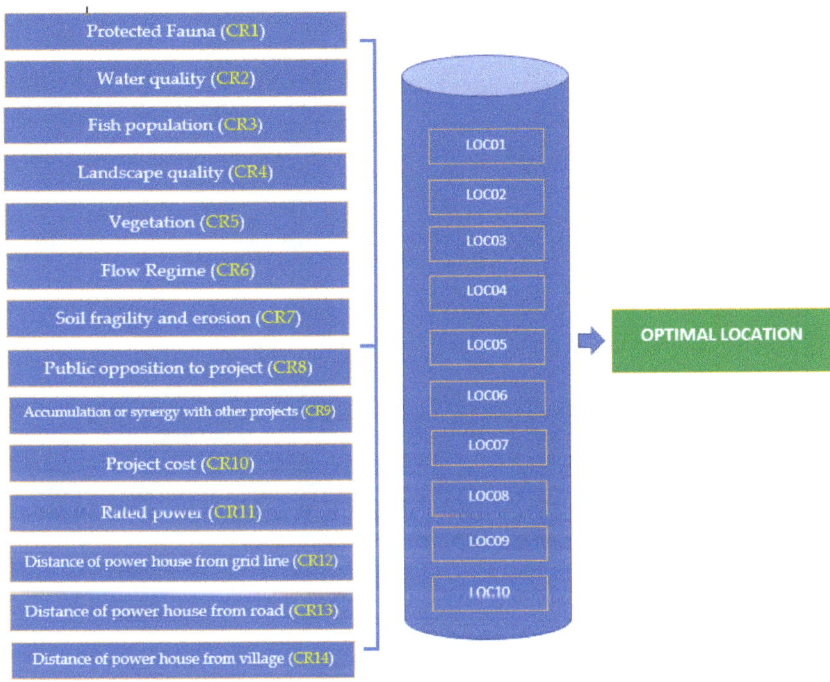

Figure 5. Criteria affecting location selection.

The input data of the FANP model are determined by the opinions of experts. The fuzzy comparison matrix of goal from the FANP model is shown in Table 2.

Table 2. Fuzzy comparison matrices for GOAL.

Criteria	A1	A2	A3	A4	A5
A1	(1,1,1)	(1,1,1)	(3,4,5)	(1,2,3)	(3,4,5)
A2	(1,1,1)	(1,1,1)	(1,1,1)	(1,2,3)	(3,4,5)
A3	(1/5,1/4,1/3)	(1/4,1/3,1/2)	(1,1,1)	(1/2,1/3,1/4)	(1,1/2,1/3)
A4	(1/3,1/2,1)	(1/3,1/2,1)	(4,3,2)	(1,1,1)	(2,3,4)
A5	(1/5,1/4,1/3)	(1/5,1/4,1/3)	(3,2,1)	(1/4,1/3,1/2)	(1,1,1)

The fuzzy numbers were converted to real numbers by using the TFN. During the defuzzification, the authors obtained the coefficients $\alpha = 0.5$ and $\beta = 0.5$. Here, α represents the uncertain environment conditions, and β represents the attitude of the evaluator is fair.

$$g0.5, 0.5(\overline{a_{A1,A4}}) = [(0.5 \times 1.5) + (1 - 0.5) \times 2.5] = 2$$

$$f0.5(LA1, A4) = (2 - 1) \times 0.5 + 1 = 1.5$$

$$f0.5(UA1, A4) = 3 - (3 - 2) \times 0.5 = 2.5$$

$$g0.5, 0.5(\overline{a_{A4,A1}}) = 1/2$$

The remaining calculations for other criteria are as in the above calculation. The real number priority when comparing the main criteria pairs is shown in Table 3.

Table 3. Real number priority.

Criteria	A1	A2	A3	A4	A5
A1	1	1	4	2	4
A2	1	1	3	2	4
A3	1/4	1/3	1	1/3	1/2
A4	1/2	1/2	3	1	3
A5	1/4	1/4	2	1/3	1

For calculating the maximum individual value as following:

$$Q1 = (1 \times 1 \times 4 \times 2 \times 4)1/5 = 2$$

$$Q2 = (1 \times 1 \times 3 \times 2 \times 4)1/5 = 1.9$$

$$Q3 = (1/4 \times 1/3 \times 1 \times 1/3 \times 1/2)1/5 = 0.43$$

$$Q4 = (1/2 \times 1/2 \times 3 \times 1 \times 3)1/5 = 1.18$$

$$Q5 = (1/4 \times 1/4 \times 2 \times 1/3 \times 1)1/5 = 0.5$$

$$\Sigma Q = 6.01$$

$$\omega_1 = \frac{2}{6.01} = 0.33$$

$$\omega_2 = \frac{1.9}{6.01} = 0.32$$

$$\omega_3 = \frac{0.43}{6.01} = 0.07$$

$$\omega_4 = \frac{1.18}{6.01} = 0.20$$

$$\omega_5 = \frac{0.5}{6.01} = 0.08$$

$$\begin{bmatrix} 1 & 1 & 4 & 2 & 4 \\ 1 & 1 & 3 & 2 & 4 \\ 1/4 & 1/3 & 1 & 1/3 & 1/2 \\ 1/2 & 1/2 & 3 & 1 & 3 \\ 1/4 & 1/4 & 2 & 1/3 & 1 \end{bmatrix} \times \begin{bmatrix} 0.33 \\ 0.32 \\ 0.07 \\ 0.20 \\ 0.08 \end{bmatrix} = \begin{bmatrix} 1.65 \\ 1.58 \\ 0.37 \\ 0.98 \\ 0.45 \end{bmatrix}$$

$$\begin{bmatrix} 1.65 \\ 1.58 \\ 0.37 \\ 0.98 \\ 0.45 \end{bmatrix} / \begin{bmatrix} 0.33 \\ 0.32 \\ 0.07 \\ 0.20 \\ 0.08 \end{bmatrix} = \begin{bmatrix} 5 \\ 4.9 \\ 5.2 \\ 4.9 \\ 5.6 \end{bmatrix}$$

Based on number of main criteria, the authors found that n = 5; λmax and CI are calculated as following:

$$\lambda_{max} = \frac{5 + 4.9 + 5.2 + 4.9 + 5.6}{5} = 5.12$$

$$CI = \frac{\lambda_{max} - n}{n - 1} = \frac{5.12 - 5}{5 - 1} = 0.03$$

To calculate the CR value, we found that RI = 1.12, with n = 5.

$$CR = \frac{CI}{RI} = \frac{0.03}{1.12} = 0.0268$$

Because CR = 0.0268 ≤ 0.1, there is no need to re-evaluate. The weights of criteria are shown in Table 4.

Table 4. The weights of criteria.

No	Symbol	Weight
1	CR1	0.0425
2	CR2	0.1017
3	CR3	0.0847
4	CR4	0.1023
5	CR5	0.1023
6	CR6	0.0534
7	CR7	0.0427
8	CR8	0.0284
9	CR9	0.0932
10	CR10	0.0415
11	CR11	0.0906
12	CR12	0.0704
13	CR13	0.0748
14	CR14	0.0715

To identify potential locations, the TOPSIS model was applied in the final phase of the study. Calculation results are shown in Table 5.

Table 5. Results from TOPSIS Model.

Alternatives	Si+	Si-	Pv_i
LOC01	0.0204	0.0177	0.4654
LOC02	0.0174	0.0161	0.4799
LOC03	0.0166	0.0169	0.5038
LOC04	0.0161	0.0195	0.5479
LOC05	0.0126	0.0247	0.6617
LOC06	0.0178	0.0184	0.5083
LOC07	0.0193	0.0201	0.5106
LOC08	0.0209	0.0129	0.3804
LOC09	0.0228	0.0173	0.4313
LOC10	0.0159	0.0210	0.5701

The TOPSIS model is based on the concept that the chosen alternative should have the shortest geometric distance from the positive ideal solution (PIS) and the longest geometric distance from the negative ideal solution (NIS), which results in Nghe An (LOC05) being the most optimal location from the potential alternatives (Table 5 and Figure 6).

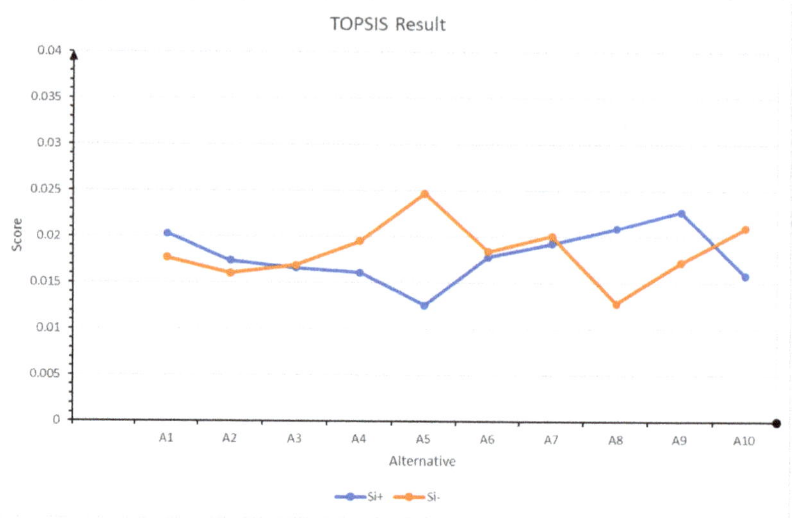

Figure 6. PIS and NIS values.

5. Discussion

Hydropower has always been an important part of Vietnam's national energy production structure and energy security. Hydropower dams are not only responsible for generating much-needed electricity—they also play an important part in controlling floods in rainy seasons and act as valued water reservoirs in dry seasons. However, while hydroelectric is a proven source of renewable energy, hydroelectric dams can cause serious environmental and social damage, such as the displacement of communities, declining number of fish, crippled food systems, and lower water quality [3].

Therefore, to ensure the goal of sustainable hydroelectric production is reached, it is necessary to improve the decision-making process by strengthening coordination among stakeholders, as well as improving quality in the participation of stakeholders. As most of the environmental and social problems of hydroelectric production are directly linked to a plant's location, it is important that economic, social, and environmental criteria are considered in the hydroelectric dam location selection process. Thus, this decision-making process can be considered as a multi-criteria decision process, in which the decision-maker must consider both qualitative and quantitative factors. In the proposed model, there are fourteen economic, social, and environmental criteria, including: fauna protection, fish population, flow regime, landscape quality, public opposition to project, water quality, vegetation, soil fragility and erosion, accumulation or synergy with other projects, project cost, rated power, distance of power house from grid line, distance of the powerhouse from village, and distance of the powerhouse from road. In the case study, input data were applied to the proposed model, which resulted in Nghe An (LOC05) being the most optimal location from the potential alternatives (Table 5).

The proposed MCDM model will assist researchers and decision-makers in identifying the optimal locations for building hydroelectric plants, while incorporating important economic, social, and environmental criteria. This research can also be used to support similar location selection processes in other countries and industries where sustainability is an important factor.

6. Conclusions

In Vietnam, hydroelectric power accounts for a high proportion of the electricity production structure. Currently, although the electricity industry has developed to diversify power sources, hydropower still accounts for a significant proportion of these.

Large hydroelectric projects can cause negative environmental and social impacts. These projects require very large reservoirs and may lead to the loss of a large area of land, most of which is agricultural. Thousands of households may need to be relocated and resettled, a cultural area within the burial area of the lake may be disturbed, and greenhouse gas emissions (mainly methane) generated by flooded organisms in the lake may be affected. To sustainably exploit and develop hydroelectricity, everything—including planning, investment projects, technical designs, construction work, and operation management—absolutely must strictly comply with specific procedures. In addition, it is necessary to ensure the quality of the construction, as well as the response scenario for dams and natural disaster mitigation for the community. Thus, hydroelectric plant location selection is an MCDM problem that decision-makers must evaluate in terms of both qualitative and quantitative factors. This is the reason why the author proposes a fuzzy MCDM model for hydroelectric plant location selection in this work. For building this proposed model, the author considered fourteen criteria that related to the decision-making process.

The research implemented fuzzy theory and the ANP and TOPSIS models for selecting the most suitable location. The implementation, using a case study, shows that the proposed model is feasible. The combined model can also be studied in conjunction with other models to diversify options.

Author Contributions: Conceptualization, C.-N.W., V.T.N. (Van Thanh Nguyen) and K.Y.C.; Data curation, F.C., C.-N.W. and V.T.N. (Viet Tinh Nguyen); Formal analysis, F.C. and V.T.N. (Van Thanh Nguyen); Funding acquisition, F.C., C.-N.W., V.T.N. (Viet Tinh Nguyen) and V.T.N. (Van Thanh Nguyen); Investigation, F.C., V.T.N. (Viet Tinh Nguyen) and K.Y.C.; Methodology, V.T.N. (Van Thanh Nguyen) and K.Y.C.; Project administration, F.C., V.T.N. (Viet Tinh Nguyen) and V.T.N. (Van Thanh Nguyen); Resources, V.T.N. (Viet Tinh Nguyen), V.T.N. (Van Thanh Nguyen) and K.Y.C.; Software, C.-N.W., V.T.N. (Viet Tinh Nguyen) and K.Y.C.; Writing—original draft, V.T.N. (Van Thanh Nguyen); Writing—review and editing, F.C. and C.-N.W. All authors have read and agreed to the published version of the manuscript.

Funding: This research received no external funding.

Conflicts of Interest: The authors declare no conflict of interest.

References

1. International Energy Agency. Fossil Fuel Energy Consumption (% of Total). 2015. Available online: https://data.worldbank.org/indicator/eg.use.comm.fo.zs?end=2015&start=1960&view=chart (accessed on 12 February 2020).

2. International Energy Agency. Electricity production from hydroelectric sources (% of total) 2019. Available online: https://data.worldbank.org/indicator/EG.ELC.HYRO.ZS (accessed on 12 February 2020).

3. Moran, E.; Lopez, M.C.; Moore, N.; Müller, N.; Hyndman, D.W. Sustainable hydropower in the 21st century. *Proc. Natl. Acad. Sci. USA* **2018**, *115*, 11891–11898. [CrossRef] [PubMed]

4. United Nations General Assembly. *Report of the World Commission on Environment and Development: Our Common Future. Transmitted to the General Assembly as an Annex to Document A/42/427–Development and International Co-Operation: Environment*; Oxford University Press: Oxford, UK, 1987; (Retrieved on 15 February 2009).

5. United Nations General Assembly. *World Summit Outcome, Resolution A/60/1, adopted by the General Assembly on 15 September 2005*; United Nations General Assembly: New York, NY, USA, 2005; (Retrieved on 17 February 2009).

6. Petrovietnam Power Corporation. Khai thác tiềm năng thủy điện: Cơ hội thúc đẩy phát triển kinh tế. Available online: https://www.pvpower.vn/khai-thac-tiem-nang-thuy-dien-co-hoi-thuc-day-phat-trien-kinh-te/ (accessed on 21 January 2020).

7. Nguyen, V.D.S. *Yếu tố kinh tế chính trị trong xây dựng đập thủy điện ở Việt Nam*; The Stimson Center: Washington, DC, USA, 2010.

8. Xu, L.; Yang, J.-B. *Introduction to Multi-Criteria Decision Making*; Trong Working Paper; Manchester School of Management: Manchester, UK, 2001.

9. Zhou, P.; Ang, B.; Poh, K. Decision analysis in energy and environmental modeling: An update. *Energy* **2006**, *31*, 2604–2622. [CrossRef]

10. Wang, C.-N.; Nguyen, V.T.; Thai, H.T.N.; Duong, D.H. Multi-Criteria Decision Making (MCDM) Approaches for Solar Power Plant Location Selection in Viet Nam. *Energies* **2018**, *11*, 1504. [CrossRef]

11. Loken, E. Use of multicriteria decision analysis methods for energy planning problems. *Renew. Sustain. Energy Rev.* **2007**, *11*, 1584–1595. [CrossRef]
12. Rouhi, F.; Ebrahimi, S.B.; Ketabian, H. Development of International facility location model applying the combination of MCDM and location covering techniques under uncertainty. *J. Ind. Manag.* **2016**, *7*, 743–766.
13. Zak, J.; Węgliński, S. The Selection of the Logistics Center Location Based on MCDM/A Methodology. *Transp. Res. Procedia* **2014**, *3*, 555–564. [CrossRef]
14. Choudhury, S.; Howladar, P.; Majumder, M.; Saha, A.K. Application of Novel MCDM for Location Selection of Surface Water Treatment Plant. *IEEE Trans. Eng. Manag.* **2019**, 1–13. [CrossRef]
15. Suder, A.; Kahraman, C. Minimizing Environmental Risks Using Fuzzy TOPSIS: Location Selection for the ITU Faculty of Management. *Hum. Ecol. Risk Assess. Int. J.* **2015**, *21*, 1326–1340. [CrossRef]
16. Hanine, M.; Boutkhoum, O.; El Maknissi, A.; Tikniouine, A.; Agouti, T. Decision making under uncertainty using PEES–fuzzy AHP–fuzzy TOPSIS methodology for landfill location selection. *Environ. Syst. Decis.* **2016**, *36*, 351–367. [CrossRef]
17. Hanine, M.; Boutkhoum, O.; Tikniouine, A.; Agouti, T. An application of OLAP/GIS-Fuzzy AHP-TOPSIS methodology for decision making: Location selection for landfill of industrial wastes as a case study. *KSCE J. Civ. Eng.* **2016**, *21*, 2074–2084. [CrossRef]
18. Erkayman, B.; Gundogar, E.; Akkaya, G.; Ipek, M. A Fuzzy Topsis Approach for Logistics Center Location Selection. *J. Bus. Case Stud. (JBCS)* **2011**, *7*, 49–54. [CrossRef]
19. Wang, C.-N.; Nguyen, V.T.; Duong, D.H.; Thai, H.T.N. A Hybrid Fuzzy Analysis Network Process (FANP) and the Technique for Order of Preference by Similarity to Ideal Solution (TOPSIS) Approaches for Solid Waste to Energy Plant Location Selection in Vietnam. *Appl. Sci.* **2018**, *8*, 1100. [CrossRef]
20. Demirel, T.; Vural, Z. Multicriteria solar energy plant using fuzzy ANP. In *Computational Intelligence*; World Scientific Publishing Pte.: Singapore, 2010; pp. 465–470.
21. Samanlioglu, F.; Ayağ, Z. A fuzzy AHP-PROMETHEE II approach for evaluation of solar power plant location alternatives in Turkey. *J. Intell. Fuzzy Syst.* **2017**, *33*, 859–871. [CrossRef]
22. Jeong, J.S.; Ramírez-Gómez, Á. Optimizing the location of a biomass plant with a fuzzy-DEcision-MAking Trial and Evaluation Laboratory (F-DEMATEL) and multi-criteria spatial decision assessment for renewable energy management and long-term sustainability. *J. Clean. Prod.* **2018**, *182*, 509–520. [CrossRef]
23. Guneri, A.F.; Cengiz, M.; Şeker, S. A fuzzy ANP approach to shipyard location selection. *Expert Syst. Appl.* **2009**, *36*, 7992–7999. [CrossRef]
24. Chang, D.-Y. Applications of the extent analysis method on fuzzy AHP. *Eur. J. Oper. Res.* **1996**, *95*, 649–655. [CrossRef]
25. Hwang, C.-L.; Yoon, K. *Methods for Multiple Attribute Decision Making*; Springer Science and Business Media LLC: Berlin/Heidelberg, Germany, 1981; Volume 186, pp. 58–191.
26. Behzadian, M.; Otaghsara, S.K.; Yazdani, M.; Ignatius, J. A state-of the-art survey of TOPSIS applications. *Expert Syst. Appl.* **2012**, *39*, 13051–13069. [CrossRef]
27. Kusumadewi, S.; Hartati, S.; Harjoko, A.; Wardoyo, R. *Fuzzy Multi-Attribute Decision Making (Fuzzy MADM)*; Graha Ilmu: Yogyakarta, Indonesia, 2006; pp. 78–79.
28. Nguyen, S. Khái quát về thủy điện Việt Nam. Vietnam Electricity. Available online: https://www.evn.com.vn/d6/news/Khai-quat-ve-thuy-dien-Viet-Nam-6-12-23805.aspx (accessed on 15 January 2020).
29. OpenDevelopmentMekong. Vietnam River Network. Available online: https://vietnam.opendevelopmentmekong.net/layers/tieng-viet-thuy-van/ (accessed on 6 January 2020).
30. Thai, V.H.; Hang, C.T.T. TIỀM NĂNG VÀ THÁCH THỨC PHÁT TRIỂN NĂNG LƯỢNG TÁI TẠO Ở VIỆT NAM [KỲ 1]. Available online: http://socongthuong.tuyenquang.gov.vn/tin-tuc-su-kien/nang-luong-moi-truong/tiem-nang-va-thach-thuc-phat-trien-nang-luong-tai-tao-o-viet-nam-ky-1!-110.html (accessed on 21 January 2020).

Article

Quadratically Constrained Quadratic Programming Formulation of Contingency Constrained Optimal Power Flow with Photovoltaic Generation

Luis M. Leon [1], Arturo S. Bretas [2] and Sergio Rivera [1],*

[1] Electrical and Electronic Engineering, Universidad Nacional de Colombia, Sede Bogotá, Bogotá 111321, Colombia; lumleongi@unal.edu.co

[2] Electrical and Computer Engineering Department, University of Florida, Gainesville, FL 32611, USA; arturo@ece.ufl.edu

*   Correspondence: srriverar@unal.edu.co

Received: 19 May 2020; Accepted: 22 June 2020; Published: 28 June 2020

Abstract: Contingency Constrained Optimal Power Flow (CCOPF) differs from traditional Optimal Power Flow (OPF) because its generation dispatch is planned to work with state variables between constraint limits, considering a specific contingency. When it is not desired to have changes in the power dispatch after the contingency occurs, the CCOPF is studied with a preventive perspective, whereas when the contingency occurs and the power dispatch needs to change to operate the system between limits in the post-contingency state, the problem is studied with a corrective perspective. As current power system software tools mainly focus on the traditional OPF problem, having the means to solve CCOPF will benefit power systems planning and operation. This paper presents a Quadratically Constrained Quadratic Programming (QCQP) formulation built within the *matpower* environment as a solution strategy to the preventive CCOPF. Moreover, an extended OPF model that forces the network to meet all constraints under contingency is proposed as a strategy to find the power dispatch solution for the corrective CCOPF. Validation is made on the IEEE 14-bus test system including photovoltaic generation in one simulation case. It was found that in the QCQP formulation, the power dispatch calculated barely differs in both pre- and post-contingency scenarios while in the OPF extended power network, node voltage values in both pre- and post-contingency scenarios are equal in spite of having different power dispatch for each scenario. This suggests that both the QCQP and the extended OPF formulations proposed, could be implemented in power system software tools in order to solve CCOPF problems from a preventive or corrective perspective.

Keywords: quadratically constrained quadratic programming; contingency constrained optimal power flow; optimal power flow

---

1. Introduction

Contingency Constrained Optimal Power Flow (CCOPF) can be studied as a reduced Security Constrained Optimal Power Flow (SCOPF). Modelling of SCOPF problems should include as many real variables as possible. As stated by Capitanescu et al. [1], SCOPF minimizes an operation cost function subject to equality and inequality constrains at three different times, in which three different limits of operation are usually met. The first of these times relates to the operation before a contingency, that means, the original power system operation.

The next group of equality and inequality constraints refers to the post-contingency at short-term time frame, which mainly focuses in keeping system stability by setting or changing system variables during a critical clearing time of the contingency. Works such as that in [2] by Tang and Sun, or that in [3] by Nguyen-Duc, Tran-Hoai, and Vol Ngoc approach the transient stability constrained optimal power flow problem in this short-term time frame.

Last set of constraints emerge when the system works while being stressed with the contingency. These medium-term time constraints could be the same as those from the original power system operation. However, in real life, the medium-term limits are more permissible than the originals, these are usually defined in specific time frames in which the violation can be endured by the equipment of the network and assuming that the contingency will not be permanent. Several medium-term time constraints can appear with, for example, time frames of 1, 5, or 30 min [1].

A contingency can be modeled as a change in the topology of the power system. This change in topology may lead to unacceptable states in the system variables, violating predefined constraints for the operation after the contingency. In order to avoid critical changes in the system, a preventive SCOPF perspective rises as a solution. As stated by Hinojosa [4], in this formulation, the post-contingency generation condition is the same as the pre-contingency generation condition; however, due to the change of topology, power flows and node voltage may have variations within the system; they must be between post-contingency limits (mid-term) in order for the SCOPF solution to actually work. PSCOPF has difficulties in the sense that not all constraints are of the same likelihood and not all have equal economic impact in the network. Furthermore, PSCOPF problem modelled with different mathematical structures must be solved by suitable optimization algorithms such as linear, non-linear or mixed-integer ones [5] or even through heuristic approaches [6] different from the methods used to solve the traditional OPF (interior point solver). As a consequence of its complexity, system compressions and linearizations have been proposed to solve these types of problems [7].

Another approach to SCOPF is that of the corrective SCOPF perspective. When constraint violations occur after a contingency, and, after the response of controls at different points in the system, those violations can be removed in a specified window of time, the system is considered as correctively secure [5]. Different corrective actions can take place depending on the equipment available in the system, e.g., modification of the power dispatch of generation machines, changes in voltage taps of transformers, use of FACTS or compensation banks, among others [8,9]. All these actions are limited by their time frame of action and/or feasibility of execution. In order to include this restriction in the formulation of the CSCOPF problem, a set of constraints could be defined for including the allowed number of corrective actions for both the short- and medium-term time frames after the contingency occurs. In this approach, some constraints, particularly on important voltages, will remain enforced as a PSCOPF approach, in order to guarantee low variations in voltage magnitudes after a contingency.

To solve the traditional SCOPF problem, it is also necessary to take into account all the contingencies that can affect the power system. The SCOPF solution would allow the system to work between specified constraints after a contingency, making this problem very complex as it was described by Gunda et al. and Fang et al. [10,11], showing that including all constraints in the traditional OPF interior point solution method could prevent convergence and, tackling the problem using metaheuristic strategies makes the problem hard to solve within short periods of time (critical clearing times of the contingency in the case of the CSCOPF) without high computational power.

More variables should be included in the problem formulation to obtain a more accurate and complete solution, such as the use and implementation of FACTS [12,13], the cost of performing regulation actions or the costs that could cause an outage originated by the contingency under study [14]. Despite a SCOPF solution that is calculated in real-time, its implementation becomes complicated since most power systems work under the unit commitment philosophy in which power dispatch decisions need to be taken well in advance, usually, the day before of the operation of the machines, which prevents regulation actions such as re-dispatches of generation.

In [15] by Gao et al. and in [16] by Cao et al. the authors explore the use and re-dispatch of Distributed Energy Resources and/or energy storage systems installed across the power network as a corrective measure to solve the CSCOPF. This is possible since inertia of DERs is small, allowing a fast change of power dispatched, when available. All these variables form a more complete view of the SCOPF.

Nevertheless, approximations and reductions of the original problem are welcomed as their results serve as a first approximation of the complex global solution. Security constraints have even been taken into account in the formulation of multi-objective management systems, for example, Amin Nasr et al. [17] propose a management system aiming to minimize active power losses and operating costs while improving the system voltage profile. This last multi-objective approach is searched while meeting voltage stability and generation contingency constraints. A more reduced version of the problem that can be considered is the limitation of contingencies for study. As discussed by Alzalg et al. [18], the SCOPF can be reduced to respond only if the OPF problem can be solved simultaneously for the original network and the network after contingency when a branch is cut. This is called a single $e - 1$ SCOPF problem formulation, CCOPF hereafter.

This paper aims to solve a CCOPF problem when a single branch contingency occurs. Motivation in this work is centered in the use of a Quadratically Constrained Quadratic Programming formulation that can represent the PCCOPF problem with a quadratic objective cost function in terms of node voltages instead of the linear methods and reductions found in the literature [7,19,20] for the solution of the PSCOPF. Moreover, as in the corrective perspective is of interest that voltage magnitudes are kept equal in important nodes after a contingency, a CCCOPF methodology is proposed to find two sets of power dispatches that ensure same voltage magnitudes in both the pre-contingency and post-contingency states of the system. This solution could give guidance on the necessary corrective actions that systems with only generation controls (with low inertia) should aim to in specific windows of time. Both strategies aim to contribute to the implementation of preventive and corrective CCOPF formulations coupled to the traditional OPF structure implemented in power system software tools, specially in the proposed CCCOPF strategy, where the optimization solver is the same one used to solve traditional OPF. Both strategies are limited to one branch contingency (complete disconnection of branch) and no short-term time frame analysis is considered, which could be more important in the CCCOPF formulation. An approximate model of renewable PV generation is included in the OPF problem as well. The specific contributions of this work towards the state-of-the-art are listed below.

- Proposition of PCCOPF formulation by using QCQP with the *matpower* environment.
- Proposition of CCCOPF formulation in the *matpower* environment that ensures the same voltage magnitude at all nodes in the system (pre- and post-contingency).

The results only apply for the operation prior to the contingency and the steady state operation after the contingency. As for the structure of the document, the problem formulation and variables included are presented in Section 2. Section 3 provides two different potential strategies for solving CCOPF problems through MATPOWER toolbox version 6.0 [21] and QCQP. Section 4 explains and compares the results obtained with both approaches and conclusions are presented in Section 5.

2. Problem Formulation

The CCOPF formulation here considered takes into account two different scenarios: one prior disconnection of a determined branch and the second one after disconnection of the branch.

The problem formulation is interpreted as in the hypothetical network of Figure 1, where the system of study has certain power dispatch (P_{i0} $i = 1, 2, 3$) for which sets of equality and inequality constraints g, h are met prior to the disconnection of one of its branches. After contingency in a selected branch k occurs, it is supposed that the system disconnects branch k and, as a consequence, the system acquires a new topology, for which a new power dispatch ($P_{ik}, i = 1, 2, 3$) needs to be regulated in order to meet with the same set of equality and inequality constraints as prior to the branch disconnection. The CCOPF formulation proposed in this work is based on the original OPF formulation for steady state conditions as described next.

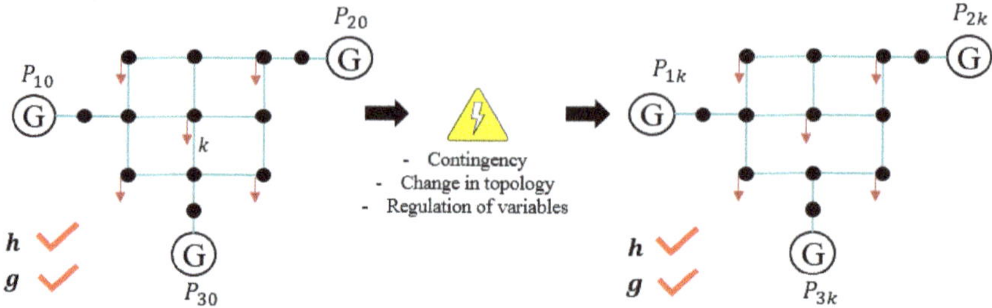

Figure 1. Interpretation of the CCOPF problem in a hypothetical network in steady state.

2.1. Objective Function

The optimal power flow can be seen as the minimization of the power system cost function maintaining some of their operation variables between specified ranges [22]. In general, the cost function of each generation agent (only with thermal generation) is defined as follows,

$$C_{G_i} = \alpha_i P_i^2 + \beta_i P_i + \gamma_i \tag{1}$$

where C_{G_i} is the cost of power injections of thermal generator i; P_i is the active power injected by thermal generator i; and α_i, β_i, and γ_i are coefficients of thermal generator i that multiply the injected power P_i when its exponent is of 2, 1 and 0, respectively. As stated by Bernal-Rubiano et al. [23], the cost function of photovoltaic (PV) generators controlled with a battery bank and whose power generation behaves with a probability density function defined with an uniform distribution could be represented as follows,

$$C_{G_{PVi}} = \alpha_i P_i^2 - \beta_i P_i + \gamma_i \tag{2}$$

with $C_{G_{PVi}}$ being the cost of power injections of PV generator i; P_i is the active power injected by PV generator i; and α_i, β_i, and γ_i are coefficients of PV generator i that multiply the injected power P_i when its exponent is of 2, 1, and 0, respectively. This has the same form as the cost function of the thermal generation, but the term β_i for every PV generator i is negative in its cost function. The negative value of this parameter is demonstrated in the mathematical uncertainty cost functions for controllable photovoltaic generators when it is considered uniform distributions for solar irradiation in a time instance [23]. The validation of this negative coefficient, it is performed comparing (in [23]) the Monte Carlo simulation with the analytic proposal where the low error in the results proved the advantages of using the analytic model due to its quadratic form and its coherence with the simulations that were performed [23].

2.2. Equality Constraints

Constraints in the OPF problem are separated into equality constraints ($g(x)$) and inequality constraints ($h(x)$). Equality constraints are, in essence, the power balance equations at each node of the system. They are represented by

$$g_P(\theta, V_m, P_g) = P_{BUS}(\theta, V_m) + P_d - C_g P_g = 0 \tag{3}$$
$$g_Q(\theta, V_m, Q_g) = Q_{BUS}(\theta, V_m) + Q_d - C_g Q_g = 0 \tag{4}$$

Where the sets g_P and g_Q are the active and reactive power balance equations at all nodes of the system: θ is the set of voltage angles; V_m is the set of voltage magnitudes; P_g and Q_g are the sets of active and reactive power generated; P_{BUS} and Q_{BUS} are the sets of net active and reactive power flows which are calculated with θ; and V_m, P_d, and Q_d are the sets of active and reactive power demanded and C_g is the set of loss coefficients at all nodes.

2.3. Inequality Constraints

Inequality constraints are separated into two sets: the first one refers to loadability line limits, typically expressed in MVA. The power flow through a branch should not exceed its loadability limits. This power flow is evaluated through the apparent power injection at the nodes that the branch connects (expressed as To and From nodes).

$$h_f(\theta, V_m) = |F_f(\theta, V_m)| \leq F_{max} \tag{5}$$

$$h_t(\theta, V_m) = |F_t(\theta, V_m)| \leq F_{max} \tag{6}$$

where h_f and h_t are the sets of inequality branch loadability constraints of the From and To nodes of each branch, respectively. F_f and F_t are the sets of calculated apparent power at the From and To nodes of each branch and F_{max} is the set of maximum apparent power flow through each branch.

The second set of inequality constraints describes voltage limits at each bus and power generation limits for each generation agent:

$$h_V = V_m^{min} \leq V_m \leq V_m^{max} \tag{7}$$

$$h_P = P_g^{min} \leq P_g \leq P_g^{max} \tag{8}$$

$$h_Q = Q_g^{min} \leq Q_g \leq Q_g^{max} \tag{9}$$

where h_V is the set of inequality voltage magnitudes at all nodes, h_P and h_Q are the sets of inequality active and reactive power generation for all generation agents, and the superscripts *min* and *max* refers to the minimum and maximum of each variable set.

With this mathematical formulation and based on the system topology, optimization takes place to find generation set-points of each agent that ensure the least operational cost of the system complying with all constraints.

2.4. Contingency Constrained Optimal Power Flow Problem

In planning and operation of actual and future power systems, a great amount of scenarios must be taken into account as technology becomes more accessible to ensure security in the power system. Even though OPF could be considered as the basic guideline for operation of power systems, network operators should extend their studies to consider CCOPF and SCOPF problems with the different and reasonable contingencies that may appear in the network, which could lead to the most expensive or damaging results.

As stated previously, the CCOPF problem differs from the traditional OPF problem because in the CCOPF, it is desired to obtain generation power dispatches that meet both pre-contingency and post-contingency constraints for both system scenarios, even if the operational cost of the system increases. In comparison, the traditional OPF only focuses on the minimization of the operational cost of the system without considering any contingency, and so, its dispatch and final values of its state variables are different from those of the CCOPF. However, many parts of the original OPF formulation can be used to build the CCOPF formulation. In this work, the original objective function will remain, i.e., trying to minimize the operation cost in the network.

The reduction of transmission line losses or improvement in bus voltages profiles both in normal and post-contingency operation have to be considered in the CCOPF formulation. These secondary objectives do not need to be included in the general objective function; however, they can be taken into account with more conservative restrictions in the problem constraints.

The traditional constraints of the OPF formulation also remain in CCOPF, in addition, as systems that search for CCOPF dispatch need to keep bus voltages and transmission line losses controlled, it is common to find that their bound limits are more restrictive that those found in traditional OPF.

CCOPF requires additional constraints to ensure that the power system will work between bounds in pre-contingency and post-contingency operation. Based on the general SCOPF given in [24], where an amount of N contingencies are included in the main problem, the following general formulation is considered and used in this work, where all constraints prior to and after the contingency need to be met in order to obtain a CCOPF solution:

$$\underset{x,u}{Min} \sum_{i=1}^{NG} C_{G_i}(x,u) \tag{10}$$

$$s.t. \quad \mathbf{g}^{(k)}(x,u); \quad k = 0,1. \tag{11}$$

$$\mathbf{h}^{(k)}(x,u); \quad k = 0,1. \tag{12}$$

where C_{G_i} refers to the operational cost of generator i, whose operation cost function could be type (1) or (2). (x,u) are independent and dependent variables of the system, respectively. NG designates the total number of generators connected to the power system. $\mathbf{g}$ and $\mathbf{h}$ represent the set of equality and inequality constraints of the problem (see Equations (3–9)). Superscript $k = 0$ refers to the pre-contingency state of the network, while $k = 1$ refers to post-contingency state of the network. As contribution to the state-of-the-art, the work explores the preventive CCOPF implementation using a QCQP model with a quadratic objective function in terms of the voltage instead of the traditional cost function using active power. The QCQP formulation could be extended to include more than just one contingency. Moreover, a corrective CCOPF implementation is explored using an extension of the traditional OPF formulation, adding a constraint to keep node voltages equal to the pre-contingency state after the contingency occurs.

3. Strategies for Solving Contingency Constrained Optimal Power Flow Problems

3.1. QCQP Formulation of Preventive CCOPF Problem

The Optimal Power Flow problem can be seen as a Quadratically Constrained Quadratic Programming model; this mainly brings out the benefit of studying OPF problems from a standard mathematical optimization perspective, encouraging participation of more research areas by tackling OPFs with different methodologies. In order for this format to work, the following are considered.

- The state variable that is iterated to find the solution is, for every node, the complex voltage at the node.
- As the injected active power P_i can be expressed as a function (f_i) of the voltage angle θ_i and voltage magnitude V_{mi} squared. High order terms of Equations (1) and (2) are omitted to keep a quadratic format in the objective function:

$$C_{G_i} = \beta_i P_i + \gamma_i = \beta_i f_i(V_{mi}^2, \theta_i) + \gamma_i \tag{13}$$

- The operation cost function (13) could also be used to represent the cost of Photovoltaic generation by changing the β_i coefficient sign to negative.
- Loadability line constraints are rewritten in terms of current square. In doing so, all constraints are quadratic as well.

The OPF of a defined power network can be converted into a QCQP instance in terms of the complex voltages of the network. This conversion is done as explained in [25]. Equation (13) for all generation agents can be expressed as

$$C_G = V^H C V + c \tag{14}$$

where C_G is the set of operational costs for all generators, V is the matrix of complex voltages at all nodes, V^H the Hermitian (complex conjugate) transpose of V, C is a matrix related to the β coefficients of the generators, and c is a constant term vector related to the γ coefficients of the generators. The minimization of this new objective function format will be subject to the same equality and inequality constraints of the OPF, this time rewritten in terms of V.

With Y being the admittance matrix of the network and Y_j being a matrix with its jth row equal to the jth row of Y and all other rows equal to the zero vector, the apparent power injected at bus j is

$$s_j = V_j I_j^H = V^H Y_j^H V \tag{15}$$

Then, by decomposing s_j into its real and imaginary parts with the help of Y_j:

$$Re(Y_j) = \Phi_j = 0.5(Y_j^H + Y_j)$$

$$Im(Y_j) = \Psi_j = 0.5(Y_j^H - Y_j)$$

$$P_j = V^H \Phi_j V$$

$$Q_j = V^H \Psi_j V$$

Moreover, by defining J_j, as the Hermitian matrix with a single 1 in the (j,j)th entry and 0 everywhere else, all the equality and inequality constraints at node j are obtained. The OPF is then formulated arranging the Φ_j, Ψ_j and J_j matrices of all nodes into the respective matrices A and B according to all equality and inequality constraints (Equations (3)–(9)) for each node. Note that all the constant terms in the equality and inequality constraints also need to be considered, these are organized into the a and b vectors for equalities and inequalities, respectively.

$$\min_{V} \quad C_G = V^H C V + c \tag{16}$$

$$\text{subject to} \quad V^H A V + a = 0 \tag{17}$$

$$V^H B V + b \leq 0 \tag{18}$$

With this QCQP formulation is possible to expand the OPF problem into a CCOPF by analyzing its different matrices and vectors on different contingencies and finding a dispatch using an accurate expanded set of matrices and vectors. First, in order to obtain the conversion of the traditional OPF structure, the conversion function developed in [26] for any power system structured in the *matpower* environment is used. The QCQP instance can be formed by hermitians, real or complex vectors, and matrices. To select the type of output desired for the A, B, and C matrices and a, b, and c vectors, the input (X) is used, which can vary from 0, to 2, to obtain the respective instance output type. The function, inputs, and outputs look as follows,

```
[nVAR, nEQ, nINEQ, C, c, A, a, B, b] = qcqp_opf(MatPowerSystem,X);
```

The following variables are obtained in the output of the *qcqp_opf* function (Table 1):

Table 1. Variables that form an OPF problem using QCQP.

Variable	Meaning
nVAR	Number of state variables (node voltages)
nEQ	Number of equality constraints
nINEQ	Number of inequality constraints
C, c	Objective function coefficient matrix, constant terms vector
A, a	Equality constraints coefficient matrix, constant terms vector
B, b	Inequality constraints coefficient matrix, constant terms vector

As observed, this OPF representation can be easily manipulated to add more restrictive constraints building a CCOPF formulation using the QCQP format, optimizing the objective function and complying with both pre-contingency and post-contingency constraints of the selected contingency cases. The methodology is explained next and shown in the flowchart of Figure 2:

1. A contingency in the test network is carried out (e.g., disconnection of specific branch)
2. The equivalent QCQP instance of the previous edited (under contingency) power system is calculated (C_{cont}, c_{cont}, A_{cont}, a_{cont}, B_{cont}, b_{cont}).
3. The original power system (without contingencies) is transformed into a QCQP instance.
4. In the QCQP instance of the original power system (C, c, A, a, B, b), equality constraints matrix and vector of the power system under contingency (A_{cont}, a_{cont}) are indexed to the original ones ($A = (A; A_{cont}), a = (a; a_{cont})$).
5. This final system is optimized and, due to its constraints, the resulting dispatch allows the system to work with their variables between limits in both cases: with and without contingencies.

This methodology takes advantage of the QCQP instance formulation to determine different constraint matrices and vectors, according to the power system topology, which can be indexed in a bigger QCQP instance to solve an OPF meeting with the constraints prior to and after a change in the power system topology. This is a CCOPF formulation, considering a contingency (or the consequence of a contingency) as a change in the topology of the network.

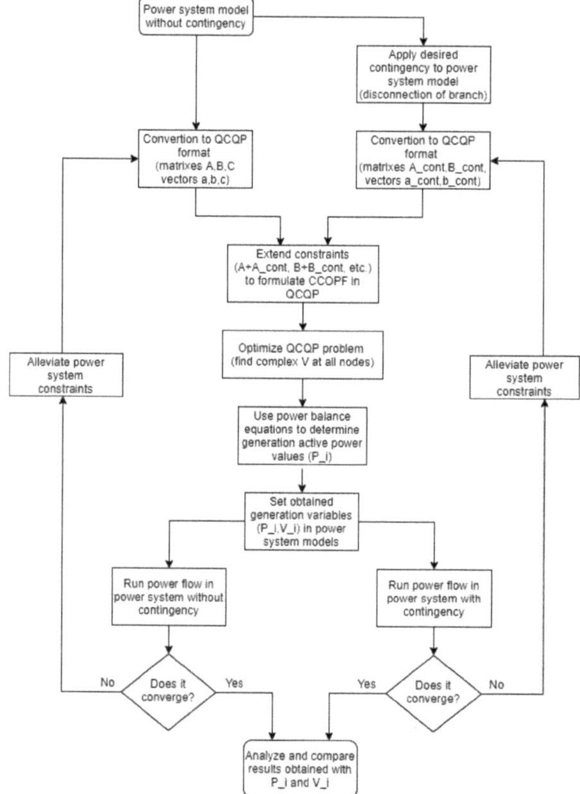

Figure 2. Flowchart of CCOPF formulation using the QCQP methodology.

3.2. Optimal Power Flow of Extended Network

One way to evaluate the generation dispatch under CCOPF can be explored by using traditional OPF formulations. For this purpose, the network of study is cloned and both power networks are connected through a high impedance transmission line that prevents the pair of networks to exchange power between them. The *from* and *to* nodes of this new line have to be chosen iteratively, ensuring no power flow across the transmission line. As shown in Figure 3, the duplicated network is subjected to a contingency, while the original system remains intact. This model allows to force different corrective CCOPF solutions by adjusting the constant terms in the inequality constraints of the stressed network, which can be done easily. The solution will be obtained through the commonly known interior point method of the traditional OPF.

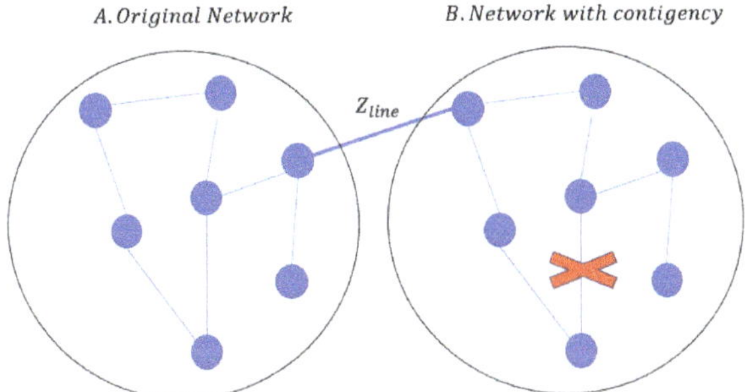

Figure 3. Topology approach to find the CCOPF through traditional OPF formulations.

This forced solution can also be extended to comply with more restrictive constraints not necessarily included in the original CCOPF problem. As an example, a constraint that forces the simulation to find a solution in which voltage magnitudes at all nodes is kept unchanged prior to ($V_{moriginal}$) and after ($V_{mcontingency}$) the contingency will be added to the extended OPF formulation, by defining the constraint as

$$|V_{moriginal}| = |V_{mcontingency}| \tag{19}$$

In order to find the power dispatch that meets the same voltage magnitudes after the contingency occurs, the generation cost coefficients in the duplicated network are suppressed, in that way, the stressed network will only focus on replicating the conditions prior to the contingency (which are obtained from the minimization of its operational cost) without considering the economical impact, avoiding possible divergences in the simulation. This condition is required for small industrial parks with small size power networks [15,16,27].

When optimizing this new extended network by using the traditional OPF solvers, two power dispatches are obtained, both ensuring same voltages at all nodes in pre- and post-contingency scenarios. In order to implement this change in the power dispatches, generation sources should have low levels of inertia, which may define this as a corrective CCOPF strategy, useful for power networks that do not allow to have any variation in the voltage magnitudes in steady state after the contingency occurs.

4. Results and Comparison

Both defined approaches are tested on the IEEE 14 bus test case network of Figure 4, composed of 14 nodes, 20 branches, and five generator agents; all of them traditional generators (all β_i coefficients are positive). In order to compare results, the linear power operation cost function (13) is used and contingencies (disconnection of branches) in branches 5, 10, and 16 are evaluated. Bus, generator, and branch test case data is summed in Tables A1–A3 located in Appendix A.

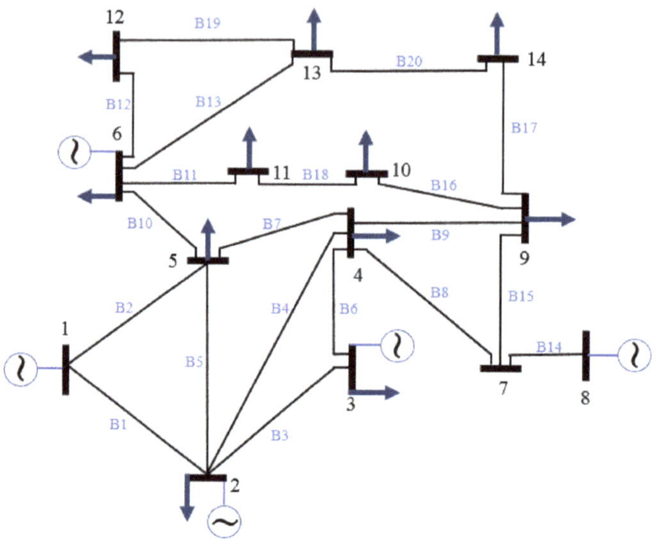

Figure 4. IEEE 14 bus test system (*Case*14) diagram indicating generators and loads positions as well as branches with their numeration.

4.1. Quadratically Constrained Quadratic Programming Formulation

4.1.1. Thermal Generation Only

In this formulation, equality and inequality constraints of the system without and with contingency are mixed in an unique set of constraints and solved in the QCQP format. The resulting voltages and power dispatch obtained in the solution are assigned to the original power network without and with the applied contingency. Three different contingencies are simulated by disconnecting branches 5, 10, and 16. After assigning generators setpoints (P and V), a power flow takes place until convergence is achieved in the network. Figures 5–7 show the final voltages and power dispatches for both scenarios. It can be observed in these three figures that the power setpoints for the generators are closely maintained after the contingency occurs. Only an increasing change is perceived in the case of generator 1, which is the slack node of the system, this suggests that branch disconnection increases active power losses in the network, which are compensated by the slack generator of the system. On the other hand, few voltage regulations are observed. In the three contingencies analyzed buses 4 and 5 present changes. When branch 5 and 16 are disconnected, voltage magnitudes at buses 4, 5, and 14 decrease, while, when branches 10 gets disconnected, the magnitude at buses 4 and 5 increases as well as in buses 10, 13, and 14.

It can be observed that the QCQP formulation only dispatches generators 1 and 2 at its maximum power in both pre- and post-contingency scenarios, so no big adjustment has to be made in the generation dispatch regulation. Voltage magnitudes at all nodes are not the same prior to and after contingency, due to the slight regulation of power of the generation agents, which presented a maximum change in power of 2.5% in generator 1 after branch 10 was disconnected. Maximum values of voltage regulation obtained were of 1.1% at node 5 after branch 5 and branch 16 were disconnected. It is important to note that even if voltages at all nodes have changes, in both scenarios voltage magnitudes are between specified bounds.

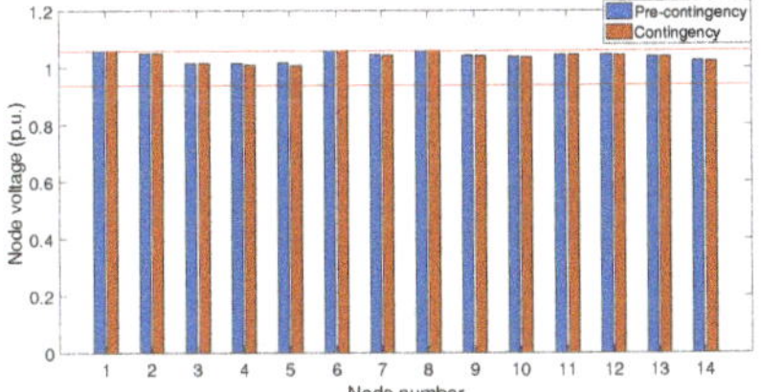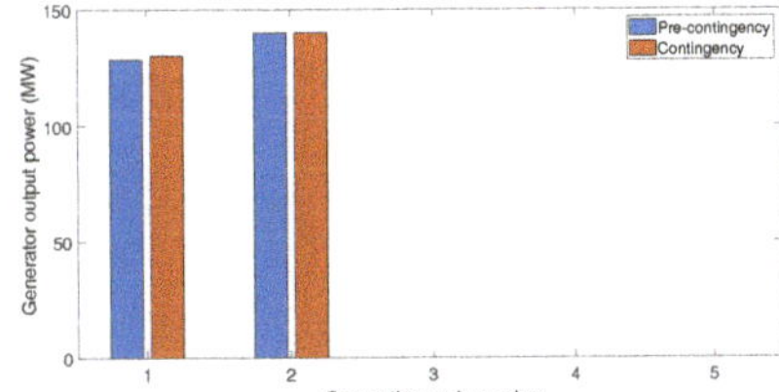

Figure 5. Voltage and power dispatched implementing the QCQP results in a power flow of the network prior to and after contingency of branch 5 (generation nodes are, respectively, 1, 2, 3, 6, and 8).

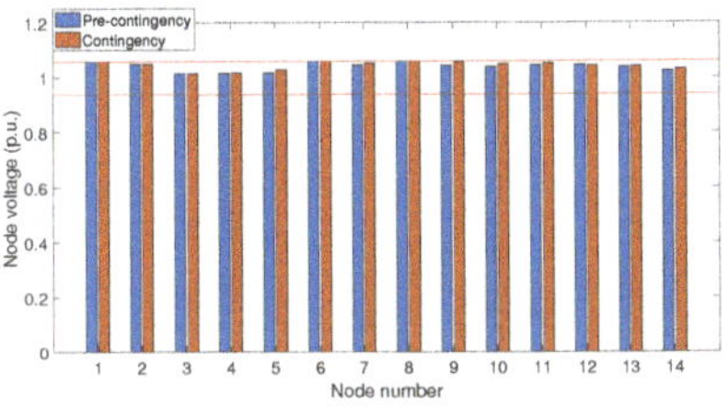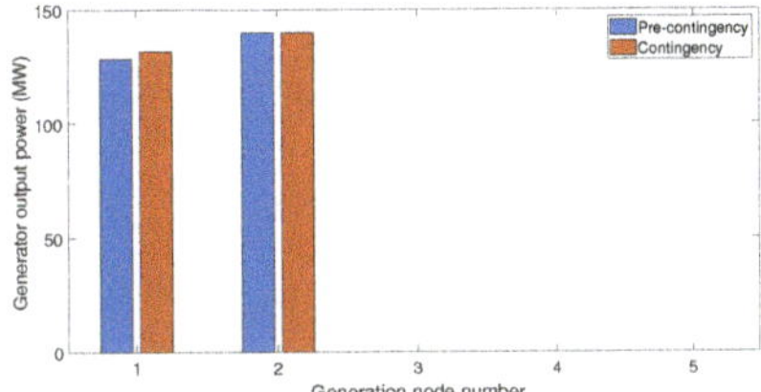

Figure 6. Voltage and power dispatched implementing the QCQP results in a power flow of the network prior to and after contingency of branch 10 (generation nodes are, respectively, 1, 2, 3, 6, and 8).

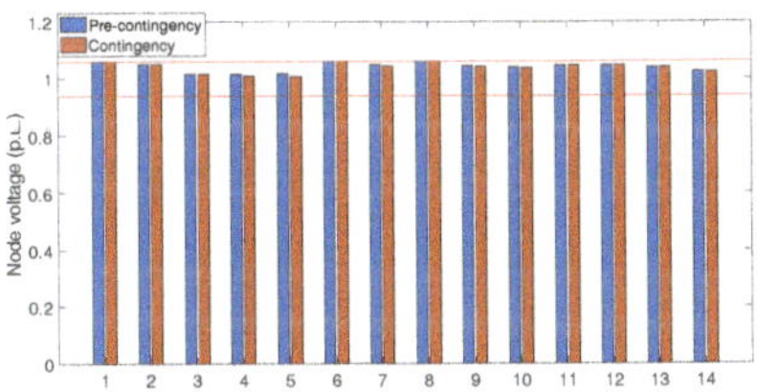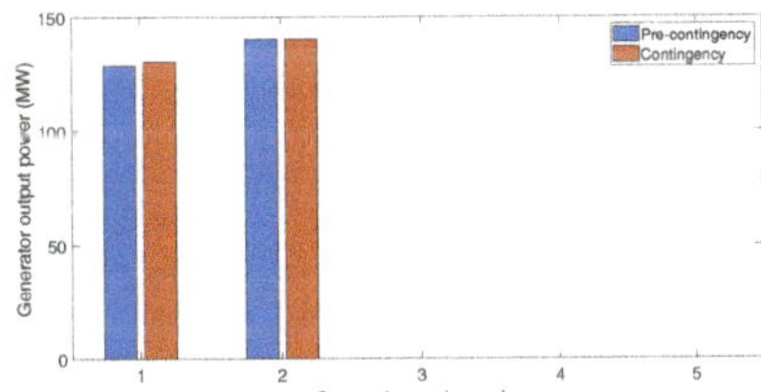

Figure 7. Voltage and power dispatched implementing the QCQP results in a power flow of the network prior to and after contingency of branch 16 (generation nodes are, respectively, 1, 2, 3, 6, and 8).

4.1.2. Thermal and Photovoltaic Generation

Photovoltaic generation with battery storage is also included in the QCQP formulation by changing the β_i coefficient of Equation (13). Generator 4, located at node 6, is selected to represent a photovoltaic generator with battery storage. Generator data is the same as presented in bus number 6 of Table A2 in Appendix A, with the only difference of having its β coefficient sign changed to negative. Figures 8–10 show the results obtained for pre- and post-contingency cases after fixing generation setpoints obtained in the QCQP formulation.

Test results show the behavior observed for the case with traditional generation in both strategies. The same power dispatch is practically kept for pre and post contingency scenarios, with a maximum percentage change of 10% experienced at generator 1 after branch 10 is disconnected. In some nodes, there is a change in the voltage magnitude, some cases greater after the contingency, some others lower. The maximum regulation observed was of 1.3% at node 10 after branch 10 was disconnected.

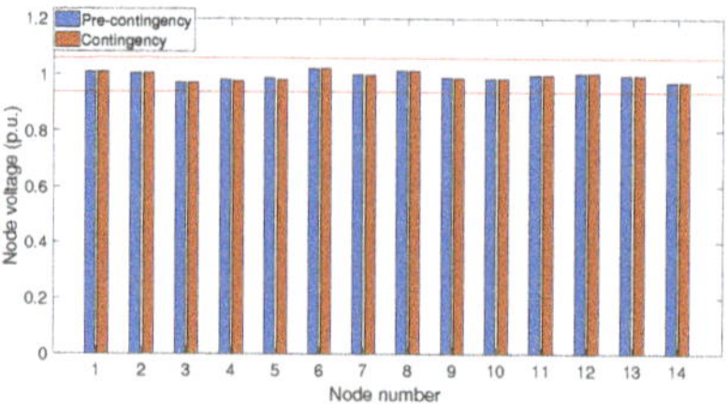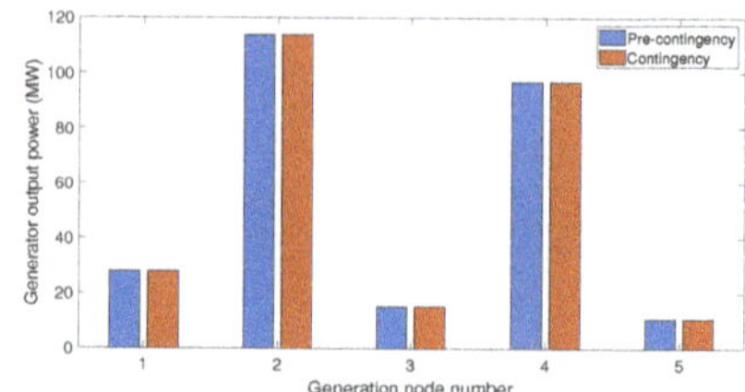

Figure 8. Voltage and power dispatched implementing the QCQP results in a power flow of the network prior to and after contingency of branch 5 (generation nodes are, respectively, nodes 1, 2, 3, 6, and 8).

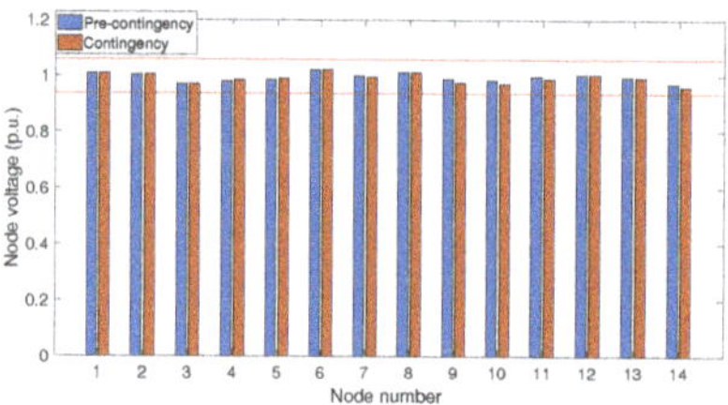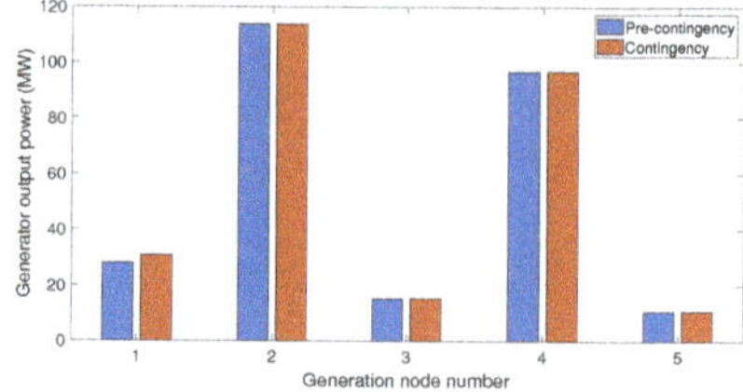

Figure 9. Voltage and power dispatched implementing the QCQP results in a power flow of the network prior to and after contingency of branch 10 (generation nodes are, respectively, nodes 1, 2, 3, 6, and 8).

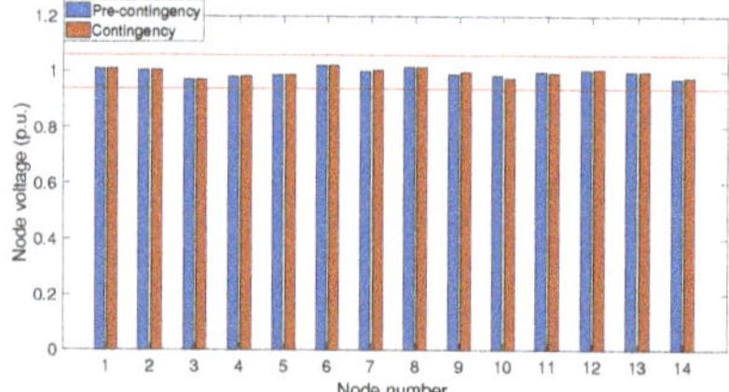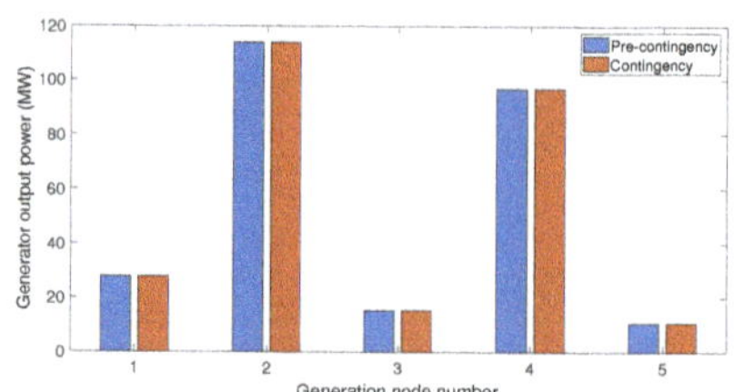

Figure 10. Voltage and power dispatched implementing the QCQP results in a power flow of the network prior to and after contingency of branch 16 (generation nodes are, respectively, nodes 1, 2, 3, 6, and 8).

When including PV generation, the QCQP strategy tends to dispatch it as much as possible, assuming the resource is available. In order to check the sensitivity of this behavior when branches connecting the generator have more restrictive limits, power flow through transmission lines connecting the renewable PV generation source is restricted to 25 MVA. This was decided because the maximum power output of generator 4 is 100 MVA and there are 4 transmission lines connected to it. A comparison of the power dispatched by the PV generator without and with this loadability restriction after disconnection of branches 5, 10, and 16 is shown in Figure 11.

As observed, there is a small reduction in the dispatched power; however, the amount dispatched is still considerably high for the generator, meaning that the strategy is in a certain way, sensitive to these loadability constraints when dispatching renewable generation. Strict power flow restriction changes and settings in the adjacent branches of PV generator after contingency will reduce its amount of active power dispatched, which entails to a more expensive operational cost of the system than without including strict loadability constraints. The changes observed in the dispatch may be due to a relaxation in the power flow of some of the adjacent connecting lines of the PV generator. It is probable that the amount dispatched by the the PV generator is not equally distributed in their 4 connecting lines, and as consequence, some of the lines are loaded to more than 25 MVA without having loadability restrictions. Now, when restrictions are applied, lines overloaded are regulated adjusting the power dispatched by the PV generator and the power difference is supported by the thermal generators, while meeting all the constraints.

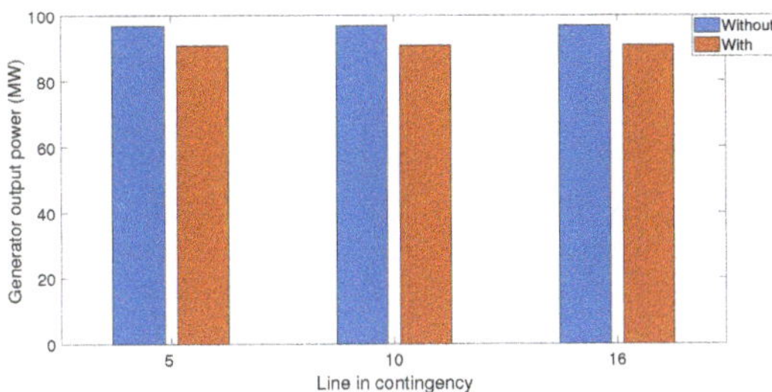

Figure 11. Power dispatch of renewable generator number 4 without and with loadability restrictions for every contingency applied (disconnection of branch).

4.2. Optimal Power Flow of the Extended Network

4.2.1. Thermal Generation Only

In this case, the system was duplicated and both networks were interconnected through a transmission line from the last node of the original system to the last node of the duplicated system. Resistance, inductance, and susceptance of the new line were adjusted in order to avoid power flow between the two networks. The choice of nodes connecting the systems is based on the amount of power flow obtained after running an OPF when interconnecting both networks without contingencies. Iteratively, power flow through the connecting transmission line is evaluated and those lines that present power flows of less than 1 MVA are considered as a good starting point for its implementation, considering that power flows of less than 0.5 MVA were found to not interfere in the power flow operation of both networks, and that changes in the new branch could reduce even more its initial power flow, i.e., increasing the impedance of the branch by adjusting its electrical parameters.

The results obtained show that all original system nodes have the same voltage than those of the duplicated system, complying with constraint (19). Due to the change in topology when the contingency is applied, a change in power dispatch is needed in order to keep the same voltage magnitudes in pre-contingency and contingency scenarios; these voltages are also within limits at all nodes. For this reason, the power dispatch may be regulated when the contingency occurs and the operation cost may change as well. Figures 12–14 show the obtained voltages and power dispatches for contingencies in lines 5, 10, and 16 compared to those when the system works without any contingency.

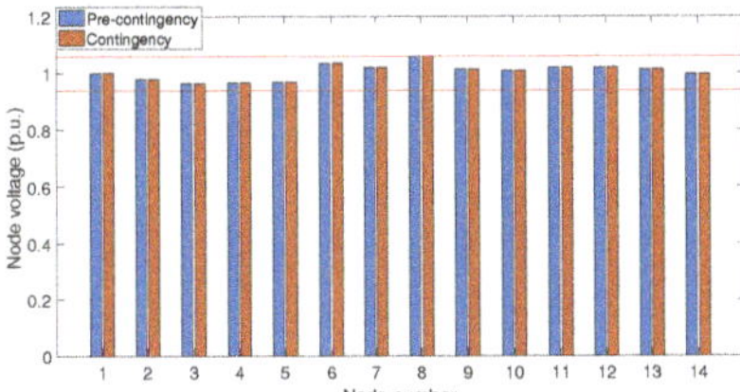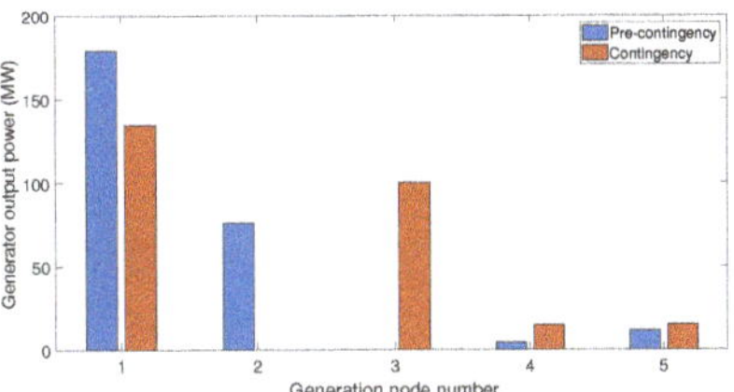

Figure 12. Voltage and power dispatched in the extended power system prior to and after contingency of branch 5 (generation nodes are, respectively, nodes 1, 2, 3, 6, and 8).

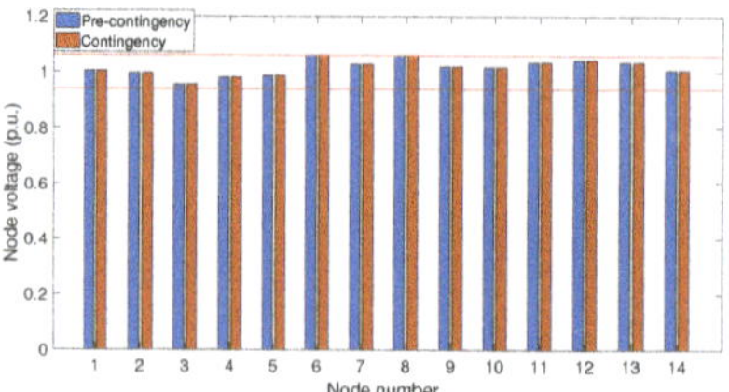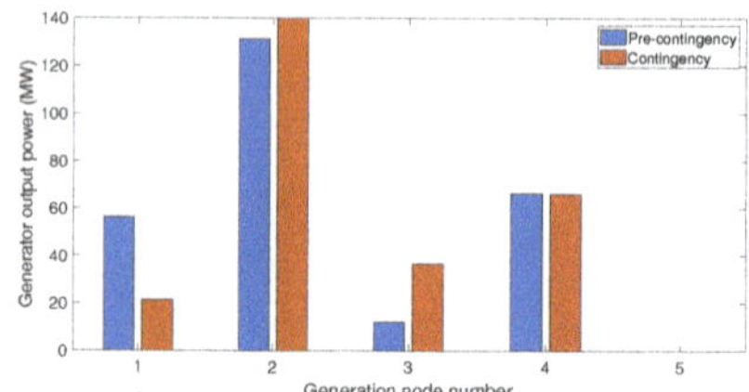

Figure 13. Voltage and power dispatched in the extended power system prior to and after contingency of branch 10 (generation nodes are, respectively, nodes 1, 2, 3, 6, and 8).

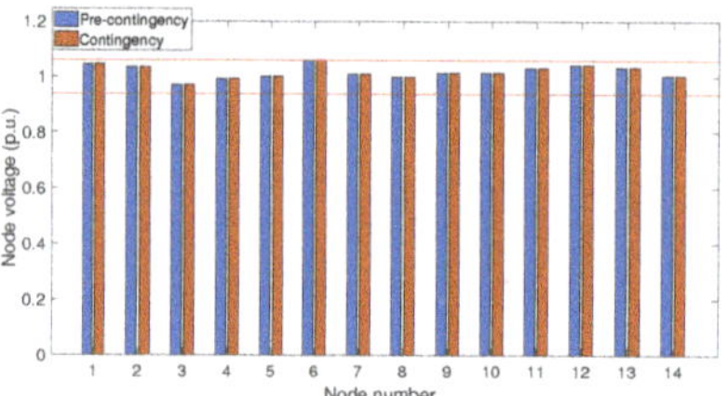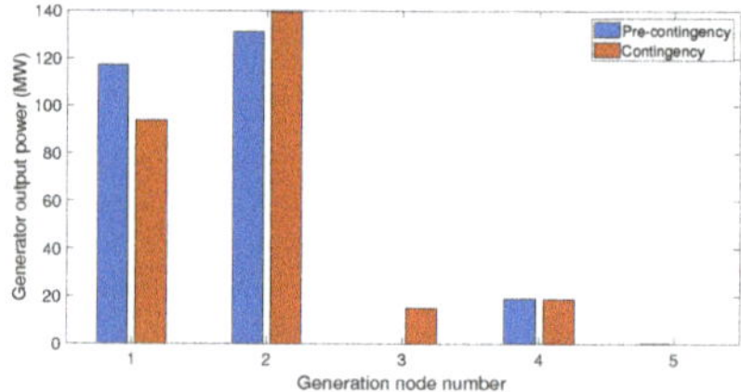

Figure 14. Voltage and power dispatched in the extended power system prior to and after contingency of branch 16 (generation nodes are, respectively, nodes 1, 2, 3, 6, and 8).

4.2.2. Thermal and Photovoltaic Generation

The strategy is now evaluated including one photovoltaic generator at node 6. To do so, the sign of the β_i coefficient of generator 4 is changed to negative. Figures 15–17 show the obtained power dispatch and voltages at all nodes following the proposed strategy. It can be observed that in all branch disconnections, voltage magnitudes are kept equal to the pre-contingency condition after their disconnection occurs. To achieve this condition, regulation in generation active power dispatched must take place. After disconnection of branches 5 and 10, the generator at node 3, which is a traditional type generator, needs to be started in order to keep voltage regulation of 0% in all nodes comparing with the pre-contingency state. Starting the generator alleviates the voltage regulation that could appear in nodes 2 and 4 and their adjacent nodes when the active power of these machines is not reduced. Generator 4, which is the PV generator with battery storage, is always dispatched in all situations, but not dispatched at its maximum in some cases. This may be due to an impossibility to meet the condition of keeping same node voltages prior to and after the contingency.

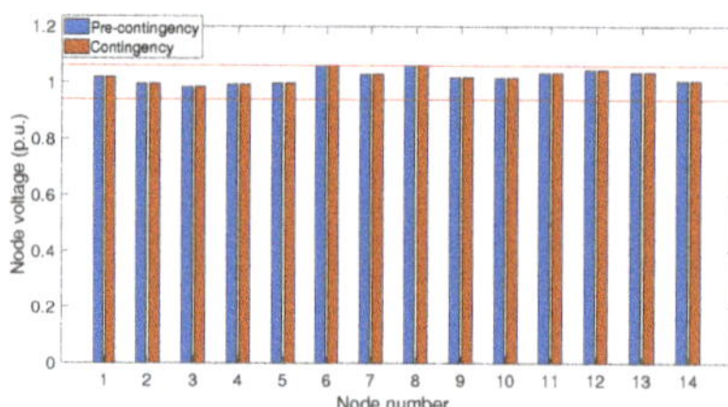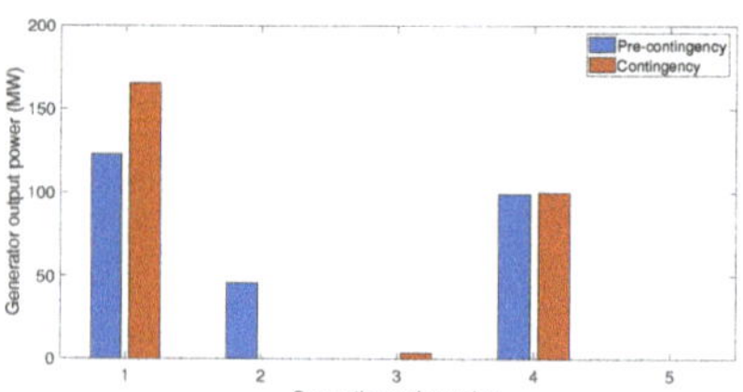

Figure 15. Voltage and power dispatched in the extended power system with PV generation prior to and after contingency of branch 5 (generation nodes are, respectively, nodes 1, 2, 3, 6, and 8).

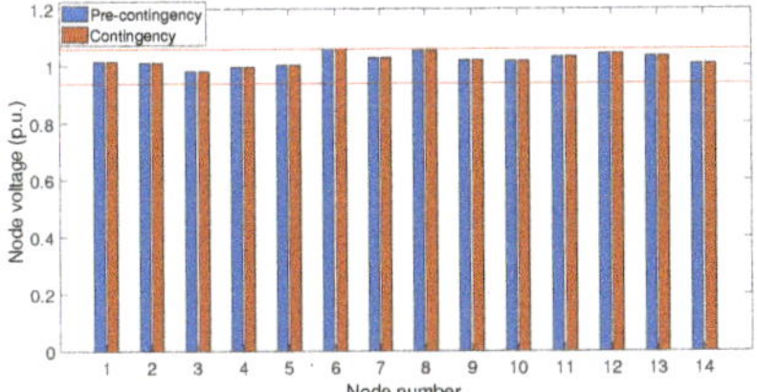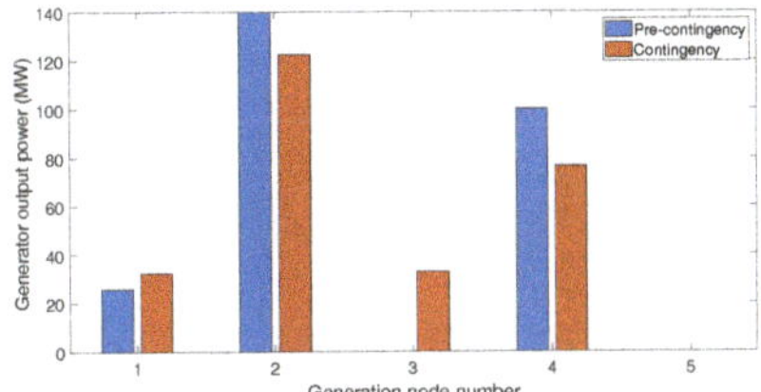

Figure 16. Voltage and power dispatched in the extended power system with PV generation prior to and after contingency of branch 10 (generation nodes are, respectively, nodes 1, 2, 3, 6, and 8).

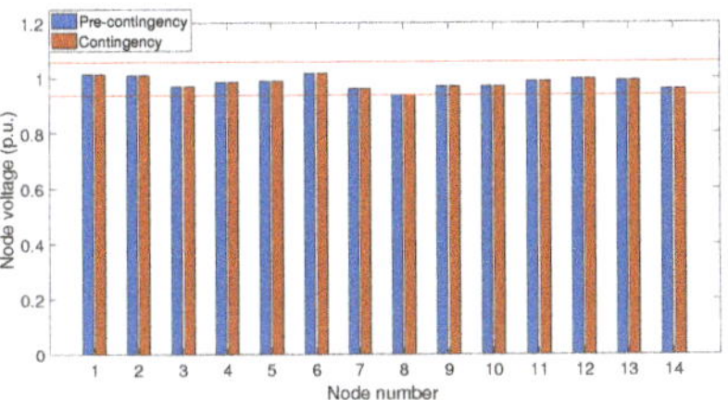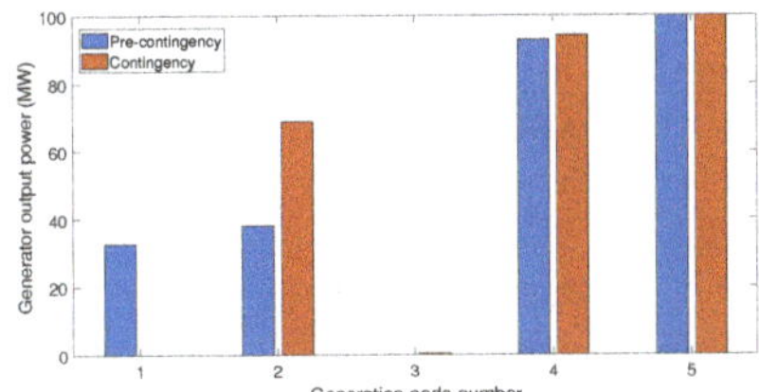

Figure 17. Voltage and power dispatched in the extended power system with PV generation prior to and after contingency of branch 16 (generation nodes are, respectively, nodes 1, 2, 3, 6, and 8).

In order to compare the sensitivity of the PV power dispatch obtained when including strict power flow constraints in branches connecting the PV generator within this extended OPF strategy, critical loadability limits (25 MVA) for the adjacent transmission lines of generator 4 are implemented and a comparison of the different dispatches with and without this constraint for disconnection of branches 10 and 16 is presented in Figure 18. Disconnection of branch 5 is not analyzed because, even though the strategy finds a solution when all branches are connected (having loadability limits or without having them), when loadability limits are applied and branch 5 is disconnected, the system did not converge. This is suspected to occur due to an impossibility of keeping voltage magnitudes equal after the disconnection of branch 5 without the PV generator loading one of its near connecting branches in more than 25 MVA. As shown in Figure 15, the dispatch of the PV generator located in node 4 without loadability limits is the nominal capacity of the generator (100 MW) and its dispatch is the same prior and after the contingency occurs. As power flow limits of the branches connected to the generator are restricted in this case, it is possible that after contingency, the generator could not be dispatched to its nominal capacity without overloading one of its connecting branches and as a consequence, some nodes may have regulation in their voltages, which prevents the system to meet the constraint of Equation (19).

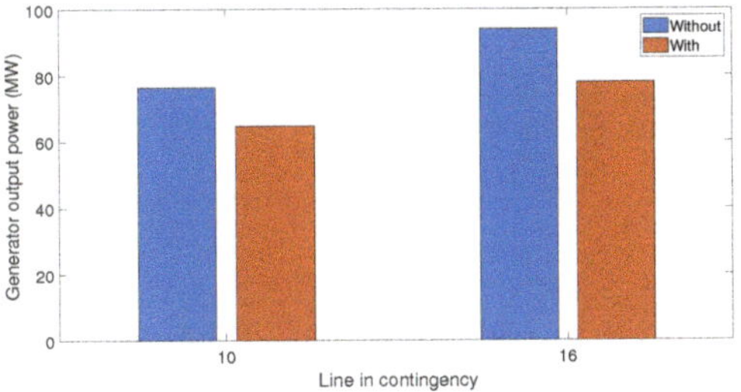

Figure 18. Power dispatch of renewable generator number 4 without and with loadability restrictions for contingencies in branches 10 and 16.

In this case, the change of power dispatch is greater than the obtained in the QCQP strategy, which presumably might be related to meeting the same voltage prior to and after the contingency condition. This is a more restrictive problem than the one formulated in the QCQP strategy, which does have some voltage regulation in their results. The changes present in the dispatch of PV generation for the extended OPF strategy are mainly due to trying to obtain the same voltage magnitude in all nodes respecting branch power flow limits which could limit the amount of power dispatched by the PV generator, and as consequence, changes the voltage magnitude of adjacent nodes. The results obtained exposes a high sensitivity of this strategy when changing loadability limits, as for one of the contingencies, the system did not even converge after the disconnection of the branch and the maximum variation of dispatched power obtained for the other contingencies is of 17% based in the original power dispatched value, while in the QCQP strategy, the maximum percentage change obtained was of near to 6%.

Sensitivity in both strategies is understand in this work as the measure of change of the behavior in the power dispatched by the PV generator implemented, when considering more severe restrictions after the contingency occurs. In the case of the renewable dispatched power, one could expect it to be dispatched at its maximum in order to reduce operational cost; however, presence of high sensitivity shows that operational cost can be highly compromised when selecting strict power flow constraints and expecting to meet all system constraints after the contingency happens.

4.3. Results Comparison

Results obtained with both approaches are not equal since the main objectives of both strategies differ. In the QCQP formulation, the generation dispatch found is supposed to be the same prior and after the contingency with the exception of the generator selected as slack. As a consequence, some voltage regulation could exist at all nodes after branch disconnection. On the other hand, in the extended OPF approach, the dispatches in all generators could change in an arbitrarily manner but the voltage magnitudes at all nodes must stay the same. This two main objectives lead to the results obtained and previously presented, where there is almost no power regulation in generation dispatches for the QCQP approach but there is presence of voltage regulation at some nodes of the system, and, in the case of the extended OPF, there is no regulation in voltage magnitudes at all nodes, but there are significant changes in the power dispatched by the generators prior and after the contingency. Even though the main focus of the two strategies vary because of the difference in their formulations and, in the case of the extended OPF strategy, the inclusion of an additional constraint, it is possible to compare the economic performance of both operational strategies from the obtained power dispatches. In Figure 19a the bar graph of operational cost for both operations (prior to and post contingency) is presented for each strategy using only thermal generation while in Figure 19b the bar graph shows the case when one PV generator with battery storage is included in the system.

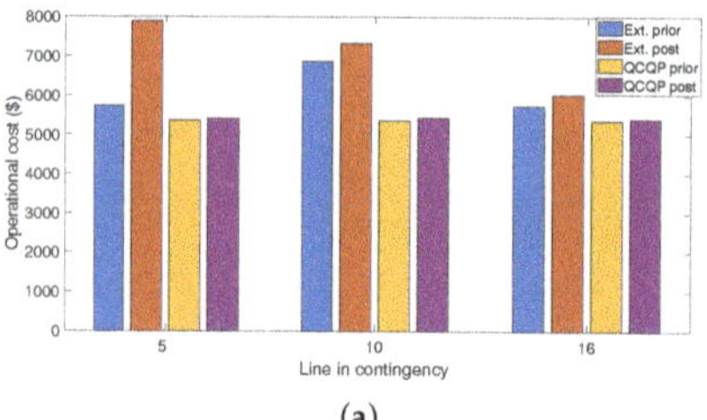
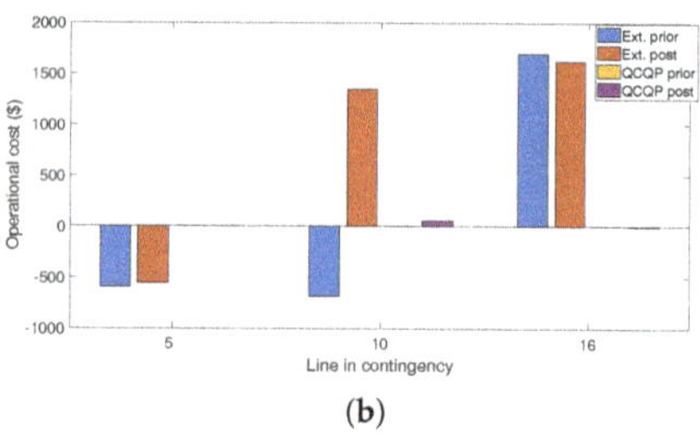

(a)   (b)

Figure 19. Operational cost for each contingency and each strategy proposed, prior to and after the contingency takes place. (**a**) On the left: only thermal generation, (**b**) on the right: thermal generation and PV generation.

The negative values obtained in the case of thermal and PV generation are due to the negative β coefficient modified from the original power system structure implemented in the *matpower* toolbox. It also has to be noted that no γ coefficients are implemented in the original structure, as shown in Table A2 in Appendix A, meaning that all linear operation cost functions (13) have their intercept in the origin and by no means could represent a real operational cost for any real generation agent. However, the β_i coefficients allow us to evaluate the operation of the system economically by only taking into account each generated power unit in all generators and, although this cost coefficient magnitudes are far from being what could be found in a real case, they help us to normalize the study conditions and to compare results of different operational strategies under the same reference frame.

Comparisons of all dispatches obtained for every contingency applied in the system without critical loadability limits are found in Figure 19. It is observed in Figure 19a that, when the system has thermal generation only, the extended OPF strategy presents a more expensive operational cost than that of the QCQP strategy prior to and after each contingency occurs. When adding PV generation in the system (Figure 19b), it is observed as well that when applying the disconnection of branch number 16, the operational cost is greater than the cost of the QCQP strategy, for both prior-post scenarios, which leads us to conclude that the behavior of the extended OPF strategy is more focused on finding dispatches that are able to meet with all constraints established (including same voltage magnitudes) in both power networks rather than an actual minimization of the operational cost in the execution of the strategy. The extended OPF strategy is proved to be more expensive than the QCQP in all disconnections performed when having thermal generation only. However, when adding one PV generator with storage, the operational cost found prior to and after disconnecting branch 5 and prior to disconnection of branch 10 had negative values which were less than the operational costs for the same cases with PV generation in the QCQP formulation, these cheaper operational costs in the extended OPF strategy are considered exceptions, as the majority of the operational costs found for the extended OPF strategy were found to be greater than the costs obtained with QCQP.

On the other hand, the QCQP results obtained (Figures 5–10) can be considered stiff with low variation in the power dispatch and voltage magnitudes for both thermal and thermal + PV generation scenarios in the three contingencies analyzed. Moreover, the operational costs results were lower than those of the extended OPF strategy for the majority of cases prior to and after the contingency (Figure 19a,b). Even though each contingency was analyzed individually, the power dispatch results in the three cases are similar. As the QCQP strategy can be easily extended to more than one contingency, it is expected that the power dispatch result of a more complete format with more than one contingency will be similar if not equal to the obtained dispatches for each contingency.

For both strategies, execution times were compared within this test system. The maximum difference in execution times observed was of approximately 750 ms in the contingency at branch 5 with thermal and PV generation. The QCQP strategy presented the slowest execution times when the system had thermal and PV generation. Simulations in both strategies ran in less than 700 ms for the cases with thermal generation only. Even though it may appear that the extended OPF strategy would be better based on its execution time, it is reminded that the result obtained with this strategy is focused on a corrective CCOPF perspective, while the QCQP strategy presents a result focused on a preventive CCOPF perspective. Execution times for all cases simulated are presented in Table 2.

Simulations in systems with more buses coincided that the QCQP strategy was much slower than the extended OPF; however, as the result of the QCQP strategy focuses on finding a preventive power dispatch for the CCOPF problem, which is supposed to remain the same after the contingencies occur. Execution time is not as critical as in the extended OPF strategy which has a corrective CCOPF perspective, where finding a power dispatch solution and making corrective actions are to be performed in a short window of time (short-term time frame analysis), this will be more challenging as systems grow in size and number of nodes due to the increase of execution time in the simulation. Selection of any methodology proposed will depend on the objectives specified by the network operator, and on the elements included in the power network with capabilities to regulate the power dispatch or voltage magnitudes in the system.

Table 2. Execution times obtained after implementation of QCQP and extended OPF strategies.

Contingency Located at	QCQP		Extended OPF	
	Execution Time (s)		Execution Time (s)	
	Thermal Generation Only	Thermal Generation + PV Generation and Battery Storage	Thermal Generation Only	Thermal Generation + PV Generation and Battery Storage
Branch 5	0.531975	1.253313	0.412515	0.507896
Branch 10	0.501589	0.939263	0.435760	0.517546
Branch 16	0.588455	0.938105	0.439151	0.469477

5. Conclusions

Two different operation strategies based on the CCOPF concept were presented and implemented in a power network with 14 nodes. Both strategies are developed in the *matpower* format and presented different results. The first one is a conception of a CCOPF problem using QCQP instances equivalent to those of an optimal power flow. The QCQP instance is modified to include contingency constraints that allow the power network to have the same power dispatch before and after a line disconnection stresses the power network. Separately, three different lines were taken out of the system and the performance of the QCQP programming strategy was evaluated through the voltage magnitudes of each node in the system prior and after the contingency happened. A maximum percentage change in the power dispatch of 2.5 % and in the voltage magnitude of 1.1 % was obtained for the QCQP strategy in a power network composed of only thermal generators, while in the case of thermal generation and one controllable photovoltaic generator with battery storage, maximum percentages of change in power dispatch and voltage of 10% and 1.3%, respectively, were obtained. It was also determined that if a PV generator is connected to the system, the strategy searches to dispatch it as much as possible. The QCQP strategy presents small sensitivity in its results when changes in the loadability constraints of lines adjacent to the PV generator are stricter.

The second strategy in this work is defined as a CCOPF corrective strategy that aims to have the same voltage magnitudes (no regulation) in steady state of a power system that is being affected by a contingency in one of its branches. As before, three different contingencies were applied to the power network when having only thermal generation and thermal generation plus one controllable PV generator with battery storage. The results obtained show that in all cases, the voltage magnitudes at all nodes are kept the same but the dispatched power prior to and after the contingency occurs can differ completely in some generators, even having to stop the operation of one of them and re-dispatch its previous power between the other machines. Furthermore, it was observed that this strategy is highly sensitive to changes of loadability constraints in transmission lines connected to the simulated controllable PV generator with energy storage. When loadability constraints changed to be stricter ones in the adjacent lines of the PV generator with battery storage, the solution to the CCOPF after disconnection of branch 5 was performed did not converge and it was impossible to find a power dispatch meeting the constraints. The extended OPF strategy is proposed to be implemented in small power systems that require fixed voltage values for its operation, and where the generators connected to these power systems have small inertia values in order to change their dispatched power rapidly after the contingency is detected.

Both methodologies can be extended to include more than one contingency. The use of these methodologies is encouraged to be implemented in power flow software tools, specially, the extended OPF approach for CCCOPF, which does not need any particular additional solver to be executed; however, this strategy does need to include the new constraint proposed in order to keep all node voltages equal after a branch disconnection. This strategy could be implemented in a more practical way after a less empirical methodology for the creation of an interconnection branch between power systems with and without contingency is established. On the other hand, the QCQP strategy does need to account with a convertion function that takes the original system structured in the power system software and transform it into a QCQP problem as discussed in Section 3 of this paper. An additional optimization solver is needed to obtain the accurate dispatch before running a power flow in the software, which makes this implementation more complicated but still promising.

Author Contributions: Conceptualization, L.M.L., A.S.B., and S.R.; methodology, L.M.L., S.R. and A.S.B.; validation, L.M.L. and A.S.B.; formal analysis, A.S.B., S.R., and L.M.L.; investigation, L.M.L., A.S.B. and A.S.; resources, L.M.L. and A.S.B.; data curation, L.M.L.; writing—original draft preparation, L.M.L., S.R. and A.S.B.; writing—review and editing, L.M.L., S.R. and A.S.B.; visualization, L.M.L., A.S.B. and S.R.; funding acquisition and supervision, A.S.B., and S.R. and A.S. All authors have read and agreed to the published version of the manuscript.

Funding: This research received no external funding.

Acknowledgments: This work was supported by Universidad Nacional de Colombia. Additionally, the authors would like to give thanks to: RED IBEROAMERICANA PARA EL DESARROLLO Y LA INTEGRACION DE PEQUENHOS GENERADORES EOLICOS (MICRO-EOLO) for their support.

Conflicts of Interest: The authors declare no conflicts of interest.

Appendix A

IEEE 14-bus test system data used in this work is found in Tables A1–A3:

Table A1. IEEE 14-bus data used.

Bus Number	P_d [MW]	Q_d [MW]	V_{max} [p.u.]	V_{min} [p.u.]
1	0	0	1.06	0.94
2	21.7	12.7	1.06	0.94
3	94.2	19	1.06	0.94
4	47.8	−3.9	1.06	0.94
5	7.6	1.6	1.06	0.94
6	11.2	7.5	1.06	0.94
7	0	0	1.06	0.94
8	0	0	1.06	0.94
9	29.5	16.6	1.06	0.94
10	9	5.8	1.06	0.94
11	3.5	1.8	1.06	0.94
12	6.1	1.6	1.06	0.94
13	13.5	5.8	1.06	0.94
14	14.9	5	1.06	0.94

Table A2. IEEE 14-generator data used.

Bus Number	Q_{max} [MVAr]	Q_{min} [MVAr]	P_{max} [MW]	P_{min} [MW]	α [$\$/MW^2$]	β [$\$/MW$]	γ [$\$$]
1	10	0	332.4	0	0	20	0
2	50	−40	140	0	0	20	0
3	40	0	100	0	0	40	0
6	24	−6	100	0	0	40	0
8	24	−6	100	0	0	40	0

Table A3. IEEE 14-branch data used.

Branch Number	From$_{bus}$	To$_{bus}$
1	1	2
2	1	5
3	2	3
4	2	4
5	2	5
6	3	4
7	4	5
8	4	7
9	4	9
10	5	6
11	6	11
12	6	12
13	6	13
14	7	8
15	7	9
16	9	10
17	9	14
18	10	11
19	12	13
20	13	14

References

1. Capitanescu, F.; Ramos, J.M.; Panciatici, P.; Kirschen, D.; Marcolini, A.M.; Platbrood, L.; Wehenkel, L. State of the art, challenges, and future trends in security constrained optimal power flow. *Electr. Power Syst. Res.* **2019**, *81*, 1731–1741. [CrossRef]
2. Tang, L.; Sun, W. An automated transient stability constrained optimal power flow based on trajectory sensitivity analysis. *IEEE Trans. Power Syst.* **2016**, *32*, 590–599. [CrossRef]
3. Nguyen-Duc, H.; Tran-Hoai, L.; Ngoc, D.V. A novel approach to solve transient stability constrained optimal power flow problems. *Turkish J. Electr. Eng. Comput. Sci.* **2017**, *25*, 4696–4705. [CrossRef]
4. Hinojosa, V.H. An improved corrective security constrained OPF for meshed AC/DC grids with multi-terminal VSC-HVDC. *Energies* **2020**, *13*, 516. [CrossRef]
5. Stott, B.; Alsaç, O. Optimal power flow: Basic requirements for real-life problems and their solutions. In Proceedings of the SEPOPE XII Symposium, Rio de Janeiro, Brazil, 20–23 May 2012.
6. Gunda, J.; Djokic, S.; Langella, R.; Testa, A. On convergence of conventional and meta-heuristic methods for security-constrained OPF analysis. In Proceedings of the 31st Annual ACM Symposium on Applied Computing, Pisa, Italy, 3-6 April 2016; pp. 109–111
7. Karbalaei, F.; Shahbazi, H.; Mahdavi, M. A new method for solving preventive security-constrained optimal power flow based on linear network compression. *Int. J. Electr. Power Energy Syst.* **2018**, *96*, 23–29. [CrossRef]
8. Cao, J.; Du, W.; Wang, H.F. An improved corrective security constrained OPF for meshed AC/DC grids with multi-terminal VSC-HVDC. *IEEE Trans. Power Syst.* **2015**, *31*, 485–495. [CrossRef]
9. Capitanescu, F.; Wehenkel, L. A new iterative approach to the corrective security-constrained optimal power flow problem. *IEEE Trans. Power Syst.* **2008**, *23*, 1533–1541. [CrossRef]
10. Gunda, J.; Djokic, S.; Langella, R.; Testa, A. Comparison of conventional and meta-heuristic methods for security-constrained OPF analysis. In Proceedings of the AEIT International Annual Conference (AEIT), Naples, Italy, 14–16 October 2015; pp. 1–6.
11. Fang, D.; Gunda, J.; Djokic, S.Z.; Vaccaro, A. Security-and time-constrained OPF applications. In Proceedings of the IEEE Manchester PowerTech, Manchester, UK, 18–22 June 2017; pp. 1–6.
12. Kaur, M.; Dixit, A. Newton's Method approach for Security Constrained OPF using TCSC. In Proceedings of the IEEE 1st International Conference on Power Electronics, Intelligent Control and Energy Systems (ICPEICES), Delhi, India, 4–6 July 2016; pp. 1–5.
13. Wu, X.; Zhou, Z.; Liu, G.; Qi, W.; Xie, Z. Preventive security-constrained optimal power flow considering UPFC control modes. *Energies* **2017**, *10*, 1199. [CrossRef]

14. Nycander, E.; Söder, L. Comparison of stochastic and deterministic security constrained optimal power flow under varying outage probabilities. In Proceedings of the IEEE Milan PowerTech, Milan, Italy, 23–27 June 2019; pp. 1–6.
15. Gao, X.; Lu, X.; Chan, K.W.; Hu, J.; Xia, S.; Xu, D. Distributed Coordinated Management for Multiple Distributed Energy Resources Optimal Operation with Security Constrains. In Proceedings of the IEEE Power and Energy Society Innovative Smart Grid Technologies Conference (ISGT), Washington, DC, USA, 18–21 February 2019; pp. 1–5.
16. Cao, J.; Liu, Y.; Ge, Y.; Cai, H.; Zhou, B.W. Enhanced corrective security constrained OPF with hybrid energy storage systems. In Proceedings of the UKACC 11th International Conference on Control (CONTROL) IEEE, Belfast, UK, 31 August–2 September 2016; pp. 1–7.
17. Nasr, M.A.; Nikkhah, S.; Gharehpetian, G.B.; Nasr-Azadani, E.; Hosseinian, S.H. A multi-objective voltage stability constrained energy management system for isolated microgrids. *Int. J. Electr. Power Energy Syst.* **2020**, *117*, 105646. [CrossRef]
18. Alzalg, B.; Anghel, C.; Gan, W.; Huang, Q.; Rahman, M.; Shum, A.; Wu, C.W. Contingency constrained optimal power flow solutions in complex network power grids. In Proceedings of the IEEE International Symposium on Circuits and Systems, Seoul, Korea, 20–23 May 2012; pp. 1636–1639.
19. Wang, Q.; McCalley, J.D.; Zheng, T.; Litvinov, E. A computational strategy to solve preventive risk-based security-constrained OPF. *IEEE Trans. Power Syst.* **2012**, *28*, 1666–1675. [CrossRef]
20. Van Acker, T.; Van Hertem, D. Linear representation of preventive and corrective actions in OPF models. In Proceedings of the Young Researchers Symposium, Eindhoven, The Netherland, 12–13 May 2016.
21. Zimmerman, R.D.; Murillo-Sánchez, C.E.; Thomas, R J. MATPOWER: Steady-State Operations, Planning and Analysis Tools for Power Systems Research and Education. *IEEE Trans. Power Syst.* **2011**, *26*, 12–19. [CrossRef]
22. Zimmerman, R.D.; Murillo-Sánchez, C.E. Matpower User's Manual, Version 7.0. 2019. Available online: https://matpower.org/docs/MATPOWER-manual-7.0.pdf (accessed on 2 May 2020).
23. Bernal, J.; Neira, J.; Rivera, S. Mathematical Uncertainty Cost Functions for Controllable Photo-Voltaic Generators considering Uniform Distributions. *WSEAS Trans. Math.* **2019**, *18*, 137–142.
24. Gaing, Z.; Lin, C. Contingency-Constrained Optimal Power Flow Using Simplex-Based Chaotic-PSO Algorithm. In *Applied Computational Intelligence and Soft Computing*; Hindawi Publishing Corporation: London, UK, 2011; p. 13.
25. Low, S.H. Convex relaxation of optimal power flow—Part I: Formulations and equivalence. *IEEE Trans. Control Netw. Syst.* **2014**, *1*, 15–27. [CrossRef]
26. Josz, C.; Fliscounakis, S.; Maeght, J.; Panciatici, P. AC Power Flow Data in MATPOWER and QCQP Format: iTesla, RTE Snapshots, and PEGASE. *arXiv* **2016**, arXiv:1603.01533.
27. Vargas, A.; Saavedra, O.; Samper, M.; Rivera, S.; Rodriguez, R. Latin American Energy Markets: Investment Opportunities in Nonconventional Renewables. *IEEE Power Energy Mag.* **2011**, *14*, 38–47. [CrossRef]

Article

Optimum Synthesis of a BOA Optimized Novel Dual-Stage $PI - (1 + ID)$ Controller for Frequency Response of a Microgrid

Abdul Latif [1,*], S. M. Suhail Hussain [2], Dulal Chandra Das [1] and Taha Selim Ustun [2]

1 Department of Electrical Engineering, National Institute of Technology Silchar, Assam 788010, India; dulal@ee.nits.ac.in
2 Fukushima Renewable Energy Institute, AIST (FREA), National Institute of Advanced Industrial Science and Technology (AIST), Koriyama 963-0298, Japan; suhail.hussain@aist.go.jp (S.M.S.H); selim.ustun@aist.go.jp (T.S.U.)
* Correspondence: abdul_rs@ee.nits.ac.in

Received: 3 June 2020; Accepted: 1 July 2020; Published: 3 July 2020

Abstract: A renewable and distributed generation (DG)-enabled modern electrified power network with/without energy storage (ES) helps the progress of microgrid development. Frequency regulation is a significant scheme to improve the dynamic response quality of the microgrid under unknown disturbances. This paper established a maiden load frequency regulation of a wind-driven generator (WG), solar tower (ST), bio-diesel power generator ($BDPG$) and thermostatically controllable load (heat pump and refrigerator)-based, isolated, single-area microgrid system. Hence, intelligent control strategies are important for this issue. A newly developed butterfly algorithmic technique (BOA) is leveraged to tune the controllers' parameters. However, to attain a proper balance between net power generation and load power, a dual stage proportional-integral- one plus integral-derivative $PI - (1 + ID)$ controller is developed. Comparative system responses (in MATLAB/SIMULINK software) for different scenarios under several controllers, such as a proportional-integral (PI), proportional-integral-derivative (PID) and $PI - (1 + ID)$ controller tuned by particle swarm optimization (PSO), grasshopper algorithmic technique (GOA) and BOA, show the superiority of BOA in terms of minimizing the peak deviations and better frequency regulation of the system. Real recorded wind data are considered to authenticate the control approach.

Keywords: isolated hybrid microgrid system ($IH\mu GS$); solar tower (ST); biodiesel power generator ($BDPG$); frequency regulation; butterfly optimization technique (BOA); microgrid energy management

---

1. Introduction

Electricity consumption is increasing in parallel with population and energy demand [1]. The increasing generation capacity with conventional energy sources has negative impacts on environment [2]. Research shows that the introduction of microgrids with renewable energy resources (RERs) is an environmentally friendly solution to this energy problem [3,4].

Microgrids can provide energy in a clean and optimal way when digital control technologies are coupled with sources such as wind and solar. However, the intermittent characteristics of RERs and low inertia of inverter-interfaced systems cause control and stability issues in microgrids [5]. Hence, combining diesel generators and RERs [6–8] is one possible solution to mitigate the detrimental effects (e.g., frequency fluctuation) of hybrid power systems. Additionally, a non-toxic bio-diesel power generator ($BDPG$) can be a more environment-conscious supplementary option for frequency regulation schemes.

To overcome the frequency fluctuation in a more reliable way, different storage devices (SDs) have been considered, such as battery (BSD), fuel cell (FCSD), ultra-capacitor (UCSD) and superconducting magnetic storage system (SMSD) [9]. There are maintenance and disposal concerns for BSD, while FCSD suffers from slow response and SMSU experiences leakage of expensive helium liquid [10]. In the following, a cost-effective, carbon-neutral-based, solid-oxide fuel cell (SOFC) can be utilized for the frequency regulation of isolated hybrid microgrids (*IHµGS*). In addition, different thermostatically controllable loads (TCLs) such as a heat pump (*HP*) and refrigerator (*RFZ*) are employed for smoothing the dynamic system responses. In the recent past, several studies have focused on the frequency regulation of *IHµGS* [11–15]. The study in [12] framed out a mathematical modeling of a system comprised of dish-stirling, solar thermal, diesel and SDs. The authors of [13] sketched a load frequency management for a hybrid power system that includes wind, solar PV and SDs. In fact, the application of a plug-in electric vehicle (PHEV)-battery (BSD) to contain frequency fluctuation is investigated in [14].

Beside the system architectures, several works have introduced a load frequency controller such as proportional-integral (*PI*) [9], proportional-integral-derivative (*PID*) [6], model predictive controller (MPC) [16], fractional order *PID* (FOPID) [17] or PIFOD [18] for the aforementioned issues. In this work, a dual-stage, proportional-integral-one plus integral-derivative *PI − (1 + ID)* controller is introduced for the provision of better system dynamics.

Several algorithmic techniques have been leveraged in order to optimally tune the controller parameters for *IHµGSs*, such as genetic algorithm (GA) [9], PSO [6], firefly technique (FA) [19], cuckoo search technique (CS) [14], mine blast technique (MBA) [20], grasshopper algorithmic technique (GOA) [21]. A comprehensive review of different algorithmic techniques for a load frequency controller is presented in [22]. In this regard, this work explores the application of a butterfly algorithmic tool (BOA) for designing the parameters of a frequency controller. This algorithm was recently developed and is considered in this paper due to its high convergence rate [23].

Therefore, the scope of this work and its contributions to the current body of knowledge can be summarized as follows:

(a) Developing a frequency controller for a *WG-ST-BDPG-SOFC-HP-RFZ*-based *IHµGS*;
(b) To establish a new transfer function model of a dual-stage *PI − (1 + ID)* controller;
(c) Comparative system dynamic analysis of different controllers such as *PI, PID* and *PI − (1 + ID)* controllers under a BOA algorithmic tool;
(d) Comparative system dynamic analysis of different algorithms (PSO, GOA and BOA), leveraging the acquired superior controller in (c);
(e) Study system dynamics under real recorded wind data and other random disturbances.

The rest of the paper is organized as follows: Section 2 details the frequency response modeling steps. Section 3 gives an overview of the BOA technique and shows its adaptation for the purposes of this work. It also presents the proposed dual-stage controller. Simulation works and their analyses are presented in Section 4. The conclusions are given in Section 5.

2. Frequency Response Modeling of the Proposed Dual-Stage Controller

The hybrid system of the proposed work consists of wind generators (1.5 MW); solar-tower-based, solar-thermal power system (1 MW), *BDPG* (800 kW), *SOFC* (200 kW); thermostatically controllable *HP*; *RFZ* elements; and demanded loads (2.2 MW). The schematic layout and abbreviation of the relevant parameters are shown in Figure 1 and Table 1, respectively.

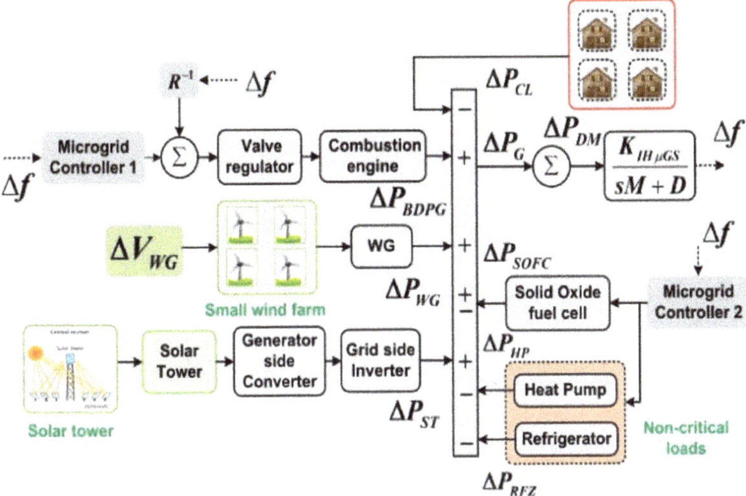

Figure 1. Schematic representation of *IHμGS*.

Table 1. Utilized symbols and abbreviations.

Symbol	Nomenclature	Value
ΔP_{CL}	Change in critical load power in p.u.	-
Δf	Aberration of frequency in Hz	-
ΔP_G	Change in net generated power	-
ΔP_{DM}	Change in net power difference from ΔP_G and ΔP_{CL}	-
K_{VR}	Gain of Valve actuator	1
K_{BE}	Engine gain	1
T_{VR}	Valve regulator delay time	0.08 s
T_{CE}	Time constant of bio-diesel power generator (*BDPG*)	0.4 s
K_{WG}	Gain of wind-driven generator (*WG*)	1
T_{WG}	Time constant of *WG*	5 s
K_{RF}, K_{RV}	Gain value of refocus and receiver	1, 1
K_G, K_T	Gain value of governor and turbine	1, 1
T_{RF}, T_{RV}	Time constant of refocus and receiver	1.33 s, 4 s
T_G, T_T	Time constant of governor and turbine	0.08 s, 1 s
K_{SOFC}	Gain of solid-oxide fuel cell (*SOFC*)	1
T_{SOFC}	Time constant of *SOFC*	0.2 s
K_{HP}	Gain of heat pump (*HP*)	1
T_{HP}	Time constant of *HP*	0.1 s
K_{RFZ}	Gain of refrigerator (*RFZ*)	1
T_{RFZ}	Time constant of *RFZ*	0.265 s
t_{sim}	Simulation time of *IHμGS*	100 s

2.1. Wind Generator (WG)

The kinetic energy of the wind converts into electrical energy through the wind generator (*WG*). As wind is a highly variable source, the power output through the *WG* depends on the instantaneous speed of the wind. Equation (1) formulates how wind energy is converted to the mechanical output power of *WG*.

$$P_{WG} = 0.5.V_{WG}^3.\rho.A_{bd}.C_P(\lambda, \beta) \tag{1}$$

where ρ, V_{WG}, A_{bd}, and C_P are, in proper order, the air density, intermittent wind speed, blade-swept area and the extractable power co-efficient. Akkanayakanpatti station's recorded wind speed data are considered and modelled in the proposed work [24]. The rate of change in real recorded wind power (ΔP_{WG}) and transfer function model of the *WG* are represented as shown in Equation (2) [25]

$$\Delta P_{WG} = \begin{cases} 0, \ V_{WG} < V_{cut-in} \, || \, V_{WG} > V_{cut-out} \\ 0, \ V_{rated} \leq V_{WG} \leq V_{cut-out} \\ (\ [0.007872 \, V_{WG}^5 - 0.23015 \, V_{WG}^4 + 1.3256 \, V_{WG}^3 \\ \ +11.061 \, V_{WG}^2 - 102.2 \, V_{WG} + 2.33]. \Delta V_{WG}), \ else \\ G_{WG}(s) = \frac{K_{WG}}{sT_{WG}+1} \end{cases} \tag{2}$$

2.2. Solar Tower (ST)

The dual-axis (vertical and horizontal) heliostats-enabled central receiver system is *ST*, placed on a surface. Here, the reflected solar radiation is focused on the central receiver of *ST* with a higher concentration ratio (500–1000) and temperature (500–850 °C) of fluid (steam or molten salt). The collected incident solar power (P_{in}) can be deliberated as

$$P_{in} = \eta_h . I . A_h \tag{3}$$

where A_h is the heliostats area, I incident solar radiation and η_h is the system constant. By solving the state equations, the linearized transfer function model of ST can be represented as [18]

$$G_{ST}(s) = \left(\frac{K_{RF}}{sT_{RF}+1}\right).\left(\frac{K_{RV}}{sT_{RV}+1}\right).\left(\frac{K_G}{sT_G+1}\right).\left(\frac{K_T}{sT_T+1}\right) \tag{4}$$

2.3. Biodiesel Power Generator (BDPG)

The combination valve regulator and combustion-engine-based biodiesel power generation (*BDPG*) was leveraged to offer support as a backup power generation. It has inherently biodegradable and non-toxic positive characteristics, which were the main reasons to incorporate it in the suggested work. Equation (5) details the transfer function of *BDPG* [21].

$$G_{BDPG}(s) = \left(\frac{K_{VR}}{sT_{VR}+1}\right).\left(\frac{K_{CE}}{sT_{CE}+1}\right) \tag{5}$$

2.4. Solid-Oxide-Based Fuel Cell (SOFC)

Through the electrochemical reaction, the fuel cell produces dc power and, by using a DC-AC converter, this power is converted into AC. With a fast charging–discharging time and higher efficiency (~80%), the *SOFC* has gained much interest in recent years among all the categories of FCSDs. In view of the above, *SOFC* was selected as the storage device in the system. Its transfer function model is given in Equation (6) [18]

$$G_{SOFC}(s) = \frac{K_{SOFC}}{sT_{SOFC}+1} \tag{6}$$

2.5. Thermostatically Controllable Loads (HP and RFZ)

In order to manage the energy consumption and to improve the system dynamic responses (by controlling operation cycles), two thermostatically controllable loads were considered, i.e., heat pump (*HP*) and *RFZ*. The transfer function models of *HP* [19] and *RFZ* [19] could be expressed as in Equations (7) and (8)

$$G_{HP}(s) = \frac{K_{HP}}{sT_{HP}+1} \tag{7}$$

$$G_{RFZ}(s) = \frac{K_{RFZ}}{sT_{RFZ}+1} \tag{8}$$

2.6. IHμGS Dynamic Model

The instantaneous change in power (ΔP_G) of the proposed *IHμGS* can be formulated as

$$\Delta P_G = \Delta P_{WG} + \Delta P_{ST} + \Delta P_{BDPG} \pm \Delta P_{SOFC} - \Delta P_{NCL} = \Delta P_{CL} \to 0 \tag{9}$$

where

$$\Delta P_{NCL} = \Delta P_{HP} + \Delta P_{RFZ}$$
$$and\ \Delta P_{DM} = \Delta P_G - \Delta P_{CL} \tag{10}$$

The equivalent dynamic model of *IHμGS* could be illustrated as

$$G_{IH\mu GS}(s) = \left(\frac{\Delta f}{\Delta P_{DM}}\right) = \frac{K_{IH\mu GS}}{D + sM} \tag{11}$$

Refer to Table 1 for the nomenclature and abbreviations used for *IHμGS* system modelling.

2.7. Objective Function Formulation

The formulation of the objective function (J) has a great impact on system dynamics and the achieved results. Therefore, the proposed work considered the integral of square error (*ISE*) objective function. This could be formulated as

$$Minimize\ J_{ISE} = \int_{0}^{t_{Sim}} (\Delta f)^2 .dt \tag{12}$$

$$Subject\ to :$$
$$\begin{cases} K_{Pi}^{min} \le K_{Pi} \le K_{Pi}^{max} \\ K_{Ii}^{min} \le K_{Ii} \le K_{Ii}^{max} \\ K_{Di}^{min} \le K_{Di} \le K_{Di}^{max} \\ K_{Ii2}^{min} \le K_{Ii2} \le K_{Ii2}^{max} \\ K_{Di2}^{min} \le K_{Di2} \le K_{Di2}^{max} \end{cases} \tag{13}$$

where $i = 1, 2$. The range of controller parameters is taken as (0–50).

3. Optimization Techniques

Three metaheuristic techniques were considered to optimally tune the controller parameters along with their comparative dynamic responses.

3.1. Particle Swarm Technique (PSO)

A swarm-based metaheuristic particle swarm technique (PSO) was developed by Eberhart and Kennedy in 1995 to solve the specified problem by improving the candidate solution with reference to the given measure quality [26]. The solution of PSO is termed as a particle. Every particle follows a track of coordinates in the problem space until the best solution is reached with respect to the suggested problem. The velocity of each particle is varied on the basis of best position (P_{best}) and I_{best} location. The optimum values obtained by optimizer are termed as I_{best} [26].

3.2. Grasshopper Algorithmic Technique (GOA)

A metaheuristic grasshopper algorithmic technique was proposed by Saremi et al. [27]. Its characteristics depend on the swarming and foraging characteristics of grasshopper, which could be modeled to form structural algorithmic techniques. The steps involved for initialization, exploitation and exploration are depicted in [27].

3.3. Butterfly Optimization Technique (BOA) and Proposed Dual-Stage Controller

Recently, since 2018, based on the food-probing approach and mating characteristics of butterflies, a natured-inspired algorithmic technique named BOA has been developed to solve several engineering problems. The unique food-probing strategy and mating characteristics of BOA are modeled in [23].

The main idea of BOA depends on three key parameters. These are sensor modality (M_s), impulsive intensity (I_s) and power component (γ). Moreover, the objective function of this technique depends on the distinction of I_s and formation of fragrance (P), which could be formulated as

$$P_i = M_s.I_s^{\gamma} \; i \in (1, \, 2...N) \tag{14}$$

where P_i is the fragrance magnitude of ith butterfly. In the following, to investigate global search stage, a dominated fitted solution q^* could be depicted as

$$\begin{aligned} m_i^{t+1} &= m_i^t + \left(n^2 \times q^* - m_i^t\right).P_i \\ where\; n &\in [0, \, 1] \end{aligned} \tag{15}$$

where m and q^* are the solution vectors of ith butterfly and current best solution among all the solutions. The formulation of the social search could be illustrated as

$$m_i^{t+1} = m_i^t + \left(n^2.m_g^t - m_h^t\right).P_i \tag{16}$$

where, m_g and m_h are the gth and hth butterflies enabled in the search space [23]. The approximate flow diagram of BOA technique is framed out in Figure 2. All the parameters considered for algorithmic techniques are given in Appendix A.

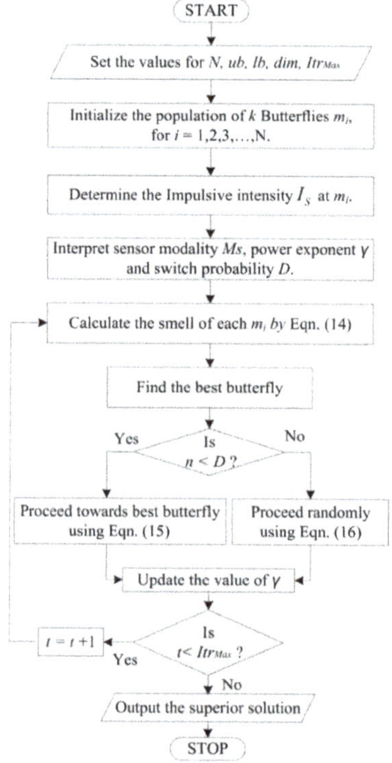

Figure 2. Flow diagram of butterfly optimization technique (BOA).

To reach above control target, a novel dual-stage proportional-integral-integral-derivative *(PI − (1 + ID))* controller is deployed, as framed in Figure 3. In the following, Δf is leveraged as an input signal, whereas *C(s)* is the output control signal of the controller. The control output signal and transfer function of the proposed controller are formulated as

$$C(s) = \Delta f . PI - (1 + ID) \tag{17}$$

$$G_{PI-(1+ID)}(s) = K_P + K_I/s - (1 + K_I/s + K_D.s) \tag{18}$$

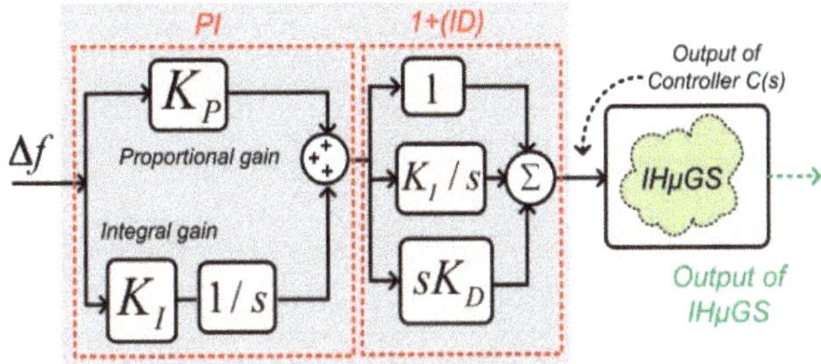

Figure 3. Proposed *(PI − (1 + ID))* controller.

4. Frequency Response Studies and Analysis

In order to verify the proposed control strategy, two scenarios are simulated in a system with Core-i7-4770 CPU under MATLAB/SIMULINK (R2013a, MathWorks, Natick, USA) was environment. Three algorithmic techniques (PSO, GOA and BOA) have been considered. Furthermore, to validate the control strategy, real recorded wind speed data have been considered.

4.1. Scenario 1: Performance Analysis of All Controllers during Non-Accessibility of All RERs

In this scenario, assume that all the RERs are unavailable due to maintenance. Therefore, the extractable power forms WG (ΔP_{WG}) and ST (ΔP_{ST}) are zero ($\Delta P_{WG} = \Delta P_{ST} = 0\%$) during the entire period. A net constant critical load demand ($\Delta P_{CL} = 30\%$) is considered from $t = 0$ s onwards. The comparative performance of different controllers such as *PI, PID* and *(PI − (1 + ID))* are displayed in Figure 4, where the tuned parameters are listed in Table 2. The system dynamics assessment of the abovementioned controllers under BOA and objective function (J_{ISE}) and figure of demerits (J_{FOD}) clearly depicts that the proposed *(PI − (1 + ID))* controller is superior to the rest. To elaborate further, the performance indicators such as peak overshoot (+O_P), peak undershoot (-U_P) and settling time (T_{ST}) are tabulated in Table 2.

Table 2. Comparative performance parameters of different controllers with optimal BOA-tuned gain values.

Controllers	PI	PID	PI − (1 + ID)
Peak Overshoot(+O_P)			
ΔF (in Hz)	0.0544	0.0136	**0.0006**
Peak Undershoot(-U_P)			
ΔF (in Hz)	0.0669	0.0389	**0.0190**
Settling Time (T_{ST})			
ΔF (in s)	3.976	4.097	**2.581**

Table 2. *Cont.*

Controllers		*PI*	*PID*	*PI − (1 + ID)*
		Minimization of J (J_{min})		
		7.79×10^{-4}	2.92×10^{-5}	2.91×10^{-5}
		Figure of Demerits (J_{FOD})		
		15.816	16.787	6.662
		Optimal Controller Parameters		
Controller-1	K_{P1}	3.010	0.502	0.325
	K_{I1}	5.103	12.11	10.702
	K_{D1}	-	0.108	-
	K_{I12}	-	-	0.513
	K_{D12}	-	-	0.118
Controller-2	K_{P2}	18.116	5.001	1.508
	K_{I2}	20.207	5.509	4.109
	K_{D2}	-	1.624	-
	K_{I22}	-	-	1.128
	K_{D22}	-	-	2.219

Bolt point out superior output.

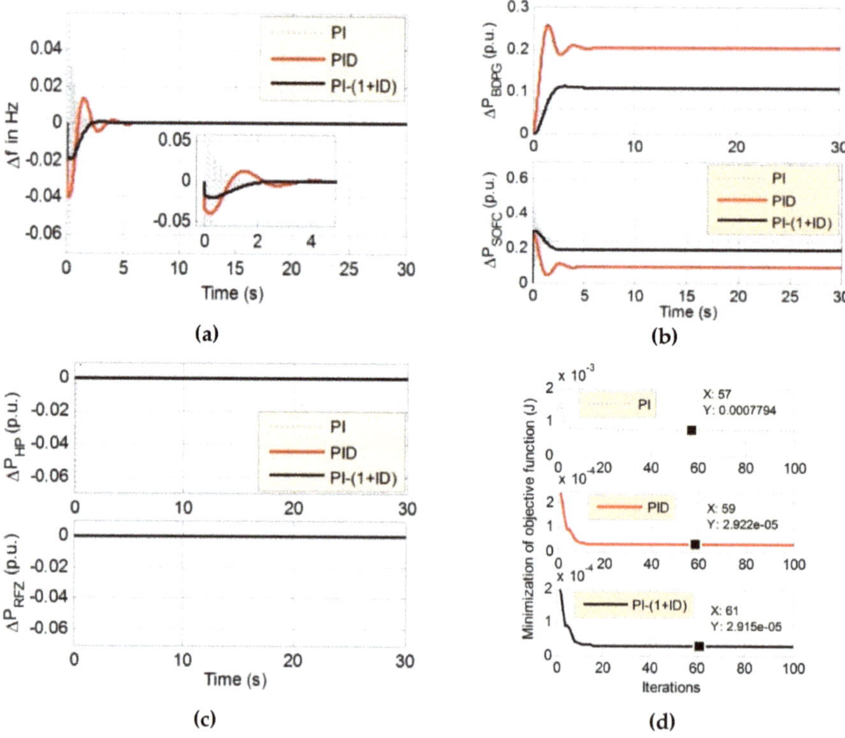

Figure 4. Comparative system dynamics analysis of different controllers (proportional-integral (*PI*), *PI*-derivative (*PID*), (*PI − (1 + ID)*)) (**a**) deviation in system frequency (Δf), (**b**) change in extractable power of bio-diesel power generator (*BDPG*) and *SOFC*, (**c**) change in extractable power of *HP* and *RFZ*, (**d**) Comparative converged objective function (J_{min}).

4.2. Scenario 2: Performance analysis of Different Algorithms Under Concurrent Random Changes of WG (Utilization of Real-Recorded Data), ST and Critical Load Demand

In this scenario, the proposed system is tested under real-recorded wind (obtained from National Institute of Wind Energy, India) [25], as displayed in Appendix B. The operating condition is illustrated with 30% average power of *ST* and 50% critical load demand for the entire time duration. The wind speed and its corresponding output power is shown in Figure 5a. The results are depicted in Figure 5b-e, showing the comparative system dynamic responses of Δf, ΔP_{BDPG}, ΔP_{SOFC}, ΔP_{HP}, and ΔP_{RFZ}. Figure 5b–e it clearly shows that the BOA-optimized *PI – (1 + ID)* controller performed better than the other suggested PSOs, GOA-tuned (*PI – (1 + ID)*) controller. The tuned values of the controller parameters are displayed in Table 3.

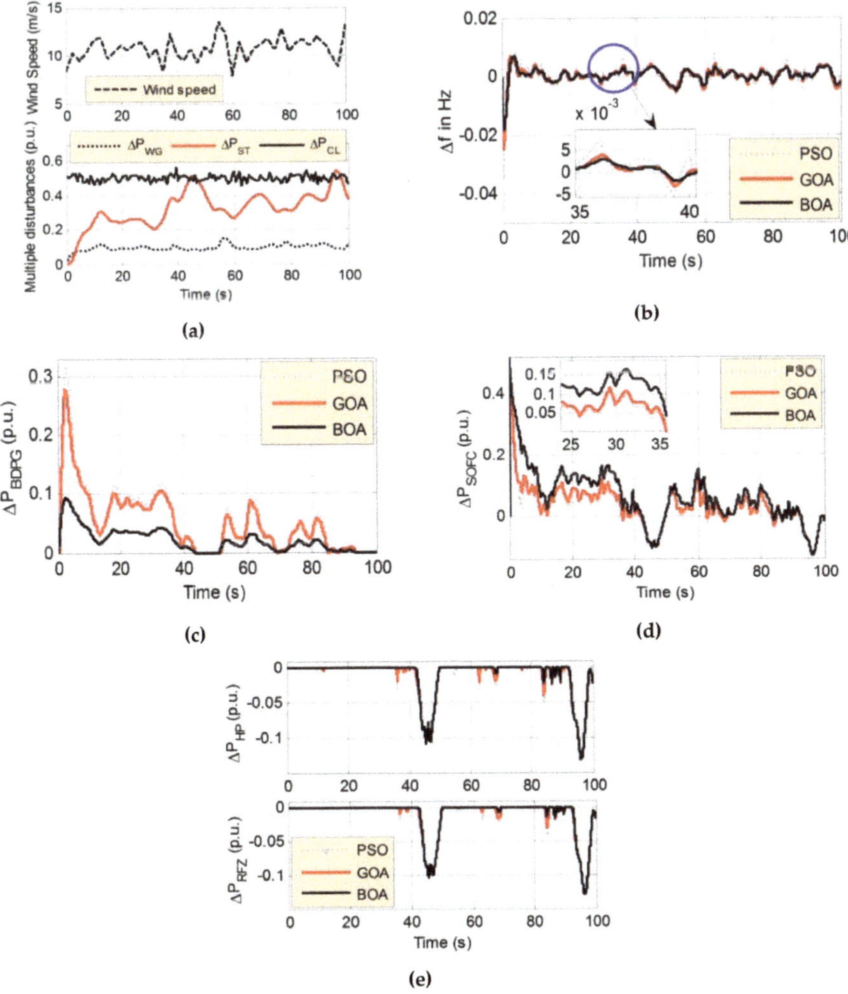

Figure 5. Comparative system dynamics analysis of different algorithmic techniques (PSO, GOA, BOA) (**a**) Real recorded wind speed and other multiple disturbances, (**b**) deviation of system frequency (Δf), (**c**) change in extractable power of *BDPG*, (**d**) change in extractable power of *SOFC*, (**e**) change in extractable power of *HP* and *RFZ*.

Table 3. Optimal values of particle swarm optimization (PSO), grasshopper algorithmic technique (GOA) and BOA techniques tuned *PI-(1+ID)* controller.

Techniques		PSO	GOA	BOA
Optimal controller parameters				
Controller-1	K_{P1}	0.3112	0.2986	0.3210
	K_{I1}	5.0070	20.051	25.053
	K_{I12}	0.5021	0.5170	0.5087
	K_{D12}	0.1085	0.1153	0.1078
Controller-2	K_{P2}	0.5170	2.5171	4.6087
	K_{I2}	4.1850	4.1751	4.1098
	K_{I22}	1.1190	1.2015	1.1069
	K_{D22}	2.2191	2.2276	2.2183

5. Conclusions

The present article develops a novel frequency regulation scheme for wind-solar-tower-biodiesel-based $IH\mu GS$. A novel dual-stage $(PI - (1 + ID))$ controller is enabled to investigate the system dynamics under different scenarios. A recently developed BOA technique is utilized to optimally tune the proposed dual stage $(PI - (1 + ID))$ controller gains and compare the system dynamics under real recorded wind data. The comparative system dynamic responses, as well as performance parameters such as peak deviation $(+O_P, -U_P)$ and settling time (T_{ST}), clearly indicate that the BOA-optimized $(PI - (1 + ID))$ controller performs better than other classical benchmark controllers. The simulation test results prove the effectiveness of the proposed control strategy. This control scheme could be further extended by integrating different RER technologies and storage devices, as well as electric vehicles, into the microgrid.

Author Contributions: Conceptualization, A.L., S.M.S.H., D.C.D. and T.S.U.; Methodology, A.L., S.M.S.H., D.C.D. and T.S.U.; Software, A.L.; Validation, A.L.; Formal Analysis, A.L., S.M.S.H., D.C.D. and T.S.U..; Writing—Original Draft Preparation, A.L. and D.C.D.; Writing—Review and Editing, S.M.S.H., and T.S.U.; Visualization, A.L.; Supervision, D.C.D.; Funding Acquisition, T.S.U. All authors have read and agreed to the published version of the manuscript.

Funding: This research received no external funding.

Acknowledgments: Authors would like to thank TEQIP-III NIT Silchar for providing technical support for this work.

Conflicts of Interest: The authors declare no conflict of interest.

Appendix A

PSO technique: Number of population: 50, Maximum iteration (Itr_{Max}): 100, Max^m weight factor (W_{max}): 0.9, Min^m weight factor (W_{min}): 0.1, Acceleration factors (C1&C2): 2.

GOA technique: Number of population: 50, Maximum iteration (Itr_{Max}): 100, Maximum coefficient factor $(C_{fmax}) = 1$, Minimum coefficient factor $(C_{fmin}) = 0.00004$, attraction intensity (f): 0.5, length scale of attractiveness (l): 1.5

BOA technique: Number of population: 50, Maximum iteration (Itr_{Max}): 100, Probability of switching (D) = 0.8, Power component $(\gamma) = 0.1$, Sensor modality $(Ms) = 0.1$.

Appendix B

WG: Date of noted data: 1st July, 2016, Minimum speed of wind: 7.4804 m/s; Maximum speed of wind: 14.08 m/s; Average speed of wind: 10.922 m/s; SD: 1.1895.

References

1. Global Energy Review 2020, International Energy Agency (IEA), Paris. 2020. Available online: https://www.iea.org/reports/global-energy-review-2020 (accessed on 2 July 2020).
2. Global Energy & CO2 Status Report 2019, International Energy Agency (IEA), Paris. 2019. Available online: https://www.iea.org/reports/global-energy-co2-status-report-2019 (accessed on 2 July 2020).
3. Xie, H.; Zheng, S.; Ni, M. Microgrid Development in China: A method for renewable energy and energy storage capacity configuration in a megawatt-level isolated microgrid. *IEEE Electr. Mag.* **2017**, *5*, 28–35. [CrossRef]
4. Ustun, T.S.; Nakamura, Y.; Hashimoto, J.; Otani, K. Performance analysis of PV panels based on different technologies after two years of outdoor exposure in Fukushima, Japan. *Renew. Energy* **2019**, *136*, 159–178. [CrossRef]
5. Dong, X.; Zhang, X.; Jiang, T. Adaptive Consensus Algorithm for Distributed Heat-Electricity Energy Management of an Islanded Microgrid. *Energies* **2018**, *11*, 2236. [CrossRef]
6. Das, D.C.; Sinha, N.; Roy, A.K. Automatic Generation Control of an Organic Rankine Cycle Solar–Thermal/Wind–Diesel Hybrid Energy System. *Energy Technol.* **2014**, *2*, 721–731. [CrossRef]
7. Dhillon, S.S.; Lather, J.S.; Marwaha, S. Multi objective load frequency control using hybrid bacterial foraging and particle swarm optimized PI controller. *Int. J. Electr. Power Energy Syst.* **2016**, *79*, 196–209. [CrossRef]
8. Ghafouri, A.; Milimonfared, J.; Gharehpetian, G.B. Fuzzy-adaptive frequency control of power system including microgrids, wind farms, and conventional power plants. *IEEE Syst. J.* **2017**, *12*, 2772–2781. [CrossRef]
9. Das, D.C.; Roy, A.K.; Sinha, N. GA based frequency controller for solar thermal–diesel–wind hybrid energy generation/energy storage system. *Int. J. Electr. Power Energy Syst.* **2012**, *43*, 262–279. [CrossRef]
10. Nadeem, F.; Hussain SM, S.; Tiwari, P.K.; Goswami, A.K.; Ustun, T.S. Comparative Review of Energy Storage Systems, Their Roles, and Impacts on Future Power Systems. *IEEE Access* **2019**, *7*, 4555–4585. [CrossRef]
11. Singh, K.; Amir, M.; Ahmad, F.; Khan, M.A. An Integral Tilt Derivative Control Strategy for Frequency Control in Multimicrogrid System. *IEEE Syst. J.* **2020**. [CrossRef]
12. Das, D.C.; Sinha, N.; Roy, A.K. Small signal stability analysis of dish-Stirling solar thermal based autonomous hybrid energy system. *Int. J. Electr. Power Energy Syst.* **2014**, *63*, 485–498. [CrossRef]
13. Hussain, I.; Ranjan, S.; Das, D.C.; Sinha, N. Performance Analysis of Flower Pollination Algorithm Optimized PID Controller for Wind-PV-SMES-BESS-Diesel Autonomous Hybrid Power System. *Int. J. Renew. Energy Res.* **2017**, *7*, 643–651.
14. Latif, A.; Pramanik, A.; Das, D.C.; Hussain, I.; Ranjan, S. Plug in hybrid vehicle-wind-diesel autonomous hybrid power system: Frequency control using FA and CSA optimized controller. *Int. J. Syst. Assur. Eng. Manag.* **2018**, *9*, 1147–1158. [CrossRef]
15. Khooban, M.H.; Gheisarnejad, M. A Novel Deep Reinforcement Learning Controller Based Type-II Fuzzy System: Frequency Regulation in Microgrids. *IEEE Trans. Emerg. Top. Comput. Intel.* **2020**. [CrossRef]
16. Pahasa, S.; Ngamroo, I. Coordinated Control of Wind Turbine Blade Pitch Angle and PHEVs Using MPCs for Load Frequency Control of Microgrid. *IEEE Syst. J.* **2016**, *10*, 97–105. [CrossRef]
17. Latif, A.; Das, D.C.; Ranjan, S.; Barik, A. Comparative performance evaluation of WCA-optimised non-integer controller employed with WPG-DSPG-PHEV based isolated two-area interconnected microgrid system. *IET Renew. Power Gener.* **2019**, *13*, 725–736. [CrossRef]
18. Latif, A.; Das, D.C.; Barik, A.K.; Ranjan, S. Maiden co-ordinated load frequency control strategy for ST-AWEC-GEC-BDDG based independent three-area interconnected microgrid system with the combined effect of diverse energy storage and DC link using BOA optimized PFOID controller. *IET Renew. Power Gener.* **2019**, *13*, 2634–2646. [CrossRef]
19. Ray, P.K.; Mohanty, A. A robust firefly–swarm hybrid optimization for frequency control in wind/PV/FC based microgrid. *Appl. Soft Comput.* **2019**, *85*, 105823. [CrossRef]
20. Atawi, I.E.; Kassem, A.M.; Zaid, S.A. Modeling, Management, and Control of an Autonomous Wind/Fuel Cell Micro-Grid System. *Processes* **2019**, *7*, 85. [CrossRef]
21. Barik, A.; Das, D.C. Expeditious frequency control of solar photobiotic/biogas/biodiesel generator based isolated renewable microgrid using grasshopper optimisation algorithm. *IET Renew. Power Gener.* **2018**, *12*, 1659–1667. [CrossRef]

22. Latif, A.; Hussain, S.M.S.; Das, D.C.; Ustun, T.S. State-of-the-art of controllers and soft computing techniques for regulated load frequency management of single/multi-area traditional and renewable energy based power systems. *Appl. Energy* **2020**, *266*, 114858. [CrossRef]

23. Arora, S.; Singh, S. Butterfly optimization algorithm: A novel approach for global optimization. *Soft Comput.* **2019**, *23*, 715–734. [CrossRef]

24. Wind Speed Data. Available online: http://niwe.res.in:8080/NIWE_WRA_DATA/DataTable_D4.jsf (accessed on 2 July 2020).

25. El-Fergany, A.A.; El-Hameed, A.M. Efficient frequency controllers for autonomous two-area hybrid microgrid system using social-spider optimizer. *IET Gener. Transm. Distrib.* **2017**, *11*, 637–648. [CrossRef]

26. Dutta, D.; Figueira, J.R. Graph partitioning by muti-objective real-valued metaheuristic: A comparative study. *Appl. Soft Comput.* **2011**, *11*, 3976–3987. [CrossRef]

27. Saremi, S.; Mirjalili, S.; Lewis, A. Grasshopper optimization algorithm: Theory and application. *Adv. Eng. Softw.* **2017**, *105*, 30–47. [CrossRef]

MDPI
St. Alban-Anlage 66
4052 Basel
Switzerland
Tel. +41 61 683 77 34
Fax +41 61 302 89 18
www.mdpi.com

Energies Editorial Office
E-mail: energies@mdpi.com
www.mdpi.com/journal/energies